Current Problems of
Lower Vertebrate
Phylogeny

Nobel Symposium 4

Current Problems of Lower Vertebrate Phylogeny

*Proceedings of the Fourth Nobel Symposium
held in June 1967 at the Swedish Museum
of Natural History* (*Naturhistoriska riksmuseet*)
in Stockholm

Edited by
TOR ØRVIG

Curator, Section of Palaeozoology,
Swedish Museum of Natural History,
Stockholm, Sweden

INTERSCIENCE PUBLISHERS
A Division of John Wiley & Sons, Inc. *New York*, *London*, *Sydney*

ALMQVIST & WIKSELL *Stockholm*

© 1968
Almqvist & Wiksell/Gebers Förlag AB, Stockholm

All rights reserved

This book or any part thereof must not be
reproduced in any form without the written
permission of the publisher

Library of Congress Catalog Card Number 68–9446

Printed in Sweden by
Almqvist & Wiksells Boktryckeri AB, Uppsala 1968

Contents

Preface 7
Sponsors 9
List of participants 10

ERIK STENSIÖ
The cyclostomes with special reference to the diphyletic origin of the Petromyzontida and Myxinoidea 13
Discussion: E. I. WHITE 70

NATASCHA HEINTZ
The pteraspid *Lyktaspis* n.g. from the Devonian of Vestspitsbergen . . . 73

ALEXANDER RITCHIE
Phlebolepis elegans Pander, an Upper Silurian thelodont from Oesel, with remarks on the morphology of thelodonts 81

MALCOLM JOLLIE
Some implications of the acceptance of a delamination principle 89

ROGER S. MILES
Jaw articulation and suspension in *Acanthodes* and their significance . . . 109

GARETH J. NELSON
Gill-arch structure in *Acanthodes* 129

ANATOL HEINTZ
The spinal plate in *Homostius* and *Dunkleosteus* 145
Discussion: E. A. STENSIÖ 148

S. E. BENDIX-ALMGREEN
The bradyodont elasmobranchs and their affinities; a discussion 153

COLIN PATTERSON
Menaspis and the bradyodonts 171

BOBB SCHAEFFER
The origin and basic radiation of the Osteichthyes 207

ERIK JARVIK
The systematic position of the Dipnoi 223

ROBERT H. DENISON
The evolutionary significance of the earliest known lungfish, *Uranolophus* . 247

GUNNAR BERTMAR
Lungfish phylogeny 259

Contents

KEITH STEWART THOMSON
A critical review of the diphyletic theory of rhipidistian-amphibian relationships 285
Discussion: H. SZARSKI 305

JEAN-PIERRE LEHMAN
Remarques concernant la phylogénie des Amphibiens 307

NATALIE S. LEBEDKINA
The development of bones in the skull roof of Amphibia 317

IRENE M. MEDVEDEVA
Die Homologie des Jacobsonschen Organs bei Anura und Urodela . . . 331

HANS C. BJERRING
The second somite with special reference to the evolution of its myotomic derivatives 341

MELVIN L. MOSS
The origin of vertebrate calcified tissues 359

TOR ØRVIG
The dermal skeleton; general considerations 373

RAINER ZANGERL
The morphology and developmental history of the scales of the Paleozoic sharks *Holmesella*? sp. and *Orodus* 399

CH. DEVILLERS & J. CORSIN
Les os dermiques crâniens des Poissons et des Amphibiens; pointe de vue embryologiques sur les "territoires osseux" et les "fusions" 413

HANS JESSEN
The gular plates and branchiostegal rays in *Amia*, *Elops* and *Polypterus* . 429

ORVAR NYBELIN
The dentition in the mouth cavity of *Elops* 439

HENRYK SZARSKI
Evolution of cell size in lower vertebrates 445

RAGNAR OLSSON
Evolutionary significance of "prolactin" cells in teleostomean fishes . . . 455

LARS BRUNDIN
Application of phylogenetic principles in systematics and evolutionary theory 473

ERIK JARVIK
Aspects of vertebrate phylogeny 497

Index of authors, genera and species 529

Preface

The origin, interrelationships and phylogeny of the vertebrates embrace problems of great significance to biologists and of interest also to the laity. In the period between the emergence of darwinism and the first world war, these problems were tackled on the basis of anatomy and embryology of recent animals, and solutions were arrived at which, to many, seemed final. Later, however, the early fossil vertebrates made their entrance on the stage, and from 1927, the year of publication of Stensiö's classical monograph on the Spitsbergen cephalaspids, Silurian and Devonian cyclostomes and fishes have gained ever increasing importance for the discussion of phylogenetic problems. Studies of early Palaeozoic vertebrates combined with reinvestigations of recent forms have raised serious doubt as to the validity of many of the views developed by anatomists of earlier days, and a reevaluation of these views, which unconsciously influence our way of thinking even today, has emerged as an urgent necessity. With these aspects in mind, it has long been planned to organize a Symposium on lower vertebrate phylogeny in Stockholm, and thanks to the courtesy of the Nobel Foundation, these plans could be realized in 1967.

On behalf of Professor Stensiö and myself, I wish to express our sincere gratitude to the Nobel Foundation for the great privilege of arranging a Nobel Symposium at the Swedish Museum of Natural History, and to the Tri-centennial Fund of the Bank of Sweden for a generous grant to the Nobel Foundation for that purpose. We also owe a debt of gratitude to the Royal Swedish Academy of Science where a reception was given on the first day of the Symposium and a concluding session and dinner in the evening of the last, to the private donor who generously defrayed the cost of that dinner, to all those who made the joint excursion to Uppsala and the Hammarby of Linnaeus on the second Symposium day such an unforgettable event, and last but not least, to the authorities of the Swedish Museum of Natural History and the entire staff of the Section of Palaeozoology for unfailing help and assistance in every way.

Nobel Symposium 4 was held June 12–16 1967 in the Section of Palaeozoology where a comprehensive material of fossils, illustrations, wax models, etc. relevant to the subjects treated, was on exhibit every day. The demonstration of this material was intended as, and turned out to be, an essential part of the agenda, and caused keen debates, notably during the well-attended evening sessions. These and other discussions have in many ways influenced the contents of this volume, and even in some cases prompted contributions only in part, or not at all, based on the papers read during the Symposium itself. The volume, incidentally, also contains papers especially prepared for the Symposium by two Russian colleagues who at the last moment were prevented from being present.

The main themes of the Symposium were the structure, interrelationships and evolutionary history of the various groups of lower vertebrates, and as is evident from this volume, these matters were dealt with from the point of view of so widely different fields of research as comparative anatomy, histology, embryology, experimental zoology, cytology, biochemistry, neurosecretion and evolutionary theory. All these contributions are related to the results gained by the study of the early fossil vertebrates, and certainly give an idea of the stimulating effect exerted by these studies in many directions. Vertebrate palaeontology today is certainly not a science of dry descriptions of fossil bones as a pastime for the select few, but an integral—and indeed essential—part of modern Life Sciences. One of the main purposes of this Nobel Symposium has in fact been to widen the interdisciplinary contacts by bringing together students from different fields of research for discussion of problems of common interest to them all.

Readers of this volume will no doubt notice the great diversity of opinion still existing as concerns most problems of vertebrate phylogeny. This, however, is no cause for dissatisfaction; the many controversial points demonstrate clearly enough that the branch of palaeontology devoted to early vertebrates, i.e. the ancestors of living forms up to and including man, is a vigorous field of research with great and yet unexplored potentialities. We all owe a great debt of gratitude to Erik Stensiö who for almost half a century has held a key position in this field, and who by his wide knowledge, ready enthusiasm and deep devotion to research has been a constant source of inspiration to us all.

Erik Jarvik

Erik Stensiö (drawn by Birger Lundquist 1946)

Sponsors

The Nobel Foundation
The Tri-centennial Fund of the Bank of Sweden
The Royal Swedish Academy of Science
The Swedish Museum of Natural History

List of participants
(present address given)

S. MAHALA ANDREWS, University Museum of Zoology, Downing Street, *Cambridge*, England

S. E. BENDIX-ALMGREEN, Mineralogical Museum, Østervoldgade 7, *Copenhagen K*, Denmark

G. BERTMAR, Section of Ecological Zoology, Dept. of Biology, Umeå University, *900 06 Umeå*, Sweden

H. C. BJERRING, Section of Palaeozoology, Swedish Museum of Natural History, *104 05 Stockholm*, Sweden

N. BONDE, Mineralogical Museum, Østervoldgade 7, *Copenhagen K*, Denmark

L. BRUNDIN, Section of Entomology, Swedish Museum of Natural History, *104 05 Stockholm*, Sweden

R. H. DENISON, Dept. of Geology, Field Museum of Natural History, Roosevelt Road at Lake Shore Drive, *Chicago*, Illinois. 60605. USA

C. DEVILLERS, Laboratoire d'Anatomie et d'Histologie Comparées, Université de Paris, 7 Quai Bernard, *Paris V*, France

D. L. DINELEY, Dept. of Geology, University of Bristol, *Bristol 8*, England

K. ENGSTRÖM, Dept. of Exhibits, Swedish Museum of Natural History, *104 05 Stockholm*, Sweden

D. GOUJET, Institut de Paléontologie, Muséum National d'Histoire Naturelle, 8 Rue de Buffon, *Paris V*, France

P. H. GREENWOOD, Dept. of Zoology, British Museum (Natural History), Cromwell Road, *London S.W. 7*, England

W. GROSS, Institut und Museum für Geologie und Paläontologie der Universität, Sigwartstrasse 10, *74 Tübingen*, Germany (DBR)

G. HAAS, Dept. of Zoology, Hebrew University, *Jerusalem*, Israel

A. HEINTZ, Palaeontological Museum, Sarsgt. 1, *Oslo 5*, Norway

NATASCHA HEINTZ, Palaeontological Museum, Sarsgt. 1, *Oslo 5*, Norway

E. JARVIK, Section of Palaeozoology, Swedish Museum of Natural History, *104 05 Stockholm*, Sweden

H. JESSEN, Section of Palaeozoology, Swedish Museum of Natural History, *104 05 Stockholm*, Sweden

A. G. JOHNELS, Section of Vertebrate Zoology, Swedish Museum of Natural History, *104 05 Stockholm*, Sweden

M. JOLLIE, Dept. of Biological Sciences, Northern Illinois University, *DeKalb*, Illinois. 60115. USA

J.-P. LEHMAN, Institut de Paléontologie, Muséum National d'Histoire Naturelle, 8 Rue de Buffon, *Paris V*, France

S. LØVTRUP, Section of Zoophysiology, Dept. of Biology, Umeå University, *900 06 Umeå*, Sweden

R. S. MILES, Dept. of Geology, The Royal Scottish Museum, Chambers Street, *Edinburgh 1*, Scotland

M. L. MOSS, Dept. of Anatomy, College of Physicians and Surgeons, Columbia University, 630 W 168th Street, *New York*, N.Y. 10032. USA

G. J. NELSON, Dept. of Ichthyology, The American Museum of Natural History, Central Park W at 79th Street, *New York*, N.Y. 10024. USA

E. NIELSEN, Mineralogical Museum, Østervoldgade 7, *Copenhagen K*, Denmark

O. NYBELIN, Natural History Museum, *400 30 Gothenburg*, Sweden

D. V. OBRUCHEV, Palaeontological Institute of the Academy of Science of the USSR, Lenin Prospekt 33, *Moscow W-71*, USSR

R. OLSSON, Dept. of Zoology, Stockholm University, Rådmansgatan 70 A, *113 86 Stockholm*, Sweden

T. ØRVIG, Section of Palaeozoology, Swedish Museum of Natural History, *104 05 Stockholm*, Sweden

Y. PAGEAU, Institut de Paléontologie, Muséum National d'Histoire Naturelle, 8 Rue de Buffon, *Paris V*, France

F. R. PARRINGTON, University Museum of Zoology, Downing Street, *Cambridge*, England

C. PATTERSON, Dept. of Palaeontology, British Museum (Natural History), Cromwell Road, *London S.W. 7*, England

T. PEHRSON, Dept. of Zoology, Stockholm University, Rådmansgatan 70 A, *113 86 Stockholm*, Sweden

A. RITCHIE, The Australian Museum ,College Street, *Sydney*, N.S.W. Australia

B. SCHAEFFER, Dept. of Vertebrate Paleontology, The American Museum of Natural History, Central Park W at 79th Street, *New York*, N.Y. 10024. USA

H.-P. SCHULTZE, Geologisch-paläontologisches Institut der Universität, Berliner Strasse 28, *34 Göttingen*, Germany (DBR)

BARBARA J. STAHL, St. Anselm's College, *Manchester*, N H. 03104. USA

E. A. STENSIÖ, Section of Palaeozoology, Swedish Museum of Natural History, *104 05 Stockholm*, Sweden

R. STRAHAN, Zoological Gardens and Aquarium, Mosman, *Sydney*, N.S.W. Australia

H. SZARSKI, Hoyer Dept. of Comparative Anatomy, Jagiellonian University, Krupnicza 50, *Kraków*, Poland

K. S. THOMSON, Division of Vertebrate Zoology, Peabody Museum of Natural History, Yale University, *New Haven*, Conn. 06520. USA

R. THORSTEINSSON, Institute of Sedimentary and Petroleum Geology, Geological Survey of Canada, 3303 33rd Street NW, *Calgary*, Alberta. Canada

G. VANDEBROEK, Institut de Zoologie, Université de Louvain, *Louvain*, Belgium

E. I. WHITE, Dept. of Palaeontology, British Museum (Natural History), Cromwell Road, *London S. W. 7*, England

EMILIA VOROBYEVA, Palaeontological Institute of the Academy of Science of the USSR, Lenin Prospekt 33, *Moscow W-71*, USSR

R. ZANGERL, Dept. of Geology, Field Museum of Natural History, Roosevelt Road at Lake Shore Drive, *Chicago*. Illinois 60605. USA

The cyclostomes with special reference to the diphyletic origin of the Petromyzontida and Myxinoidea

By Erik Stensiö

Section of Palaeozoology, Swedish Museum of Natural History, Stockholm, Sweden

The provisional classification of the Ostracoderms used in this paper agrees with the one earlier suggested by the writer (Stensiö, 1958; 1964), but with the exception that three orders, Lyktaspida N. Heintz (in this Volume), Hibernaspida Obruchev and Olbiaspida Obruchev, are added to the Heterostraci, which hence are a very multiform group. On the basis of our present knowledge the classification of the Cyclostomi would hence be somewhat as follows.

I. Cephalaspidomorphi
    A. Osteostraci
        Orthobranchiata, Oligobranchiata, Nectaspiformes
    B. Anaspida
    C. Petromyzontida

II. Pteraspidomorphi
    A. Heterostraci
        Astraspiformes
            Astraspida, Eriptychiida
        Pteraspiformes
            Drepanaspida, Pteraspida, Traquairaspida, Lyktaspida
        Cyathaspiformes
            Cyathaspida
        Corvaspiformes
            Corvaspida
        Amphiaspiformes
            Amphiaspida, Hibernaspida, Olbiaspida
        Cardipeltiformes
            Cardipeltida
        Turiniiformes
            Turiniida
    B. Myxinoidea

Incertae sedis
        Fam. *Polybranchiaspididae*

Thelodonti
A. Phlebolepida
B. Thelodontia

The still imperfectly known genus *Polybranchiaspis* which Liu (1965) classified with the Heterostraci, differs considerably from these in several morphologically important characters. Since at least in the majority of these characters it is reminiscent of the Osteostraci, *Polybranchiaspis* may very well turn out to be an aberrant representative of the latter group of Ostracoderms. That being so, *Polybranchiaspis* is left out of consideration in the present paper.

On the account that they are a provisional group comprising slightly known forms which may belong to both the Cephalaspidomorphi and Pteraspidomorphi, the Thelodonti are not either discussed in the subsequent account.

Since in 1927 the present writer has maintained that already at their first appearance in the geological sequence, the cyclostomes had divided into two principal stocks, the Cephalaspidomorphi and Pteraspidomorphi, and that the Petromyzontida have evolved from some primitive cephalaspidomorphs, whereas the Myxinoidea are derived from some early Heterostraci forms. Thus I regard the Petromyzontida and Myxinoidea as diphyletic groups far apart. The opinion just referred to concerning the affinities of the Heterostraci to the Myxinoidea and the diphyletic origin of the Petromyzontida and Myxinoidea has been strongly contested by several writers above all on the account that both the Petromyzontida and Myxinoidea possess a complicate rasping tongue.

On purpose to elucidate the reasons of my opinion of the evolution of the Petromyzontida and Myxinoidea it may be appropriate first to list some of the characters common to both fossil and recent cyclostomes and then, after this, to deal with the petromyzontid characters of the Osteostraci and Anaspida and the myxinoid characters of the Heterostraci.

General cyclostome characters

Out of the numerous characters of this category (Stensiö, 1927; 1932; 1958; 1964; Holmgren & Stensiö, 1936; Jarvik, 1964; 1965) only the following fourteen (A–N), thirteen of which (A–M) refer to the visceral skeleton and the branchial apparatus, will here be specially mentioned.

(A) The fact that the visceral arches are not segmented off from the endocranium but are directly continuous with this (Figs. 1, 2, 8, 17, 21, 22).

In the Osteostraci and in all the better known Heterostraci this was true of all the visceral arches. This condition still is persisting to a great extent in the Petromyzontida, and it is clearly traceable in the Myxinoidea too, where, however, the vast majority of the visceral arches are lost. In this particular respect the Anaspida may be supposed to have been nearest as the Petromyzontida (Stensiö, 1958; 1964).

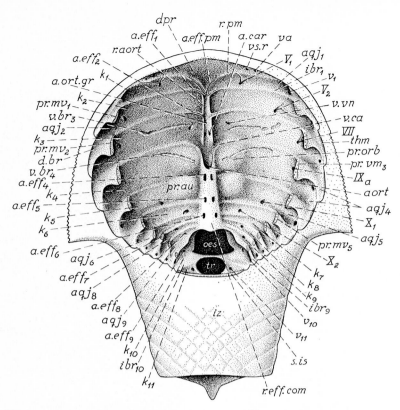

Fig. 1. *Kiaeraspis auchenaspidoides* Stensiö. Cephalic shield in ventral view. Hypotrematic division of visceral endoskeleton removed. From Stensiö, 1958; 1964.

a.car. canal for internal carotid artery; *a.eff*$_1$, *a.eff*$_2$, grooves for efferent branchial arteries 1 and 2; *a.eff*$_4$–*a.eff*$_9$, foramina and canals for efferent branchial arteries 4–9; *a.eff.pm*, groove for premandibular efferent branchial artery; *aort.gr*, aorta groove; *aqj*$_1$–*aqj*$_9$, extrabranchial atria where the epitrematic and hypotrematic extrabranchial spaces in the inidvidual branchial compartments joined; *h.br*, impressions of external branchial ducts; *dpr*, supraoral area (dorsal boundary of oral cavity; roof of stomodaeum invagination); *ibr*$_1$–*ibr*$_{10}$, interbranchial ridges 1–10, showing the position of the epitrematic halves of the mandibular, hyoid, glossopharyngeal and vagal arches in the endoskeleton of the cephalic shield; *iz*, interzonal part of exoskeleton of cephalic shield; k_1–k_{11}, branchial fossae (k_1, premandibular fossa; k_{11}, hindmost vestigial fossa without gills); *oes*, oesophagal foramen in postbranchial wall; *pr.au*, otic prominence; *pr.mv*$_1$–*pr.mv*$_9$, medio-ventral processes of *ibr*$_1$–*ibr*$_9$, for the connection with the hypotrematic division of the visceral endoskeleton; *pr.orb*, orbital prominence; *r.aort*, aortic ridge; *r.eff.com*, subaortic ridge; *r.pm*, premandibular ridge, showing the position of the epitrematic halves of the premandibular arches of both sides; *s.is*, section area of foremost median part of ventral exoskeletal wall of *iz*; *tr*, ventral foramen in postbranchial wall, for the truncusarteriosus; v_1–v_{11}, canals for lateral transverse superficial veins 1–11; *vbr*$_3$, *vbr*$_4$, grooves caused by transverse superficial veins (probably v_3 and v_4); *v.ca*, paired groove for anterior cardinal vein; *v.vn*, canal for the preorbital anastomosis between the facial vein and the anterior cardinal veins; *va*, vascular canal; *vs.r*, canal for rostral vein sinus; V_1, canal for visceral trunk of n. trigeminus I (premandibular nerve); V_2, canal for visceral trunk of n. trigeminus II (mandibular nerve); *VII*, canal for facialis nerve; IX_a, canal for n. glossopharyngeus (external glossopharyngeus canal leading out from labyrinth cavity); X_1, X_2, canals for branchial trunks of n. vagus.

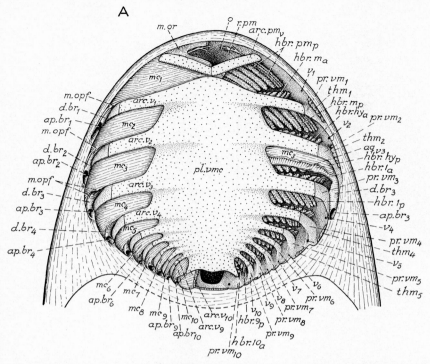

Fig. 2. *Mimetaspis* sp. *A*, hypothetical restoration of the hypotrematic parts of the visceral endoskeleton in position relative to cephalic shield. Ventral view. The branchial constrictor muscles and the external branchial ducts shown on the left side of the figure. On the right side the branchial constrictor muscles and the walls of the extrabranchial spaces removed except outside the branchial compartment 3 where they have been laid in to some extent. *B*, same hypothetical restoration as in *A*, but showing the extrabranchial spaces and the extrabranchial atria 2–10. From Stensiö, 1964.

$ap.br_1$–$ap.br_{10}$, external branchial orifices 1–10; aqj_2–aqj_4, soft tissues bounding extra-branchial atria; aq_{v2}–aq_{v5}, soft tissues bounding extrabranchial spaces; $arc.pm_v$, hypotrematic half of premandibular arch; $arc.v_1$, $arc.v_2$, $arc.v_3$, hypotrematic halves of mandibular arch, hyoid arch and glossopharyngeal arch, respectively; $arc.v_4$–$arc.v_{10}$, hypotrematic halves of vagal visceral arches; $d.br_1$–$d.br_{10}$, external branchial ducts; $hbr.m_a$, anterior hemibranch of mandibular arch; $hbr.m_p$, posterior hemibranch of mandibular arch; $hbr.hy_a$, $hbr.hy_p$, anterior and posterior hemibranchs of hyoid arch; $hbr.pm_p$, posterior hemibranch of premandibular arch

(B) The presence of unmodified premandibular and mandibular arches and the absence of true jaws.

The said two visceral arches still were fairly unchanged in the Osteostraci (Figs. 1, 2, 8, 9 A, B) where, however, the epitrematic divisions of the two participate in the formation of the cephalic shield. In the Heterostraci (Figs. 15, 16, 17 A, 21) the premandibular arch apparently had undergone certain changes, whereas in the Petromyzontida and Myxinoidea (Fig. 22) extensive modifications have taken place in the two arches concerned but in different ways. In the Anaspida, where they are unknown, these two visceral arches, no doubt also were much modified (see below).

(C) The position of the gills on branchial septa of soft tissues internal to the visceral endoskeleton.

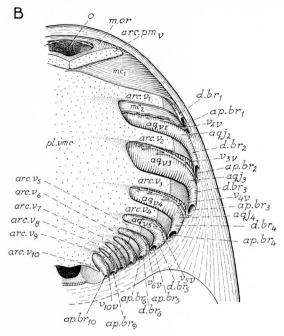

(the only hemibranch of this arch); $hbr.1_a$, $hbr.1_p$, anterior and posterior hemibranchs of glossopharyngeal arch; $hbr.9_p$, posterior hemibranch of visceral arch 9; $hbr.10_a$, anterior hemibranch of visceral arch 10; mc_1–mc_{10}, branchial constrictor muscles 1–10 (represented also by superficial portions; mc_1, the muscle of the premandibular compartment); $m.opf$, hypothetical muscle fibres effecting a dilatation of the external branchial ducts; $m.or$, probable oral muscle (entirely hypothetical), differentiated from mc_1; o, oral cavity represented solely by the stomodaeum; $pl.vmc$, probable, median ventral cartilaginous plate arisen by fusions of the inferior parts of the visceral arches (posterior to $arc.pm_v$); $pr.mv_1$–$pr.mv_{10}$, medio-ventral processes of the interbranchial ridges passing into the cartilaginous hypotrematic halves of the visceral arches; thm_1–thm_5, supratrematic thymus glands; v_{2v}–v_{10v}, superficial ventro-lateral veins.

This state of affairs occurring in the fossil cyclostomes still is persisting in the Petromyzontida (Fig. 3 A) and in principle in the Myxinoidea too. In the lower gnathostomes, on the other hand, the brancial septa lie lateral to the branchial arches (Fig. 3 B).

(D) The position of the gills on the individual branchial septa, more exactly the presence of an anterior and a posterior hemibranch on each septum and thus internally to the visceral arches (Figs. 2 A, 3 A, 8, 17).

This position of the gills was typical of all the fossil forms and is still retained in the Petromyzontida and Myxinoidea.

(E) The subdivision of the branchial apparatus by the branchial septa into branchial compartments in open communication with the pharynx throughout their height and the presence in each branchial compartment of two hemibranchs, the posterior hemibranch of the septum nearest in front and the anterior hemibranch of the septum nearest behind (Fig. 3 A).

These conditions existed in Osteostraci (Figs. 1, 2, 8, 9, A, B, 11) and in the better known Heterostraci (Pteraspida, Cyathaspida, Turiniida; Figs. 17, 21), and they are also persisting in the larvae of the Petromyzontida. On the other hand in the adult Petromyzontida and in the Myxinoidea the branchial compartments have modified to form gill pouches connected with a special subpharyngeal duct (Petromyzontida) or with the pharynx (Myxinoidea) by internal branchial ducts. Judging from the external branchial openings, the persisting branchial compartments in the adult Anaspida may also very well have tended to modify into branchial pouches.

(F) The position of the two hemibranchs in each branchial compartment in that the posterior hemibranch of the branchial septum in front extended obliquely backwards and outwards, whereas the anterior hemibranch of the septum behind had an almost transverse position relative to the longitudinal axis of the head.

That this was so in the Osteostraci and Heterostraci (Pteraspida and Cyathaspida) is proved by impressions of the anterior hemibranch in the individual branchial compartments (Figs. 2 A, 8, 17). The character is also persisting in the larvae of the Petromyzontida (Fig. 3 A) and can still be traced both in the adult Petromyzontida and in the Myxinoidea.

(G) The existence of premandibular and mandibular (spiracular) branchial compartments equipped with gills and thus with full respiratory functions. The gills in the premandibular compartment consisted of the posterior hemibranch of the premandibular arch (the only hemibranch of this arch) and the anterior hemibranch of the mandibular arch, whereas the mandibular compartment housed the posterior hemibranch of the mandibular arch and the anterior hemibranch of the hyoid arch.

Up to now this primitive stage has been met with only in the Osteostraci (Figs. 1, 2 A, 8, 11). In the Heterostraci (Figs. 16 A, 17 A, 21) the premandibular compartment had been incorporated into the oral cavity and had lost its gills (Stensiö, 1958; 1964). In the larvae of the Petromyzontida both the premandibular and mandibular compartments are persisting without gills (Fig. 3 A), and in the adult Petromyzontida they form part of the oral cavity. Judging from the position far backwards of the external branchial openings, the reduction of the branchial compartments in the Anaspida probably had proceeded even somewhat further caudally than in the adult Petromyzontida so that not only the premandibular and mandibular compartments were lost but possibly also the hyoid compartment. Finally, in the Myxinoidea both the vagal compartments and the branchial compartments anterior to them all can be recognized as distinct evaginations of the pharynx in the young embryos. Fairly soon during the further ontogenetic development, however, the premandibular, mandibular, hyoid and glossopharyngeal evaginations disappear altogether. Consequently only the vagal evaginations develop to form the gill pouches in the adult Myxinoidea.

(H) The presence of numerous branchial compartments, at the most 14–15, innervated by the vagus nerve.

14–15 vagal branchial compartments still are persisting in certain *Bdellostoma* species, and this seems also to have been the case in some Anaspida forms (*Pharyngolepis*). Generally, however, the vagal compartments were or are fewer: 5–10 in the

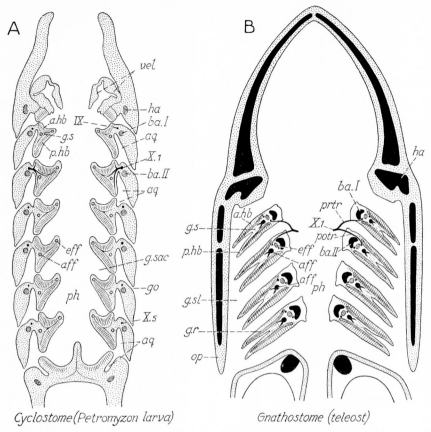

Cyclostome (Petromyzon larva)   Gnathostome (teleost)

Fig. 3. Diagrammatic horizontal sections of the head ventral to the endocranium. The two sections show the fundamental differences between cyclostomes and gnathostomes in the position, innervation and blood vascular system of the gills. A, *Petromyzon* larva; B, teleost. From Jarvik, 1964.

aff, afferent branchial artery; *a.hb*, anterior hemibranch; *ba.I*, *ba.II*, branchial arches 1 and 2 (innervated by n. glossopharyngeus and foremost branchial trunk of n. vagus, respectively); *aq*, extrabranchial spaces; *eff*, efferent branchial artery; *go*, external gill opening; *g.r*, gill ray; *gs*, interbranchial gill septum; *g.sac*, branchial compartment; *g.sl*, gill slit; *ha*, hyoid arch; *op*, gill cover; *ph*, pharynx; *phb*, posterior hemibranch; *potr*, *prtr*, posttrematic and pretrematic branches of visceral nerve trunk; *IX*, *X.1–X.5*, visceral trunks of glossopharyngeus and vagus nerves, respectively; *vel*, velum.

Heterostraci (Figs. 17 A, 21) and in the majority of the Myxinoidea, 6 in the Osteostraci (Figs. 1, 2, 8, 9 A), 5 in the Petromyzontida and probably 5–9 in the majority of the Anaspida.

(I) The mode of opening to the exterior of the respiratorily functioning branchial compartments by means of separate, paired external (efferent) branchial ducts.

This condition was persisting in the Osteostraci (Figs. 2, 10 A, D) and Anaspida and is still retained in the Petromyzontida and in the bdellostomids. On the other hand, in the vast majority of the Heterostraci the external branchial ducts of each side open to the exterior through a single external branchial opening (Figs. 17 A, 19 D, 20 C, D, 21), and, as is well known, this is also the case in the myxinoids (*Myxine* and the other representatives of this family).

(J) The strikingly broad (deep) shape of the gill lamellae in the individual hemibranchs.

Such a shape of the gill lamellae is distinctive of the recent cyclostomes, both the Petromyzontida and Myxinoidea. That the gill lamellae had a similar shape in the ostracoderms is shown by impressions of the hemibranchs on the roof of the oralobranchial chamber in the Osteostraci (Fig. 8) and on the inside of the exoskeletal armour in the Pteraspida and Cyathaspida (Fig. 17).

(K) The presence of two separate metameric trigeminus nerves, trigeminus I (premandibular nerve) and trigeminus II (mandibular nerve) both represented by dorsal and branchial trunks (Stensiö, 1927; 1958; 1963 b; 1964).

The n. trigeminus I and n. trigeminus II were persisting as such only as long as the premandibular and mandibular branchial compartments possessed respiratory functions. As pointed out under *G*, that was still the case in the Osteostraci, where the presence of the two branchial trunks can be established (V_1, V_2, Figs. 1, 8, 11). In the Heterostraci where the premandibular branchial compartment had lost its respiratory function and had been incorporated in the oral cavity the branchial trunk of trigeminus I (V_1) naturally must have undergone a considerable modification. In the Anaspida this must have been true not only of the branchial trunk of trigeminus I but also that of trigeminus II. In recent cyclostomes the two branchial nerve trunks concerned have even undergone such an extensive modification that their original nature can be made out only on the basis of the conditions in the Osteostraci.

(L) The innervation of the branchial apparatus solely by posttrematic nerves situated in the outermost parts of the branchial septa slightly antero-internal to the visceral endoskeleton and the breaking up of each of these nerves into minor branches for the corresponding branchial septum and its anterior and posterior hemibranchs (the posterior hemibranch of the branchial compartment nearest in front and the anterior hemibranch of the branchial compartment nearest behind).

Such an innervation of the branchial septa and gills apparently existed in all the better known fossil cyclostomes (Figs. 8, 11) and is still persisting both in the Petromyzontida and Myxinoidea (*IX, X.1–X.5*, Fig. 3 A). In contrast to this each branchial septum and its gills in the lower gnathostomes are innervated by two nerve branches, the posttrematic branch of its own branchial nerve and the pretrematic branch of the branchial nerve nearest behind (*prtr, potr*, Fig. 3 B). Moreover, these two nerve branches run downwards lateral to the branchial arches.

(M) The position of the afferent and efferent branchial arteries in the outer and inner parts, respectively, of the individual branchial septa and thus internal to the visceral arches between two consecutive branchial compartments. Each afferent branchial artery gave off branches to the hemibranchs on both sides of the corresponding branchial septum, whilst each efferent branchial artery arose by the confluence of branches from the same two hemibranchs.

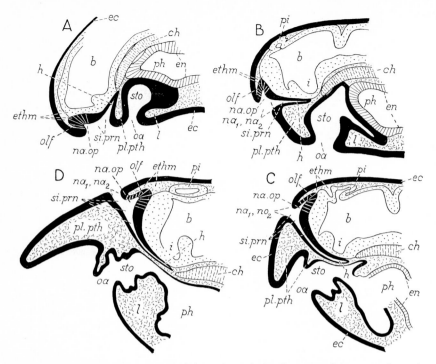

Fig. 4. *Petromyzon* sp. Four stages of the embryonic development of the snout shown in median, longitudinal sections. *A*, the youngest stage, *D*, the oldest stage. From Goodrich, 1909.

b, brain; *ch*, notochord; *ec*, ectoderm (solid black); *en*, entoderm; *ethm*, ethmoidal part of cranial division of head (superior, cranial component of definitive snout; embryonic snout corresponding to the definitive snout in the gnathostomes); *h*, adeno-hypophysis and its cavity; *i*, hypothalamus with neuro-hypophysis; *l*, lower lip; na_1, na_2, naso-hypophysial opening (opening of prenasal sinus); *na.op*, nasal opening proper; *oa*, future oral opening; *olf*, olfactory organ; *ph*, pharynx; *pi*, parietal and pineal organs; *pl.pth*, posthypophysial fold ("upper lip"; inferior, visceral component of definitive snout); *si.prn*, prenasal sinus (extracephalic space secondarily enclosed in definitive snout); *sto*, stomodaeum invagination.

The characters just mentioned existed in the Osteostraci (Figs. 1, 8, 11; Stensiö, 1964, fig. 8), the better known Heterostraci and presumably also in the Anaspida, and they are still retained in the Petromyzontida (Fig. 3 A). In the Myxinoidea, on the other hand, each afferent artery leads directly into the capillary system of a single gill pouch and the efferent artery (single or double) arises directly from the capillary system of the same gill pouch. Thus in the Myxinoidea each efferent branchial artery carries blood to the two hemibranchs in each original branchial compartment (the posterior hemibranch of the branchial septum in front and the anterior hemibranch of the septum behind), while the corresponding efferent branchial artery drains these two hemibranchs. As may readily be gathered, one is here concerned with a secondary condition arisen as a result of the special, complicate structure of the branchial pouches.

(N) Up to now the olfactory organ in the recent cyclostomes has generally been interpreted as an unpaired formation, but there have been, and still are, divergent opinions whether or not this state of affairs should be regarded as a primitive character. On the basis of the embryological investigations published after the turn of the century (Lubosch, 1905; Sewertzoff, 1916; Neumayer, 1938; Damas, 1944; Holmgren, 1946; Johnels, 1948; etc.) and with the guidance of the conditions in the Osteostraci, however, a different interpretation of the olfactory organ in the cyclostomes can now be arrived at.

In the lower gnathostomes each external endocranial nasal opening (fenestra endonarhina) pierces the morphologically anterior wall of the nasal capsule and is separated from the corresponding opening of the opposite side by a more or less thick internasal wall which consists of bone, cartilage or soft tissue, and which in the latter case may even be as thin as to be represented only by a septum. Such a paired external nasal opening (fenestra endonarhina) also exists in the recent cyclostomes (Lubosch, 1905; Neumayer, 1938), where, however, it has not been recognized as such, apparently for the reason that it leads out into the "hypophysial sac" or "duct" (see Stensiö, 1927; 1932; 1958; 1964). Because, as we shall see, it is in fact an extracephalic space secondarily enclosed in the snout, this "sac" or "duct" is referred to in the sequel as the *prenasal sinus*, whereas the term *naso-hypophysial aperture* (Stensiö, 1958; 1964) is retained for its unpaired external opening through which the water was conveyed inwards into, and outwards from, the paired external nasal opening proper.

The primordium of the olfactory organ in the lower gnathostomes is a paired placode. On the other hand, the primordium of the olfactory organ in the Petromyzontida is said to be an unpaired placode, which during its further development behaves in such a way that, as maintained by several writers, it must originally have been paired. Thus fairly soon during the course of the ontogenetic development this unpaired placode becomes bilobed and by an evagination on each side of the median line gives rise to a right and a left nasal sac (*sac.n*, Fig. 5; *olf*, Fig. 4) separated by a descending, internasal septum (*s.in*, Fig. 5) of soft tissues (Kupffer, 1894; Lubosch, 1905; Peter, 1906; Goodrich, 1909; Matthes, 1934; Hagelin & Johnels, 1955). During its ontogenetic development the nasal sac in the Petromyzontida hence is a paired organ, and this condition is also retained after the metamorphosis (Lubosch, 1905). The presence of the internasal septum effects that the nasal sac of each side possesses an external opening and that hence already prior to the development of the cartilaginous nasal capsule there has arisen a paired external nasal opening (*na.op*, Figs. 4, 5) which leads out into the prenasal sinus (*si.prn*) both downwards and forwards. Since after the origin of the cartilaginous nasal capsule it lies in the anterior wall of its own half of this capsule the paired external nasal opening is in fact a fenestra endonarhina which corresponds perfectly to the similarly termed, likewise paired, opening in the gnathostomes. Finally, it is also clear now (Kupffer, 1895, fig. 24; Damas, 1944; Johnels, 1948: 173, fig. 14), that the nasal capsule in the Petromyzontida arises embryologically from a paired primordium and that, therefore, it must originally have been a paired

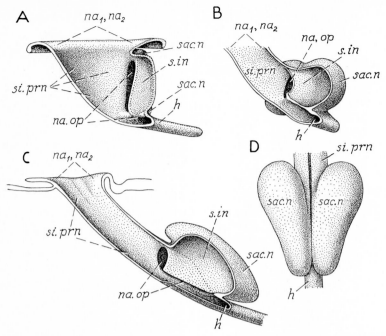

Fig. 5. *Petromyzon planeri*. Drawings of the olfactory organ in larvae at three different stages of development. *A*, the youngest stage; *C*, *D*, the oldest stage. In *A–C* the left half of the prenasal sinus and the principal part of the lateral wall of the left nasal sac removed; lateral view. *D*, nasal sacs of both sides in dorsal view. From Lubosch, 1905.

h, adeno-hypophysis with its cavity; na_1, na_2, naso-hypophysial opening; *na.op*, nasal opening proper (fenestra endonarhina) of right side; *sac.n*, nasal sac; *s.in*, internasal septum; *si.prn*, prenasal sinus.

formation, just as in the gnathostomes. The olfactory organ in the Myxinoidea develops from a seemingly unpaired placode in essentially the same way as in the Petromyzontida so that there arise two nasal sacs (*olf*, Figs. 6, 7; *sac.n*, Fig. 6 C), a right and a left one, separated by an internasal septum of soft tissues (*s.in*, Figs. 6 C, 7; Neumayer, 1938; Marinelli & Strenger, 1958). Each nasal sac leads out into the prenasal sinus (*si.prn*) through an external opening of its own (*na.op*, Figs. 6, 7), an opening which in the adults pierces the anterior wall of the corresponding half of the nasal capsule. Consequently the nasal sac and the external nasal opening both are paired formations in the Myxinoidea too. Judging from Neumayer (1938, pls. 7–10) those of the minor embryonic cartilages in *Bdellostoma* which fuse to form the nasal capsule have a bilaterally symmetrical arrangement indicating that this capsule in the Myxinoidea originally also must have been paired. The olfactory nerve is paired in both the Petromyzontida and Myxinoidea, but in the latter group it is represented by fila olfactoria (*ff.olf*, Fig. 22 B) which also is the case in many gnathostomes. Moreover, the group of the fila olfactoria of each side in the recent cyclostomes,

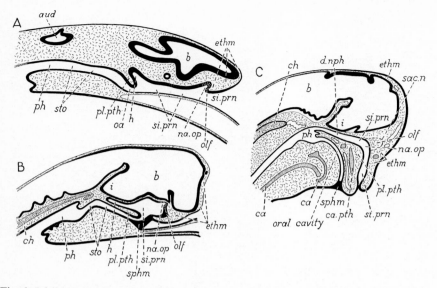

Fig. 6. *Bdellostoma stouti* Lockington. Three stages of the embryonic development of the anterior part of the head shown in median, longitudinal sections. *A*, youngest stage; *C*, oldest stage. From Kupffer, 1900; 1906.

aud, auditory organ; *b*, brain; *ca*, cartilages; *ca.pth*, cartilage in post-hypophysial fold; *ch*, notochord; *d.nph*, naso-pharyngeal duct; *ethm*, ethmoidal part of cranial division of head (superior, cranial component of definitive snout; embryonic snout, corresponding to the definitive snout in the gnathostomes); *h*, adeno-hypophysis with its cavity; *i*, hypothalamus; *na.op*, nasal opening proper; *oa*, future oral opening; *olf*, olfactory organ; *ph*, pharynx; *pl.pth*, posthypophysial fold ("upper lip"; inferior, visceral component of definitive snout); *sac.n*, nasal sac (position denoted in *C* with an interrupted line); *si.prn*, prenasal sinus (extracephalic space secondarily enclosed in definitive snout); *sphm*, secondary membrane closing oral and naso-hypophysial openings; *sto*, stomodaeum.

both the Petromyzontida and Myxinoidea, consist of the neurites from the olfactory sac of the same side (Holmgren, 1946, fig. 34). This state of affairs is of importance in that it corroborates the conclusion here arrived at that, like the gnathostomes, the recent cyclostomes actually also possess a paired nasal sac. Finally, it should be stressed here that the olfactory centre in the forebrain is paired in both the Petromyzontida and Myxinoidea.

It is clear from what has just been set out that the olfactory organ in the recent cyclostomes actually is paired and of the same fundamental type as in the gnathostomes but has undergone a considerable modification to which there is no parallel in the gnathostomes. More precisely, this modification refers to the following conditions: (1) the early development from an at least seemingly unpaired placode; (2) the fusion of the originally separate nasal capsules of both sides; (3) the presence in the snout of an extracephalic space, a prenasal sinus ("hypophysial duct", "hypophysial sac") which opens directly to the exterior; and (4) the position of the paired external nasal opening in that this opening leads out into the prenasal sinus.

In order to elucidate the nature of the prenasal sinus and the reason why the paired external nasal opening leads out into this sinus it is necessary to deal here with the ontogenetic development of the snout in the recent cyclostomes.

Already at the early embryonic stage at which the first differentiation of the olfactory organ has taken place there develops in the Petromyzontida and Myxinoidea a median, antero-ventrally directed fold slightly in front of, and ventral to, the anterior end of the notochord. This fold, the posthypophysial fold (*pl.pth*, Figs, 4 A, 6 A; Stensiö, 1958; 1964; "upper lip" in the older literature), which on each side is directly continuous with the adjacent parts of the head and latero-ventrally passes into the lateral wall of the stomodaeum invagination (*sto*), is lined with epidermis on its exposed faces and on its anterior margin. To be more exact, the fold under discussion arises immediately posterior to the adeno-hypophysial invagination (*h*) between this and the stomodaeum invagination (*sto*) so as to form the anterior part of the roof of the latter invagination. With the proceeding development of the embryos the posthypophysial fold grows out forwards (*pl.pth*, Figs. 4 B, C, 6 B) below the embryonic snout which is formed by the ethmoidal part of the cranial division (*ethm*, Figs. 4 A–C, 6 A, B) of the head and therefore corresponds to the snout in the gnathostomes. Since however, the median part of the fold fails to fuse with the suprajacent inferior wall of the embryonic snout there is retained between the two an anteriorly open extracephalic space (*si.prn*, Figs. 4, 5, 6 A, B) into which both the cavity of the adeno-hypophysial invagination (*h*) and the paired external nasal opening (*na.op*) proper lead out. *It is this extracephalic space that is now termed the prenasal sinus (si.prn)*, whereas its external opening is referred to as the naso-hypophysial aperture (na_1, na_2, Figs. 4, 5).

After the stages just dealt with the ontogenetic development of the snout in the Petromyzontida and Myxinoidea takes place in somewhat different ways. Thus in the Petromyzontida (Fig. 4 D) the postero-ventral part of the prenasal sinus represented by the persisting cavity of the adeno-hypophysial invagination (*h*) does not lengthen backwards but terminates blindly in the fossa hypophyseos. On the other hand, the anterior half of the posthypophysial fold in the Petromyzontida (*pl.pth*, Fig. 4 C, D) undergoes a considerable modification in that after an excessive increase in size it swings upwards-forwards in such a way as to form the anterior, principal part of the definitive snout. In consequence the embryonic snout (the ethmoidal part of the cranial division of the head; *ethm*, Fig. 4) in the Petromyzontida is pushed backwards and suppressed to such an extent that the prenasal sinus (Figs. 4 C, D, 5 A–C) leads upwards and the naso-hypophysial aperture (na_1, na_2, Figs. 4 D, 5 A–C) acquires a dorsal position far behind the anterior margin of the head. Turning to the Myxinoidea we notice here first of all that the cavity of the adeno-hypophysial invagination lengthens backwards so as to open posteriorly into the antero-dorsal

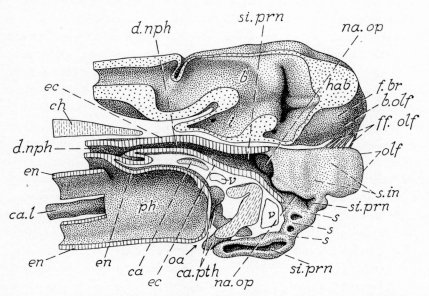

Fig. 7. *Bdellostoma stouti* Lockington. Restoration of left half of head of a somewhat advanced embryo in lateral view. The section passes slightly lateral to the median line so that the internasal septum is shown. From Neumayer, 1938.

b, brain; *b.olf*, olfactory bulb; *ca, ca.l*, cartilages; *ca.pth*, cartilages in posthypophysial fold ("upper lip"); *ch*, notochord; *d.nph*, nasopharyngeal duct; *ec*, ectoderm; *en*, entoderm; *f.br*, forebrain; *ff.olf*, fila olfactoria; *hab*, habenular ganglion; *i*, hypothalamus with neuro-hypophysis; *na.op*, nasal aperture proper of left nasal sac; *oa*, future oral opening; *olf*, olfactory organ of left side (cavity concealed by internasal septum); *ph*, pharynx; *s*, secondary septa; *s.in*, internasal septum; *si.prn*, prenasal sinus; *v*, blood vessels in posthypophysial fold ("upper lip").

part of the pharynx immediately in front of the velum. Consequently this cavity in the Myxinoidea modifies to form the naso-pharyngeal duct (*d.nph*, Figs. 6 C, 7) which, as is well known, has a respiraratory function in that by the pumping movements of the velum it conveys the water backwards to the branchial apparatus through the prenasal sinus and the naso-hypophysial aperture. Furthermore, contrary to the case in the Petromyzontida, the posthypophysial fold in the Myxinoidea (*pl.pth*, Fig. 6) grows out forwards only to a point morphologically below the anterior end of the embryonic snout (the ethmoidal part of the cranial division of the head; *ethm*, Fig. 6). That being so and since its anterior half does not increase appreciably in size, the posthypophysial fold retains throughout its position morphologically ventral to the embryonic snout, which therefore does not undergo any backward shift or suppression whatever but forms the superior component of the definitive snout forwards right to the anterior end of the head. As a result the prenasal sinus in the Myxinoidea (*si.prn*, Figs. 6, 7, 22) is strikingly long and runs morphologically forwards between the two components of the definitive snout—the ethmoidal part of the cranial division of the head and the posthypophysial fold—so that the naso-hypophysial aperture lies on the morphologically anterior face of the definitive snout. However, at later ontogenetic stages the two components of the snout bend strongly downwards causing the prenasal sinus to make a corresponding bend (Figs. 6 C, 7; Holmgren, 1946, fig. 28, Lindström, 1947, fig. 1). During a long period of the ontogenetic development the naso-hypophysial aperture in the Myxinoidea therefore is situated on the ventral side of the head closely in front of the oral opening, a fact, which, as we shall see, is of interest for the interpretation of the snout in the Pteraspida. In the adult Myxinoidea where the bend just dealt with is straightened out, on the other hand, the naso-hypophysial aperture has a subterminal position and is directed much more forwards. It should be added here that the prenasal sinus and the naso-hypophysial duct both are strikingly wide (Fig. 22; Goodrich, 1909, fig. 30; Marinelli & Strenger, 1956; figs. 88, 89, 119), a condition which naturally enables a considerable quantity of water to be pumped backwards to the branchial apparatus.

According to its position relative to the future endocranium and to its innervation (Sewertzoff, 1916; Neumayer, 1938; Johnels, 1948; Lindström, 1949, etc.) the anterior, principal part of the posthypophysial fold belongs to the visceral division of the head arisen from the epitrematic halves of the premandibular metameres of both sides, which hence undergo an extensive forward shift. All the subnasal and prenasal cartilages that are found in the definitive snout in the Petromyzontida therefore are derived from the epitrematic halves of the premandibular arches of both sides and from the skeletogenous tissue between and nearest behind (Stensiö, 1927; 1932; 1958; 1964; Holmgren & Stensiö, 1936; Damas, 1944; Johnels, 1948). This is also true of the subnasal and prenasal cartilages of the snout in the Myxinoidea (see Neumayer, 1938), but with the exception that the ring-shaped, cartilaginous elements supporting the prenasal sinus probably develop chiefly or entirely within the ethmoidal part of the cranial division of the head.

In summing up what has been stated above under point N we find the following. (a) Apart from the fusions of the nasal capsules of both sides the olfactory organ in the Petromyzontida and Myxinoidea is a paired organ. (b) That being so, the recent cyclostomes must have evolved from early agnathous forms which possessed a paired olfactory organ of the same fundamental type as the gnathostomes. This leads us to the opinion that a paired olfactory organ existed already in the earliest agnathous craniate vertebrates from which both the cyclostomes and gnathostomes have evolved. (c) The snout in the ancestors of the recent cyclostomes was originally formed by the ethmoidal part of the cranial division of the head, just as is the case in the gnathostomes. (d) In consequence of a considerable forward shift of the oralo-branchial apparatus, however, the antero-median supraoral part of this apparatus containing the epitrematic divisions of the premandibular visceral arches of both sides became

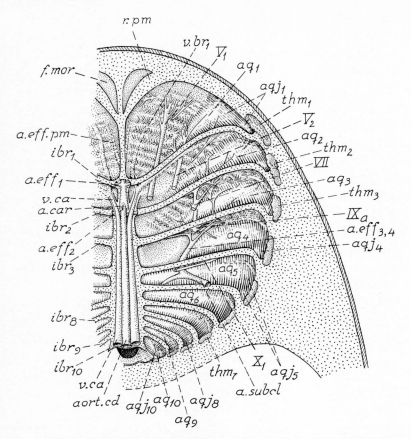

Fig. 8. "*Cephalaspis*" *signata* Wängsjö. Right half of cephalic shield in dorsal view, but with the skeleton removed in such a way that only the aorta ridge and the interbranchial ridges are left. The thin layers of soft tissues lining the epitrematic halves of the individual branchial compartments internal to the perichondral bone of the oralo-branchial space restored, showing the position of the extrabranchial spaces relative to the gills in the individual branchial compartments (see Fig. 2). From Stensiö, 1964.

a.car, internal carotid artery; *a.eff*$_1$, *a.eff*$_2$, efferent branchial arteries of the mandibular and hyoid arches, respectively; *a.eff*$_{3,4}$, efferent branchial artery formed by confluence of *a.eff*$_3$ and *a.eff*$_4$; *a.eff.pm*, efferent branchial artery of premandibular arch; *aort.cd*, unpaired cephalic division of dorsal aorta; aq_1–aq_{10}, extrabranchial spaces; aqj_1–aqj_{10}, extrabranchial atria; *a.subcl*, subclavian artery; *f.mor*, parathoral concavity; ibr_1–ibr_{10}, interbranchial ridges showing the position of the epitrematic halves of the post-premandibular visceral arches; *r.pm*, premandibular ridge, showing the position of the epitrematic half of the premandibular arch; thm_1–thm_7, position of epitrematic thymus glands; $v.br_1$, foremost, transverse, dorsal extrabranchial vein; *v.ca*, anterior cardinal vein; V_1, branchial trunk of premandibular nerve (trigeminus I); V_2, branchial trunk of mandibular nerve (trigeminus II); *VII*, facial nerve; IX_a, glossopharyngeus nerve; X_1, foremost branchial trunk of vagus nerve.

incorporated into the snout already in the ancestors common to the Petromyzontida and Myxinoidea. Thus, the definitive snout in the recent cyclostomes is a secondary, compound formation made up of a dorsal, cranial component and a ventral visceral component, the latter of which in the Myxinoidea has the same forward extent as the cranial one, whereas in the Petromyzontida it extends considerably further forwards and upwards in front of the cranial component so as also to form the anterior, dorsal, principal part of the definitive snout. A certain parallel to this condition is met with in many arthrodires (Stensiö, 1963a; 1968) and in the teleostomians and tetrapods (Jarvik, 1954), where, however, only the infrapharyngeal and suprapharyngeal segments of the premandibular arch have fused with the posterior part of the ethmoidal region. (e) Prior to the origin of the visceral component of the definitive snout the paired external nasal opening led directly to the exterior, just as in the gnathostomes. During the proceeding phylogenetic development of the recent cyclostomes this opening retained its original position, and hence it leads out into the extracephalic interspace between the cranial and visceral components of the definitive snout, i.e. into the prenasal sinus ("hypophysial duct"; "hypophysial sac"). (f) Because this space has been mistaken for an unpaired external nasal aperture, the recent cyclostomes have up to now been misinterpreted as monorhinal animals.

With regard to the definitive snout and the olfactory organ the recent cyclostomes have hence reached an advanced stage of specialization to which there is no correspondence in the gnathostomes. A snout and an olfactory organ of fundamentally the same type as in the Petromyzontida, however, existed already in the Silurian Osteostraci and Anaspida. Since, as has been maintained previously (Stensiö, 1958; 1964) and as will be further shown below, a myxinoid snout had evolved already before the Middle Ordovician, it is clear that the definitive, compound snout distinctive of the cyclostomes must have arisen still earlier and that consequently in principle this type of snout has been persisting unchanged for more than 400 million years.

Petromyzontid characters of the Osteostraci and Anaspida

As has long been known (Kiaer, 1924; Stensiö, 1927; 1932; 1958; 1963b; 1964; Heintz, 1939), the Osteostraci and Anaspida are much closer akin to each other and to the Petromyzontida than what they are to the Myxinoidea and Heterostraci. The Osteostraci, Anaspida and Petromyzontida therefore in fact represent a major stock of cyclostomes of their own, the Cephalaspidomorphi (Stensiö, 1927). However, this classification has not gained general acceptance (see Romer, 1945; 1966) for in mistaking the naso-hypophysial aperture for the external nasal opening and in disregarding the fundamental structural differences of the snout and rasping tongue in the Petromyzontida and Myxinoidea (see Jarvik, 1964; 1965) several writers still group the Myxinoidea too with the cephalaspidomorph stock which then generally is named Monorhina. Under these circumstances it may be appropriate to list here the principal

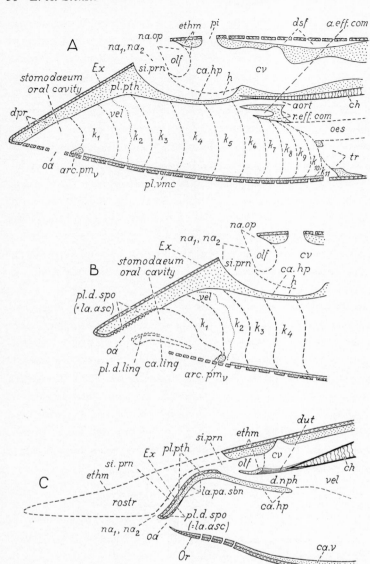

Fig. 9. *A*, diagrammatic longitudinal section of the head of an orthobranchiate Osteostraci form with the approximate position of the olfactory organ and branchial compartments denoted with interrupted lines. *B*, diagrammatic longitudinal section of the anterior part of an oligobranchiate Osteostraci form drawn in the same way as *A*, but with a possible small lingual cartilage and lingual tooth plate also laid in. *C*, diagrammatic longitudinal section of the head of a pteraspid, a section which is intended to show the different composition and structure of the snout in the Heterostraci and Osteostraci (see also Figs. 13 A, B, 16). Sections of cartilage stippled; sections of dermal bones cross-hatched.

Ex, exoskeleton on morphologically dorsal (originally external) face of inferior, visceral component of definitive snout (in the Osteostraci situated on the prenasal part of the dorsal face of the cephalic shield); *Or*, oral plate; *a.eff.com*, canal for arteria branchialis efferens communis (closed posteriorly); *aort*, canal for cephalic division of dorsal aorta and for paired anterior cardinal vein (see Fig. 8); *arc.pm$_v$*, probable position of the connection between the

Petromyzon-like characters of the Osteostraci and Anaspida. These characters, in which the Cephalaspidomorphi s.str. (Stensiö, 1927; 1932; 1958; 1963b; 1964) also differ from the Myxinoidea and Heterostraci, are briefly as follows.

(1) The Osteostraci possessed a prenasal sinus (*si.prn*, Fig. 9 A, B; Stensiö, 1963b, *hy.s*, figs. 7, 9; 1964, *hy.s*, figs. 17, 18) which, bending strongly upwards, opened to the exterior through a naso-hypophysial aperture (na_1, na_2) situated on the dorsal side of the head far behind the anterior margin of the cephalic shield. Moreover, the prenasal sinus ended blindly in the fossa hypophyseos, where apparently it also included the cavity of the adeno-hypophysial invagination. Finally, it is also clear that the olfactory organ led out into the prenasal sinus. As can be inferred from the position of the naso-hypophysial aperture, the prenasal sinus and the olfactory organ in the Anaspida must also have been essentially as in the Osteostraci.

(2) Merely on the basis of the conditions mentioned in the preceding point it is clear that the prenasal part of the snout in the Osteostraci and Anaspida is of the petromyzontid type, that this part of the snout is of a visceral nature and that is must have arisen embryologically from an excessively developed posthypophysial fold which in swinging forwards-upwards has caused an extensive suppression of the ethmoidal division of the cranial division of the head. As can be inferred from the position of the premandibular ridge (*r.pm*, Figs. 1, 8; see also Figs. 2, 11) and of the interbranchial ridge nearest behind (ibr_1) on the roof of the oralo-branchial space ("chamber"), the prenasal part of the endoskeleton of the cephalic shield in the Osteostraci comprises the epitrematic divisions of the premandibular and mandibular visceral arches together with the interjacent skeletogenous tissues. Consequently the forward shift of the

---

hypotrematic halves of the premandibular arches of both sides (see Figs. 2, 11); *ca.hp*, original ventral wall of fossa hypophyseos (hypophysial cartilage); *ca.ling*, possible lingual cartilage; *ca.v*, probable median ventral cartilaginous plate arisen by fusions of the lowermost parts of the visceral arches of both sides (=*pl.vmc*); *ch*, notochord; *cv*, cranial cavity; *d.nph*, nasopharyngeal duct; *dpr*, supraoral field (dorsal boundary of oral cavity); *dsf*, dorsal sensory field; *dut*, dural tissue forming a secondary ventral wall of the fossa hypophyseos; *ethm*, superior, ethmoidal component of definitive snout (embryonic snout); *ext*, exoskeleton on external face of inferior, visceral component of definitive snout; *h*, space for adeno-hypophysis; k_1–k_{10}, approximate positions of branchial compartments 1–10; k_{11}, vestigial branchial compartment; *la.pa.sbn*, palato-subnasal lamina (endoskeleton of visceral component of snout); na_1, na_2, nasohypophysial aperture (external opening of *si.prn*); *na.op*, external nasal opening proper (fenestra endonarhina); *oa*, oral opening; *oes*, position of oesophagus; *olf*, position of olfactory organ; *pi*, pineal opening; *pl.d.ling*, possible dental plate of tongue; *pl.d.spo*, supraoral dental "plate" continuous with exoskeleton of external side of rostral part of cephalic shield (=ascending postrostral bone lamina, *la.asc*, in the Heterostraci; see B, C and Figs. 13 B, 16 A); *pl.pth*, inferior, visceral component of definitive snout formed by posthypophysial fold ("upper lip"); *pl.vmc*, probable median ventral cartilaginous plate arisen by fusions of the inferior parts of the branchial arches of both sides (=*ca.v*); *r.eff.com*, subaortic ridge; *rostr.* rostrum; *si.prn*, prenasal sinus; *tr*, foramen below *oes*, for tr. arteriosus and probably also for an anterior part of the heart; *vel*, velum (in the Osteostraci of the same simple type as in the *Petromyzon*-larvae).

visceral division of the head in the Osteostraci was not restricted solely to the premandibular metamere but has also influenced the mandibular one, and in this particular respect the Osteostraci hence are even somewhat more specialized than the Petromyzontida. Whether this also applies to the Anaspida cannot be decided for the time being. However, since most certainly they had acquired a long complicated rasping tongue the representatives of the latter group may very well be imagined to have been approximately as the Petromyzontida in that the mandibular arch probably lay as far caudally as not to participate in the formation of the snout. In view of what has been stated concerning the embryological development and nature of the visceral component of the definitive snout in the Petromyzontida it is self-evident that the corresponding component of the definitive snout in the Osteostraci and Anaspida, like the other parts of the head and the entire trunk, is strengthened with exoskeleton.

(3) The cranial cavity in the Osteostraci is of such a shape that it can only be imagined to have housed a brain of a petromyzontid type (Stensiö, 1927; 1958; 1963b; 1964).

(4) The presence of a special bulge in the roof of the cranial cavity to the right of the median line and closely behind the pineal canal, a bulge which indicates that, like in *Petromyzon*, the right habenular ganglion was considerably larger and extended much more upwards than the left one (Stensiö, 1927; 1958; 1963b; 1964).

(5) The separate course of the dorsal and ventral roots of the individual spino-occipital and spinal nerves in the Osteostraci and (6) the alternating position of these nerves of both sides in the Osteostraci (Stensiö, op.cit.).

(7) The condition of the membraneous labyrinth in the Osteostraci which consisted of an extensive vestibular division and two semicircular canals, anterior and posterior, each with an extensive ampulla.

(8) Several details in the blood vascular system of the Osteostraci. (a) The presence of a facial artery and a facial vein in the preorbital division of the cephalic shield. (b) The conditions of the anterior cardinal veins of both sides which during their passage backwards through the postbranchial wall joined to form an unpaired anterior cardinal vein which had such an asymmetrical course to the right of the median line as clearly to show that it emptied into the right ductus Cuvieri and that hence the left ductus Cuvieri had disappeared, just as in the adult Petromyzontida. (c) The presence of longitudinal and transverse superficial branchial veins outside the branchial compartments.

Since the endoskeleton has not been found preserved in the Anaspida nothing definitive can at present be stated concerning characters 3–8 in this group of cephalaspidomorphs.

(9) The facts that except in *Nanpanaspis* (Liu, 1965; provisionally classified with the Osteostraci) the eyes were relatively large in the adults of both the

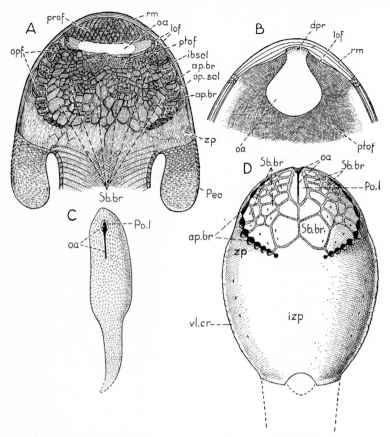

Fig. 10. Oral openings of four fossil cyclostomes. Ventral view. *A*, *Hirella gracilis* (Kiaer); *B*, *Ateleaspis* ("*Aceraspis*") *robusta* Kiaer; *C*, *Phlebolepis elegans* Pander (thelodont; for a new interpretation see Ritchie, in this Volume). *D*, *Tremataspis* sp. From Stensiö, 1958; 1964.

Pec, pectoral fin; *Po.l*, lateral oral plates (concerning *C* see Ritchie, in this Volume); *Sb.br*, subbranchial scales and bone plates; *ap.br*, external branchial openings; *dpr*, supraoral area (roof of oral cavity); *ibscl*, interbranchial scales; *izp*, interzonal part of cephalic shield; *lof*, lateral oral field of scales or plates; *oa*, oral opening; *opf*, opercular valves (of the inidvidual external branchial openings) strengthened with minute scales; *op.scl*, minute scales on opercular folds; *prof*, preoral field of scales; *ptof*, postoral field of scales; *rm*, ventral rim of cephalic shield; *vl.cr*, ventro-lateral crest; *zp*, zonal part of cephalic shield.

Osteostraci and Anaspida and that, therefore, as could be made out in the Osteostraci, the eye muscles and their nerves generally had not disappeared. In the Osteostraci the superior oblique eye muscle took origin in a conspicuous myodome of its own situated in the postero-dorsal corner of the orbital cavity immediately anterior to the auditory capsule, and as a result the trochlear nerve (IV) emerged into this myodome and thus far backwards. With regard to the superior oblique eye muscle and the exit of the trochlear nerve from the endocranium the Osteostraci hence were as the Petromyzontida. In the lower gnathostomes, on the other hand, the superior oblique eye muscle takes origin

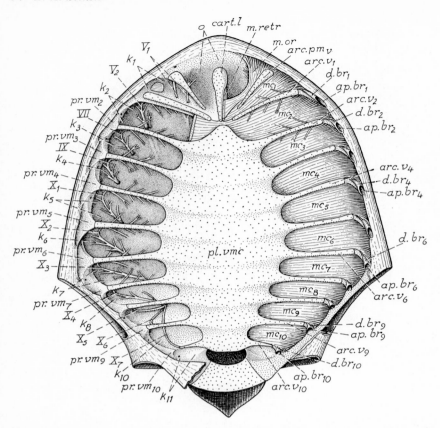

Fig. 11. *Nectaspis areolata* Wängsjö, Cephalic shield in ventral view with a hypothetical restoration of the hypotrematic division of the visceral endoskeleton. The branchial compartments and nerves are shown on the left side of the figure whilst the branchial constrictor muscles and the external branchial openings are laid in on the right side. From Stensiö, 1964.

$ap.br_1–ap.br_{10}$, external branchial openings; $arc.pm_v$, hypotrematic half of premandibular arch; $arc.v_1$, $arc.v_2$, $arc.v_3$, hypotrematic halves of mandibular, hyoid and glossopharyngeal branchial arches, respectively; $arc.v_4–arc.v_{10}$, hypotrematic halves of vagal branchial arches; $cart.l$, possible lingual cartilage; $d.br_1–d.br_{10}$, external branchial ducts; k_1, k_2, premandibular and mandibular branchial compartments, respectively; k_3, k_4, hyoid and glossopharyngeal branchial compartments, respectively; $k_5–k_{10}$, vagal branchial compartments; k_{11}, vestigial branchial compartment without gills; $mc_1–ms_{10}$, branchial constrictor muscles; $m.or$, $m.retr$, possible muscles differentiated from mc_1, for moving a small rasping tongue; o, oral cavity; $pl.vmc$, probable cartilaginous branchial plate formed by the fusion of $arc.v_3–arc.v_{10}$; $pr.vm_2–pr.vm_{10}$, medio-ventral processes of interbranchial ridges; V_1, branchial trunk of premandibular nerve (trigeminus I); V_2, branchial trunk of mandibular nerve (trigeminus II); VII, facial nerve; IX, glossopharyngeus nerve (branchial trunk); $X_1–X_7$, branchial truks of vagus nerve.

in the antero-dorsal part of the orbital cavity above the inferior oblique muscle, a state of affairs which causes the trochlear nerve to emerge from the endocranium a fair distance anterior to the auditory capsule. Hence, it is clear that the eye muscles in the Osteostraci were of the petromyzontid type. In all probability this was also true of the Anaspida.

(10) The presence of a non-biting mouth in the Osteostraci (Figs. 2, 9 A, B, 10 A, B, D, 11) and Anaspida (Stensiö, 1958; 1964, figs. 22, 23, 25, 28). In the Osteostraci this mouth has all prerequisites for modifying into one of the suctorial type similar to that in the Petromyzontida; in the Anaspida there is strong reason to believe that such a modification had already taken place and that then in this group there may have existed an "oral sucker" of fundamentally the same type as in the adult Petromyzontida.

(11) As can be inferred from the exoskeleton on its inferior (external) face (Fig. 10 A, B, D), the ventral wall of the oralo-branchial space (chamber) in the Osteostraci was flexible in such a way as to make possible constrictions and expansions of the oral cavity and branchial apparatus. This condition naturally means that the branchial constrictor musculature (mc_1–mc_{10}, Figs. 2, 11) was persisting in full extent but was restricted to the hypotrematic division of the visceral endoskeleton. More exactly, this musculature, which took origin on the most ventro-lateral part of the epitrematic division of the visceral endoskeleton close to the exoskeleton on the ventral rim of the cephalic shield (Figs. 1, 2, 11), was situated internal to the hypotrematic division of the visceral endoskeleton between this and the extrabranchial atria. Consequently the respiration (the pumping of water through the branchial apparatus) in the Osteostraci took place by contractions and expansions of the hypotrematic division of the branchial region in fundamentally the same way as in the Petromyzontida. This was still more the case in the Anaspida, where, like in the Petromyzontida, the constrictor muscles of the persisting part of the branchial apparatus apparently also possessed epitrematic divisions (Stensiö, 1964, fig, 28). Owing to the complete incorporation of the epitrematic divisions of the visceral arches into the cephalic shield the corresponding divisions of the branchial constrictor muscles in the Osteostraci had lost their function and had therefore disappeared.

As may readily be gathered from points 1–11, the Osteostraci and Anaspida agree with the Petromyzontida not only in their general plan of organization but also in numerous remarkable details. It is therefore clear that the Osteostraci, Anaspida and Petromyzontida all belong to one and the same principal stock of the cyclostomes, the Cephalaspidomorphi. However, both the Osteostraci and Anaspida are too much specialized to have given rise to Petromyzontida; these must therefore have evolved from some early, much more primitive cephalaspidomorphs which also comprised the ancestors of the Osteostraci and Anaspida.

Since they differ a good deal with regard to the oral opening, oral cavity and foremost part of the branchial apparatus the Osteostraci have provisionally been classified into three principal subgroups (Stensiö, 1958; 1964), the Orthobranchiata, Nectaspiformes and Oligobranchiata.

In the Orthobranchiata, where the oral opening is a transverse more or less broad slit (Figs. 9 A, 10 A; see also Fig. 1), the oral cavity lay entirely in front of and between the premandibular branchial compartments of both sides and was so short and small that it corresponded solely to the stomodaeum invagination in the embryos of the Petromyzontida and to the prelingual part ("the oral sucker"; Goodrich, 1909) of the oral cavity in the adult Petromyzontida. That being so, the Orthobranchiata naturally must have been primarily without a rasping tongue. The interbranchial ridge 1 (representing the epitrematic division of the mandibular arch; ibr_1, Fig. 1) is of the same type as the interbranchial ridges following further backwards, but that this is so does not necessarily mean that a velum was absent. In fact, it is very likely that the Orthobranchiata possessed a simple velum (*vel*, Fig. 9 A) of the same type as in the *Petromyzon* larva (*vel*, Fig. 3 A).

In the Nectaspiformes, where the oral opening still is unknown, the oral cavity (Fig. 11) was not restricted solely to the stomodaeum invagination but extended also backwards between the premandibular branchial compartments of both sides so that these are small and lie far apart. Judging from the conditions in the recent cyclostomes, the backward extension of the oral cavity just referred to most certainly indicates that the Nectaspiformes had acquired a small, primitive rasping tongue with a musculature of its own differentiated from the antero-inferior portions of the premandibular branchial constrictor muscles of both sides. Moreover, the interbranchial ridges 1 (ibr_1) of both sides in the Nectaspiformes have fused dorso-medially to form a conspicuous, backward arching, transverse velar crest (Stensiö, 1958; 1964, fig. 15) to which a well developed velum must have been attached, a velum which probably was essentially as in the *Petromyzon* larva (*vel*, fig. 3 A).

In the Oligobranchiata the oral cavity also had undergone a backward lengthening (Figs. 8, 9 B), a lengthening which, however, has taken place in a way different from that in the Nectaspiformes in that the three foremost interbranchial ridges (premandibular, mandibular and hyoid ridges; *r.pm*, ibr_1, ibr_2, Fig. 8) have been pushed backwards in their dorsal and ventral parts and, therefore, together with the corresponding parts of the visceral arches have acquired a conspicuous oblique position relative to the longitudinal axis of the head. Furthermore, at least in some oligobranchiate forms the anterior part of the roof of the oral cavity is covered with tubercle-bearing exoskeleton (a "maxillary tooth plate"; a process from the exoskeleton on the ventral rim of the cephalic shield or separate minor plates; *pl.d.spo*, fig. 9 B; see also Stensiö, 1964, fig. 116) against which a moveable median ventral apparatus may have worked. The two conditions just mentioned render it very likely that a small, primitive rasping tongue existed in the Oligobranchiata too, but had evolved independently of that in the Nectaspiformes. In harmony with this opinion is the fact that at least in the better known Oligobranchiata the oral opening is a longitudinal slit (Fig. 10 B, D) which apparently could be much expanded and thus made possible a protraction of a rasping tongue. Like in the Orthobranchiata, the interbranchial ridges 1 of both sides (the epitrematic divisions of the mandibular arches) do not show anything concerning the presence of a velum in the Oligobranchiata, which therefore probable also possessed a velum of the same type as in the *Petromyzon* larva.

The Anaspida (Stensiö, 1964, figs. 23, 25, 28, 29 A, B) have a circular oral opening which in all respects is such as to indicate an oral apparatus of a suctorial type reminiscent of that in the adult Petromyzontida. Another distinctive feature of the Anaspida is that the external branchial openings all lie closely together in a backward-downward sloping row a fair distance posteroventral to the orbital opening and thus far behind the oral opening. Moreover, the exoskeleton on the lateral and ventral sides of the head and branchial region in this group of cephalaspidomorphs has undergone an extensive disintegration into minor plates and scales. The aforementioned conditions can be understood only on the basis of the assumption that there existed a long rasping tongue fundamentally similar to that in the adult Petromyzontida or even longer. The evolution of such a rasping tongue would naturally have caused a long backward extension of the oral cavity, a total reduction of the gills in the two or three foremost branchial compartments (the premandibular and mandibular compartments and possibly also the hyoid compartment), a considerable backward shift of the mandibular arch, and an extensive modification of the original constrictor musculature in at least the premandibular and mandibular metameres, a modification which in turn would have brought about extensive changes in the entire preotic part of the visceral endoskeleton. The moveability of a suctorial mouth and of a long, complex rasping tongue must also have necessitated a great flexibility of the surrounding parts of the skin so that these could not contain any more extensive dermal bones. In fact, the presence of such a mouth conjointly with a long rasping tongue explain in a satisfactory way the conditions with regard to the exoskeleton of the head and branchial region in the Anaspida, viz. that these parts of the animals are covered solely with minor non-overlapping, thin bone plates and scales which in some forms even had disappeared more or less completely. Further at least in one genus, *Pharyngolepis*, there has been found a "mandibular" plate (Kiaer, 1924), which, judging from its shape, most certainly was attached to the anterior part of a rasping tongue in such a way that it worked against a "maxillary tooth plate" of the same nature as the similarly termed bone plate in the oligobranchiate Osteostraci (Stensiö, 1958; 1964). According to Ritchie (1964) the "mandibular" plate in *Pharyngolepis* would have been wedged in between the exoskeletal elements ventral to the oral opening and oral cavity. If this were true there would naturally exist other, somewhat extensive ventral plates to which the "mandibular" plate was attached. However, this is not the case in that the exoskeletal elements posterior to the oral opening in *Pharyngolepis* all are small, vestigial scales situated losely in the skin. That being so and since it does not give any clue to the extensive modifications which apparently has taken place in the oral and branchial apparatus, Ritchie's interpretation of the "mandibular" plate seems unlikely from all points of view.

By the above comments on the oral apparatus I have intended to stress that in the cephalaspidomorphs there was a general trend towards the evolution of a rasping tongue and that this condition decidedly favours the opinion that the rasping tongue in the Petromyzontida has arisen independently of that in the Myxinoidea, an opinion which is corroborated by the great anatomical differences between these two groups not only with regard to the rasping tongue itself but also in most other respects.

Myxinoid characters and other special conditions in the Heterostraci

As may readily be gathered from what has been stated above, the Heterostraci have so many general characters in common with the Cephalaspidomorphi and Myxinoidea, that, like these, they have to be considered as true cyclostomes.

The endoskeleton of the cranium and branchial apparatus in the Heterostraci consisted of cartilage and could therefore not be preserved in the fossils. However, the cartilage was of a fairly firm nature so that not infrequently it caused more or less extensive impressions on the inner faces of the armour, while, on the other hand, in the unarmoured form, as *Turinia* for example, its general external characters are fortuitously shown to some extent in relief. In some of the armoured forms the inner faces of the armour also exhibit impressions of the gills and extrabranchial atria. Since the conditions just mentioned make an investigation of the internal details of the endocranium and oral apparatus impossible, the Heterostraci naturally are much less known from an anatomical point of view than the Osteostraci. All the same it is clear that with regard to the following characters the Heterostraci were fundamentally as the Myxinoidea.

(I) With the exception of the Lyktaspida and Drepanaspida, both of which are separately dealt with in the sequel, the dorsal wall of the prepineal part of the head is not pierced by any opening or openings leading inwards to the olfactory organ or pharynx (Figs. 9 C, 12, 13, 14, 15 A, 16, 17 A, 19, 20). This means that, including both exoskeleton and endoskeleton, the dorsal, prepineal part of the snout in the Heterostraci cannot possibly comprise any premandi-

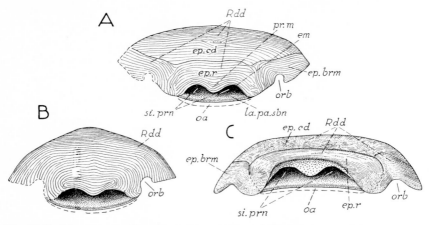

Fig. 12. Sketches showing the writer's interpretation of the superior, cranial component of the definitive snout in three different cyathaspids without a rostrum. The rostro-dorsal disc in anterior view. *A, Dikenaspis yukonensis* Denison; *B, Vernonaspis sekwiae* Denison; and *C, Aequiarchegonaspis schmidti* (Geinitz). The sketches all are based on drawings and photographs published by Denison, 1964. Cartilage shaded with stippling.

Rdd, rostro-dorsal disc; *em*, paired emargination in antero-ventral margin of *Rdd*; *ep.brm*, prebranchial marginal epitegum; *ep.cd*, central dorsal epitegum; *ep.r*, rostral epitegum; *la.pa.sbn*, palato-subnasal lamina (endoskeleton of posthypophysial fold); inferior, visceral component of definitive snout; *oa*, position of oral opening; *orb*, orbital notch; *pr.m*, descending median process of antero-inferior margin of *Rdd*; *si.prn*, prenasal sinus and its external opening (the naso-hypophysial opening).

bular visceral component but must be formed solely by the ethmoidal part (*ethm*) of the cranial division of the head, and that hence it corresponds alone to the cranial component of the definitive snout in the recent cyclostomes. Thus, contrary to the case in the Osteostraci, Anaspida and Petromyzontida the cranial component of the snout in the Heterostraci had not undergone any backward shift and suppression as a result of an excessive development and forward-upward swing of the anterior dorso-medial part of the premandibular, visceral division of the head but retained its full forward extent and position in the same way as in the Myxinoidea and gnathostomes.

(II) As is clear from the position of the pineal pit and from other impressions on the basal face of the dorsal armour in the Pteraspida and Cyathaspida (Figs. 13 A, B, 14 D, 15 A, 16 A, B, 17; see also Stensiö, 1964, figs. 57, 59, 81, 82, 97, 102–107, 109–112, 114, 122 A), the cranial cavity in the Heterostraci extended only a short distance forwards into the ethmoidal part of the cranial division of the head (Stensiö, 1963*b*). Since the basal face of the dorsal armour does not show any impression whatever of the roof of this ethmoidal part of the cranial cavity, the telencephalon in the Heterostraci must have lain somewhat lower down than the other divisions of the brain (Figs. 13 A, B, 15, 16 A, B), and in some forms (Fig. 16 B) it may also, like in the somewhat advanced embryos of the Myxinoidea (Figs. 6 C, 7; Stensiö, 1963*b*), have been bent down to such an extent as to have lain chiefly below the nasal capsule. In analogy with the condition in the other cyclostomes (the Cephalaspidomorphi and Myxinoidea) the paired olfactory organ (*olf*, Figs. 13 A, B, 16, 21) may be assumed to have been situated closely in front of the telencephalon. That this was so is proved by the impression of the nasal capsule in some Pteraspida forms (Fig. 15 A; Stensiö, 1964, figs, 59, 112, 122 A), an impression which also shows that the nasal capsule was secondarily unpaired and that it did not extend forwards beyond the posterior third or thereabouts of the total length of the ethmoidal part of the cranial division of the snout. The opinion maintained by several writers that the nasal capsule in the Heterostraci would have been distinctly paired and would have been situated much further forwards in the snout hence is untainable from all points of view.

(III) From what has been set out under points I and II it naturally follows that the paired external nasal opening (*na.op*, Figs. 13 A, B, 16, 21) which pierced the anterior wall of the nasal capsule did not lead directly to the exterior but opened into a broad, forward or forward-downward extending interspace between the ethmoidal part of the cranium (*ethm*) and the anterior dorso-median part of the visceral division of the premandibular metamere (*pl.pth*, *la.pa.sbn*), i.e. into a prenasal sinus (*si.prn*) homologous with that in the Cephalaspidomorphi and Myxinoidea.

In order to elucidate this condition we shall now enter somewhat in detail upon the structure of the snout in the Cyathaspida and Pteraspida, and in so doing we shall first turn to two Cyathaspida forms, *Vernonaspis sekwiae* and *Dikenaspis yukonensis* (Denison, 1964). In these two forms, which lack a rostrum and where the rostro-dorsal disc of the armour extends somewhat downwards on to the anterior face of the ethmoidal part of the cranial division of the head, the antero-inferior margin of this disc (Fig. 12 A, B) shows two

---

Fig. 13. *A*, diagrammatic, hypothetical, longitudinal section of head in a cyathaspid without rostrum. Sections of cartilage stippled; sections of dermal bones cross-hatched. *B*, corresponding hypothetical section of a cyathaspid with a short rostrum (*Poraspis*) drawn in the same way as *A*. *C*, attempted hypothetical restoration of a cyathaspid with a short rostrum in anterior view. *B* and *C* from Stensiö, 1958; 1964 (*B* slightly altered).

Br, branchial plate; *Ifo*, infraorbital plate; *Og*, orogonial plate; *Or*, oral plates; *Pl.d.spo*, probable superior labial dental plates; *Rdd*, rostro-dorsal disc; *Vd*, ventral disc; *ca.hp*, hypophysial cartilage constituting hindmost part of *la.pa.sbn*; *ca.r*, rostral cartilage; *ca.sor*, probable cartilaginous suboral plates to which the oral plates were attached; *ca.v*, median, ventral part of visceral endoskeleton; *ch*, notochord; *cv*, cranial cavity; *d.nph*, naso-pharyngeal duct; *ethm*, superior, cranial component of definitive snout (= embryonic snout); *la.asc*, ascending postrostral lamina; *la.pa.sbn*, palato-subnasal lamina (endoskeleton of inferior, visceral component of definitive snout + *ca.hp*); *la.sbr*, subrostral lamina of rostral plate; na_1, na_2, naso-hypophysial opening (external opening of *si.prn*); *na.op*, nasal opening proper; *oa*, oral opening; *olf*, olfactory organ; *pi*, pineal pit; *pl.pth*, endoskeleton of posthypophysial fold; *pr.m*, descending median process of antero-inferior margin of *Rdd*; *rm* median ridge of *la.sbr*; *rostr*, rostrum; *si.prn* prenasal sinus (in *C* its anterior bilobated opening also shown); t_1, t_2, possible tentacles; *vel* velum (most certainly of a myxinoid type).

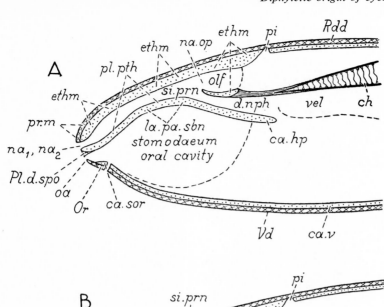

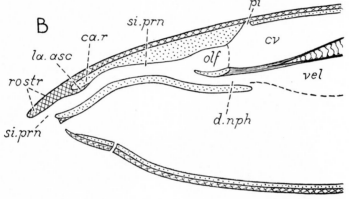

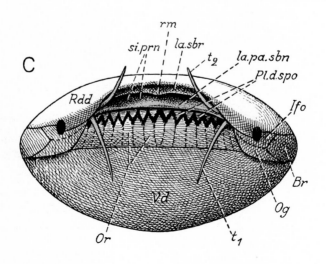

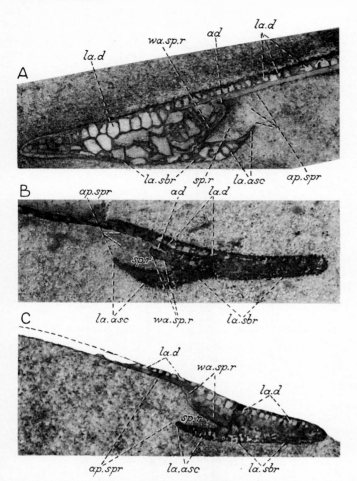

Fig. 14 A–C. *A*, median longitudinal sections of anterior part of rostral plate in three different pteraspids. *A, B*, indeterminable forms; *C, Zascinaspis* sp. From Stensiö, 1958; 1964.

ad, antero-doral angle of rostral space; *ap.spr*, posterior opening of rostral space; *la.asc*, ascending postrostral lamina of rostral plate; *la.d*, dorsal lamina of rostral plate; *la.sbr*, subrostral lamina of rostral plate; *sp.r*, rostral space; *wa.sp.r*, anterior wall of rostral space.

conspicuous notches (*em*), a right and a left, separated by a rounded-off, descending, median process (*pr.m*). These two notches and the interjacent process jointly formed the dorsal, exoskeletal boundary of the bilobated external opening of a broad prenasal space (*si.prn*) into which the paired external nasal opening led out. That this space cannot possibly be the oral cavity is immediately clear from the fact that it lay entirely above the anterior mediodorsal part of the visceral division of the head which in all cranial vertebrates, both cyclostomes and gnathostomes, forms not only the dorsal boundary of the oral opening but also the roof of the stomodaeum invagination. Hence we find that the space concerned (*si.prn*) in *Vernonaspis sekwiae* and *Dikenaspis*

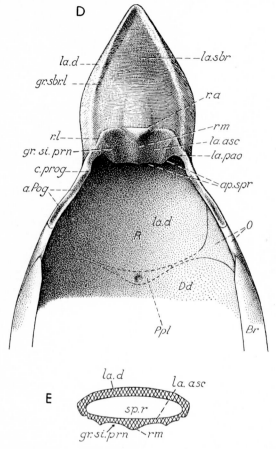

Fig. 14 D, E. *Althaspis? spatulirostris* Stensiö. D, Anterior part of dorsal armour in ventral view. From Stensiö 1958; 1964. E, transverse section of rostrum passing through ascending postrostral lamina (*la.asc*) of rostral plate.

Br, branchial plate; *Dd*, dorsal disc; *O*, orbital plate; *Ppl*, pineal plate; *R*, rostral plate; *a.Pog*, area of attachment of the preorogonial and orogonial plates; *ap.sp.r*, posterior opening of rostral space; *c.prog*, preorogonial angle; *gr.sbr.l*, lateral subrostral groove; *gr.si.prn*, paired groove on postero-ventral face of *la.asc* (anterior part of roof of prenasal sinus; see Fig. 12 A, B); *la.asc*, ascending postrostral lamina of *R*; *la.d*, dorsal lamina of *R*; *la.pao*, para-oral lamina of *R*; *la.sbr*, subrostral lamina of *R*; *r.a*, *r.l*, transverse ridge between *la.sbr* and *la.asc*; *rm*, median ridge separating the grooves *gr.si.prn* of both sides; *sp.r*, rostral space.

yukonensis was a broad prenasal sinus of the same nature as the similarly termed space in the Cephalaspidomorphi and Myxinoidea and that it was bounded off ventrally from the oral opening and oral cavity by the anterior medio-dorsal part of the visceral division of the premandibular metamere (*la.pa.sbn*, Figs. 12 A, B, 13 A). Consequently the definitive snout in the two Cyathaspida species under discussion (Fig. 13 A) was made up of two components, a superior cranial component (*ethm*) and an inferior visceral component (*la.pa.sbn*, *pl.pth*), the latter of which, like in the Myxinoidea, extended forwards

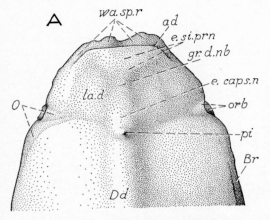

Fig. 15 A. Undeterminable pteraspid. Cast of internal face of anterior half of dorsal armour (rostrum broken off). From Stensiö, 1958; 1964.

Br, branchial plate (impression); *Dd*, dorsal disc (impression); *O*, orbital plate (impression); *ad*, antero-dorsal angle of rostral space (see Fig. 14 A, B); *e.caps.n*, impression of roof of nasal capsule; *e.si.prn*, impression of roof of prenasal sinus; *gr.d.nb*, impression of roof of prenasal sinus immediately in front of nasal capsule; *la.d*, dorsal lamina of rostral plate (impression); *orb*, orbital opening; *pi*, pineal pit (impression); *wa.sp.r*, anterior wall of rostral space (impression; see Fig. 14 A–C).

only as far as to a point below the anterior end of the cranial one. Both the prenasal sinus itself and its external opening—the naso-hypophysial aperture (na_1, na_2)—thus are strikingly broad in the two species concerned, a state of affairs which, as we have seen above, also is met with to some extent in the Myxinoidea (Fig. 22). Moreover, the bilobated shape of the naso-hypophysial opening may very well mean that the prenasal sinus was subdivided in places into right and left halves by a secondary incomplete septum of soft tissues. What has just been stated concerning *V.sekwiae* and *D.yukonensis* also applies to all those Cyathaspida forms (e.g. *Aequiarchegonaspis*, Fig. 12 C) where the rostrum is absent and where the rostro-dorsal shield does not extend downwards on to the anterior face of the snout. In the latter forms, however, the naso-hypophysial aperture lay somewhat antero-ventral to the anterior margin of the rostro-dorsal disc so that it was without an exoskeletal boundary on its dorsal side. Thus in all the Cyathaspida without a rostrum the naso-hypophysial opening had a terminal position or almost so.

Somewhat different conditions are met with in the Pteraspida all of which possess a more or less long rostrum formed in the fossils by the rostral plate. The anterior part of this exoskeletal rostrum (Figs. 14–16, 21) is a solid mass of bone (Fig. 14 A–C), whereas the posterior part contains a broad, transverse cavity, the rostral space (*sp.r*, Fig. 14), open postero-dorsally throughout its breadth. More exactly, this space is bounded by three bone laminae: dorsally by the anterior part of the dorsal, principal lamina (*la.d*) of the rostral plate,

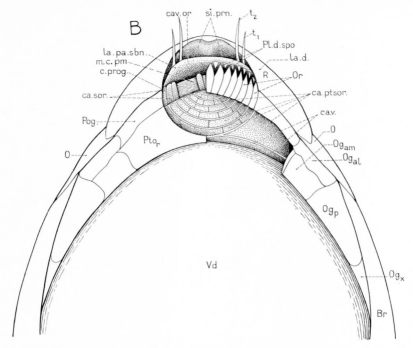

Fig. 15 B. *Zascinaspis* sp. Hypothetical restoration of oral apparatus in ventral view. Cartilage and ascending postrostral lamina of *R* (exoskeleton) stippled. Postoral plates removed with the exception of Pto_r. From Stensiö, 1958; 1964.

Br, branchial plate; *O*, orbital plate; Og_{al}, Og_{am}, Og_p, Og_x, orogonial plates; *Or*, oral plates; *Pl.d.spo*, probable superior labial dental plates; *Pog*, preorogonial plate; Pto_r, postoral plate of right side (that of left side removed); *R*, rostral plate; *Vd*, ventral disc; *ca.ptsor*, probable separate cartilages to which the postoral plates were attached; *ca.sor*, probable cartilaginous bars to which the oral plates were attached; *cav.or*, oral cavity; *ca.v*, anterior median ventral part of visceral endoskeleton; *c.prog*, preorogonial angle; *la.d*, dorsal lamina of *R*; *la.pa.sbn*, palato-subnasal lamina (endoskeleton of posthypophysial fold); *m.c.pm*, inferior part of original premandibular constrictor muscle much modified to effect the biting movements of the jaw apparatus; *si.prn*, prenasal sinus and its external opening (the naso-hypophysial opening); t_1, t_2, possible tentacles.

ventrally by a subrostral lamina (*la.sbr*) extending backwards from the anterior marginal part of the former lamina, and postero-ventrally by an ascending postrostral lamina (*la.asc*) continuous with the subrostral one from which it stretches obliquely backwards-upwards. However, the ascending postrostral lamina does not extend as far upwards as to close the rostral space, and the posterior opening of this space therefore is a broad, transverse slit directed backwards. Since the dorsal, principal lamina of the rostral plate naturally rested on the ethmoidal region of the endocranium (Fig. 16 A, B), the rostral space must have housed the thickened, foremost part of this region (*ca.r*, Fig. 16 A, B). The rostrum in the Pteraspida was hence formed entirely by the ethmoidal division of the head (*ethm*) and consisted of both endoskeleton and exoskeleton, the latter of which also comprises inferior and infero-posterior parts which originally lay on the upper part of the morphologically anterior face of the ethmoid region. Like the external faces of the dorsal, principal and subrostral bone laminae, the external (postero-ventrally directed) face of the ascending, postrostral bone lamina (*la.asc*) is ornamented with dentine, a state of affairs which means that this bone lamina was covered with epidermis, just as the outer faces of the armour as a whole. Moreover, the bone lamina under discussion (*la.asc*) also exhibits two dorso-ventrally running, broad grooves, a right and a left (*gr.si.prn*, figs. 14 D, E, 15 B), separated by a more or less rounded off, median ridge (*r.m*, Figs. 14 D, E, 15 B) so that in transverse or obliquely transverse sections of the rostrum (Fig. 14 E) it has fundamentally the same shape as the antero-inferior margin of the rostro-dorsal disc in *Vernonaspis sekwiae* and *Dikenaspis yukonensis* (Fig. 12 A, B). That being so, it is clear that the prenasal sinus in the Pteraspida (*si.prn*) lay immediately posterior (morphologically postero-ventral) to the ascending postrostral bone lamina of the rostral plate, extending from here far upwards-backwards (Figs. 16, 21), and that it was separated from the oral opening and oral cavity by the anterior dorso-median part of the visceral division of the premandibular metamere (*pl.pth, la.sbn*). Consequently, like in the somewhat advanced embryos of the Myxinoidea (figs. 6 C, 7), the external opening of the prenasal sinus—the naso-hypophysial aperture (na_1, na_2)—in the Pteraspida had a ventral position closely in front of the oral opening. Owing to the relief of the ascending postrostral bone lamina of the rostral plate described above both the outer part of the prenasal sinus itself and the naso-hypophysial aperture are of a broad, bilobated shape, a condition which may mean that the sinus was subdivided to some extent into right and left halves by an imperfect, secondary septum of soft tissues approximately as in the non-rostrate Cyathaspida. Thus we find that the definitive snout in the Pteraspida also was made up of a superior, cranial component and an inferior, visceral component the former of which, however, is produced forwards-downwards so as to form a rostrum situated in front of the latter.

Besides the Cyathaspida already discussed, however, there are also certain other Cyathaspida forms, as, for instance, *Poraspis* (Fig. 13 B, C; Stensiö, 1958; 1964), which possess a short pteraspid-like rostrum. These Cyathaspida forms were fundamentally as the Pteraspida where the naso-hypophysial aperture, the prenasal sinus and the structure of the definitive snout as a whole are concerned. However, contrary of the case in the Pteraspida, the ascending postrostral bone lamina is lower and formed by the rostral part of the rostro-dorsal disc, not by any separate rostral plate.

Owing to the fact that it was an extracephalic space between the cranial and visceral components of the definitive snout (see p. 25), the prenasal sinus in the Cyathaspida and Pteraspida must originally have been lined with exoskeleton (scales or minor bone plates). The morphologically dorsal part of this exoskeleton situated on the cranial component of the snout is represented by the ascending, postrostral bone lamina (*la.asc*) of the rostral plate in the Pteraspida (Fig. 16 A, B) and by the similarly termed bone lamina of the rostrodorsal disc in the rostrated Cyathaspida (Fig. 13 B, C). There is reason to believe, however, that minor bone plates or scales situated losely in the skin also were persisting on the other walls of the prenasal sinus in the Pteraspida and Cyathaspida above all on the morphologically inferior wall (*Pl.d.pth*, Fig. 16 A) formed by the visceral component of the snout. That this may have been so is strongly supported by the conditions both in the Osteostraci, where the corresponding visceral part of the cephalic shield is covered with exoskeleton (*Ex*, Fig. 9 A, B; see also Fig. 9 C), and in the Anaspida where the prepineal, dorsal face of the visceral part of the snout generally also is provided with dermal bone plates.

It has repeatedly been stressed that the naso-hypophysial aperture (na_1, na_2) and the outer (anterior) part of the prenasal sinus in the Cyathaspida and Pteraspida were very broad. In fact, the naso-hypophysial aperture (*si.prn*, Figs. 12, 13 C, 15 B, 21) was a transverse slit almost as broad as the oral opening, and it is clear that the anterior part of the prenasal sinus must have been approximately of the same breadth. As we shall see, these conditions mean that the Cyathaspida and Pteraspida had acquired a myxinoid mode of respiration in that the prenasal sinus was continued caudally by a naso-pharyngeal duct and that the water was conveyed backwards to the gills through the naso-hypophysial aperture, prenasal sinus and naso-pharyngeal duct by the pumping movements of a velum.

(IV) The inferior, visceral component of the definitive snout in the Cyathaspida and Pteraspida (*la.pa.sbn*, *pl.pth*, Figs. 12, 13, 15 B, 16, 21) was the partition between the prenasal sinus on the one hand and the oral opening and oral cavity on the other. Since naturally it developed embryologically from a posthypophysial fold which, like in the Myxinoidea (Fig. 6), did not grow out forwards-upwards in front of the ethmoidal part of the cranial division of the head, the component of the snout under discussion must have been supported internally by the epitrematic halves of the premandibular arches of both sides. Under all circumstances these halves of the premandibular arches cannot possibly have had any other position relative to both the endocranium and to the oral cavity. In consequence of the mechanism of the oral apparatus as a whole (see point VI below), however, the movements in the inferior part of this apparatus apparently effected such a strain on the supraoral part that the epitrematic halves of the premandibular arches of both sides must have fused along their morphologically anterior margins to form an extensive, cartilaginous palatinosubnasal lamina (*la.pa.sbn*, Figs. 12, 13, 15 B, 21) which was continuous postero-medio-dorsally with the ventral endocranial wall (*la.hp*, Figs. 13 A, B 16),

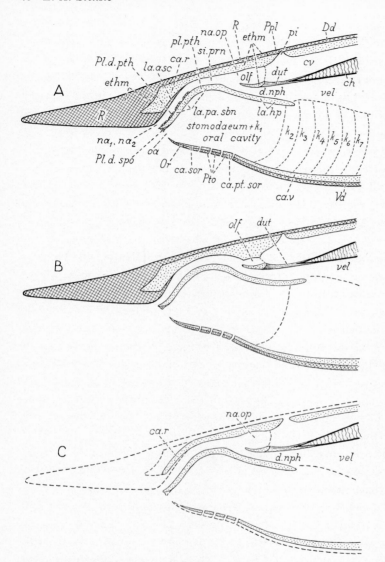

Fig. 16. *A, B*, two hypothetical diagrammatic median longitudinal sections of heads of a rostrate Pteraspida form showing possible different position of olfactory organ. In *A*, the probable exoskeletal elements (*Pl.d.pth*) on the morphologically dorsal side of the inferior, visceral component of the definitive snout also laid in. *C*, a corresponding section as in *A* but drawn under the assumption that the exoskeleton has disappeared and that the rostral cartilage (*ca.r*) is represented only by a thin lamina. Modified in this way the definitive snout in the Pteraspida would be essentially as in the Myxinoidea. Sections of cartilage stippled; sections of dermal bones cross-hatched.

Dd, dorsal disc; *Or*, oral plate; *Pl.d.pth*, probable exoskeletal plates on morphologically dorsal face of visceral component of definitive snout (these possible plates correspond to the exoskeleton on the dorsal side of the prenasal part of the cephalic shield in the Osteostraci; *Ex*, Fig. 9 A, B; see also Fig. 9 C); *Pl.d.spo*, probable supralabial dental plate or plates (=*pl. d.spo* in the Osteostraci; Fig. 9 B); *Ppl*, pineal plate; *Pto*, postoral plates; *R*, rostral plate; *Vd*, ventral disc; *ca.pt.sor*, suboral cartilaginous plates to which the postoral plates were attached; *ca.r*, rostral cartilage; *ca.sor*, suboral cartilaginous plates to which the oral plates were attach-

and postero-dorso-laterally with the epitrematic halves of the mandibular arches of both sides.

(V) The eyes in the Heterostraci were under reduction so that generally they were strikingly small (Figs. 12, 13 C, 15 A, 17 A, 18 A, 19 A, D, 20 A–C). Above all in the Drepanaspida (Fig. 18 B, C), but also in several representatives of other orders, e.g. many Pteraspida forms, the eyes seem even to have been so small that both their muscles and the nerves for these may have been completely lost. In one of the Hibernaspida forms, *Eglonaspis* (Fig. 19 B), which lacks orbital openings, even the eye-ball and the optic nerve may have disappeared altogether. Because of the conditions just dealt with the Heterostraci must have searched for food chiefly or entirely with the aid of the olfactory organ and organs of taste the latter of which at least in some forms presumably were situated on tentacles. Amongst the recent cyclostomes a parallel to this is met with in the Myxinoidea, which, as is well known, are practically blind.

(VI) The Pteraspida and Cyathaspida have a biting mouth, and most certainly this was also the case in the Traquairaspida and some other, still slightly known Heterostraci forms. The biting took place by means of a kind of "lower jaw" (Figs. 13, 15 B, 16, 21; Stensiö, 1964, figs. 31, 42 B, 43 B, 45 A, 58, 92 B, 94, 95, 99 C, D) made up in the fossils of a transverse series of separate, minor dermal bone plates, the oral plates (*Or*, Figs. 13, 15 B, 16), posterior to which in many forms there follow other, minor bone plates (*Pto_r*, *Pog*, Og_{am}, Og_{al}, Fig. 15 B) differing in number and arrangement in the individual genera. Together with the oral plates these minor, dermal bone plates—the postoral plates—formed a superficial, ventral covering of the "lower jaw". With the exception only of the freely forward or forward-upward projecting, biting parts of the oral plates all the dermal bone plates concerned must originally have been attached to the suprajacent cartilaginous hypotrematic halves of the right and left premandibular arches; these halves of the premandibular arches must therefore be assumed to have been subdivided into numerous, minor cartila-

---

ed; *ca.v*, median ventral cartilage formed by fusions of the hypotrematic halves of the visceral arches; *ch*, notochord; *cv*, cranial cavity; *d.nph*, naso-pharyngeal duct; *dut*, probable membranous median part of ventral cranial wall below hypophysis (this probable membranous part formed by dura mater); *ethm*, superior, ethmoidal component of definitive snout; k_1, premandibular branchial compartment incorporated in oral cavity and without gills; k_2, k_3, mandibular and hyoid branchial compartments; k_4, glossopharyngeal branchial compartment; k_5–k_7, anterior vagal branchial compartments; *la.asc*, ascending postrostral bone lamina of rostral plate (exoskeleton on morphologically inferior face of superior, cranial component of definitive snout); *la.hp*, hypophysial cartilage (original ventral wall of hypophysial fossa); *la.pa.sbn*, palatino-subnasal lamina (endoskeleton of inferior, visceral component of definitive snout) continuous posteriorly with *la.hp*; na_1, na_2, naso-hypophysial opening (external opening of *si.prn*); *na.op*, nasal opening proper; *oa*, oral opening; *olf*, position of olfactory organ; *pi*, pineal pit; *pl.pth*, endoskeleton of posthypophysial fold; *si.prn*, prenasal sinus; *vel*, position of velum.

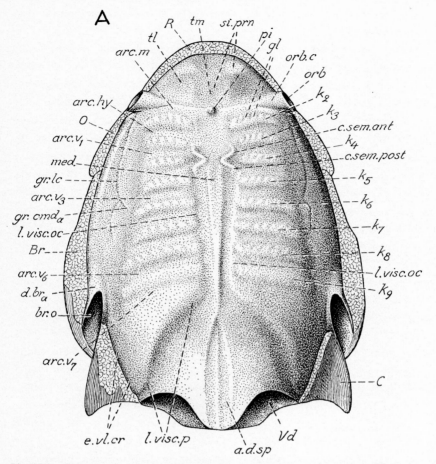

Fig. 17 A. *Simopteraspis primaeva* (Kiaer). Cast of basal internal face of dorsal wall of armour (see also Fig. 21). Sections of armour denoted with reticulation. From Stensiö, 1958; 1964.

Br, branchial plate (section); *C*, cornual plate (partly in section); *O*, orbital plate (section); *R*, rostral plate (section); *Vd*, ventral disc; *a.d.sp*, impression of basal face of dorsal spine; *arc.hy*, *arc.m*, positions of epitrematic halves of hyoid arch and mandibular arch, respectively; $arc.v_1$–$arc.v_7$, position of the epitrematic halves of the other visceral arches in the anterior division of the branchial region (in front of the external branchial opening); *br.o*, external branchial opening; *c.sem.ant*, *c.sem.post*, impressions of the cartilaginous ridges above the anterior and posterior semicircular canals; *d.bra*, position of hindmost part of anterior paraatrial branchial duct; *e.vl.cr*, part of visceral endoskeleton continuous with endoskeleton in proximal part of cornual plate (endoskeleton of the anterior part of an original paired fin fold); *gl*, impressions of gill lamellae belonging to the posterior hemibranchs of the visceral arches; *gr.lc*, $gr.cmd_a$, grooves caused by sensory canals; k_2–k_8, epitrematic halves of open gill slits in endoskeleton (corresponding to branchial compartments 2–8); k_9, impression of thin cartilaginous wall still separating branchial compartment 9 from exoskeleton (this branchial compartment, like those in the posterior division of the branchial region, was closed internal to the armour); *l.visc.oc*, *l.visc.p*, steps marking medial and backward extent of branchial region; *med*, impression of roof of myelencephalic division of cranial cavity; *orb*, orbital opening; *orb.c*, orbital cavity (impression of roof); *pi*, impression of pineal pit; *si.prn*, impression of anterior part of dorsal wall of prenasal sinus; *tl*, *tm*, impressions of thickening on basal face of rostral plate.

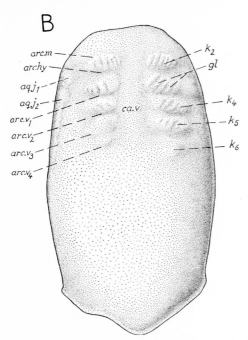

Fig. 17 B. *Poraspis pompeckji* Brotzen. Cast of basal (internal) face of ventral disc. From Stensiö, 1958; 1964.

arc.hy, *arc.m*, positions of hypotrematic halves of hyoid and mandibular arches, respectively; $arc.v_1$–$arc.v_4$, positions of hypotrematic halves of four consecutive visceral arches; aqj_1, aqj_2, impressions of extrabranchial atria 1 and 2; *gl*, impressions of gill lamellae belonging to the posterior hemibranchs of the visceral arches; *ca.v*, area marking the position of the median ventral cartilaginous plate of the visceral endoskeleton (arisen by fusions of branchial arches); k_2–k_5, hypotrematic halves of open gill slits in endoskeleton, corresponding to branchial compartments 2–5; k_6, impression of thin endoskeletal wall of hypotrematic half of branchial compartment 6 (as the other, further backwards following branchial compartments, this had not acquired any opening internal to the armour).

ginous plates more or less moveable relative to each other (*ca.sor*, *ca.pt.sor*, Figs. 13 A, B, 15 B, 16, 21). As a whole this unique "lower jaw" was raised and lowered by special muscles (*m.c.pm*, Fig. 15 B; Stensiö, 1964, fig. 94 A) which could only have arisen from the original constrictor muscles of the premandibular branchial compartments of both sides. Being of a visceral nature the aforementioned special musculature must have been situated internal to the visceral endoskeleton so that it inserted on the upper (morphologically internal) faces of the probable cartilaginous plates which represented the hypotrematic halves of the premandibular arches. As a result of the modifications just dealt with the premandibular branchial compartments of both sides had lost their gills and had become incorporated into the oral cavity (Figs. 13 A, B, 16, 21) which hence, unlike in the Osteostraci (Fig. 9 A, B), was not restricted to the stomodaeum invagination but extended a fair distance further backwards, in fact as far as right to the mandibular arches of both sides. It may be added here, that, as can be seen from impressions on the internal side of the armour, the mandibular arches (arc.m, Fig. 17 A, 21), which lay postero-ventral to the orbital cavities, had not undergone any more appreciable modifications except that, as we shall see, jointly they must have given rise to the skeleton of an extensive velum (see points VII, VIII).

The oral opening in the Pteraspida and Cyathaspida (and naturally also in the other orders with a biting mouth) was a broad transverse slit (oa, Figs. 12, 13 A, B, 15 B, 16) immediately below the antero-inferior margin of the palatino-subnasal lamina (the endoskeleton of the visceral component of the definitive snout), more exactly between the said margin of the palatino-subnasal lamina and the oral plates. As already pointed out, the oral cavity extended as far backwards as also to comprise the original premandibular branchial compartments of both sides, and, that being the case, its endoskeletal boundary must naturally have been formed anteriorly and dorsally by the palatino-subnasal lamina (la.pa.sbn+la.hp) and ventrally by the probable minor cartilaginous plates (ca.sor, ca.pl.sor) arisen by the subdivision of the hypotrematic halves of the right and left premandibular arches. Since the dentiferous bones of the lower jaw, or of the lower jaw apparatus, in lower vertebrates never bite (or bit) against cartilage but only against dermal plates or teeth attached to the epitrematic halves of the foremost one or the two foremost visceral arches, this must also have been the case in the Cyathaspida, Pteraspida and Traquairaspida. Hence we are led to the conclusion that the oral plates in these three orders also bit against small *supra-labial tooth plates* (Pl.d.spo, Figs. 13, 15 B, 16 A, B) attached to the inferior part of the morphologically posterior (oral) face of the palatino-subnasal lamina. It is true that these supra-labial tooth plates have not been found preserved as yet (see discussion by White, in this Volume). That nevertheless they existed is clear, however, from the conditions in certain Pteraspida forms, e.g. *Mylopteraspis* (Stensiö, 1964, fig. 44), where the upward bending, much thickened, anterior parts of the oral plates exhibit distinct, plane biting areas worn off in such a way that they can only have arisen as a result of biting against bone plates situated on the palatino-subnasal lamina. Moreover, since it developed ontogenetically from the stomodaeum invagination lined with epidermis (Figs. 6, 7) the anterior part of the oral cavity may be assumed originally to have contained tuberculated scales not only on its dorsal face (the morphologically inferior face of the palatino-subnasal lamina) but also on its other faces. The supra-labial tooth plates would then represent an anterior dorsal part of the original exoskeleton of the stomodaeum invagination. Remains of this exoskeleton are still persisting on the inferior face of the foremost part of the oral cavity in the oligobranchiate Osteostraci (Fig. 9 B; see point III).

By the decaying of the cartilaginous palatino-subnasal lamina the supra-labial tooth plates in the Pteraspida and Cyathaspida naturally became detached, and in consequence they generally were lost prior to the embedding of the specimens in the bottom sediments. Since, as a rule, they may have been quite small, the tooth plates concerned may also easily be overlooked even in such cases where still they happen to be preserved but lie scattered in the rock inside the carapace. In these cases they can only be identified with full certainty by the aid of serial grindings at close intervals between the consecutive sections. At any event the grinding of crossopterygian skulls carried out in this way have made it possible to find and to interpret practically all of the minute tooth plates in the mouth and on the branchial skeleton, which would have been impossible by ordinary mechanical and chemical methods of preparation.

The oral opening in the craniate vertebrates originally lay between the premandibular arches of both sides. This state of affairs still is met with in the Osteostraci, Anaspida, Petromyzontida and Myxinoidea, and it is also persisting to some extent in the gnathostomes, where, however, as a result of a reduction of the hypotrematic halves of the premandibular arches, the posterodorsal and inferior parts of the oral opening are bounded by the mandibular arches, more exactly by the epimandibular segments and the hypotrematic halves of the mandibular arches (Jarvik, 1954; Stensiö, 1963a; 1968). In those forms where there evolved an actively biting jaw apparatus the biting on each side therefore took place in the cyclostomes by means of the hypotrematic half of the premandibular arch against the epitrematic half, but in the gnathostomes by means of the hypotrematic half of the mandibular arch against the persisting epitrematic half of the premandibular arch and against the epal segment of the mandibular arch. Thus, under no circumstances the jaw apparatus in craniate vertebrates bit or bites directly against the ethmoidal division of the head. The opinion still maintained by several writers that the oral plates in the Cyathaspida and Pteraspida would have bitten against the ascending postrostral bone lamina (*la.asc*) of the ethmoidal division of the cranium then is not only in contrast to what could actually be established concerning the structure of the definitive snout in these two groups, but it is also contradicted by the general anatomical conditions in all craniate vertebrates. For these reasons and because, as already stressed, the Heterostraci are true cyclostomes, the opinion that the oral plates in the Cyathaspida and Pteraspida would have bitten against the ascending postrostral bone lamina hence is untainable from all points of view.

As may be inferred from the above discussion of its anatomic structure, the oral apparatus in the Cyathaspida, Pteraspida and certain other forms was much more specialized than that in the Osteostraci in the two respects that it comprised the premandibular branchial compartments of both sides and that it was decidedly of a biting type. Such an oral apparatus cannot possibly have given rise to the oral apparatus either in the Osteostraci or in the Anaspida and

Petromyzontida. On the other hand, taking everything into consideration, it is very likely that, correlative with the evolution of a rasping tongue an oral apparatus of the cyathaspid-pteraspid type may have modified to form a biting mouth of a myxinoid type.—As will be shown below (point VIII), the oral apparatus in several Heterostraci was of a much modified, non-biting type.

(VII) In the Pteraspida, Traquairaspida, Cyathaspida and Corvaspida where they were attached to each other and to the dorsal disc chiefly only by ligaments and skin, the branchial plates and the extensive ventral disc of the armour were slightly moveable relative to the dorsal disc so that the exoskeleton of the branchial region was not rigid altogether. However, this region of the exoskeleton could be only very slightly expanded and constricted. Since, as can be inferred from impressions on its interior faces, the armour was firmly attached to the entire endoskeleton (Fig. 17), both the endocranium and the visceral endoskeleton, it is clear that regarded as a whole (including both its cartilage and dermal bones) the skeleton of the branchial region was so stiff that the respiration in the four orders listed must have taken place in a way entirely different from that in the Cephalaspidomorphi. In those forms, e.g. *Simopteraspis*, *Seretaspis* and *Poraspis*, where the lateral cartilaginous walls of the anterior branchial compartments had disappeared between the visceral arches there are distinct impressions of gills and gill lamellae (Figs. 17, 21) on the internal faces of both the dorsal and ventral shields (discs), a condition which proves that the superficial, principal portions of the branchial constrictor muscles originally situated immediately lateral to the gills and internal to the visceral endoskeleton were lost altogether, just as in the Myxinoidea. One therefore arrives at the conclusion that the respiration in the Pteraspida, Traquairaspida, Cyathaspida and Corvaspida must have been effected by a powerful myxinoid velum (*vel*, Figs. 13 A, B, 16; see also Fig. 9 A, B) which pumped the water backwards to the gills through the naso-hypophysial opening, the prenasal sinus and through a naso-hypophysial duct of the same nature as in the Myxinoidea (pp. 26–27).

A myxinoid respiration naturally had also evolved in those Heterostraci orders in which the armour was so rigid that it could not be expanded or constricted at all. More exactly, this applies to the Drepanaspida, where all the bone components of the armour are firmly attached to each other by sutures, and to the Amphiaspida, Hibernaspida and Olbiaspida (Figs. 19, 20), where because of a complete fusion of all its original bone components, the armour is one rigid unit without any traces of sutures. In the four orders just listed, and in the Lyktaspida as well, the ingestion of the food took place in connection with the respiration, and as we shall see (point VIII), this mode of feeding had caused considerable changes of the originally biting oral apparatus and of its relations to the prenasal sinus and to the naso-pharyngeal duct.

Owing to the total reduction of the external, principal portions of the branchial constrictor muscles the endoskeleton was not subjected to any powerful strain for expansions and compressions, and as a result the original scales and minor bone plates on its outside usually fused to form larger plates or a continous rigid cuirass. The prerequisite condition for the evolution of a non-expansive and non-constrictive armour in the Heterostraci hence was the disappearance of the superficial portions of the branchial constrictor muscles. Under these circumstances there is reason to believe that in the Cardipeltida where the antero-dorsal, lateral and ventral walls of the armour secondarily had disintegrated into minor plates and scales or perhaps had disappeared altogether (*Aspidosteus*), the respiration also took place in the same way as in the Myxinoidea. The Cardipeltida may therefore also have possessed a naso-pharyngeal duct. On the other hand in *Turinia*, whose armour, like the exoskeleton of the other parts of trunk, is represented only by small synchronomorial scales one would rather be inclined to assume that the superficial portions of the branchial constrictor muscles still might have been persisting. However, there are no other conditions indicating that this was so.

Consequently we find that the Heterostraci, or at any event the vast majority of these, were fundamentally as the Myxinoidea where the constrictor musculature of the branchial apparatus, the presence of a naso-pharyngeal duct and the mode of respiration are concerned.

(VIII) In addition to the orders with a biting mouth dealt with under point VI the Heterostraci comprise several other orders where the oral apparatus was adapted for feeding on detritus, on planktonic organisms or on both. More exactly this applies to the orders Lyktaspida, Amphiaspida, Hibernaspida, Olbiaspida and Drepanaspida.

Lyktaspis ("*Doryaspis*"; see also N. Heintz, in this Volume), the only genus of the Lyktaspida known at present, is decidedly akin to the Pteraspida. However, it differs considerably from these above all in the following respects (see Figs. 15 B, 18 A). The original rostrum has completely disappeared, but has been replaced by a very long, narrow sub-oral pseudo-rostrum (*Psr*, Fig. 18 A) consisting of a single plate attached posteriorly to the anterior margin of the ventral disc (*Vd*). According to its position this pseudo-rostrum may be assumed to have evolved from the median oral and median postoral bones which, after fusing with each other and with a couple of adjacent medial oral and postoral plates, have undergone an excessive forward lengthening. The remaining more laterally situated postoral bones and the orogonial bones of each side (Fig. 15 B; see also Stensiö, 1958; 1964, figs. 43 B, 45 A, 58, 92 B, 95) are represented by an extensive lateral postoral plate (*Pto.l*, Fig. 18 A) attached posteriorly to the ventral disc, medially to the posterior part of the pseudo-rostrum (*Psr*) and latero-dorsal to the orbital plate (*O*). Finally, the original lateral oral bones on each side of the pseudo-rostrum have also fused to form a com-

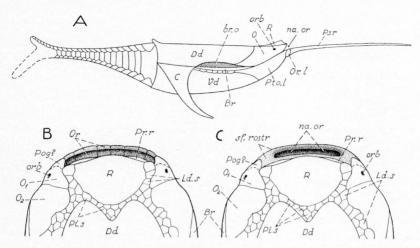

Fig. 18. *A*, *Lyktaspis nathorsti* (Lankester). Restoration in lateral view made by N. Heintz and published by Obruchev, 1964. *B, C, Drepanaspis gemuendenensis* Schlüter. *B*, skeleton of anterior part of head in dorsal view. *C*, same part of head in dorsal view but with the soft foremost part of the snout and the naso-oral opening tentatively restored. *B*, chiefly from Obruchev, 1943.

Br, branchial plate; *C*, cornual plate; *Dd*, dorsal disc; *Ld.s*, longitudinal dorsal field of minor, polygonal plates or scales; *O*, orbital plate; O_1, O_2, anterior and posterior orbital plates; *Or*, oral plates; *Or.l*, lateral oral plate; *Pi.s*, transverse pineal field of minor, polygonal plates or scales; *Pog?*, possible preorogonial plate; *Pto.l*, compound lateral postoral plate; *Pr.r*, prerostral plate; *Psr*, pseudorostral plate; *R*, rostral plate; *Vd*, ventral disc; *br.o*, external branchial opening; *na.or*, naso-oral opening (in *Drepanaspis* restored on the basis of the assumption that it had fundamentally the same position as in *Lyktaspis*; it is also possible, however, that this opening in *Drepanaspis* had a more or less pronounced terminal position); *orb*, orbital opening; *sf.rostr*, soft rostral tissue (denoted with delicate stippling).

pound minor bone which may be referred to as the lateral oral plate (*Or.l*, Fig. 18 A); this plate is connected medially with the pseudo-rostrum, posteriorly with the lateral postoral plate. On the foremost prepineal part of the dorsal side of the head *Lyktaspis* possesses a broad transverse upward-forward directed opening (*na.or*, Fig. 18 A), which is bounded posteriorly (morphologically dorsally) by the rostral plate (*R*), laterally by the pointed antero-dorsal corner of the orbital plate (*O*) and anteriorly (morphologically ventrally) by the pseudo-rostrum and the lateral oral plates (*Or.l*) of both sides. Because of its position relative to the bone plates just referred to the opening in question (*na.or*) apparently corresponds to both the naso-hypophysial and oral openings in the Cyathaspida and Pteraspida, and it is therefore termed here the naso-oral foramen (*na.or*). The presence of this compound foramen must mean, that the partition between the prenasal sinus and the oral cavity, i.e. the visceral component of the snout, was represented only by a posterior part (see Figs. 16 and 20 D) so that the naso-pharyngeal duct was short and that in consequence there had arisen an extensive secondary communication between the prenasal sinus and the oral cavity (*si.prn + oral cavity*, Fig. 20 D). In *Lyktaspis* there hence

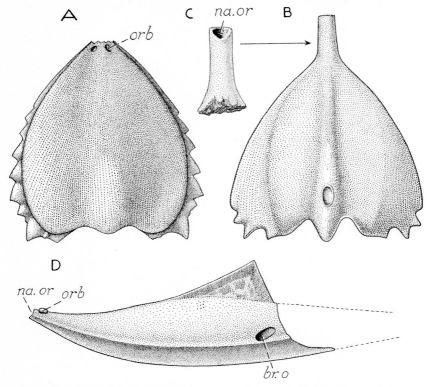

Fig. 19. *A*, *Hibernaspis macrolepis* Obruchev. Armour in dorsal view. *B*, *C*, *Eglonaspis rostrata* Obruchev. *B*, armour in dorsal view. *C*, rostral tube-shaped process in ventral view. *D*, *Olbiaspis coalescens* Obruchev. Armour in lateral view. From Obruchev, 1964.

br.o, external branchial opening; *na.or*, naso-oral opening; *orb*, orbital opening.

most certainly existed an oralo-prenasal cavity of such a kind that the velum had acquired both a respiratory and feeding function. To be more exact, the velum in *Lyktaspis* must have pumped in a mixture of water and food (planktonic organisms, detritus or both) into the oralo-prenasal cavity, where in some way or other the food was separated from the water that was conveyed to the gills. These conditions naturally had also brought about an extensive, or perhaps even total, reduction of the originally pteraspid-like oral musculature. Concerning the oral apparatus *Lyktaspis* hence had reached a much more advanced stage of specialization than, for example, the Cyathaspida and Pteraspida. In fact *Lyktaspis* had undergone a modification so as to become an inactive feeder, whereas the Cyathaspida and Pteraspida with their biting oral apparatus were active feeders. As may readily be gathered from the above analysis, *Lyktaspis* differs so considerably from the Pteraspida that it has to be regarded as a representative of an order of its own.

As has already been mentioned (point VII), the three orders Amphiaspida,

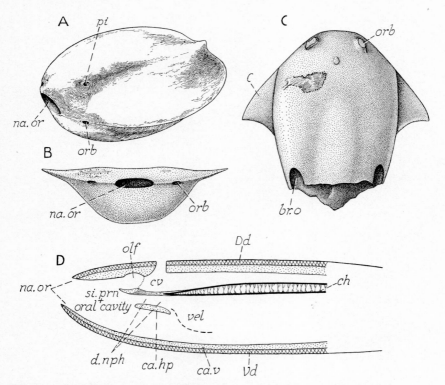

Fig. 20 *A, B, Siberiaspis plana* Obruchev. Armour in oblique antero-dorsal and anterior aspects, respectively. *C, Angaraspis urvantzevi* Obruchev. Armour in dorsal view. From Obruchev, 1964. *D*, hypothetical, diagrammatic, median longitudinal section of the head in a Heterostraci form somewhat similar to *Siberiaspis*. Sections of cartilage stippled; sections of dermal bones cross-hatched (cf. Figs. 13 A, B, 16 A, B).

C, cornual process; *Dd, Vd*, dorsal and ventral walls of armour; *br.o*, external branchial opening; *ca.hp*, hypophysial cartilage (probable persisting part of endoskeleton in inferior, visceral component of definitive snout); *ca.v*, probable persisting median ventral part of visceral endoskeleton (See Figs. 13 A, B, 16 A, B); *ch*, notochord; *cv*, cranial cavity; *d.nph*, nasopharyngeal duct; *na.or*, naso-oral opening; *olf*, olfactory organ; *orb*, orbital opening; *pi*, pineal opening; *si.prn*, prenasal sinus; *vel*; probable position of a powerfully developed velum.

Hibernaspida, and Olbiaspida (Obruchev, 1964) comprise Heterostraci forms where the originally separate exoskeletal bones of the head and branchial region have fused to form a single unit, a completely rigid armour without any traces of sutures (Figs. 19, 20 A–C). These Heterostraci forms all have a terminal naso-oral foramen (*na.or*, Figs. 19, 20 A–C) which in the genus *Eglonaspis* (Fig. 19 B, C) even lies on the truncate anterior end of a fairly long, tube-shaped rostral process. The presence and position of this foramen prove that, like in *Lyktaspis*, the visceral component of the snout in the Amphiaspida, Hibernaspida and Olbiaspida was persisting only most posteriorly (Fig. 20 D) and that the oral cavity and the prenasal sinus therefore had fused to a considerable extent to form an oralo-prenasal cavity. As the Lyktaspida, the Amphiaspida, Hibernaspida and Olbiaspida hence were inactive feeders in that the food, consisting of planktonic organisms, detritus or both, was carried backwards into the oraloprenasal cavity together with the water for the respiration by the pumping movements of a velum. With regard to this mode of feeding, however, the three orders under discussion most certainly had reached a still more advanced stage of specialization than the Lyktaspida, *inter alia* in the respect that the original oral and postoral plates may have been completely incorporated into the ventral wall of the carapace to form the anterior part of this wall immediately below the naso-oral foramen.

What has just been set out concerning the Lyktaspida, Amphiaspida, Hibernaspida and Olbiaspida also contributes in throwing new light on the oral apparatus and feeding of the Drepanaspida.

In the Drepanaspida the oral and postoral plates (Fig. 18 B) are found as separate bones, but contrary to the case in the Pteraspida and Cyathaspida, these plates all are rigidly attached by sutures both to each other and to the rigid armour as a whole. This state of affairs shows clearly that the Drepanaspida are derived from ancestors which possessed a biting mouth of the pteraspid-cyathaspid type. The thick, upward bending oral plates extend somewhat forwards beyond the dorsal wall of the armour in such a way that in the fossils there is a broad, forward-upward facing opening, a naso-oral foramen, situated approximately as the similarly termed foramen in *Lyktaspis*. To be more exact, this foramen in the Drepanaspida (Fig. 18 B) is bounded ventrally by the upper margins of the oral plates (*Or*), laterally by the antero-dorsal pointed ends of the anterior orbital plates of both sides ($O_1 + Pog$?) and dorsally by the rostral and prerostral plates (*R, Pr.r*). In the living animals, however, where the armour was completed to some extent forwards by soft tissues (*sf. rostr*, Fig. 18 C), the naso-oral foramen was most certainly somewhat narrower than in the fossils. Moreover, it is also possible that in the living animals it was dorsal or even had a terminal position on the snout (Fig. 18 C). Be this as it may, it is practically certain that where the structure of the oral apparatus and the communication of this apparatus with the prenasal sinus are concerned the Drepanaspida were fundamentally as the Lyktaspida so that the feeding was combined with the respiration. In view of the broad flat shape which shows that they were bottom dwellers the Drepanaspida must have fed chiefly or entirely on detritus.

Summing up what has been set out above under point VIII, we find that at least five Heterostraci orders, the Drepanaspida, Lyktaspida, Amphiaspida, Hibernaspida and Olbiaspida, were planktonic feeders, detritus feeders or both, and that the food in these five orders was carried backwards into the pharynx by the current of water caused by the velum conjointly with the respiration. Judging from the conditions in the Drepanaspida and Lyktaspida, all the five orders may originally have had a biting oral apparatus of a cyathaspid-pteraspid type.

(IX) As is clear from what has been stated under point I on p. 19, the external (efferent) branchial ducts in the cyclostomes originally possessed separate external openings. Judging from the restoration reproduced in Fig. 18 A, this may very well still have been the case to some extent in *Lyktaspis* where the branchial plate (*Br*, Fig. 18 A) is so low and has such a position that there is a long opening between it and the dorsal disc (*Dd*). Moreover, it has also to be taken into consideration that the external (efferent) branchial ducts still might have retained their separate external openings in some of the slightly known Heterostraci orders too, as the Turiniida, for example. On the other hand, in the Drepanaspida, Pteraspida, Traquairaspida, Cyathaspida, Corvaspida, Amphiaspida, Hibernaspida, Olbiaspida and Cardipeltida the external (efferent) branchial ducts of each side led out into a common external branchial opening (*br.o*, Figs. 17 A, 19 D, 20 C, 21). In this particular respect the nine orders just listed were nearest as in *Myxine*, *Notomyxine*, *Neomyxine* and *Nemomyxine*, i.e. the recent genera of the family Myxinidae.

(X) Like the Cephalaspidomorphi, the Heterostraci also possessed extra-branchial atria ($a.qj_1$–$a.qj_{14}$, Figs. 17 B, 21; see also Stensiö, 1958; 1964) situated external to the branchial compartments and internal to the visceral endoskeleton. Being completely enclosed in the visceral endoskeleton, which was firmly attached to the armour, the extra-branchial atria must have been almost rigid, and with the exception of possible weak sphincter muscles they cannot therefore have contained any musculature. This condition harmonizes with the conclusion (see points VI–VIII) that the respiration in the Heterostraci was effected, like in the Myxinoidea, by the velum and to some minor extent also by the persisting internal portions of the individual branchial constrictor muscles which, like in the Myxinoidea (Gustafson, 1936) effected the expulsion of the coarser objects which every now and then happened to get stuck in the branchial apparatus.

(XI) As is clear from the impressions of their endoskeletal walls on the internal (basal) face of the dorsal wall of the armour, the extrabranchial atria in the Heterostraci (Figs. 17 B, 21; Stensiö, 1964, figs. 107, 109–111) generally extended so far laterally that the interspace between them and the lateral wall of the armour was fairly narrow. In those forms, e.g. *Turinia*, where they possibly opened separately to the exterior, the external (efferent) branchial ducts must hence have been short, approximately as short as in the Osteostraci and Petromyzontida. As opposed to this, the external (efferent) branchial ducts had undergone considerable modifications in the other Heterostraci which have only one, paired external branchial opening situated more or less far caudally in the lateral wall of the armour. Turning first to the Traquairaspida, Cardipeltida

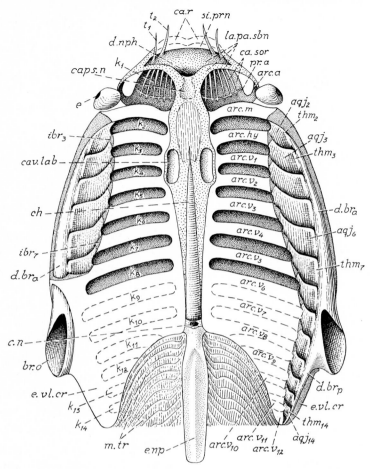

Fig. 21. *Simopteraspis primaeva* (Kiaer). Hypothetical, tentative restoration of the endoskeleton situated inside the armour, based on Fig. 17 A. Dorsal view. The principal upper parts of the endocranium and the lateral parts of the visceral endoskeleton removed. Extrabranchial atria, para-atrial branchial ducts and membranous floor of fossa hypophyseos shaded with lines; sections of cartilage stippled. From Stensiö, 1958; 1964, but altered in some details.

aqj_2–aqj_{14}, extrabranchial atria; *arc.a*, part of probable cartilage derived from premandibular arch; *arc.hy*, *arc.m*, epitrematic halves of hyoid and mandibular arches, respectively; *arc.v₁*, epitrematic half of glossopharyngeal arch; $arc.v_2$–$arc.v_{12}$, epitrematic halves of vagal visceral arches; *br.o*, external branchial opening; *caps.n*, inferior part of nasal capsule; *ca.r*, position of rostral cartilage (in rostral space); *ca.sor*, suboral cartilaginous bars to which the oral plates were attached (see Figs. 13, 15 B, 16); *cav.lab*, inferior part of labyrinth cavity; *ch*, notochord; *c.n*, neural canal; $d.br_a$, $d.br_p$, anterior and posterior para-atrial branchial ducts, respectively; *d.nph*, naso-pharyngeal duct (anterior opening); *e*, eye-ball; *e.np*, notopinnary component of occipital region (endoskeleton of foremost part of dorsal fin fold); *e.vl.cr*, endoskeleton of pectoral part of paired fin fold (extending slightly outwards into cornual plate); ibr_3, ibr_7, interbranchial ridges (sections); k_1, premandibular branchial compartment (incorporated into oral cavity and therefore without gills); k_2, mandibular branchial compartment; k_3, hyoid branchial compartment; k_4, glossopharyngeal branchial compartment; k_5–k_{14}, vagal branchial compartments (k_9–k_{14} with complete cartilaginous walls between the branchial arches); *la.pa.sbn*, palato-subnasal lamina (endoskeleton of posthypophysial fold(= inferior visceral component of definitive snout); *m.tr*, anterior dorsal part of trunk musculature; *pr.a*, possible anterior process of the median ventral part of the visceral endoskeleton; *si.prn*, prenasal sinus; t_1, t_2, possible tentacles; thm_2–thm_{14}, spaces occupied by thymus glands.

and the vast majority of the Pteraspida and Cyathaspida, i.e. those forms where the external branchial opening lies a fair distance in front of the posterior margin of the armour, we find here the following conditions (Figs. 17 A, 21): (a) the branchial region extends a fair distance backwards beyond the external branchial opening ($br.o$); (b) from a descriptive point of view the branchial region may therefore be regarded as consisting of two divisions, anterior and posterior, situated in front of and behind the external branchial opening, respectively; (c) these two divisions both contained branchial compartments with respiratory functions (k_2–k_{14}). As a result of these three conditions the external (efferent) branchial ducts in front of the external branchial opening naturally had undergone a backward lengthening, whilst the external branchial ducts behind the said opening had undergone a forward lengthening. Owing to the narrow interspace between the extrabranchial atria and the lateral exoskeletal wall of the branchial region, however, the external (efferent) branchial ducts of each side in all the forms concerned could hardly have been persisting as separate canals, and concurrently with their lengthening they may therefore have fused to form two compound para-atrial, efferent branchial ducts, anterior and posterior ($d.br_a$, $d.br_p$, Fig. 21), which jointly opened to the exterior through the external branchial opening of their own side. As may readily be gathered, the posterior one ($d.br_p$) of these probable para-atrial branchial ducts, which ran almost straight forwards, discharged the water from the hindmost part of the oesophagus and from all the 4–6 branchial compartments of its own side situated in the posterior division of the branchial region, and consequently it was also in communication with the oesophagus by means of all these 4–6 branchial compartments. Somewhat different conditions are met with in the Drepanaspida, Amphiaspida, Hibernaspida and Olbiaspida, in some Pteraspida forms, e.g. *Protaspis* (=? *Glossoidaspis*) and in some Cyathaspida, as, for instance *Allocryptaspis* and *Ctenaspis*, all of which possess a long, or at least not particularly short, branchial region. In the latter forms the external branchial opening of each side ($br.o$, Figs. 19 D, 20 C) lies slightly anterior to or in the posterior margin of the armour and thus opposite the hindmost part of the branchial apparatus. Under these circumstances the probable posterior para-atrial, efferent branchial duct ($d.br_p$) in the forms just listed must have been short and must have run almost transversally outwards towards the external branchial opening. Moreover, this duct could only have been in communication with the oesophagus by means of the two or three posterior branchial compartments, the hindmost one of which perhaps may even be imagined to have lost its respiratory function. Finally, in the Cyathaspid genus *Seretaspis* (Stensiö, 1964, fig. 109) where in consequence of reduction from behind-forwards the branchial region is very short (comprising eight branchial compartments in all) the probable posterior para-atrial, efferent branchial duct can only have discharged the water from the hindmost part of the oesophagus and from a few branchial compartments, the hindmost one of which may have been under reduction and tended to lose, or perhaps had already lost, its respiratory function.

As can be inferred from its position, course and internal communications with the oesophagus the probable posterior para-atrial, efferent branchial duct in the Heterostraci tended to modify into a paired oesophago-cutaneous duct of the same nature as the similarly termed duct in the Myxinoidea which, however, is persisting only on the left side of the branchial region and hence is secondarily an unpaired, asymmetrical canal.

Being hypothetical to a great extent the above account of the modifications of the external (efferent) branchial ducts may perhaps be regarded as questionable, even to the extent of being rejected without any discussion. It should be kept in mind, however, that in those Heterostraci forms where the paired external branchial opening lies a fair distance in front of the posterior margin of the armour, the outflow of the water from the hindmost part of the oesophagus and from the long, posterior division of the branchial apparatus took place in a forward direction, thus in a direction reverse to that from the anterior division. Consequently in the Heterostraci just referred to there existed all the prerequisites necessary for the evolution of a right and a left oesophago-cutaneous duct. The reduction of the posterior division of the branchial apparatus in a direction from behind-forwards (which apparently has taken place in several of those Heterostraci forms where the paired, common external branchial opening lies in the hindmost part of, or just behind the carapace) naturally caused not only a decrease in the outflow of water from the said division of the branchial apparatus but also that this outflow became directed straight laterally and outwards, just as in the single, persisting oesophago-cutaneous duct in the Myxinoidea.

Summing up briefly what has been stated above under points Nos. I–XI we find that the Heterostraci agree with the Myxinoidea in the following characters. (a) The composition and structure of the definitive snout in that the visceral component of this does not extend forwards-upwards in front of the cranial component and (b) that, therefore, except in the forms adapted to an inactive feeding (on plankton, on detritus or both), the naso-hypophysial aperture had a terminal or ventral position. (c) The considerable breadth both of the naso-hypophysial aperture and of the adjacent principal part of the prenasal sinus. (d) The presence in the actively feeding forms, e.g. the Cyathaspida and Pteraspida, of a biting (non-suctorial) oral apparatus and the circumstance that a biting oral apparatus of an essentially similar type originally may also have existed in the ancestors of all the Heterostraci. (e) The retrogressive development of the eyes which generally were so small that they may have been without muscles and the fact that at least one form (*Eglonaspis*) even was com-

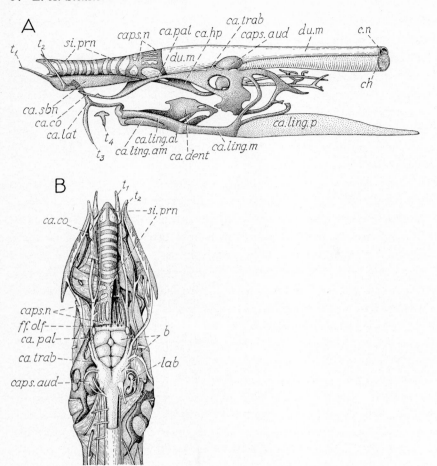

Fig. 22. *Myxine glutinosa*. A, skeleton of head of an adult specimen in lateral view. B, head in dorsal view dissected so as to show prenasal sinus and olfactory organ; dorsal wall of right half of nasal capsule removed. Cartilage stippled. From Marinelli & Strenger, 1956.

b, brain; *ca.co*, cornual cartilage; *ca.dent*, dentiferous cartilage (of rasping tongue); *ca.hp*, hypophysial cartilage; *ca.lat*, lateral labial cartilage (tentacular cartilage); *ca.ling.al*, anterior lateral lingual cartilage; *ca.ling.am*, anterior median lingual cartilage; *ca.ling.m*, middle lingual cartilage; *ca.ling.p*, posterior lingual cartilage; *ca.pal*, palatine cartilage; *caps.aud*, auditory (otic) capsule; *caps.n*, nasal capsule; *ca.sbn*, subnasal cartilage (belonging to inferior, visceral component of definitive snout); *ca.trab*, trabecular cartilage; *ch*, notochord; *c.n*, neural canal; *du.m*, dural membrane; *ff.olf*, fila olfactoria; *lab*, membranous labyrinth (upper part); *si.prn*, prenasal sinus (its cartilages probably derived from the skeletogenous tissue in the superior, cranial component of the definitive snout); t_1–t_4, cartilaginous tentacular processes.

pletely blind. (f) A total absence of the superficial, principal portions of the branchial constrictor muscles, which means (g) that the pumping of water to the gills took place by the movements of a powerful velum and (h) that the Heterostraci therefore must have possessed a naso-pharyngeal duct leading backwards from the prenasal sinus to the pharynx. That this duct actually existed is proved by the forms feeding on plankton and detritus, where the food could have been conveyed to the pharynx solely by the backward directed current of water effected by a velum of fundamentally the same type as in *Myxine* and *Bdellostoma*. (i) The fact that in all the better known forms the external (efferent) branchial ducts of each side joined to form two major para-atrial ducts, anterior and posterior, which in turn fused so as to lead to the exterior through a single, wide external branchial aperture. And finally (j) the condition of the posterior para-atrial duct which together with the posterior branchial compartments opening into it was fundamentally as a paired oesophago-cutaneous duct.

As is apparent from the characters discussed and listed, the Heterostraci had acquired a decided myxinoid anatomical organization and together with the Myxinoidea they hence represent an old stock of cyclostomes of their own, the Pteraspidomorphi (Stensiö, 1927). In several respect, but above all with regard to the oral apparatus and branchial apparatus, the Heterostraci had reached a more advanced stage of specialization than the Osteostraci and cannot possibly have given rise either to the Osteostraci or to the other Cephalaspidomorphs (Anaspida and Petromyzontida). Taking all known facts into consideration one hence arrives at the conclusion that the cyclostomes are represented by two separate major stocks, the Cephalaspidomorphi and Pteraspidomorphi, evolved long before the Middle Ordovician, in the Cambrian or even earlier, and that consequently the recent cyclostomes are diphyletic.

Under these circumstances it naturally follows that the rasping tongue in the Myxinoidea has evolved independently of that in the Petromyzontida, a conclusion which harmonizes with the fact stressed by many writers (Müller, 1836: 281; Holmgren & Stensiö, 1936; Holmgren, 1946; Luther, 1938; Marinelli & Strenger, 1954; 1956; Dawson, 1963; Jarvik, 1964; 1965), that this tongue differs radically in the two groups where both its anatomy and function are concerned. The rasping tongue in the Myxinoidea can only have evolved in forms possessing a fairly modified, biting oral apparatus of fundamentally the same type as in the Pteraspida and Cyathaspida, an oral apparatus which could not any longer give rise to a "sucker". As opposed to this, the rasping tongue in the Petromyzontida apparently evolved in forms with a more primitive oral apparatus of a type similar to that in the Osteostraci, an oral apparatus which also could give rise to a "sucker". It should also be added here that with regard to their anatomy in general the Myxinoidea and Petromyzontida differ so considerably that a comparison between them is difficult and in part even impossible (see Brodal & Fänge, 1963; Jarvik, 1964; 1965).

Amongst the Heterostraci a rasping tongue naturally could not arise in the forms adapted for feeding on detritus and minute organisms (including plankton). Notwithstanding the presence of all the necessary prerequisites a rasping tongue, even at an incipient stage, seems not to have existed either in those of the better known forms which possessed a biting oral apparatus or in the forms adapted to an inactive mode of feeding. The remaining Heterostraci forms still are entirely unknown where their oral apparatus and mode of feeding are concerned. However, it is not impossible that the disintegration of the antero-dorsal, lateral and ventral parts of the armour in the Cardipeltida may have been caused by the presence of a small, primitive rasping tongue. A small rasping tongue may also be imagined to have existed in the Turiniida and those Thelodonti suspected by the writer to be Heterostraci, where the entire armour is represented solely by small synchronomorial scales (in some forms partly also lepidomoria) situated losely in the skin. Be this as it may, no definitive indications of the existence of a rasping tongue have hitherto been found in the Heterostraci. At least to some extent, however, this may very well be due to fact that the visceral endoskeleton still is slightly known in these fossil cyclostomes. On the other hand, in the Osteostraci, as we have seen, there was a general trend towards the evolution of a small primitive rasping tongue. Since the Anaspida most certainly possessed a long rasping tongue similar to that in the Petromyzontida, but probably somewhat longer, a trend towards the evolution of such a tongue must even have existed already in some of the common ancestors of the Osteostraci and Anaspida.

Summary and conclusions

The principal results arrived at may be briefly summarized as follows.

The definitive snout in the adult Petromyzontida and Myxinoidea is made up of two different components, a superior, cranial component which is the ethmoidal part of the cranial division of the head and an inferior, visceral component which, being formed by the foremost upper part of the visceral division of the head, contains the derivatives of the epitrematic halves of the premandibular arches of both sides. During the embryonic development the latter component grows out forwards as a fold below the superior, cranial component in such a way that the external, upward-backward extending incurvity between the two components becomes enclosed in the definitive snout, where its median part is persisting as an outward opening space in front of the olfactory organ and the cavity of the adeno-hypophysial invagination. *This space, in all previous publications termed the "hypophysial duct" or "hypophysial sac", hence is an extracephalic cavity in a secondary position inside the definitive snout.* That being the case and since it has nothing to do with the olfactory organ except that this organ opens into it, *the space under discussions has been termed here the prenasal sinus*, whereas its external opening has been referred to under the old term, the naso-hypophysial aperture (Stensiö, 1958; 1964). A prenasal sinus of the same nature as in the Petromyzontida and Myxinoidea also existed in all the fossil cyclostomes.

Two different types of definitive snouts, myxinoid and petromyzontid are met with in the cyclostomes. In the definitive myxinoid snout the two components have either the same forward extent or else the inferior, visceral component is shorter than superior, cranial one, and as a result the naso-hypophysial opening has a terminal, subterminal or ventral position. This type of definitive snout is distinctive of the Myxinoidea and must also have existed in the Heterostraci. In the definitive petromyzontid snout the inferior, visceral component has undergone such an excessive evolution in a forward-upward direction as to form alone the anterior, principal part of the definitive snout. This has caused an extensive suppression of the superior, cranial component because of which the prenasal sinus bends upwards in such a way that the naso-hypophysial opening has acquired a dorsal position far behind the anterior end of the head. The latter type of definitive snout is distinctive of the Petromyzontida, Osteostraci and Anaspida.

With regard to the structure and composition of their definitive snout the cyclostomes hence are much specialized but in a way different from that in the gnathostomes.

The Petromyzontida and Myxinoidea are diplorhinal vertebrates, just as the gnathostomes. The originally separate nasal capsules have fused to form an unpaired capsule, a condition which has caused the original internasal wall to be represented solely by a thin septum of soft tissues. The nasal sac of each side has an external opening of its own, a fenestra endonarhina, situated in the anterior wall of the secondarily unpaired nasal capsule and separated from the corresponding opening of the other side by the internasal septum. This paired external nasal opening hence has the same position relative to the olfactory organ and to the endoskeleton as the fenestra endonarhina in fishes, and it must therefore be homologous with this. The fact that it leads out into the prenasal sinus means that it has retained its original external position relative to the cranial and visceral divisions of the head. What has just been stated concerning the olfactory organ in the recent cyclostomes also applies to all the fossil ones. Consequently we find that, notwithstanding the advanced stage of specialization of the snout, the olfactory organ has undergone only comparatively slight changes and that, apart from the secondary unpaired nasal capsule, it was, and still is, of the paired gnathostome type. *Thus we cannot any longer speak of the cyclostomes as Monorhini or monorhinal vertebrates.*

The Heterostraci had not only acquired a naso-pharyngeal duct of the same nature and with the same function as in the Myxinoidea, but, as has been stressed, they also possessed many other important myxinoid characters. It is therefore clear that the Heterostraci are so closely akin to the Myxinoidea that jointly with these they represent a principal branch of cyclostomes of their own, the Pteraspidomorphi (Stensiö, 1927; 1932; 1958; 1963b; 1964). *Hence it follows that the recent cyclostomes—the Myxinoidea and Petromyzontida—are diphyletic* (see also the physiological and biochemical papers in Brodal & Fänge, 1963).

The oral apparatus in the Heterostraci originally was of a biting type with a kind of moveable "lower jaw". An oral apparatus of this kind still was persisting in the Pteraspida, Cyathaspida and presumably also in some other slightly known forms. To be more exact, a biting oral apparatus existed as long as the feeding was active. However, many Heterostraci, as, for example, the Drepanaspida, Lyktaspida, Amphiaspida, Hibernaspida and Olbiaspida had adapted themselves to an inactive feeding on minute organisms, detritus or both. In the five latter orders the food was conveyed into the pharynx with the water pumped backwards to the gills by the velum, and in consequence the oral apparatus had undergone considerable changes and was no longer of a biting type.

The recent Myxinoids must have evolved from some early Heterostraci possessing a biting oral apparatus.

References

Bertmar, G. (1959). On the ontogeny of the chondral skull in the Characidae, with a discussion on the chondrocranial base of the visceral chondrocranium in fishes. *Acta zool. Stockh.*, **40**: 203–364.
Brodal, A. & Fänge R. (ed.) (1963). *The Biology of Myxine*. Oslo: Universitetsforl.
Cole, F. J. (1905–1925). A monograph on the general morphology of myxinoid fishes, based on a study of *Myxine*. *Trans. R. Soc. Edinb.*, **41**: 749–788; **45**: 683–757; **46**: 669–681; **49**: 215–230; 293–344; **54**: 309–342.
Damas, H. (1944). Recherches sur le dévelopement de *Lampetra fluviatilis* L. Contribution à l'étude de la cephalogenèse des vertébrés. *Arch. Biol. Liége*, **29**: 1–284.
Dawson, J. A. (1963). The oral cavity, the "jaws" and the horny teeth of *Myxine*. In *The Biology of Myxine*, ed. Brodal, A. & Fänge, R.,: 231–235. Oslo: Universitetsforl.
Denison, R. H. (1964). The Cyathaspididae. A family of Silurian and Devonian jawless vertebrates. *Fieldiana: Geol.*, **13**, 307–473.
Goodrich, E. S. (1999). Cyclostomes and fishes. In *A Treatise on Zoology*, ed, Lankester, E. R., **9**: Vertebrata Craniata, fasc. 1, London: A. & C. Black.
Göppert, E. (1906). Die Entwickelung des Mundes und der Mundhöhle mit Drüsen und Zunge; die Entwickelung der Schwimmblase, der Zunge und des Kehlkopfes bei den Wirbeltieren. A. Die Entwickelung des Mundes, der Mundhöhle und ihrer Organe. In *Handbuch der vergleichenden und experimentellen Entwicklungslehre der Wirbeltiere*, hrsg. Hertwig. O., **2**:**2**: 1–80. Jena:Fischer.
Gustafson, G. (1936). On the Biology of *Myxine glutinosa*. *Ark. Zool.*, **28** *A*: 1–8.
Hagelin, L.-O. & Johnels, A. G. (1955). On the structure and function of the accessory olfactory organ in lampreys. *Acta zool. Stockh.*, **36**: 113–125.
Heintz, A. (1939). Cephalaspida from Downtonian of Norway. *Skr. norske Vidensk-Akad. Oslo, Mat.-naturv. Kl.*, **1939**: 1–119.
Heintz, A. (1962). Les organes olfactifs des Heterostraci. *Colloques int. Cent. natn. Rech. Scient.*, **104**: 13–29.
Heintz, A. (1963). Phylogenetic aspects of myxinoids. In *The biology of Myxine*, ed. Brodal, A. & Fänge, R.,: 9–21. Oslo: Universitetsforl.

Herre, W. (1964). Zum Herstammungsproblem von Amphibien und Tylopoden sowie über Parallelbildungen und zur Pholyphyliefrage. *Zool. Anz.*, **173**: 66–98.

Holmgren, N. (1946). On two embryos of *Myxine glutinosa*. *Acta zool. Stockh.*, **27**: 1–90.

Holmgren, N. & Stensiö, E. A. (1936). Kranium und Visceralskelett der Akranier, Cyclostomen und Fische. In *Handbuch der vergleichenden Anatomie der Wirbeltiere*, hrsg. Bolk, L., Göppert, E., Kallius, E. & Lubosch, W., **4**: 233–500. Berlin & Wien: Urban & Schwarzenberg.

Jarvik. E. (1954). On the visceral skeleton in *Eusthenopteron* with a discussion of the parasphenoid and palatoquadrate in fishes. *K. svenska VetenskAkad. Handl.*, (4) **5**: 1–104.

Jarvik. E. (1964). Specializations in early vertebrates. *Annls Soc. R. zool. Belg.*, **94**: 11–95.

Jarvik, E. (1965). Die Raspelzunge der Cyclostomen und die pendactyle Extremität der Tetrapoden als Beweise für monophyletische Herkunft. *Zool. Anz.*, **175**: 101–143.

Johnels, A. G. (1948). On the development and morphology of the skeleton of the head in *Petromyzon*. *Acta zool. Stockh.*, **29**: 139–279.

Kaensche, C. C. (1890). Beiträge zur Kenntniss der Metamorphose des *Ammocötes branchialis* in *Petromyzon*. *Zool. Beitr.*, **2**: 219–250.

Kiaer, J. (1924). The Downtonian fauna of Norway. 1. Anaspida with a geological introduction. *Skr. norske VidenskSelsk. Oslo, Mat.-naturv. Kl.*, **1924**: 1–139.

Kiaer, J. (1928). The structure of the mouth of the oldest known vertebrates, pteraspids and cephalaspids. *Palaeobiologica*, **1**: 117–134.

Kupffer, C. W. v. (1893–1900). *Studien zur vergleichenden Entwicklungsgeschichte des Kopfes der Kranioten*. München & Leipzig: Lehmann.

Kupffer, C. W. v. (1894). Ueber Monorhinie und Amphirhinie. *Sber. bayer. Akad. Wiss.*, **24**: 51–60.

Kupffer, C. W. v. (1906). Die Morphogenie des Centralnervensystems. In *Handbuch der vergleichenden und experimentellen Entwicklungslehre der Wirbeltiere*, hrsg. Hertwig, O., **2:3**: 1–272. Jena: Fischer.

Lindström, T. (1949). On the cranial nerves of the cyclostomes with special reference to N. trigeminus. *Acta zool. Stockh.*, **30**: 315–458.

Liu, Y.-H. (1965). New Devonian agnathans of Yunnan. *Vertebrata Palasiatica*, **9**: 125–134.

Lubosch, W. (1905). Die Entwicklung und Metamorphose des Geruchsorgans von *Petromyzon* und seine Bedeutung für die vergleichende Anatomie des Geruchsorganes. *Jena. Z. Naturw.* (2) **33**: 95–148.

Luther, A. (1938). Die Visceralmuskulatur der Acranier, Cyclostomen und Fische. A. Acranier, Cyclostomen, Selachier, Holocephalen und Dipnoer. In *Handbuch der vergleichenden Anatomie der Wirbeltiere*, hrsg. Bolk L., Göppert, E., Kallius, E. & Lubosch, W., **5**: 467–542. Berlin & Wien: Urban & Schwarzenberg.

Marinelli, W. & Strenger, A. (1954). *Vergleichende Anatomie und Morphologie der Wirbeltiere*. 1. *Lampetra fluviatilis*. Wien: F. Deuticke.

Marinelli, W. & Strenger, A. (1958). *Vergleichende Anatomie und Morphologie der Wirbeltiere*. 2. *Myxine glutinosa* (L.). Wien: F. Deuticke.

Matthes, E. (1934). Geruchsorgan. In *Handbuch der vergleichenden Anatomie der Wirbeltiere*, hrsg. Bolk, L., Göppert, E., Kallius, E. & Lubosch, W., **2:2**: 881–886. Berlin & Wien: Urban & Schwarzenberg.

Müller, J. (1836). Vergleichende Anatomie der Myxinoiden, der Cyclostomen mit durchbohrtem Gaumen. *Abh. preuss. Akad. Wiss.*, **1836**: 65–340.

Neumayer, L. (1938). Die Entwicklung von *Bdellostoma St. L. Arch. ital. Anat. Embriol.*, **40** (*Suppl.*): 1–222.

Obruchev, D. V. (1964). Podklass Heterostraci (Pteraspides). Inopantsernuie (pteraspidui). In *Osnovui paleontologii* (*Fundamentals of Palaeontology*), ed. Orlov, I. A., **11**: 45–82. Moscow: Acad. Sci. U.S.S.R. In Russian.

Peter, K. (1906). Die Entwicklung des Geruchsorganes und Jacobson'schen Organ in der Reihe der Wirbeltiere. In *Handbuch der vergleichenden und experimentellen Entwicklungslehre der Wirbeltiere*, hrsg. Hertwig. O., **2:2**: 1–78. Jena: Fischer.

Ritchie, A. (1964). New light on the morphology of the Norwegian Anaspida. *Skr. norske VidenskAkad. Oslo, Mat.-naturv. Kl.*, **1964**: 1–22.

Romer, A. S. (1945). *Vertebrate Paleontology*. 2nd Ed. Chicago: University of Chicago Press.

Romer, A. S. (1966). *Vertebrate Paleontology*. 3rd Ed. Chicago: University of Chicago Press.

Sewertzoff, A. N. (1916). Études sur l'évolution des vertébrés inférieurs. 1. Morphologie du squelette et de la musculature de la tête des cyclostomes. *Arch. russ. Anat. Hist. Embr.*, **1**: 1–104.

Shipley, A. E. (1887). On some points in the development of *Petromyzon fluviatilis*. *Q.J.micr.Sci.*, (n.s.) **27**: 325–370.

Stensiö, E. A. (1927). The Downtonian and Devonian vertebrates of Spitsbergen. 1. Family Cephalaspidae. *Skr. Svalbard Nordishavet*, **12**: i-xii, 1–391.

Stensiö, E. A. (1932). *The cephalaspids of Great Britain*. London: Br. Mus. (nat. Hist.).

Stensiö, E. A. (1958). Les cyclostomes fossiles ou ostracodermes. In *Traité de Zoologie*, ed. Grassé, P.-P., **13:1**: 173–425. Paris: Masson.

Stensiö, E. A. (1963a). Anatomical studies on the arthrodiran head. 1. Preface, geological and geographical distribution, the organisation of the arthrodires, the anatomy of the head in the Dolichothoraci, Coccosteomorphi and Pachyosteomorphi. Taxonomic appendix. *K. svenska VetenskAkad. Handl.*, (4) **9**: 1–419.

Stensiö, E. A. (1963b). The brain and the cranial nerves in fossil, lower craniate vertebrates. *Skr. norske VidenskAkad. Oslo, Mat.-naturv. Kl.*, **1963**: 1–120.

Stensiö, E. A. (1964). Les cyclostomes fossiles ou ostracodermes. In *Traité de Paléontologie*, ed. Piveteau, J., **4:1**: 96–382. Paris: Masson.

Stensiö, E. A. (1968). Les arthrodires. In: *Traité de Paléontologie*, ed. Piveteau, J., **4:2**: Paris: Masson. In press.

Strahan, R. (1958a). The velum and respiratory current of *Myxine*. *Acta zool. Stockh.*, **39**: 227–240.

Strahan, R. (1958b). Speculation on the evolution of the agnathan head. *Proc. cent. & bicent. Congr. Biol. Singapore*: 83–94.

Watson, D. M. S. (1954). A consideration of ostracoderms. *Phil. Trans. R. Soc.*, (B) **238**: 1–25.

Discussion

E. I. White

While Stensiö is to be congratulated on the great interest and importance of his paper, there is one point on which I have to register my continued disagreement, on the existence of an upper dentition in the pteraspids. Whatever the postulates of Stensiö's theory may be, the fact is that such a dentition has never been discovered, and the extraordinarily fine and varied conditions of preservation under which remains of the mouth region have been found make it clear that it never existed—in Kiaer's (1928: 119, pl. 12) original material from Spitsbergen the mouth region is beautifully preserved in place and clearly shown in section; in the Wayne Herbert (Herefordshire) fossils the same area is shown as impressions in undistorted detail (White, 1935: 409, figs. 41–44); and finally the specimens from the Clee Hills (White, 1961: 251, pls. 37, 38) are lower "teeth" in the round free of matrix, with no sign of wear such as must have been evident had they worked against an upper dentition. Pteraspids fed on detritus either by scooping or sucking.

Kiaer, J. (1928). The structure of the mouth of the oldest known vertebrates, pteraspids and cephalaspids. *Palaeobiologica*, **1**: 117–134.
White, E. I. (1935). The ostracoderm *Pteraspis* Kner and the relationships of the agnathous vertebrates. *Phil. Trans. R. Soc.*, (B) **225**: 381–457.
White, E. I. (1961). The Old Red Sandstone of the Brown Clee Hill and adjacent area. 2. Palaeontology. *Bull. Br. Mus. nat. Hist.: Geol.*, **5**: 243–310.

The pteraspid *Lyktaspis* n.g. from the Devonian of Vestspitsbergen

By Natascha Heintz

Palaeontological Museum, Oslo, Norway

The Devonian sediments of Old Red type in Dickson Land, Andrée Land, Reinsdyrflya, and several other places in Vestspitsbergen, contain rich faunas of ostracoderms and fishes, on the stratigraphic distribution of which the subdivision of these deposits is mainly based. For the so-called Wood Bay Formation (Føyn & A. Heintz, 1943; Friend, N. Heintz & Moody-Stuart, 1966) a rather aberrant type of pteraspid, described herein as *Lyktaspis* n.g., has proved to be an excellent guidefossil.

Material of this genus has been assembled over a long period of time. The first specimens were found by Nathorst in the Dicksonfjorden area in 1882, and much more material was assembled by later Swedish, Norwegian and British expeditions. Vogt's Expedition in 1925, the English-Norwegian-Swedish Expedition in 1939 and A. and N. Heintz in 1959, 1960, and 1964, have brought back the largest and most valuable parts of the collections at my disposal.

Previous investigations

In his description of some Devonian lower vertebrates from Spitsbergen, Lankester (1884) introduced the name *Scaphaspis nathorsti* for a species which Woodward later (1891; 1900) referred to the family Pteraspididae as *Pteraspis nathorsti*. On the basis of a more comprehensive material, Kiær took up an extensive study of the species, but unfortunately did not complete this work before his death. In an unpublished manuscript he pointed out that because of its unusual ornamentation, large cornual plates and long, slim rostrum, the species in question had to be placed in a separate genus which he proposed to call *Doryaspis*. This generic name was formally introduced by White (1935) who figured a ventral disc as *Doryaspis nathorsti* (Lankester), and has later received mention by Føyn & A. Heintz (1943), Tarlo (1962), Stensiö (1964), and Obruchev (1964). Since, however, *Doryaspis* is preoccupied (Denison, 1967), it is here replaced by *Lyktaspis* n.g. with the type species *L. nathorsti* (Lankester).

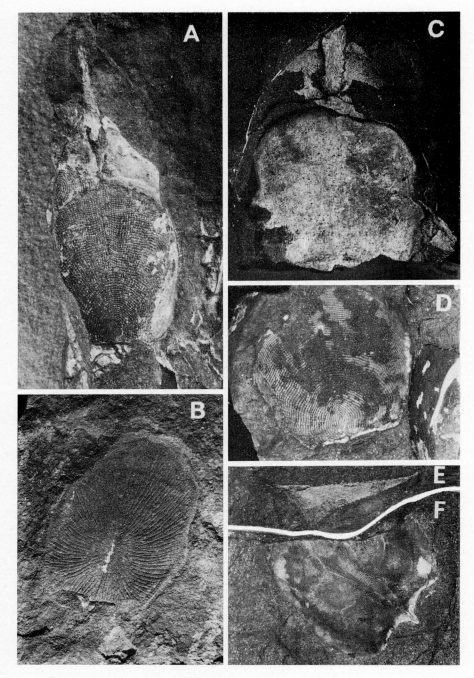

Fig. 1. Different plates of *Lyktaspis nathorsti* (Lankester), all approx. nat. size. (PMO refers here, as in Fig. 4, to specimens belonging to the Palaeontological Museum, Oslo). *A*, ventral side of the carapace in ventral view; PMO A 27942. *B*, external mould of dorsal disc; PMO D 5208. *C*, specimen showing in its most anterior part ventral view of the anterior part of the ventral disc, the lateral plates, the antero-lateral plates and the posterior part of the pseudo-

Description

The following brief description of *Lyktaspis* n.g. is based mainly on the type species *L.nathorsti*, by far the most common species of the genus in the Devonian of Vestspitsbergen. Two other species *L.dani* n.sp. and *L.minor* n.sp. also exist; these which are comparatively rare, will be defined and described by the writer in a subsequent paper.

General remarks. The dorsal side of the carapace in *Lyktaspis* n.g. (Fig. 2) is made up of nine plates, viz. one dorsal disc (d) with a small dorsal spine (ds) posteriorly; two rather slim branchial plates (b); two large cornual plates (c); two orbital plates (o); a small, circular pineal plate (p); and a short, blunt rostrum, semicircular in outline (r). The ventral side (Fig. 3) consists generally of six plates, viz. one ventral disc (v); two lateral plates (l); two small antero-lateral plates (a); and a very long, slender pseudorostrum (s). In a few cases a row of six or seven small, scale-like plates are found ventrally to the branchial plate along the anterior part of the ventral disc (Fig. 4 A).

The trunk behind the carapace is covered by comparatively large scales. The three or four anterior rows contain a large number of scales whereas the following each consist of only six scales. Median fulcra-like scales (f, Figs. 2, 3) form one row along the dorsal side of the trunk and another along the ventral; of these the dorsal ones continue posteriorly to the very tip of the tail (Fig. 2). The length of the hypocercal tail is about one third of that of the scale-covered part of the trunk.

All the plates of the carapace are thin, averaging a little less than 1 mm in thickness. As normal in pteraspids, they are built up of three layers, but the cavities of the middle layer are comparatively smaller than normal in other Heterostraci (Fig. 1 F). The highly characteristic ornamentation of the plates was described by Lankester (1884) as consisting of concentric and radiating ridges; it is more adequate in this respect, however, to speak of concentric ridges subdivided by a system of radiating grooves (Fig. 1 A, B, D). The dorsal and ventral discs, the branchial plates, pineal plate and rostrum show each one centre of growth. The orbital plate, on the other hand, exhibits three such centres, one in the normal position and two others postero-medially to the orbit. The lateral plates and the pseudorostrum seem each to possess two centres of growth, whereas in the small antero-lateral plates and the cornual plates

---

rostrum, and in its posterior part ventral view of the dorsal disc and a fragment of the right cornual plate; PMO A 28581. *D*, ventral disc in ventral view; laterally in this plate are seen sensory canals and their tubuli; PMO A 32717. *E*, cornual plate; PMO A 32718. *F*, nearly complete dorsal disc with posterior spine; sensory canals are seen in the cancellous layer; PMO A 32719.

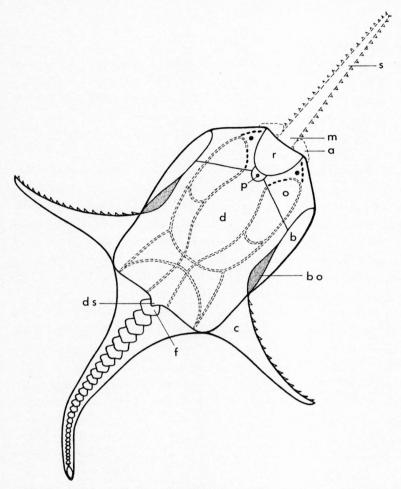

Fig. 2. Restoration of the dorsal side of the carapace in *Lyktaspis* n.g. Sensory lines indicated by double interrupted lines, presumed extension of these canals by dotted lines.

a, antero-lateral plate; *b*, branchial plate; *bo*, branchial opening; *c*, cornual plate; *d*, dorsal disc; *ds*, dorsal spine; *f*, fulcra scales; *l*, lateral plate; *m*, mouth; *o*, orbital plate; *p*, pineal plate; *r*, rostrum; *s*, pseudorostrum; *v*, ventral disc.

no typical growth centres are encountered since the ornamentation is here made up of parallel, length-wise arranged, ridges. The anterior margin of the cornual plates and the lateral margins of the pseudorostrum are set with small spines (Figs. 2, 3).

With certain deviations to be dealt with elsewhere, the sensory canal system of the carapace follows the usual pteraspid pattern. These canals are rather fine, varying from 0.5 to 1 mm in width, and are generally finer in the anterior part of the carapace than posteriorly. So far no traces of sensory canals have been found in the squamation.

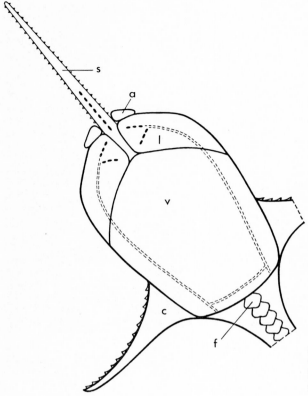

Fig. 3. Restoration of the ventral side of the carapace in *Lyktaspis* n.g. Lettering as in Fig. 2.

The rostral region of the carapace. The rostrum is semicircular in outline, comparatively very short, blunt and transversely truncated anteriorly (Figs. 2, 4 B). Laterally and postero-laterally it is adjoined by the orbital plates and posteriorly by the small pineal plate. Its anterior margin, which is thickened, does not protrude forwards beyond the corresponding margins of the orbital plates. In an anterior direction, the rostrum is strongly bent upwards (cf. reconstruction in lateral view by the writer, reproduced in Obruchev, 1964, fig. 44). The dorsal side of the rostrum shows the typical ornamentation of dentine ridges whereas its ventral side right up to the anterior margin is smooth, indicating that the latter was directly adjoined by soft tissues other than those of the skin. In shape and position, the rostrum differs markedly from that of all other members of the family Pteraspididae where the mouth is always situated ventrally, although lying in cases quite far anteriorly. In the other pteraspids, a larger or smaller part of the ventral side of the rostrum is covered by ornamentation and was thus in the living state covered by skin (Stensiö, 1958; 1964). In *Lyktaspis* n.g. on the other hand, the shape and location of the rostrum indicates that the mouth was situated fully on the antero-dorsal side of the animal (*m*, Fig. 2).

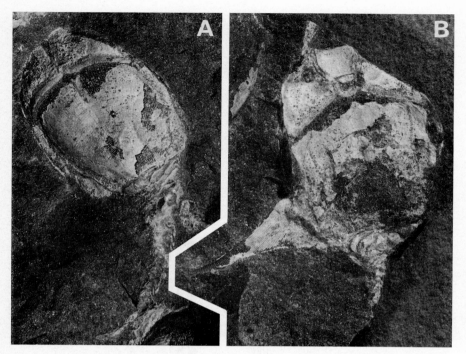

Fig. 4. *Lyktaspis nathorsti* (Lankester). *A*, Dorsal view of ventral disc, lateral plates and trunk squamation; along its left side the ventral disc is adjoined by a series of six small, scale-like plates; PMO A 27938; approx. nat. size. *B*, ventral view of dorsal side of carapace and a few trunk scales. PMO A 27942; approx. nat. size.

The long, spear-like plate which in *Lyktaspis* n.g. lies in an antero-medial position between the lateral plates and in front of the ventral disc (*s*, Fig. 3; cf. Fig. 1 A, C) has been tentatively referred to as a pseudorostrum, more as a means of indicating its position than to show its possible relation to the antero-ventral plates in other pteraspids. The plate in question is a little longer than the total length of the carapace following behind, and protrudes anteriorly far in front of the dorsally situated proper rostrum. It bears the typical ornamentation of dentine ridges both dorsally and ventrally, a circumstance which shows that it was originally covered by skin on both sides. In his unpublished manuscript Kiær interpreted the plate here named the pseudorostrum as the real rostrum, in consequence of which he failed to recognize what was the dorsal and what the ventral side of the animal, and orientated it upside-down. Postero-laterally to the pseudorostrum, and in front of the lateral plates, there are two small plates, one on each side, which are tentatively named antero-lateral plates (*a*, Fig. 3; cf. Fig. 1 A, C); at present, however, I find it difficult to compare them with any special plates in the oral region of other pteraspids (see also Stensiö, in this Volume). Even these plates show the characteristic external ornamentation of ridges both dorsally and ventrally.

Systematic position

Although Kiær, as mentioned, misinterpreted the pseudorostrum of *Lyktaspis nathorsti* as a true rostrum, and believed that one is here concerned with much the same condition in the rostral region of the carapace as in pteraspids in general, he nevertheless placed this form in a separate genus within the group because of the characteristic ornamentation, large cornual plates and other features as well. Similar conclusions were reached by White (1935), according to whom "*Scaphaspis nathorsti* ... is clearly not a species of *Pteraspis*, although a Pteraspid ..."

The new interpretation of the mouth region presented here and the recognition of a pseudorostrum, implies that *Lyktaspis* n.g. is clearly different from all other members of the family Pteraspididae. In recent years particularly Stensiö (1958; 1964) has emphasized that the configuration of the mouth region in pteraspids, especially the ventral side of the rostrum, should be ascribed considerable systematic significance. On the basis of such characters, Stensiö established several new genera of pteraspids.

The mouth region in *Lyktaspis* n.g. is without doubt highly specialized as compared with that of any other hitherto described pteraspid genus. This, and the fact that so far *Lyktaspis* n.g. has only been found within a rather limited stratigraphic sequence in Vestspitsbergen, indicate that this genus belongs to a lineage which has long been separated from other representatives of the pteraspid stock; in these circumstances the genus *Lyktaspis* n.g. should be placed in a new family Lyktaspididae (=Doryaspididae) or perhaps even in a new pteraspid suborder Lyktaspidida (=Doryaspidida) as in fact both Obruchev (1964) and Tarlo (1962) have maintained.

References

Denison, R. H. (1967). A new *Protaspis* from the Devonian of Utah, with notes on the classification of Pteraspididae. In *Fossil Vertebrates*, ed. Patterson, C. & Greenwood, P. H., *J. Linn. Soc. (Zool.)*, **47**: 31–37.

Føyn, S. & Heintz, A. (1943). The Downtonian and Devonian vertebrates of Spitsbergen. 8. The English-Norwegian-Swedish Expedition 1939. Geological results. *Skr. Norg. Svalbard og Ishavsunders.*, **85**: 1–51.

Friend, P. F. (1961). The Devonian stratigraphy of North and Central Vestspitsbergen. *Proc. Yorks. geol. Soc.*, **33**: 77–118.

Friend, P. F., Heintz, Natascha & Moody-Stuart, M. (1966). New unit terms for the Devonian of Spitsbergen and a new stratigraphical scheme for the Wood Bay Formation. *Norsk Polarinst. Årb.*, **1965**: 59–64.

Lankester, E. R. (1884). Report on fragments of fossil fishes from the Palaeozoic strata of Spitzbergen. *K.svenska VetenskAkad. Handl.*, **20**: 1–7.

Obruchev, D. V. (1964). Podklass Heterostraci (Pteraspides). Inopantsernuie (pteraspidui). In *Osnovui paleontologii (Fundamentals of Palaeontology)*, ed. Orlov, I. A., **11**: 45–82. Moscow: Acad.Sci. U.S.S.R. In Russian.

Stensiö, E. A. (1958). Les cyclostomes fossiles ou ostracodermes. In *Traité de Zoologie*, ed. Grassé, P.-P., **13:1**: 173–425. Paris: Masson.

Stensiö, E. A. (1964). Les cyclostomes fossiles ou ostracodermes. In *Traité de Paléontologie*, ed. Piveteau, J., **4:1**: 96–382. Paris: Masson.

Tarlo, L. B. H. (1962). The classification and evolution of the Heterostraci. *Acta palaeont.polon.*, **7**: 249–290.

White, E. I. (1935). The ostracoderm *Pteraspis* Kner and the relationships of the agnathous vertebrates. *Phil. Trans. R. Soc.*, (B) **225**: 381–457.

Woodward, A. S. (1891). The Devonian fish-fauna of Spitzbergen. *Annls Mag. nat. Hist.*, (6) **8**: 1–15.

Woodward, A. S. (1900). Notes on fossil fish-remains collected in Spitzbergen by the Swedish Arctic Expedition, 1898. *K. svenska VetenskAkad. Handl.*, **25**: 1–7.

Phlebolepis elegans Pander, an Upper Silurian thelodont from Oesel, with remarks on the morphology of thelodonts

By Alexander Ritchie

The Australian Museum, Sydney, N.S.W. Australia

Recent research on articulated material of Silurian and Devonian thelodonts from Europe has provided much new information concerning the morphology of this, the most poorly known group of fossil agnathans. Important additions to our knowledge of thelodont scales, their histology and their stratigraphic significance have been made by Gross (1947; 1967), Ørvig (1951) and others. The Scottish, English and Norwegian thelodonts will be covered in full in a separate paper. During the course of the present research the writer has had the opportunity of examining, in Oslo, a considerable number of specimens of *Phlebolepis elegans* Pander, an intriguing little thelodont from the Upper Silurian of the island of Oesel in the Baltic. Features were observed in this material which help to resolve earlier conflicting interpretations and bring *Phlebolepis* into line with other thelodont genera.

Phlebolepis elegans Pander was originally described from isolated scales (Pander, 1856) with a figure showing a basal section The first articulated material, consisting of over sixty individuals, was discovered by A. Luha in 1929––30. The specimens were sent to Prof. J. Kiaer in Oslo for examination and description but he died before completing the study. His successor, Prof. A. Heintz, using Kiaer's prepared plates and preliminary notes, published a detailed account of the material (Kiaer & Heintz, 1932).

Heintz was able to show that the form found by Luha was a small agnathan, up to 7 cm long, entirely covered with rhomboidal, imbricating scales; in addition to the hypocercal caudal fin there was a low, rounded anal fin. Dorsal and paired fins appeared to be lacking. The small rhomboidal scales each bore a prominent, median ridge, produced posteriorly into a spine, flanked on either side by several finer ridges. Heintz was unable to section any of the scales since, as he explained, it was "practically impossible to find any scale which could be used for this purpose"; most of the specimens are preserved as detailed external moulds. In the absence of sections Heintz assumed that the scales were of the normal thelodont type with enamel-coated dentine and a pulp cavity. From the apparent resemblance of the form to *Thelodus* and *Lanarkia* Heintz named it *Coelolepis luhai* since he did not recognize, in Luha's articulated material, the form *Phlebolepis elegans* Pander which was based on scales alone.

Hoppe (1931; 1933) prepared both horizontal and (the more informative) vertical sections of *Coelolepis luhai* scales and discovered that, in most respects, they corresponded closely to those of *Phlebolepis elegans*. His sections, like those of Pander, revealed no trace of a pulp cavity, a characteristic feature of thelodont scales.

Buistrov (1949) showed, however, that either Hoppe's material or his preparations were defective since his own sections demonstrated convincingly the presence of a well-developed pulp cavity which originated on the smooth, flat base of the scale and passed upward and backward into the long posterior projection. The rather coarse, branching canals which penetrate the dentine from the basal plate have been figured (and variously interpreted) by Hoppe (1933: 127, figs. 1, 3 b), Gross (1947: 111, fig. 6; 1967, fig. 15 E–Q), Buistrov (1949: 247, fig. 2) and Ørvig (1951: 385, fig. 10 D–G). The presence of the pulp cavity removes any doubt as to the thelodont relationships but Buistrov (1949) considered that the relatively small size of the pulp cavity indicated a primitive condition when compared with the rather larger cavity in numerous other thelodonts.

Westoll (1945: 347–348) remarked that, as described, "in body-form, fin-details and size and place of orbits and mouth *Phlebolepis* has a most surprising similarity to Anaspida... It is true that there are, according to Kiaer & Heintz, no signs of gill openings, but it seems not impossible that such openings might have been obliterated during fossilization. The possibility of relationships to anaspids seems sufficiently strong for a close re-examination of *Phlebolepis* to be very desirable."

Stensiö (1958: 418, fig. 110 C; 1964: 372, fig. 4 C) figured a specimen from the Museum of Comparative Zoölogy, Harvard, in which he interpreted the mouth as a longitudinal slit terminated anteriorly by a larger, oval structure; the latter, bordered by two crescentic plates, was interpreted as a "prenasal sinus".

The writer has had the opportunity of examining, and making latex casts of, material of *Phlebolepis* during visits to the Paleontologisk Museum, Oslo, in 1962 and 1966. The specimens are closely, and apparently randomly, scattered over the bedding surface of large limestone slabs; preserved as external moulds, in places they partly overlap one another. Although earlier reconstructions give the impression that *Phlebolepis* is laterally compressed throughout (Buistrov, 1949, fig. 1; Westoll, 1945, fig. 5 A; Obruchev, 1964: 43, fig. 10) examination of the total collection reveals that many individuals are seen from the dorsal or ventral surface. Kiaer & Heintz (1932: 5) pointed this out, remarking that "in specimens which are compressed from above the front part of the head is roundish, quite broad, reminding one of the head of *Thelodus* or *Lanarkia*... On the contrary, in specimens pressed from the sides the front part of the head is more or less sharply pointed".

Fig. 1. *Phlebolepis elegans* Pander. Latex cast, whitened with ammonium chloride, of individual in dorsal view. Specimen figured by Kiaer & Heintz (1932, pl. 4, fig. 1); ×4.
o, orbit; *o.m.*, oral margin; *p.f.*, pectoral fin.

In one of these dorso-ventrally compressed individuals (Fig. 1; cf. Kiaer & Heintz, 1932, pl. 4, fig. 1), probably seen in dorsal view, the anterior part of the animal is roughly oblong in outline with a rather wide, slightly concave anterior margin. In this and other specimens the anterior margin is composed of several rows of small, rhomboid scales with their long axes arranged transversely; a symmetrical arrangement may be preserved with an anterior, subtriangular area somewhat reminiscent of the "rostral epitegum" in *Anglaspis* (Stensiö, 1964, fig. 70). Immediately posterior to this (Fig. 1) there is a median indentation which may possible represent the position of the pineal organ. Kiaer & Heintz observed features which they interpreted as orbits in several specimens (1932, pl. 3, figs. 1–4, pl. 4, fig.1). The orbits are laterally placed, immediately posterior to the antero-lateral corners of the head, and are bordered by at least two, long, crescentic scales; these are the scales which Stensiö (1964, fig. 40) has interpreted as forming the borders of a large "prenasal sinus" situated immediately anterior to the "oral opening", which is figured as a long, narrow, rostro-caudal slit. The writer prefers the explanation of Kiaer & Heintz (1932: 5) that the mouth was a terminal cleft; although the arrangement varies greatly in different specimens "the brim of small, roundish scales along the front of the head is always present. It is very probable that this brim formed the upper (and under) limit of the mouth opening".

The most important of the new observations concern mainly the number, nature and position of the fins. The asymmetrical caudal fin has been generally accepted to be hypocercal in form —as in other thelodonts, anaspids and most pteraspidomorphs—and Kiaer & Heintz (1932, pl. 4, fig. 2) figured a particularly well-preserved specimen in which a low rounded, anal fin is also clearly visible. The epichordal lobe of the caudal fin in this specimen shows a peculiar notch which has been incorporated into later reconstructions, obviously derived from Kiaer & Heintz's plate (Buistrov, 1949, fig. 1; Obruchev, 1964: 43, fig.10). The writer has been unable to detect this feature in other, well-preserved individuals (Fig. 2a) and suggests that it is the result of damage or faulty preservation in this particular individual only.

Although a dorsal fin has not been previously recognised in *Phlebolepis* the writer has observed it in at least three individuals. In one (Fig. 2a) the caudal fin is undistorted but incomplete distally. A well-preserved dorsal fin is present as a low rounded structure, covered with scales like those over the trunk and situated almost opposite but slightly anterior to the ventral fin (Fig. 3a). In two of the specimens figured by Kiaer & Heintz (1932, pl. 3, figs. 2, 3) the crumpled nature of the caudal fin makes it difficult to be absolutely certain of the orientation but the writer believes that in both cases the fin labelled *Af*. is actually the dorsal fin (Fig. 2b). *Phlebolepis* thus possesses the same complement of median fins as *Logania* (*Thelodus*) *scotica* and *Lanarkia*, both of which have

Fig. 2. *Phlebolepis elegans* Pander. Latex casts, whitened with ammonium chloride, of two individuals in lateral view. *a*, × 1¾; *b*, specimen figured by Kiaer & Heintz (1932, pl. 3, fig. 3); × 2.
a.f., anal fin; *d.f.*, dorsal fin; *p.f.r.*, basal ridge of pectoral fin.

fins in approximately the same positions; this has recently been established in material collected by the writer, which will be described in a later paper.

In the earlier accounts of *Phlebolepis* it was observed that paired lateral fins appeared to be absent—somewhat surprisingly in view of their presence in the other thelodonts (*Logania, Lanarkia, Turinia*). Of the sixty or so specimens seen by the writer only one individual shows convincing evidence for the presence of such fins in *Phlebolepis*. This specimen (Fig. 1; cf. Kiaer & Heintz, 1932, pl. 4, fig. 1) has, along the right margin, a narrow strip of fine, slender scales. From a point shortly behind the carbonaceous stain of the right eye the strip extends posteriorly for about 12 mm reaching a maximum width of 3 mm just before it is abruptly truncated. It is more easily observed on a whitened latex cast, as here, than on the actual specimen in the rock. Without any doubt this structure represents the remains of a well-developed, triangular, pectoral flap or fin of a type similar to that found in *Logania* and *Lanarkia*. Its preservation in only one of the numerous articulated specimens must surely indicate that it was rather flimsy and easily lost after death. Several individuals seen in lateral view (Fig. 2 a, b) have a slender, longitudinal raised ridge over approximately the same stretch of trunk and this may represent a slightly stiffened fin-base. It should be noted that the lateral fins are not always visible in much larger and otherwise well-preserved individuals of *Logania scotica* (Traquair) from Scotland.

The writer, like earlier workers, has not observed any evidence of branchial

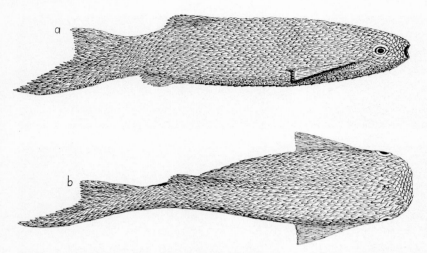

Fig. 3. *Phlebolepis elegans* Pander. Reconstructions in lateral view (*a*) and in dorsal view (*b*); ×1¼.

apertures in *Phlebolepis*. This is scarcely surprising in view of the size of the animal and the nature of the squamation; any apertures would be very small and easily obscured by slight scale movement during compression. This does not mean, however, that the Thelodonti, unlike the Heterostraci, lacked gill openings (Kiaer & Heintz 1932: 8). The writer has examined many hundreds of specimens of *Logania* and *Lanarkia* from the Scottish Silurian inliers and has only been able to discern the actual branchial apertures in a single specimen of each genus. The branchial apertures form a straight, condensed row, probably eight in number, situated opposite and ventral to the lateral fins; in this the thelodonts differ markedly from the Heterostraci and it must be questioned whether they are as closely related to that group as has been suggested in the

Fig. 4. *Phlebolepis elegans* Pander. Detail of scales from anterior margin and orbital region (*a*) and from posterior trunk region (*b*). Latex casts whitened with ammonium chloride.

past. The thelodonts are proving to be remarkably uniform in general morphology with a dorso-ventrally flattened, cephalo-thoracic region, triangular pectoral flaps, both dorsal *and* anal fins and a hypocercal caudal fin. In view of this it would seem probable that *Phlebolepis*, too, had separate branchial apertures ventral to the paired fins, but only additional material can settle this aspect. Such a condensed row of branchial apertures need not necessarily indicate any relationship with anaspids, as suggested by Westoll (1945, fig. 5) since, in the latter group, the paired fins are of an entirely different type, situated posterior to the last branchial aperture (Ritchie, 1964, fig. 1).

As in other thelodonts the scales of *Phlebolepis* display considerable variation in shape and ornamentation in different parts of the trunk and fins (Fig. 4a, b). This aspect has been covered by Gross (1967: 42–44, fig. 15 E–Q, pl. 7, figs. 11–15) and others, and the histological studies in particular have led to a useful revision of thelodont genera and species. The writer's research, on articulated material, confirms Gross's findings that the thelodonts (*Thelodus, Logania, Lanarkia, Phlebolepis, Turinia* etc.) form an apparently natural group of agnathans whose relationships with the other agnathans are as yet uncertain.

This study was supported by grants from the Research Grant Fund, University of Edinburgh, and the British Council.

References

Buistrov, A. P. (1949). *Phlebolepis elegans* Pander. *Dokl. Acad. Sci. U.S.S.R.*, (N.S.), **64**: 245–247. In Russian.
Gross, W. (1947). Die Agnathen und Acanthodier des obersilurischen Beyrichienkalks. *Palaeontographica*, (A) **96**: 91–158.
Gross, W. (1967). Über Thelodontier-Schuppen. *Palaeontographica*. (A) **127**: 1–67.
Hoppe, K.-H. (1931). Die Coelolepiden und Acanthodier des Obersilurs der Insel Ösel. Ihre Paläobiologie und Paläontologie. *Palaeontographica*, **76**: 35–94.
Hoppe, K.-H. (1933). *Phlebolepis elegans* Pander aus dem Obersilur von Ösel. (Bemerkungen zu dem von A. Heintz beschriebenen vollständigen Exemplare von "*Coelolepis luhai*"). *Centralb. Min. Geol. Paläont.*, **1933**:*B*: 124–130.
Kiaer, J. & Heintz, A. (1932). New coelolepids from the Upper Silurian of Oesel (Esthonia). *Eesti Loodustead. Arh.*, (1) **10**: 1–8.
Obruchev, D. V. (1964). Podklass Thelodonti (Coelolepides). In *Osnovui paleontologii* (*Fundamentals of Paleontology*), ed. Orlov, I. A., **11**: 39–44. Moscow: Acad. Sci. U.S.S.R. In Russian.
Ørvig, T. (1951). Histologic studies of placoderms and fossil elasmobranchs. 1. The endoskeleton, with remarks on the hard tissues of lower vertebrates in general. *Ark. Zool.* (2) **2**: 321–454.
Pander, C. H. (1856). *Monographie der fossilen Fische des silurischen Systems der Russisch-Baltischen Gouvernements*. St. Petersburg.
Ritchie, A. (1964). New light on the morphology of the Norwegian Anaspida. *Skr. norske VidenskAkad. Oslo, Mat.-naturv. Kl.*, **1964**: 1–35.

Stensiö, E. A. (1958). Les cyclostomes fossiles ou ostracodermes. In *Traité de Zoologie*, ed. Grassé, P.-P., **13**:**1**: 173–425. Paris: Masson.

Stensiö, E. A. (1964). Les cyclostomes fossiles ou ostracodermes. In *Traité de Paléontologie*, ed. Piveteau, J., **4**:**1**: 96–382. Paris: Masson.

Westoll, T. S. (1945). A new cephalaspid fish from the Downtonian of Scotland, with notes on the structure and classification of ostracoderms. *Trans. R. Soc. Edinb.*, **61**: 341–357.

Some implications of the acceptance of a delamination principle

By Malcolm Jollie

Dept. of Biological Sciences, Northern Illinois University, DeKalb, Illinois. USA

This review was stimulated by, and is based on, comments made by Jarvik (1964). I do not propose to argue for or against the acceptance of the delamination principle nor to concern myself with what is delamination or how this occurs. I am assuming that "delamination" occurs and am concerned with some of the implications of this. I do not propose to discuss all of the possible alternative implications but only those which seem to me to be most probable. Further, it is recognized that the ideas expressed here must be tested in terms of views relative to the origin of skeletal tissues, both in phylogeny and ontogeny, and to the origin of the vertebrate head and its metamery.

The delamination principle is based on Holmgren's (1940) account of the successive contributions of neural crest mesenchyma to the meninges and blastematic cranium and visceral arches of the shark. Jarvik (1959: 44–45) identified and defined this principle as follows: "The repeated formation and sinking in of laminae of skeletogenous tissue in the head and in the fins ... must be manifestations of a common morphogenetic principle. This principle which applies to the endoskeleton as well as the exoskeleton I propose to call Holmgren's principle of delamination".

Further, he stated, "Delamination may be said to be the capacity of the outermost parts of the undifferentiated ectomesenchyme, or—in later ontogenetic stages and the adult—the outermost parts of the corium, to a repeated production of laminae" [of cells] "with potentialities to form skeletal structures." As an implication, he pointed out (1959: 46), "Hence it follows that there cannot be any fundamental differences between the structures into which these laminae develop, that is between the endoskeleton and the exoskeleton or between the various types of exoskeletal and endoskeletal formations".

Historically a "delamination principle" was suggested quite early, and reviewed by Goodrich (1909: 66), from which the following remarks are taken. From Hertwig (1874*a*; 1874*b*; 1876) on there has been general agreement that dermal bones arose by the coalescence of denticles and that such aggregates have tended in evolution to extend or sink deeper into the tissues. Here the agreement ended and there were two lines of general thought. According to Gegenbaur a first generation of such ossifications became associated with the chondrocranium and later became ossifications of it. In contrast Hertwig and Kölliker (1859a; 1859b) maintained that dermal bones may cover over a cartilage but always lie outside the perichondrium; that cartilage bones can never be traced back to denticles and are developed entirely in direct relation to the cartilaginous skeleton.

Current belief, while recognizing delaminations of some sort as a phylogenetic event and the contribution of ectomesoderm (or neural crest) to the teeth, scales, visceral skeleton and the trabeculae of the cranium, nevertheless, views bone and cartilage as "mesodermal tissues", derived primarily from sclerotome (Balinsky, 1965). The "in vogue" view stresses the idea that there is no real difference between dermal and chondral bone and suggests that there is some sort of interchangeability of these. It is assumed that what was formerly a chondral bone may arise in ontogeny in a dermal fashion. The possibility of fusion of these kinds of bones or their combination is not denied while new bones or combinations are thought to arise or disappear rather irregularly. There has been comparison of the distribution of neural crest tissue with the cephalopectoral shield of fossil forms and Holmgren assumed that in the shark this mesectoderm represented former dermal bone. The phylogenetic origin of this neural crest tissue and its skeletogenous capability (said to be lacking in neural crest more posteriorly, but obviously only in forms without "bony" body scales) is unknown.

The discussion of skeletal tissues often becomes confused because of failure to separate *process* from *structure* or *phylogeny* from *ontogeny* even though one may believe that these are but parts of a spectrum of events. By process I refer to the formation of matrix (and fibers) or the deposition of hydroxyapatite crystals. By structure I mean the relatively consistent pattern of skeletal units, body scales or plates (scutes), etc. Jarvik's remark, above, to the effect that exoskeletal and endoskeletal structures do not differ may be accurate in terms of process and some aspects of ontogeny—they are of neural crest origin, or competence, and ossify or calcify in similar ways—but not in terms of structure (details of shapes and interrelationships of units, position in body, thickness and histological details, etc.) or phylogeny. We must assume that the genetic systems which produce the different structural end products, or generations of end products (scales lying over head bones) seen in various vertebrates, *are* different. Process is of little concern here since phylogenetic divergence can only be indicated by details of structure and structural ontogeny.

The implication, from Hertwig, that dermal bones can be traced back to the coalescence of denticles, is not an ontogenetic fact (to my knowledge) but a phylogenetic speculation in need of clarification. It may be that the first step in evolution was the coalescence of denticles but this was followed by the development of new growth systems. Many or most of these involved a center of reticulate bone which extended out peripherally while thickening. Denticles as such are not observed except as later acquisitions—the odontodes of Ørvig (in this Volume).

Further what constitutes a bone (i.e. a comparable, homologous, named skeletal unit) is never actually defined nor, perhaps, considered definable. Phylogenetically we are uncertain as to whether the various kinds of skeletal tissues are of equal age (did bone preceed cartilage or vice versa?), whether cartilage was invented as an ontogenetic model for bone or whether, in contradiction to the expressed belief in a delamination principle, the primitive vertebrate was armored or naked.

Thus our current state of beliefs is rather confused. This is due in part to the attempt to credit all views and viewpoints (or to discredit certain of them). We do not always attempt to sort fact from fiction, the untenable view from the well documented; in short we fail to develop the necessary consistency of argumentation along with a balance of facts. The identification of a delamination principle by Jarvik gives us a tool with which to reconsider some of the elements of this situation, a tool which may make possible additional theory and lead to revelation of the history of the vertebrate head and skeletal tissues.

Jarvik (1954) has already used this principle in an analysis of the parasphenoid. I propose to apply it in an attempt to interpret the interrelationship between the agnaths and gnathostomes, a relationship which has been variously described and variously related to functional explanation. A brief review of these opinions (which does not attempt to account for all of the nuances of belief) follows.

Huxley (1876), Fürbringer (1875), Parker (1883), Dohrn (1884), and Ayers (1921; 1931) believed (hypothesis A of Fig. 1) that the cyclostomes were degenerate or modified gnathostomes. This view was based on the apparent correspondence of parts of the cranial and visceral skeletons, on the similarities of cranial nerves, etc. It suggests a specialization of an originally "jawed" mouth for sucking (and "biting" by means of a "tongue" apparatus). In contrast Sewertzoff (1916; 1917; 1927; 1928), Tretjakoff (1927; 1929a; 1929b), Stensiö (1927; 1932), and Holmgren & Stensiö (1936) viewed the cyclostomes (hypothesis B) as representative of a vertebrate stem which existed prior to the development of jaws. This view leads to the assumption now held by the majority of zoologists that some early agnath was directly ancestral to the gnathostomes. A third hypothesis (C), suggested by Stensiö (1958), Jollie (1962) and Jarvik (1964), is that both agnaths and gnathostomes are derived from a common ancestor as much like one as the other. This ancestor is "agnathous" but lacks the obvious specializations of the known fossil or living agnaths, an area that I will return to.

Much of the above speculation, leading to a particular phylogenetic conclu-

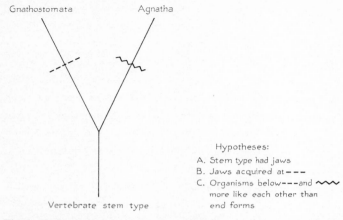

Fig. 1. Phylogenetic scheme for relationship between agnaths and gnathostomes.

sion, was based upon the nature and development of the head skeleton. The possibly different nature of the visceral arches of the cyclostomes has been repeatedly remarked. As early as 1832, Rathke suggested the branchial arches of *Petromyzon* were homologous with the extrabranchials of sharks. Balfour (1881), Allis (1923; 1924) and Damas (1944) also held this view. Goette (1901) stated that the branchial arches of *Petromyzon* were fundamentally different because of their superficial position. Johnels (1948: 261) described them as "almost exclusively extrabranchial." Müller (1839) said the skeleton of *Myxine* was of a "different type" than that of the gnathostome. Cole (1905) viewed the skeleton of *Myxine* as a "neomorph" and therefore of late and independent origin; however, Holmgren (1942; 1946) described both internal and external arches in *Myxine* and suggested this might apply to cyclostomes generally. Huxley (1876), Parker (1883), Dohrn (1884), Howes (1892), Gaupp (1904), and Sewertzoff (1916; 1917) viewed the arches of agnaths as equivalent to those of gnathostomes. Similarly Goodrich (1909: 38) and Jarvik (1954: 91; 1964), after recounting the strikingly different relationships to the axial portion of the skeleton, pharynx wall, muscles, blood vessels, heart and nerves, maintained that the arches of agnaths are homologous to those of gnathostomes. Jollie (1962) suggested essentially independent origins of the head skeletons of agnaths and gnathostomes since homologies of the endocranial parts in the two groups do not appear to exist.

The key to the above divergence of viewpoints lies in the interpretation of the visceral skeleton. It seems to me that an understanding of these seemingly contradictory viewpoints hinges on the functional evolution of the mouth and pharynx as suggested by the delamination process. As an approach to the problem of the primary radiation of the vertebrates let us begin with picturing a series of hypothetical stages and see if they complement each other and show

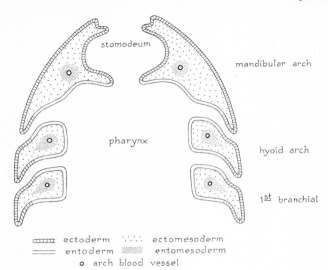

Fig. 2. Diagrammatic protovertebrate as seen in frontal section.

the necessary consistency to assure us of their probability. Alternate hypotheses can be proposed but their degree of probability seems less and space limits their discussion here.

The ancestral vertebrate (Fig. 2) was small, as are the living protochordates, with nose, eye, otic vesicle and lateral-line system. The original mouth margin was formed by the edges of the oral plate (velum) between stomodeum and pharynx. The stomodeum was formed by outgrowth of an "oral hood". This organism was predaceous, feeding on smaller organisms sucked into the pharynx (it was not a sedentary filter feeder—nor derived from one). Its sensory structures helped it locate and capture prey with quick, swimming dashes. The pharynx had a number of slits to the exterior which allowed for flow of water through the mouth and pharynx and aided in the capture of prey. The slits were not equipped with gills since the body surface and pharyngeal lining served this need. With increase in size, development of denticulate body surface and more efficient and sustained swimming ability a respiratory function for the slits was gradually acquired. In terms of skeleton, other than denticles, "dermal" fibrous connective tissue layers were present along with concentrations of sclerotomic mesenchyme forming an axial system. The former tissue was particularly well developed as a cephalopectoral shield but was present on the trunk and tail as well and its cells arose as neural crest.

One line of development from this ancestor gave rise to the agnaths (Fig. 3). In this the predatory capacity was not greatly increased (probably the living lamprey and hagfishes are the most active predators ever developed in this group). Early groups tended to be bottom foragers and scavengers (see Denison, 1961). In these exoskeletal development lead to unjointed cephalopectoral

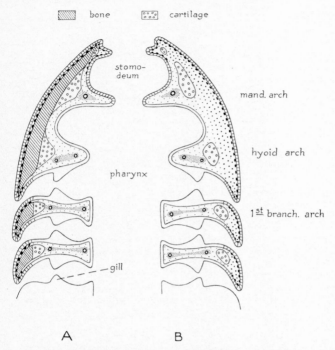

Fig. 3. Diagrammatic early agnathan as seen in frontal section. *A*, armored form; *B*, denticle covered form. Tissues as indicated or as in Fig. 2.

shields and a trunk-tail cover of large articulated scale-plates in two of the several known kinds. The outer pharyngeal openings were restricted to round, valved openings aiding the pharyngeal pumping action and as a result any development of gill filaments must be inward. The mouth margins were now formed by the edges of the oral hood, and in some the opening of the first branchial pouch (hyoid) was lost, and the pharyngeal openings tended to be shifted posteriorly (in the living forms the posterior shifting of pouches represents the extreme known for the group). The gills were housed in sac-like spaces with constricted inner and outer openings. Concurrent with this was the development of a pumping velum in some forms or a velum acting as a valve (other pumping mechanisms were developed in these early forms: the floor of the pharynx in the cephalaspids, the lateral walls of the stomodeum in the anaspids (?), the floor of the mouth in pteraspids or a velum according to Denison, 1964). Some sort of "tongue" structure was present in all and played a part in the operation of the mouth.

The second line of development, the gnathostome, continued in the direction of active predation (Fig. 4). The mouth formed by the oral hood margins was enlarged and an angle between upper and lower lips accentuated and extended posteriorly; the mandibular arch in turn was moved back to contact

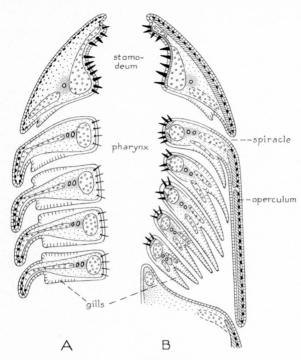

Fig. 4. Diagrammatic early gnathostome as seen in frontal section. *A*, denticle covered stage without operculum; *B*, armored stage with operculum. Tissues and arches as indicated in Figs. 2, 3.

the hyoid arch thus reducing the hyoid pouch to a spiracle (this state may have been achieved in the common ancestor of agnath and gnathostome). The mouth margin was covered by enlarged denticles, now identified as teeth. These teeth extended into the stomodeum and finally into the pharynx. Teeth in the mouth and pharynx helped hold and subdue increasingly larger and more active prey. The movements of the mouth and pharyngeal wall were used in sucking in prey and moving it into the esophagus. With increasing size and activity the members of this group also developed gills but these necessarily extended outward since the inner denticulate surface of the arch was involved in food handling. The outer margin of the arch either formed a denticle studded flap or such a flap from the hyoid arch extended back to enclose a branchial chamber into which naked gill filaments projected. This hyoid operculum served both a protective function as well as acting as a valve for maintaining the directional flow of water. This flow had both a feeding and a respiratory function.

In the gnathostomes the sensory organs underwent further development as a result of search-locate-and-capture operations while in the agnath the eye, ear and lateral-line system were less developed, less strongly localized, and sometimes, as in the living cyclostomes, apparently degenerate.

From the above, it is suggested that, although both groups shared an inherited tendency for skeletal formation, the skeleton of each developed independently in relation to functional needs. The cephalopectoral shield developed as a single continuous element in the agnaths and as separate, moveable, head and pectoral girdle in the gnathostome. Not all agnaths or gnathostomes carried the development of the dermal skeleton to a "shield" stage, but in those that did this fundamental difference, perhaps independently arrived at in several lines of each of these major groups, supports the basic argument. Endoskeletal development was carried much further in the gnathostomes as a correlate with a more active life. The visceral arches of the two groups developed differently and this difference deserves further comment.

While both groups required some sort of valving of the pharyngeal slits, for feeding and later for respiration, and tended to develop external supporting elements, the arches of the gnathostomes were also involved in food handling and developed a strong internal skeletal system for the support of tooth plates and/or gill rakers. In both groups gills developed along with the skeletal supports of the arch system. The gills are essentially analogous since Stensiö (1927: 169) pointed out that those of agnaths are covered with entoderm while those of gnathostomes are covered with ectoderm.

In the fossil agnaths, the arches are unknown as separate elements comparable to those of the gnathostomes. In the cephalaspids the dorsal parts form ridges on the interior of the cephalopectoral shield. The ventral arch elements are assumed to have been formed of cartilage. The nature of the arches in the armored heterostracans can only be guessed, but are usually described as cartilaginous. The loss of pharyngeal wall movement with the development of armor is a modified state as compared with the assumed primitive condition and may have been accompanied by lack of development of visceral arches. In living agnaths the visceral arches are only slightly developed, unjointed, and fused with the axial skeleton.

In the early gnathostomes the visceral arches were rotated posteriorly from their original transverse position. Further the arches were subdivided into articulating parts: infrapharyngo- (or pharyngo-), epi-, cerato-, and hypo- elements. Such units are definitive of the gnathostome arch and are unknown for agnaths (Fig. 5). A suprapharyngeal (or supra-) unit was probably a later development while the basi- unit is a midline brace of secondary origin.

The dorsal parts of the skeletogenous arches of the agnaths and gnathostomes were assumed by Säve-Söderbergh (1936: 143-144) and Jarvik (1954: 93) to be comparable, therefore this area should be given closer scrutiny. The supra- and infra- (pharyngo) elements can be viewed as delamination products, like all of the arch parts, related to dental plates. The supra-element has been associated with the teeth extending dorsally in each pharyngeal pouch (and also the stomodeal pouch) lateral to the dorsal end of the arch. However, the

Implications of delamination 97

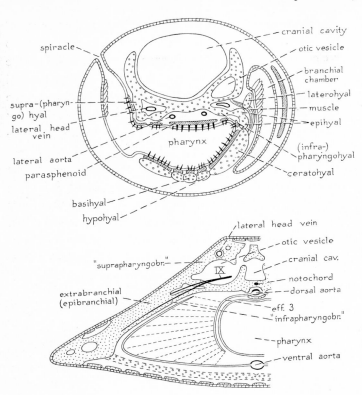

Fig. 5 (above). Diagrammatic transverse section of a gnathostome in the region of the hyoid arch; the left side somewhat anterior to the right. The cartilaginous precursors of the arch elements are indicated in relation to some of the other head structures. The epihyal is shown as a double element, the lower part giving rise to the symplectic bone.

Fig. 6 (below). Diagrammatic transverse section of head region of a cephalaspid showing supposed homologs of parts of the visceral arch (after Jarvik, 1954, fig. 46).

supra-element might also be viewed as a dorsally extending brace for the medial end of the moveable epal element without any relation to dental plates and thus a tissue stress response. The infra-element is just the anteromedial continuation of the arch proper. These elements act to anchor, or form a base for, the dorsal end of the epal element. There is an anteroposterior gradient in the completeness of this support function (i.e. jaws, hyoid arch, and branchial arches).

The presumed arches (Fig. 6), fused to the cephalopectoral shield in the cephalaspids, do not show the necessary kind of agreement to prove any homology. Further, the ossification of the pharynx roof, below the dorsal aorta, in these agnaths does not resemble, functionally or in relationship to the blood vessel, the infra-elements of gnathostomes and is best assumed to be an independent dermal skeletogenous development.

7 – 689869 *Nobel Symposium 4*

As a digression, a comment on the structure of the "premandibular arch" of gnathostomes is in order. From the above it follows that there can be no suprapremandibular element since no pouch extends upwards lateral to the dorsal end of this "arch" (except the fold between the lip and toothed margin of the jaw). The lamina orbitonasalis, which has been identified as the supra-element, is generally not a dorsally oriented rod but rather a transversely oriented one, attaching medially to the trabecula. The search for another infra-element has not revealed anything more probable than a vague blastema, or the ventral aspect of the trabecular commissure (the ethmoid plate), or the paraseptal cartilage of mammals. These structures are associated with the vomer which is considered to be their dental plate. The apparent lack of any other infra-element is best explained as the result of an incorrect assumption. If the lamina orbitonasalis and perhaps a blastematic continuation below the nasal septum is viewed as the infra-element then it is the supra-element which is lacking—as would be assumed applying the delamination principle.

Returning to a consideration of the conditions in agnaths, the assumption that the living cyclostomes have trabeculae is a part of the problem of comparison of arches. Such an assumption would imply that we are dealing with an infrapharyngeal of the mandibular arch taken into the axial skeleton. Although these so-called trabecles, and the nasal capsule, have been described as ectomesodermal by Damas (1944), Koltzoff (1902) and de Beer (1937) indicated that they are derived in *Petromyzon* from the sclerotomes of the three preotic segments and are thus equivalent to the anterior part of the gnathostome parachordals. Johnels (1948) indicated the parachordal (i.e. sclerotome) nature of these structures and suggested that the dorsal plates (anterior to the nasal capsule) might represent the trabecles.

Jarvik (1964: 26) states that "there is thus no true trabecula in cephalaspids", however, he assumed that the parachordal commissure of *Petromyzon*, identified as ectomesenchymatic by Johnels (1948: 173–175, 256–260), represented these structures. This commissure lies below and somewhat anterior to the pituitary anlage and as such possibly represents the dorsomedial ends of the mandibular arch skeletogenous tissue. Can this be called trabecular? I doubt this since it represents a cell organization the ancestry of which preceded anything as definitive as a trabecula.

The assumption that the "pedicel", an anterolateral projection of the anterior "basicapsular" region, may represent a "basitrabecular process" is thus impossible. This posterior end of the subocular arch possibly represents a piece of mandibular arch material which parallels the line of attachment of the velum to the roof of the pharynx. At its posterior end it later gives rise to the styliform process which might be considered an internal mandibular arch, but in no way does this material resemble the corresponding arch of a gnathostome. Thus from what we know it is evident that there are no trabeculae in agnaths nor homologs of other parts of the visceral arch system of gnathostomes.

Again as a digression, one might question whether the trabecles of gnathostomes are really infrapharyngomandibulars since both Platt (1894) and Stone (1929) described those of amphibians as formed of both neural crests and (posteriorly) somitic tissue. Bertmar (1959: 297) has suggested that the posterior part of what has been called the trabecle of amphibians is in fact the "polar cartilage", a separate anterior bit of "parachordal tissue". This does not change the fact that these two elements appear fused together and the basitrabecular connection with the palatoquadrate cartilage lies within the so-called polar cartilage area. Even though Bertmar considered the trabeculae of dipnoans, actinopterygian fishes and amphibians to be completely ectomesodermal and thus of arcual origin, he assumed that in sharks a mixture of tissues occurs.

If a mixture occurs in any vertebrate this could be rationalized as due to the fact that cells of both origins, with comparable skeletogenous capabilities, occur in this region. This state might be a secondary one or, more likely, an indication that the homology does not exist. Therefore, it could be strongly argued that no gnathostome ever had an infrapharyngomandibular and that we are dealing with a blastematic precursor, not a formed element.

It has been remarked that the trabecles of sharks are very short and thus appear to differ from those of most gnathostomes. Certainly the relationship of the palatoquadrate to the cranium, the peculiar position of the basitrabecular process (or its lack) are noteworthy, as are the differences in the arches generally. One might leap to the conclusion that the development of the jaw suspension and visceral arches in sharks was, at least in part, independent of their development in the methyostylic line (i.e. plagiostomes vs. teleostomes).

Returning to the idea of the independent origin of the head skeleton in the agnaths there are other details seen in the development of the lamprey which are of interest. The parachordals appear as separate anterior and posterior parts (de Beer, 1937; Johnels, 1948) but these are not equivalent to the polar cartilages and parachordals of gnathostomes. The posterior part of the lamprey parachordal is joined to the first branchial arch. The nasal capsule develops early and is separate from the rest of the cranium.

Strahan (1964) has compared the adult head skeletons of the lamprey and hagfish and from his diagrams there appears to be some similarity in the two living types but little agreement with gnathostomes (the use of such terms as "palatoquadrate" and "extrahyal", suggesting gnathostome homology, is confusing). As a final remark on the lamprey the association of elastic connective tissue, or mucocartilage, with skeletogenous tissue has hindered understanding of its skeletal development.

From all of this, and more could be added, it is my opinion that the most

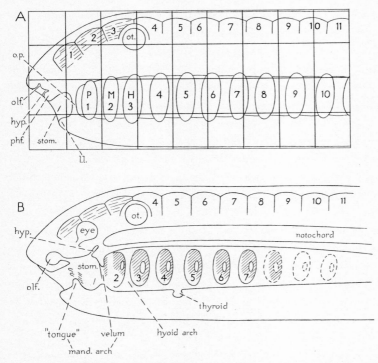

Fig. 7. Fundamental pattern of vertebrate head region, seen in sagittal section, as based on the lamprey larva. *A*, "basic agnathan embryo" from Strahan (1958, fig. 1). *B*, basic vertebrate as proposed here. Somite and gill pouches numbered.

H, hyoid gill pouch (apparently behind the hyoid arch); *hyp*, hypophysis rudiment (Rathke's pouch); *l.l.*, lower lip; *M*, mandibular gill pouch; *o.p.*, oral plate (velum); *olf.*, olfactory rudiment (nasal sac or vesicle); *ot.*, otic capsule (vesicle); *P*, premandibular gill pouch; *phf.*, posthypophysial fold; *stom.*, stomodeum.

exacting comparisons of the arch elements, or the head skeleton generally, of agnaths and gnathostomes will not reveal any convincing grounds for homologies beyond those of the parachordals (which probably extend farther forward han in the gnathostome) and the otic capsule.

Another approach to the ancestry of gnathostomes and agnaths is suggested in a paper by Strahan (1958) concerning the early radiation of the agnaths. In this he has described the origin of the head form of the various types as determined by D'Arcy Thompson's method of coordinate deformation, using a hypothetical ammocoete larva as a starting point (as suggested by Watson, 1954).

Starting from Strahan's base figure (Fig. 7 A), I have produced diagrams of the various agnaths. From Strahan's figures it is clear that there is confusion as to the relationship of the stomodeum to the pharynx and its visceral arches. It is now generally assumed that the velum (oral plate) marks the division between stomodeum and pharynx and that it is associated with the mandibular arch (Stensiö, 1927: 347). Thus the suggestion that there is a premandibular pouch is untenable; this is in fact the so-called "mandibular pouch" or stomodeum (Fig. 7 B). For definition purposes the pouch of a particular arch is viewed as lying anterior to the arch. The hyoid (spiracular) pouch is the first real pharyngeal pouch. Further the mouth of the hypothetical ancestor should show a premaxillary "arch" in the oral hood margin and have upper and lower lips. I have suggested in this figure that the mandibular arch has been shifted posteriorly and subdivided into a tongue tubercle and velar portions. Although the several fossil types may have had a hyoid pouch opening to the exterior, as shown in Figs. 8–11, neither type of living cyclostome possesses a hyoid pouch or spiracle.

Of these diagrams, that of the heterostracan (Fig. 9) purports to be least modified from the basic ancestral type. Evidence of such a primitive position is slim. The nasal openings were probably paired and opened inside the mouth and this can be viewed as primitive. The general agreement of the other groups of agnaths and the living forms in having a hypophyseal duct (opening on the top of the head or the tip of the snout) suggests that the heterostracan should basically agree, as Stensiö (1932; 1958) has argued.

The following facts suggest the heterostracan is divergent: the gill pouches having a common external opening, the tendency for a heavy cephalopectoral armor formed principally of aspidin, the obviously reduced swimming ability coupled with a hypocercal tail, and the form of the lateral-line system and inner ear. The cranial endoskeleton of this group was cartilaginous and continous with the dermal bony cover and branchial arches were possibly present (and unjointed?). The rigid "carapace" of some forms necessitated some sort of pumping mechanism.

At this point, reference should be made to the thelodont *Phlebolepis elegans* (Ritchie, in this Volume). This species has the body shape, pectoral fin folds, and dorsal and anal fins of a possible common ancestor of the other agnaths. The hypocercal tail and lack of pelvic fin folds, however, removes it and other thelodonts from the position of ancestor of the gnathostomes.

Fig. 12 is an attempt to treat the gnathostome in the same way as the several kinds of agnaths. This figure indicates that, although the mouth is different, the interrelationships of parts are as much like the hypothetical ancestor as some agnaths. This is especially suggested by the hypophyseal invagination (Rathke's pouch) being in the mouth, the paired nasal sacs opening near the mouth margin, and the retention of the spiracle. The similarities of brain structure and cranial nerve distribution in agnaths and gnathostomes also suggests that this gnathostome pattern is near the primitive one.

As a last comment, relative to these figures, Strahan accepted the proposition of Stensiö that the myxinoid fishes were derived from the heterostracan. I can-

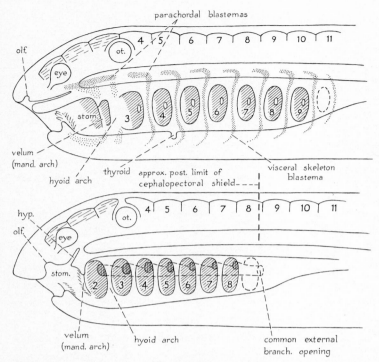

Fig. 8 (above). Diagrammatic sagittal section of theoretical "lamprey larva" based in part on Damas (1944).

Fig. 9 (below). Diagrammatic sagittal section of theoretical heterostracan, based in part on Denison, 1964.

not. However, there can be little doubt in the minds of those who have considered the details of structure of the living cyclostomes that there are greater anatomical differences between them than separate the most diverse gnathostome fishes. Therefore, the suggestion that the living cyclostomes are representatives of a single "class' is hard to accept. The argumentation behind such a viewpoint is of interest.

According to Romer (1965: 144), "It is agreed by most workers on vertebrate classification and evolution that the Cyclostomata, including the lampreys, such as *Petromyzon*, on the one hand, and the various hagfishes (*Myxine, Bdellostoma*, etc.) on the other, are a natural group of jawless vertebrates. They have in common numerous unique characters, not merely in their general body shape and round, jawless mouth, but in a long series of special features such as the monorhine condition of the nose, the pouch-like gill slits, an internal ear with but two or one semicircular canals, and, most notable, the peculiar rasping tongue-like structure formed from modified gill bars. Certain of these features might be attributed to degeneracy; some might perhaps be truly primitive vertebrate characters. But it is difficult to believe that the total set of characters

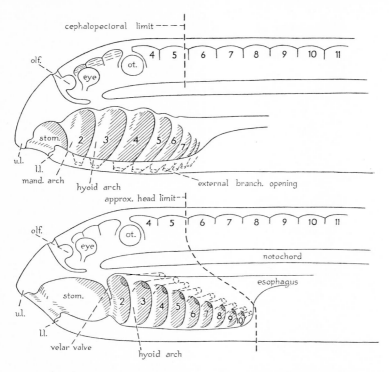

Fig. 10 (above). Diagrammatic sagittal section of theoretical cephalaspid, based on Stensiö's several papers.

Fig. 11 (below). Diagrammatic sagittal section of theoretical anaspid based on Stensiö, 1958.

could have been achieved by two distinct lines of descent from ancestral types far removed from one another."

Most, if not all, of the points made in this quotation have been discussed at one time or another and further comment is not needed (see Jarvik, 1965). It is the last sentence that brings the reader around to the sematic nature of the problem.

As an example of this problem of words, Romer remarked in disbelief that, "...under Stensiö's hypothesis, the lampreys and hagfishes despite their numerous resemblances, would have had discrete lines of descent from the very dawn of vertebrate history." Depending upon what is meant by dawn, Romer's disbelief could be real or feigned. Stensiö's suggestion that the lines giving rise to the petromyzontids and myxinoids were already present at the time of the osteostracans, heterostracans, anaspids and thelodonts (is this the dawn?) strikes me as almost necessary to account for the differences between the two living types. Going further, certainly Stensiö never thought of the ancestral types of the living forms as being "far removed from one another", nor have I.

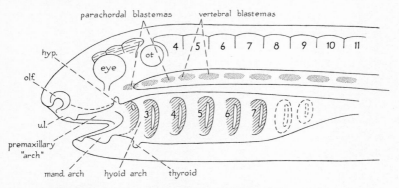

Fig. 12. Diagrammatic sagittal section of theoretical gnathostome.

These ancestral forms did differ in specified ways and were distinct as defined but otherwise were good, interrelated agnaths.

This whole discussion of the origin of the cyclostomes has become reduced to an acceptance or rejection of Stensiö's views. This should not be the heart of the argument since other hypotheses or alternatives are available which give full recognition not only to the basic agnath similarities of the lampreys and hagfishes, but which also recognize the great gulf of dissimilarities. The assumption of a long independent history perhaps from the middle Ordovician, is not impossible nor is there anything wrong with admitting that the true affinities of the myxinoids remain obscure. The possible presence of paired nasal sacs in the Heterostraci does not greatly alter the situation since it is well known that by Silurian times the monorhinal condition had already been achieved in two groups; the myxinoid ancestor might have been a third such group.

In summary, the problems and disagreements concerning the primary radiation of vertebrates can be reconsidered in the light of the delamination principle as defined by Jarvik. From this one can assume that skeletal development in agnaths and gnathostomes occurred independently, in part in parallel and in part in divergent fashion. Among the gnathostomes the differences seen in the "plagiostomes" as compared with the "teleostomes" suggest a basic divergence of skeletal development. The different styles can be related to functional situations of feeding, food handling and respiratory needs.

The fact that the ancestral vertebrates was naked, or essentially so, and without a bony (or cartilaginous) endoskeleton, can be accepted with little reservation. The delamination principle suggests that in the head at least, the exoskeleton and endoskeleton developed in continuity, bone characterizing the vascular (higher oxygen tension) outer layer and cartilage the largely non-vascular (relatively lower oxygen tension) deeper axial region. Caudal extension of these skeletal developments occurred later, the dermal and axial skeletons in the trunk and tail appearing to be quite separate from their inception.

The early phase of vertebrate development into agnath and gnathostome lines was apparently accompanied by the development of some heavily armored types in both. Functional specialization of skeletal elements, particularly of the visceral endoskeleton occurred at the same time as these experiments in armor and eventually gave rise to the types seen later in the fossil record.

Unanswered at this time is the problem of whether the neural crest was originally the only tissue capable of producing skeleton or whether this capacity also resided in sclerotome tissue, or mesoderm generally. The availability of neural crest cells throughout the body for skeletal formation is suggested by the distribution of pigment cells. Neural crest cells could have invaded the various blastemas to lend skeletal competence under the inductive influence of the nervous system or other sources.

References

(* not seen)

Allis, E. P. Jr. (1923). Are polar cartilages and trabecular cartilages of vertebrate embryos the pharyngeal elements of the mandibular and premandibular arches? *J. Anat.*, **58**: 37–51.
*Allis, E. P. Jr. (1924). On the homologies of the skull of the Cyclostomata. *J. Anat.*, **58**: 256–264.
Ayers, H. (1921). Vertebrate cephalogenesis. 5. Origin of jaw apparatus and trigeminus complex—*Amphioxus, Ammocoetes, Bdellostoma, Callorhynchus. J. comp. Neurol.*, **33**: 339–404.
Ayers, H. (1931). Vertebrate cephalogenesis. 6. A. The velum—its part in head building—the hyoid. The velata. The origin of vertebrate head skeleton. B. Myxinoid characters inherited by the Teleostomi. *J. Morph.*, **52**: 309–371.
*Balfour, F. M. (1881). *A treatise on comparative embryology.* London: Macmillan.
Balinsky, B. I. (1965). *An introduction to embryology.* Philadelphia & London: Saunders.
Bertmar, G. (1959). On the ontogeny of the chondral skull in Characidae, with a discussion on the chondrocranial base and the visceral chondrocranium in fishes. *Acta zool. Stockh.*, **40**: 203–364.
Cole, F. J. (1905). A monograph on the general morphology of the myxinoid fishes, based on a study of *Myxine.* 1. The anatomy of the skeleton. *Trans. R. Soc. Edinb.*, **41**: 749–788.
Damas, H. (1944). Recherches sur le développement de *Lampetra fluviatilis.* Contribution à l'étude de la céphalogenèse des vertébrés. *Arch. Biol. Liége,* **55**: 1–284.
de Beer, G. R. (1937). *The development of the vertebrate skull.* Oxford: Oxford University Press.
Denison, R. H. (1961). Feeding mechanisms of agnatha and early gnathostomes. *Am. Zool.,* **1**: 177–181.
Denison, R. H. (1964). The Cyathaspididae. A family of Silurian and Devonian jawless vertebrates. *Fieldiana: Geol.,* **13**: 307–473.
*Dohrn, A. (1884). Studien zur Urgeschichte des Wirbelthierkörpers. 5. Zur Entstehung und Differenzirung der Visceralbogen bei *Petromyzon planeri. Mitt. zool. Stn. Neapel,* **5**: 102–150, 152–161.

Fürbringer, P. (1875). Untersuchungen zur vergleichenden Anatomie der Musculatur des Kopfskelets der Cyclostomen. *Jena. Z. Naturw.*, **9**: 1–93.
*Gaupp, E. (1904). Das Hyobranchialskelet der Wirbelthiere. *Ergeb. Anat. EntwGesch.* **14**: 808–1048.
Goette, A. (1901). Über die Kiemen der Fische. *Z. wiss. Zool.*, **69**: 532–577.
Goodrich, E. S. (1909). Cyclostomes and fishes. In *A Treatise on Zoology*, ed. Lankester, E. R., vol. **9**: Vertebrata Craniata, fasc. 1. London: A. & C. Black.
*Hertwig, O. (1874a). Über Bau und Entwickelung der Placoidschuppen und der Zähne der Selachier. *Jena. Z. Naturw.*, **8**: 331–402.
*Hertwig, O. (1874b). Über das Zahnsystem der Amphibien und seine Bedeutung für die Genese des Skelets der Mundhöhle. Eine vergleichend anatomische, entwicklungsgeschichtliche Untersuchung. *Arch. mikr. Anat.*, **11** (*Supplh.*): 1–208.
Hertwig, O. (1876). Über das Hautskelet der Fische. *Morph. Jb.*, **2**: 329–391.
Holmgren, N. (1942). Studies on the head of fishes. An embryological, morphological and phylogenetic study. 3. The phylogeny of elasmobranch fishes. *Acta zool. Stockh.*, **23**: 129–261.
Holmgren, N. (1946). On two embryos of *Myxine glutinosa. Acta zool. Stockh.*, **27**: 1–90.
Holmgren, N. & Stensiö, E. A. (1936). Kranium und Visceralskelett der Akranier, Cyclostomen und Fische. In *Handbuch der vergleichenden Anatomie der Wirbeltiere*, hrsg. Bolk, L., Göppert, E., Kallius, E. & Lubosch, W., **4**: 233–500. Berlin & Wien: Urban & Schwarzenberg.
*Howes, G. B. (1892). On the affinities, inter-relationships and systematic position of Marsipobranchii. *Proc. Lpool biol. Soc.*, **6**: 122–147.
*Huxley, T. H. (1876). On the nature of the craniofacial apparatus of *Petromyzon. J. Anat. Physiol. Lond.*, **10**: 412–429.
Jarvik, E. (1954). On the visceral skeleton in *Eusthenopteron* with a discussion of the parasphenoid and palatoquadrate in fishes. *K. svenska VetenskAkad. Handl.*, (4) **5**: 1–104.
Jarvik, E. (1959). Dermal fin-rays and Holmgren's principle of delamination. *K. svenska VetenskAkad. Handl.*, (4) **6**: 1–51.
Jarvik, E. (1964). Specializations in early vertebrates. *Annls Soc. R. zool. Belg.*, **94**: 11–95.
Jarvik, E. (1965). Die Raspelzunge der Cyclostomen und die pentadactyle Extremität der Tetrapoden als Beweise für monophyletische Herkunft. *Zool. Anz.*, **175**: 101–143.
Johnels, A. G. (1948). On the development and morphology of the head of *Petromyzon. Acta zool. Stockh.*, **29**: 139–279.
Jollie, M. (1962). *Chordate morphology.* New York: Reinhold Publ. Corp.
Kölliker, A. (1859a). On the different types in the microscopic structure of the skeleton of osseous fishes. *Proc. R. Soc. Lond.*, **9**: 656–668.
*Kölliker, A. (1859b). Ueber die verschiedenen Typen in der mikroskopischen Structur des Skeletes der Knochenfische. *Verh., phys.-med. Ges. Würzb.*, **9**: 257–271.
*Koltzoff, N. K. (1902). Entwicklungsgeschichte des Kopfes von *Petromyzon planeri. Bull. Soc. Imp. Nat. Moscou*, **15**: 259–589.
*Müller, J. (1839). Über den eigentümlichen Bau des Gehörorganes bei den Cyclostomen, mit Bemerkungen über die ungleiche Ausbildung der Sinnesorgane bei den Myxinoiden. Fortsetzung der vergleichenden Anatomie der Myxinoiden. *Abh. preuss. Akad. Wiss.*, **1839**: 15–48.
Parker, W. K. (1883). On the skeleton of marsipobranch fishes. (Abstr.). *Proc. R. Soc. Lond.*, **34**: 447–449.
Platt, Julia B. (1894). Ontogenetic differentiations of the ectoderm in *Necturus.* Second preliminary notice. *Anat. Anz.*, **9**: 51–56.

Romer, A. S. (1965). Possible polyphylety of the vertebrate classes. *Zool. Jb.* (*Syst.*), **92**: 143–156.

Säve-Söderbergh, G. (1936). On the morphology of Triassic stegocephalians from Spitsbergen, and the interpretation of the endocranium in the Labyrinthodontia. *K. svenska VetenskAkad. Handl.*, (3) **16**: 1–181.

*Sewertzoff, A. N. (1916). Études sur l'évolution des vertébrés inférieurs. 1. Morphologie du squelette et de la musculature de la tête des cyclostomes. *Arch. russ. Anat. Hist. Embr.*, **1**: 1–104.

*Sewertzoff, A. N. (1917). Études sur l'évolution des vertébrés inférieurs. 2. Organisation des ancêtres des vertébrés actuels. *Arch. russ. Anat. Hist. Embr.*, **1**: 425–572.

*Sewertzoff, A. N. (1927). Études sur l'évolution des vertébrés inférieurs. 3. Intérprétation morphologique des parties du squelette des arcs viscéraux. *Publ. Staz. zool. Napoli*, **8**: 475–554.

Sewertzoff, A. N. (1928). Einige Bemerkungen über die systematischen Beziehungen der Anaspida zu den Cyclostomen und Fischen. *Paläont. Z.*, **10**: 111–125.

Stensiö, E. A. (1927). The Downtonian and Devonian Vertebrates of Spitsbergen. 1. Family Cephalaspidae. *Skr. Svalbard Nordishavet*, **12**: i–xii, 1–391.

Stensiö, E. A. (1932). *The cephalaspids of Great Britain*. London: Br. Mus. (nat. Hist.).

Stensiö, E. A. (1958). Les cyclostomes fossiles ou ostracodermes. In *Traité de Zoologie*, ed. Grassé, P.-P., **13**:**1**: 173–425. Paris: Masson.

Stone, L. S. (1929). Experiments showing the role of migrating neural crest (mesectoderm) in the formation of head skeleton and loose connective tissue in *Rana palustris*. *Arch. EntwMech. Org.*, **118**: 40–77.

Strahan, R. (1958). Speculations on the evolution of the agnathan head. *Proc. cent. & bicent. Congr. Biol., Singapore*: 83–94.

Strahan, R. (1964). In *Parker & Haswell: A Text-Book of Zoology*, ed. Marshall, A. I., vol. 2. London: Macmillan.

*Tretjakoff, D. K. (1927a). Das periphere Nervensystem des Flussneunauges. *Z. wiss. Zool.*, **129**: 359–452.

*Tretjakoff, D. K. (1927b). Das Gefässsystem im Kiemengebiet des Neunauges. *Morph. Jb.*, **58**: 209–264.

*Tretjakoff, D. K. (1929a). Die schleimknorpeligen Bestandteile im Kopfskelett von *Ammocoetes*. *Z. wiss. Zool.*, **133**: 470–516.

*Tretjakoff, D. K. (1929b). Ursprung der Chordaten. *Z. wiss. Zool.*, **134**: 558–640.

Watson, D. M. S. (1954). A consideration of ostracoderms. *Phil. Trans. R. Soc.*, (B) **238**: 1–25.

*Woodland, W. N. F. (1913). On the supposed gnathostome ancestry of the Marsipobranchii, with a brief description of some features of the gross anatomy of the genera *Geotria* and *Mordacia*. *Anat. Anz.*, **45**: 113–153.

Jaw articulation and suspension in *Acanthodes* and their significance

By Roger S. Miles

Royal Scottish Museum, Edinburgh, Scotland

Acanthodes bronni Agassiz has principally been of interest over the last thirty years as the only acanthodian from which strong, direct evidence can be drawn in support of Watson's (1937) aphetohyoid concept.

It is surprising therefore that a detailed account of the relationships between the palatoquadrate and braincase has not been given, although material for such an account exists in some abundance. It is surprising also that the functional aspect of the aphetohyoid problem in acanthodians has been ignored, for a consideration of function is clearly necessary for a satisfactory interpretation of morphology.

In a preliminary note (Miles, 1964), I have attempted to give a reinterpretation of the visceral skeleton of *Acanthodes*, which can now be refined and extended. The sketch restoration I have published of the brain-case and jaws of *Acanthodes* (Miles, 1964, fig. 1 B, C; 1965, fig. 1 B, C) is a little-modified version of Watson's (1937, fig. 18) restoration. It now proves to be inaccurate in important respects, although I find no reason to alter my concept of *Acanthodes* as a fish with a modified spiracular gill-slit, that is to say as a non-aphetohyoid form.

In this paper the abbreviation H.U. is used for the Palaeontological Institute of the Humboldt University, Berlin, and D.M.S.W. for the collection of Prof. D.M.S. Watson, now housed in the University Museum of Zoology, Cambridge.

Morphology

Published sources of information on the cranial morphology of *Acanthodes bronni* include the papers of Reis (1895; 1896), Jaekel (1899; 1906; 1925, and others), Dean (1907), Watson (1937) and Miles (1964; 1965).

Palatoquadrate. Three separate ossifications can normally be distinguished (Figs. 1, 2), the autopalatine (*Aup*), metapterygoid (*Mpt*) and quadrate (*Qu*). However in large (i.e. old) individuals there is frequently some perichondral

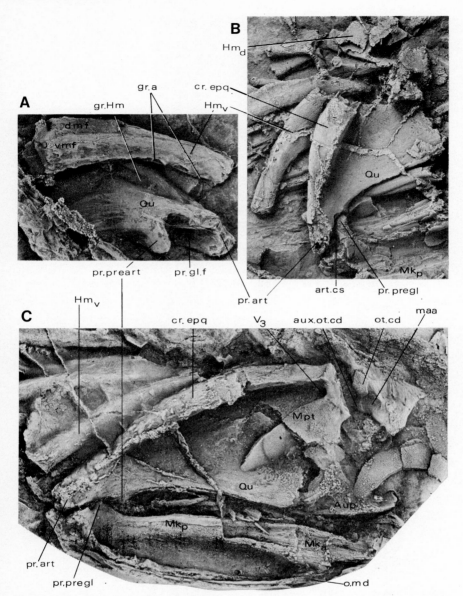

Fig. 1. *Acanthodes bronni* Ag. L. Permian, Lebach, Saarland. Casts of *A*, H. U. MB 3b; approx. ×2; *B*, Univ. of Bonn, unreg.; approx. ×4/3; *C*, H. U. MB 14a; approx. ×5/3.

Aup, autopalatine ossification; Hm_d, Hm_v, dorsal and ventral ossification of hyomandibula respectively; Mk_a, Mk_p, anterior and posterior ossification of Meckel's cartilage respectively; *Mpt*, metapterygoid ossification; *Qu*, quadrate ossification; V_3, foramen for r. mandibularis trigemini; *art.cs*, articular cotylus of mandible; *aux.ot.cd*, auxiliary otic condyle; *cr.epq*, extrapalatoquadrate ridge; *dmf*, dorsomesial face of hyomandibula; *gr.a*, attachment areas for hyomandibular gill-rakers; *gr.Hm*, groove for hyomandibula; *maa*, possible attachment area for slip of m. add. mandibulae; *ot.cd*, otic condyle; *o.md*, mandibular bone; *pr.art*, articular process of quadrate ossification; *pr.gl.f*, preglenoid fossa of quadrate ossification; *pr.preart*, prearticular process of quadrate process of quadrate ossification; *pr.pregl*, preglenoid process of Meckel's cartilage; *vmf*, ventromesial face of hyomandibula.

co-ossification of these regions. There is a prominent extra-palatoquadrate ridge (*cr.epq*, Figs. 1 B, C, 4) on the lateral surface of the palatoquadrate, which dorsally demarcates the deep adductor muscle fossa. On the mesial surface there is a broad groove for the hyomandibula (*gr.Hm*, Figs. 1 A, 2, 4 B; Miles, 1964, fig. 1; 1965, fig. 1). This groove descends over the dorsal surface of the articular process of the quadrate ossification, so that on this process it is more correctly described as being on the posterior margin of the palatoquadrate.

The autopalatine ossification bears the basal process (*pr.b*, Figs. 2, 4). The mesial surface of this process (Fig. 2) may be roughened where it participates in the basal articulation with the brain-case. The metapterygoid ossification is pierced by a foramen for the ramus mandibularis trigemini (V_3, Figs. 1 C, 2, 4). The anterior face of this ossification bears a dorsally-situated, well-developed cotylus (*ot.cs*, Figs. 2, 3 C) for the otic articulation with the brain-case. Slightly below this otic cotylus there is a dorso-ventrally elongated, shallow cotylus (*aux. ot.cs*, Figs. 2, 3 C; Miles, 1964, fig. 1 A; 1965, fig. 1 A) for the auxiliary otic articulation with the brain-case. The quadrate ossification is drawn out posteriorly to form the articular process of the palatoquadrate (*pr.art*, Figs. 1, 2, 4), and at the same time transversely expanded so that a broad articular surface is developed on its ventral margin (Fig. 1 A). The articular process is preceded by a large fossa on the mesial surface (*pr.gl.f*, Figs. 1 A, 2; Reis, 1896, pl. 1, fig. 6; Miles, 1964, fig. 1 A; 1965, fig. 1 A), which may be termed the preglenoid cotylus ("Praeglenoid Hemmgrube", Reis). This cotylus is anteriorly delimited by the stout, broad-based prearticular process of the palatoquadrate (*pr.preart*, Figs. 1, 2, 4).

Lower jaw. Meckel's cartilage is perichondrally ossified in two sections, an anterior or mentomandibular ossification (Mk_a, Fig. 1 C, 2) and a posterior or articular ossification (Mk_p, Figs. 1 B, C, 2). The two regions may be perichondrally co-ossified in large individuals (Fig. 1 C). A long, slender, dermal mandibular bone (extramandibular spine, mandibular splint) is present along the lower part of the lateral face of the jaw (*o.md*, Figs. 1 C, 2, 4 A; Miles, 1966: 154). The posterior ossification of Meckel's cartilage bears virtually the whole of the large, laterally-situated, clearly-defined fossa for the adductor muscles (Figs. 1 C, 2, 4 A). At its posterior end this ossification is strongly expanded laterally, and on its dorsal surface it bears the articular cotylus (*art.cs*, Fig. 1 B, C) for the reception of the articular process of the palatoquadrate. A well-developed preglenoid process (*pr. pregl*, Figs. 1 B, C, 2, 4 A; auxiliary articular facet, coronoid elevation, mandibular knob) arises from the anteromesial angle of the articular cotyle. This process articulates with the auxiliary articular cotylus of the palatoquadrate (Figs. 1 B, C, 2), thus forming the auxiliary jaw articulation.

Fig. 2. *Acanthodes bronni* Ag. L. Permian, Lebach, Saarland. Cast of H. U. MB 17 b; approx. × 3/2.

Bh, basihyal; Hb_a, Hb_p, anterior and posterior ossification of hyoid bar respectively; *pr.b*, basal process of autopalatine ossification; *gr.hb*, depression on mesial face of Meckel's cartilage for hyoid bar; *k*, anterior knob of hyoid bar for articulation with basihyal. Other abbreviations as in Fig. 1.

The nature of the mandibular symphysis is obscure. There is no definite evidence of a basimandibular element (cf. Dean, 1907: 212, fig. 12) but there is some indication of a ligamentous connection between the anterior ends of the mesially-curved mandibular rami (Watson, 1937: 103).

The lower part of the mesial surface of the posterior ossification bears a broad, shallow depression (*gr. hb*, Fig. 2), in which lies the posterior part of the hyoid bar. This depression continues up along the posterior margin of the mesial surface of the mandible, and ends immediately posteroventral to the jaw articulation, subjacent to the groove on the palatoquadrate for the hyomandibula.

Hyoid arch. Dorsally the hyoid arch is developed as a hyomandibula (Miles, 1964; 1965), in which there is a small dorsal or anterior ossification (Hm_d, Figs. 1 B, 4 A) and a large ventral or posterior ossification (Hm_v, Figs. 1, 4 A).

The two ossifications are not known to be perichondrally co-ossified, and they seem always to have been separated by a moderately extensive cartilaginous region. The mesial surface of the ventral ossification (Fig. 1 A) is angled so that dorsomesial (*dmf*) and ventromesial (*vmf*) faces can be recognized, except most ventrally, where the element tends to be circular in section. The anterior edge of the ventral element is about 1 mm broad in adult specimens; it bears up to a dozen or so dorso-ventrally orientated, oval areas lacking perichondral bone (*gr.a*, Fig. 1 A), on which were situated the gill-rakers.

Ventrally in the hyoid arch there is a stout, median basihyal element (*Bh*, Fig. 2; Watson, 1937: 105). There is also a large lateral element which is typically represented by separate anterior ($Hb._a$, Figs. 2, 4 A) and posterior ($Hb._p$, Figs. 2, 4 A) ossifications. These ossifications are usually identified as the hypohyal and ceratohyal respectively (e.g. Watson, 1937; but cf. Jollie, 1962, fig. 5–21 C), but they are perichondrally co-ossified in large specimens, and this proves that they are separately ossifying regions of one long element. This element should almost certainly be termed the ceratohyal, the hypohyal not being present; but in default of a detailed discussion I shall refer to it here as the hyoid bar. The hyoid bar is mesially curved on its anterior half, and anteriorly bears an expanded knob (*k*, Fig. 2) for articulation with the basihyal. The mesial face of the bar bears a broad, moderately deep grove (Fig. 2; Jaekel, 1925, fig. 12), which is conceivably for, *inter alia*, the insertion of the anterior interhyoideus musculature.

Gill-rakers. These structures have been of some significance in the interpretation of the hyoid arch and jaw suspension in *Acanthodes*, and they require brief notice here. They have been described by Reis (1896, pl. 1, figs. 3, 18) and Watson (1937). So far as we can tell they are purely dermal structures, like gill-rakers of other fish. They have a ribbed ornamentation and an expanded base for articulation with the gill-arches; the base of each raker is pierced by at least one fine foramen. The rakers articulate with more or less transversely placed ridges on the gill-arches and hyoid bars (Watson, 1937: 105, pl. 13, fig. 1), but as noted above conditions are slightly different on the hyomandibula. The gill-rakers which articulate with the upper surface of the hyoid bar are very long, as described by Watson, but those of the hyomandibula (on the ventral ossification only) appear to be much less well developed, and in H.U. MB 8a, for example, they are only half as long as those of the first epibranchial (note however that the growth and replacement of the rakers cannot yet be taken fully into account).

The conflicting restorations of the gill-rakers published by Reis and Watson have been the source of some confusion. Reis (1896, pl. 1, fig. 3) restored gill-rakers in both anterior and posterior rows on the ventral elements of all the arches but the hyoid, where only the posterior row was shown, and the last, where only the anterior row was shown. Anterior gill-rakers only, however, were shown on the hyomandibula (Reis, 1896, pl. 1, fig. 5). Watson (1937, fig. 18) restored an anterior row only, on all the elements of the hyoid and succeeding arches. Holmgren (1942: 145) has noted this discrepancy and suggested that the single row of gill-rakers on the hyoid arch "answers to the posterior branchial series (the anterior being annihilated by the closure of the gill-slit)".

I find that two rows of gill-rakers are present on the first four posthyoid branchial arches (as shown by Reis) *in juvenile individuals only* (length of mandibular bone ca. 15 mm). In adults the posterior row is never present on the first, second or third gill-arch, and it is only infrequently retained on the fourth. There seems to be little doubt that the rakers on the hyoid and succeeding arches in adults are equivalent to those of the anterior rows of the juvenile gill-arches.

Articular surfaces of endocranium. D.M.S.W. P. 495 is a negatively prepared nodule containing a large brain-case of *Acanthodes bronni*. If it is treated as a two-piece mould, perfect life-size replicas of the original, major ossification of the brain-case can be obtained by the use of flexible casting media. Such casts, from which Fig. 3 A and B have been prepared, confirm and extend our knowledge of the brain-case derived from less well-preserved specimens.

In the otic region of the brain-case (Figs. 1 C, 3 A, 4), immediately below the level of the dorsolateral ridge (*cr.dl*) and a short distance behind the postorbital process (*pr.po*), lies the well-developed otic condyle (*ot.cd*) which is received by the otic cotylus of the metapterygoid ossification of the palato-quadrate. The major axis of the elongated otic condyle is directed anteroventrolaterally with reference to the anteroposterior axis of the head.

Anteroventrally to the otic condyle, on the lateral wall of the braincase immediately behind the orbit, lies a second, less well-formed yet nevertheless prominent condyle (*aux.ot.cd*, Figs. 1 C, 3 A, 4). This condyle, which seems to be rather variable in its development, functions as a bearing surface for the shallow auxiliary otic cotylus of the metapterygoid ossification, and may be termed the auxiliary otic condyle.

I now regard the anterior basal ossification (Watson, 1937) of the brain-case as a purely endocranial element, with structures comparable with those found in the corresponding region of the brain-case in early bony fish (Miles, 1965: 236).

At each side, the anterior basal ossification bears a basipterygoid process (*pr.bpt*, Fig. 3 A, B; Miles, 1964, figs. 1 A, 2 B; 1965, figs. 1 A, 2 B). The lateral face of the basipterygoid process has a rough, "unfinished" surface (for a capsular ligament?) where it participates in the palato-basal articulation with

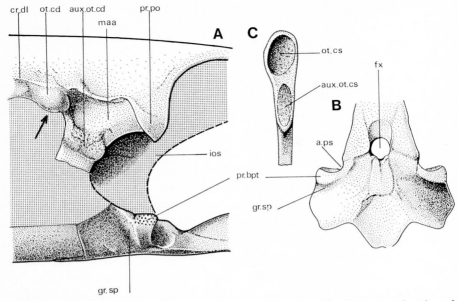

Fig. 3. *Acanthodes bronni* Ag. *A*. Restoration of anterior otic and orbitotemporal region of the brain-case in lateral view; approx. × 5/2. Arrow indicates position of smooth concavity where brain-case wall is deeply undercut. Regular stipple represents cartilaginous regions. *B*, posterior part of "anterior basal" ossification in ventral view; approx. × 7/4. *C*, anterior face of metaterygoid ossification of palatoquadrate; approx. × 7/4. *A*, *B*, D.M.S.W. P. 459; *C*, H.U. MB 14 a.

a.ps, notch for efferent pseudobranchial artery; *aux.ot.cd*, auxiliary otic condyle; *aux.ot.cs*, auxiliary otic cotylus; *cr.dl*, dorsolateral ridge of brain-case; *fx*, foramen usually determined as the hypophysial foramen; *gr.sp*, spiracular groove; *ios*, interorbital septum; *maa*, possible attachment area for slip of m. add. mandibulae; *ot.cd*, otic condyle; *ot.cs*, otic cotylus; *pr.bpt*, basipterygoid process; *pr.po*, postorbital process.

the palatoquadrate. There is a broad sulcus immediately behind the basipterygoid process, which I have identified as the spiracular groove (*gr.sp*, Fig. 3 A, B; Miles, 1964; 1965). This groove compares well in its morphological relations with the ventral part of the spiracular groove in palaeoniscoids such as *Pteronisculus* (Nielsen, 1942, fig. 6). In bony fishes there is of course also a parasphenoid bone, which however is accurately moulded to and reflects the relief of the basal region of the brain-case (see also Holmgren, 1942: 132).

Restoration. It is an easy matter to articulate the palatoquadrate with the brain-case through the well-formed otic joint, both graphically and in an accurate life-size model, so that the auxiliary otic cotylus is brought into contact with its condyle and the mesial surface of the basal process is placed against the basipterygoid process (Fig. 4). The otic articulation is clearly the most highly developed and mechanically most important joint between the upper jaw and brain-case. The auxiliary otic articulation does not appear to involve a very tight relationship between the metapterygoid and brain-case, and only a certain amount of sliding movement of the auxiliary cotylus over the surface of the auxiliary condyle is indicated.

The relationship between the basal process and the basipterygoid process was clearly a loose one and the two structures were probably only in ligamentous connection. This is the relationship between the orbital process and the brain-case in *Chlamydoselachus* (Smith, 1937: 355) and hexanchid sharks (Huxley, 1876: 44; Daniel, 1934: 45). It is also of interest to remind ourselves of conditions in some palaeoniscoids, as described by Nielsen (1942: 144) for *Pteronisculus*. Here (as in *Acanthodes*) there is no specially developed articular fossa on the basal process of the palatoquadrate, and it seems likely that the basal process rested against the distal end or the dorsal face of the basipterygoid process, and that during the abduction of the palatoquadrate the two contact faces simply slid against each other.

The hyomandibula clearly rests in the broad groove on the mesial face of the palatoquadrate, and the posterior part of the hyoid bar rests against the depression on the mesial face of the posterior ossification of Meckel's cartilage, at least when the jaws are adducted (Fig. 4 A). The relationship between the groove and depression just noted (see Figs. 2, 4 A) would seem to indicate quite clearly that the hyomandibula and hyoid bar articulated with each other immediately behind the jaw articulation, either directly through cartilaginous extensions or indirectly by the intervention of an additional element (Miles, 1965: 241).

It is of interest to note the similar relations to each other of the hyoid and mandibular arches in some other long-jawed fishes.

In the hexanchid shark *Heptranchias perlo*, for example, the hyomandibula and the ceratohyal are lodged in excavations in the mesial surface of the palatoquadrate and Meckel's cartilage respectively, and Gadow (1888: 425) records that the "hyoid-hyomandibular junction rests against the hinder surface of a prominent knob, which forms the median or inner quadratomandibular articulation". Very similar relations obtain between the jaw articulation and hyomandibula-ceratohyal joint in *Chlamydoselachus* (Goodey, 1910, pl. 42, fig. 5; Allis, 1923, pl. 10, figs. 15, 17, 18–20) and *Xenacanthus* (Hotton, 1952, fig. 1).

In the Crossopterygii there may also be a distinct groove on the mesial face of the palatoquadrate, which is best described for *Eusthenopteron* (spiraculohyomandibular recess, Jarvik, 1954: 31–32, figs. 14, 16 C, 24). Dorsally it was apparently occupied in life by the dorsal diverticulum of the spiracular tube, but ventrally where it is especially conspicuous (Thomson 1967: 230) it houses the hyomandibula, which articulates with the ceratohyal (through the stylohyal) immediately behind the jaw articulation. Principally the same relations are found between the hyoid and mandibular arches in other early bony fishes (see e.g. *Pteronisculus*, Nielsen, 1942, figs. 35, 36).

By restoring the upper jaw apparatus in *Acanthodes* it is possible to determine where the dorsal end of the hyomandibula meets the lateral surface of the brain-case (Fig. 4 A; Miles, 1964; 1965). It does so, with the jaws adducted, slightly behind and ventral to the otic condyle in the otic region. The lateral wall of the brain-case is not perichondrally ossified in this region, however, and this might be regarded as unusual if we suppose that a conspicuous articular fossa was developed for the hyomandibula. It is possible therefore that the hyomandibula had no direct articular relations with the brain-case, but was bound to it by ligaments as is the case in hexanchid sharks (Gadow, 1888; Daniel, 1934), and as probably was the case in most cladodont-level sharks (Hotton, 1952; Schaeffer, 1967: 9–10).

We have no good evidence of ligaments binding the mandibular and hyoid arches in the region of the jaw articulation in *Acanthodes*, although we may almost certainly assume that such were developed.

Musculature. There are, as Reis (1896) notes, clearly defined muscle attachment areas in *Acanthodes* which enable us to restore the cranial musculature in outline. Reis's own account, however, is unnecessarily complicated by his view of the roles of the "premandibular" element and "extra-mandibular-spine" in the jaw mechanism.

The main point I wish to make here on the mandibular and hyoid muscles is that a cranial musculature of generalized type for fishes can be restored in *Acanthodes* without difficulty.

It seems clear that the intermandibularis musculature was present between the lower jaw rami, and that the m. adductor mandibulae took origin entirely or almost entirely on the concave lateral face of the palatoquadrate. It is just possible however that a slight anterodorsal slip of muscle originated on the brain-case between the postorbital process and otic condyle (*maa*, Fig. 3 A). The range of possible movements of the lower jaw is limited by the double jaw articulation (*vide infra*), and it is doubtful whether there was any marked differentiation of the m. adductor mandibulae. The possible presence of a m. preorbitalis (suborbitalis) of elasmobranch type is considered below. The m. constrictor dorsalis probably took origin on the brain-case under the dorsolateral ridge, and inserted principally on the mesial face of the metapterygoid ossification of the palatoquadrate. This ossification may be supposed, therefore, to have formed a processus oticus externus (Holmgren, 1943: 62). A slip of muscle may have extended ventrally to insert on the autopalatine ossification, posterior to the basal process.

It seems beyond question that a major part of the m. constrictor hyoideus formed a sheet of muscle in the gill-cover or operculum (*sensu* Eaton, 1939: 42). For this and other reasons (Miles, 1964) I regard the main gill-cover of acanthodians as belonging to the hyoid and not to the mandibular arch (cf. Watson, 1937). There is some evidence (*vide supra*) that the m. constrictor hyoideus ventralis was differentiated into anterior and posterior interhyoideus muscles. There may be some differentiation of the constrictor hyoideus dorsalis into a levator hyomandibularis, inserting on the broad dorsomesial face of the hyomandibula. However the close relationship of the hyomandibula to the palatoquadrate, the presence of a simple series of opercular bones (the branchiostegal rays), the apparent lack of an opercular process on the hyomandibula and the closely circumscribed movements of the palatoquadrate and gill cover (*vide infra*), indicate only slight differentiation of the constrictor hyoideus dorsalis, and the absence of a differentiated dilatator operculi in the mandibular musculature.

Discussion

Comparison with elasmobranchs. There are no living acanthodians and it is necessary to interpret the jaw mechanism of *Acanthodes* by analogy with other fish groups.

Many workers have recognized a similarity between the palatoquadrates of hexanchid sharks and *Acanthodes* (e.g. Reis, Jaekel, Holmgren). The similarity depends above all on the dorsal articulation with the brain-case. However the otic process of the palatoquadrate in hexanchids articulates with the postorbital process of the brain-case (Huxley, 1876, fig. 9; Goodrich, 1909, fig. 59 A; Daniel, 1934, fig. 48), and the articulation is notably loose in *Hexanchus*. This is in contrast to the tight cup-and-ball otic articulation of *Acanthodes*, where the otic condyle is situated distinctly behind the postorbital process on the lateral wall of the brain-case. There is therefore a suggestion that the otic articulations in the two cases have been independently acquired (cf. Holmgren, 1942: 136, 137). Palaeontological evidence goes a long way to confirming this suggestion. An otic articulation closely comparable with that of hexanchids is found in cladodont-level sharks (Schaeffer, 1967: 9, fig. 5). This suggests that such an articulation is a primary feature of elasmobranchs, despite suggestions to the contrary derived from the embryology of hexanchids (e.g. Edgeworth, 1935:64). In sharp contrast, the most primitive known, articulated acanthodians (from the Lower Old Red Sandstone of Scotland) appear to lack an external otic process of the palatoquadrate and, therefore, the otic articulation of the upper jaw with the brain-case (e.g. *Climatius*, Watson, 1937: 118). The otic articulation of Acanthodii appears to be a specialization of the Acanthodiformes (Miles, 1966), correlated with their mode of feeding (*vide infra*), and not a primary character of the class. An otic articulation is present in the Crossopterygii (Stensiö, 1932; Jarvik, 1954), but has again been acquired independently of that of acanthodians.

There is some difference of opinion over the nature of the ventral articulation of the palatoquadrate with the brain-case in the posteroventral region of the orbit in *Acanthodes*. Watson (1937), Nielsen (1949) and Miles (1964; 1965) identify it as the basal articulation, but Holmgren (1942: 139) claims that it is an orbital articulation of elasmobranch type, citing as his evidence, "the almost identical organization of the palato-quadrate in Acanthodians and in squaloid and notidanid sharks".

There are two points of contention here: (1) the relationship between the orbital articulation of elasmobranchs and the basal articulation of other vertebrates, and (2) the nature of the articulation in *Acanthodes*.

With regard to point (1), Holmgren's (1942; 1943) embryological researches on sharks would seem to confound de Beer's (1937: 419) view that the two articulations in question are homologous. Holmgren (1942: 138, 139) summarizes his results in this way: "In a squaloid embryo the palato-quadrate is attached to the basicranial region by two connections, viz., an anterior with the trabecula and a posterior with the polar cartilage. The anterior develops into the articulation of the orbital process, the posterior is quite rudimentary and disappears. The latter never has the character of an articulation in sharks, whether embryonic or adult... The posterior connection is present, so far as I know, in Palaeoniscids, Crossopterygians and Dipnoi. It may also be present in embryonic ganoids". Confirmation for Holmgren's view comes from arthrodires, if the homologies are correct, for Stensiö (1963) describes well-developed basal and orbital articulations existing side by side in some forms.

Which articulation is present in *Acanthodes*? Holmgren (1942: 138–140) attempts to justify his view by reference to the variable position that the orbital articulation has in modern sharks, noting in particular its posterior position in the orbits of squaloids and notidanids.

However the orbital articulation seems generally to be anteriorly situated in cladodont-level sharks (see e.g. Holmgren, 1942: 140), and it is instructive to compare Goodrich's (1909, fig. 159) sketch of "an acanthodian" with an orbital articulation (in the anterior part of the orbit) with the different conditions that actually obtain in *Acanthodes*. I believe that Nielsen (1949: 109) is perfectly correct in holding that the palatoquadrate of *Acanthodes* articulates in exactly the same place as the palato-basal articulation in palaeoniscoids; and I believe that the basipterygoid process of *Acanthodes* has exactly the same morphological relations to structures such as the postorbital process, spiracular groove and efferent pseudobranchial artery (see further Miles, 1964) as the basipterygoid process of palaeoniscoids. For these reasons I hold that the palato-basal, and not the more anterior orbital articulation is present in *Acanthodes*.

Closely bound up with the problem just discussed is the possible existence of an anterior, symphysial relationship between the right and left palatoquadrates in *Acanthodes*, for Holmgren (1942: 140) admits that if a symphysial connection is absent then the articulation between the palatoquadrate and brain-case in the posterior region of the orbit is the basal articulation. It is clear that the palatoquadrate did not have a long palatine process in *Acanthodes* (Miles, 1965: 238) and largely on this evidence I agree with Watson (1937) that there is no palatoquadrate symphysis. Holmgren (1942: 142, 146), however, suggests that a broad ligamentous symphysis was present, and that the "anterior ossicles" (Watson, 1937) are dermal structures which lay in the symphysial band, as do several rows of teeth in *Heptranchias* and *Notorhynchus*. However I find that the "anterior ossicles" are endocranial structures, normally perichondrally co-ossified with the anterior basal ossification; and there is no evidence from the specimens that a broad ligamentous band attached to the autopalatine ossifications and crossed the ventral surface of the anterior basal ossification.

Most of the elements of the jaw and suspensory apparatus in *Acanthodes* are perichondrally ossified in two or more sections, and Nelson (in this Volume) shows that this is also true of the epibranchials of the gill-arches. Attempts have been made to compare the ossifications of the palatoquadrate and mandible with the centres of chondrification found in these elements in elasmobranch embryos (e.g. Holmgren, 1942: 139, 140), but the value of such comparisons has been diminished by Nielsen's (1942) demonstration that similarly situated centres of ossification may be found in young individuals of some bony fish.

I (Miles, 1964; 1965) have compared the dorsal and ventral ossifications of the hyomandibula in *Acanthodes*, with the laterohyal and epihyal elements respectively of the developing bony fish hyomandibula. But two centres of chondrification are found in the developing selachian hyomandibula (Holmgren, 1943), and my comparison is of uncertain value. The main point I wish to make, however, is that a straightforward interpretation of the hyomandibula in terms of the dorsal elements of the gill-arches may be of limited validity, for the embryology of recent fishes teaches us that hyoid gill-ray material can play an important part in the composition of the suspensorium (Holmgren, 1943; Bertmar, 1959).

From the above discussion I suggest that the similarities between the early elasmobranch type of palatoquadrate and that of *Acanthodes* are of a convergent nature and without phylogenetic significance, notwithstanding the striking similarity in the structure of the double jaw joint. Nielsen (1949: 109

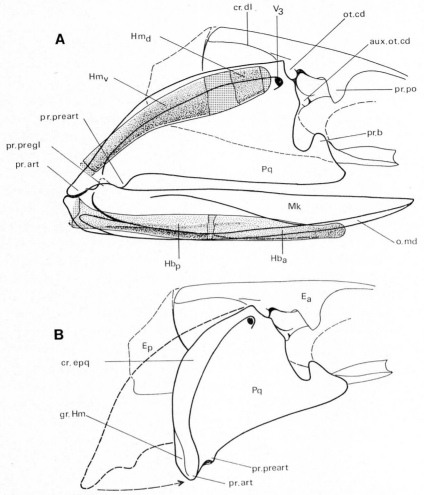

Fig. 4. *Acanthodes bronni* Ag. Restoration of skull and jaw apparatus. *A*, with hyoid arch shown; *B*, with palatoquadrate fully abducted. Regular stipple indicates cartilaginous regions.
 E_a, E_p, anterior and posterior units of endocranium respectively; Hb_a, Hb_p, anterior and posterior ossification of hyoid bar respectively; *Mk*, Meckel's cartilage; *Pq*, palatoquadrate; *cr.dl*, dorsolateral ridge; *pr.b*, basal process of autopalatine ossification; *pr.po*, postorbital process. Other abbreviations as in Fig. 1.

has listed features in which the palatoquadrate of *Acanthodes* agrees with those of typical palaeoniscoids, of which the possession of a palato-basal articulation would seem to be the most important.

Movement of jaws. The double articulation between the mandible and palatoquadrate in *Acanthodes* permits simple up-and-down movements of the mandible to take place, but probably severely restricts any fore-and-aft movement of the mandible relative to the palatoquadrate. Clearly the auxiliary jaw articulation does not serve as a locking mechanism for the lower jaw in the way suggested by Reis (1896).

The palatoquadrate of *Acanthodes* can rotate laterally about the otic articulations, through perhaps some 30° (Fig. 4 B). The otic articulations are so well defined that it is inconceivable that the palatoquadrate did not make such movements in life (cf. the relatively fixed position of palatoquadrate in hexanchid sharks). It seems impossible, however, that the upper jaw could have been protruded as in galeoid and squaloid sharks (Luther, 1909; Schaeffer, 1967).

There is no very clear evidence as to the mechanism responsible for the swinging out of the palatoquadrate. This could not have been accomplished by the constrictor dorsalis musculature if it is distributed as I have suggested above (cf. Reis, 1896). One possibility is that the movement was brought about by the contraction of a preorbitalis muscle of selachian type (see e.g. Luther, 1909; Edgeworth, 1935). This muscle plays an important part in the jaw protrusion mechanism of galeoids and squaloids, but in *Acanthodes* the palatoquadrate would pivot about the otic condyles and swing out as a m. preorbitalis contracted. This is not an entirely satisfactory hypothesis, however, as *Acanthodes* is a large-eyed, microsmatic form like early actinopterygians, and there is no evidence that a preorbitalis muscle took origin on the nasal capsule and passed back through the orbit to the palatoquadrate.

Perhaps a more satisfactory suggestion is that the basic mechanism for abducting the jaws in actinopterygians (see Schaeffer & Rosen, 1961) was also employed in *Acanthodes*. In this mechanism the hyoid bar acts as a lever, expanding the oralobranchial chamber as it pivots about its anterior articulation with the basihyal element, carring the upper jaw and mandible laterally. The lower jaw can be dropped so that the mouth is widely opened at the same time.

The strong development of the hyoid bar, and its well formed articulation with the basihyal element suggest that this is the likely mechanism in *Acanthodes*. If so the hyomandibula may be important in integrating the lateral movements of the palatoquadrate with those of the hyoid bar and mandible.

The morphological relationships of the mandibular and hyoid arches, and the apparently simple development of the bones and muscles of the gill cover, suggest that the wide abduction of this cover could only occur as the palatoquadrate was abducted.

It is clear that during the abduction of the palatoquadrate there must either have been some sliding movement of the hyomandibula in the groove on its mesial surface, or else the head of the hyomandibula underwent fairly marked movement relative to the neurocranium. In the analogous case of the Rhipidistia, Thomson (1967: 234) suggests the first of these alternatives, but I think the second is more likely in *Acanthodes*, considering the intimate relations between the hyoid and mandibular arches. An analogy is suggested with *Chlamydoselachus*, where the articulation of the hyomandibula with the brain-case is said to be a sliding joint of very loose construction, aiding greatly in the range of movement of the hyomandibula (Smith, 1937: 354).

Nature of jaw suspension. It is usual to consider the significance of the hyomandibula in fishes in terms of its suspensory functions. It has been suggested above that the relations of the hyomandibula to the brain-case and the palatoquadrate might be closely similar in *Acanthodes* and hexanchid sharks, opening up the possibility of interpreting the jaw mechanism of the former by analogy with the latter. Here, however, we meet a difficulty in that there is some difference of opinion as to whether or not the hyomandibula of hexanchids and also *Chlamydoselachus* and cladodont-level sharks, has any suspensory functions (Hotton, 1952; Schaeffer, 1967).

In his paper of 1876 which first defined the major modes of jaw suspension, Huxley noted that the slender hyomandibula of *Heptranchias* was attached to the otic region of the brain-case by ligaments and was closely bound by ligaments to the mandibular arch near to the articulation of the mandible. He concluded (1876: 45), however, that the hyomandibula contributed little or nothing to the support of the palatoquadrate. The same was probably true of *Acanthodes*.

The hyoid arch in *Acanthodes* is clearly not an unmodified branchial arch, and a more valuable concept of the hyomandibula can probably be formulated in terms of whether or not it has any mechanical or dynamic role to play in the jaw mechanism. I have indicated above that the hyomandibula might have an important part to play in *Acanthodes* in jaw abduction. *Chlamydoselachus* is an excellent example of a fish where the hyomandibula has been considered as not forming a true suspensorium (Daniel, 1934: 63) yet it is known to play a significant role in the jaw mechanism. Smith (1937: 354) states that the wide expansion of the oralobranchial chamber in this fish "is made possible by the length and mobility of the hyomandibular".

The jaw suspension of *Acanthodes* is amphistylic according to Huxley's (1876) original classification, and amphyostylic according to Gregory's (1904) more detailed analysis. I am not sure that it has a place in de Beer's (1937: 425) scheme, because although the mode in *Acanthodes* may closely resemble the amphistyly of hexanchids, I believe it has been independently acquired.

The double jaw articulation of *Acanthodes* quite strikingly resembles that of early elasmobranchs (Hotton, 1952; Schaeffer, 1967). The double joint appears to be the primary condition in elasmobranchs and it is retained in living hexanchids, *Chlamydoselachus* and *Heterodontus*. There is no evidence however that it is a primary character of acanthodians (see *Mesacanthus*, *Ischnacanthus*, Watson, 1937). Hotton (1952: 494) suggests that the function of the preglenoid process of Meckel's cartilage (mandibular knob) is directly related to the degree of functional amphistyly, and that it acts as an auxiliary jaw articulation in forms with a non-suspensory hyomandibula and an accompanying close articulation between the posterior end of the palatoquadrate and the brain-case. With the exception of the last point, this view appears to be sound. The presence of the auxiliary joint may be correlated with a long gape and the power to swing out the palatoquadrate (secondarily lost in some forms; Schaeffer, 1967: 9–11) rather than with the maintenance of close relations between the posterior end of the palatoquadrate and the brain-case. In such cases there is presumably a need for some mechanism to guard against the disarticulation of the lower jaw.

Biological note. *Acanthodes* is an anguilliform fish with an elongated branchial region (Watson, 1937, fig. 21). Nelson (1966) has demonstrated that there is a correlation between the anguilliform habit and the posterior displacement of the branchial arches in teleosts of the order Anguilliformes. In *Acanthodes*, however, there appears to be a more extensive correlation between body form, elongated branchial region, lack of teeth, large gape, presence of well-developed gill-rakers and microphagous mode of feeding. All of the known Acanthodiformes lack teeth and were presumably typically microphagous fish. There is a clearly defined trend in this order to the elongation of the branchial region, which can be followed from the Lower Devonian *Mesacanthus* through the Middle Devonian *Cheiracanthus* to the Carboniferous and Lower Permian *Acanthodes*.

I picture *Acanthodes* swimming slowly during feeding, with the oralobranchial chamber widely expanded, the mouth wide open and the gill-cover fully abducted (i.e. with the palatoquadrate turned out about the otic articulations), straining off moderately small organisms as a continuous stream of water passes through the oralobranchial chamber past the gill-rakers. In this way the otic articulations can be related to the mode of life.

This concept is supported by the gut contents of some fossils (e.g. "*Estheria*" in *Acanthodes*, Kner, 1868, pl. 5, fig. 1; Watson, 1937: 115; see also *Homalacanthus*, Miles, 1966: 181, pl. 3). That larger animals were occasionally ingested is indicated by the presence of a palaeoniscoid within the body cavity of a specimen of *Acanthodes sulcatus* (Watson, 1937: 115), but it seems most improbable that the toothless *Acanthodes* was habitually an active predator. This palaeoniscoid does attest however to the wide gape of *Acanthodes*.

Phylogenetic note. Watson (1937) defined the aphetohyoid condition as the presence of a complete spiracular gill-slit. He attempted to demonstrate the aphetohyoid condition in *Acanthodes* through the structure of the hyoid arch which was said to be an unmodified branchial arch with gill-rakers crossing a complete spiracular gill-slit, and through the presence of a mandibular operculum covering the complete slit. The Aphetohyoidea, which comprise all of the Acanthodii and Placodermi, were said to occupy an intermediate position between the Agnatha and Pisces in vertebrate phylogeny.

There are, however, (Miles, 1964; 1965) good reasons for regarding the main gill-cover of acanthodians as a hyoid gill-cover of normal fish type. In addition the close morphological relations that exist between the hyoid and mandibular arches in *Acanthodes*, particularly in the region of the jaw articulation, would seem to leave little reason for supposing that a complete spiracular gill-slit existed. Further, it is now clear that the hyoid arch is not developed as a typical gill arch, and it seems possible that the hyomandibula plays a functional role in the jaw mechanism. The presence of a spiracular groove on the brain-case suggests that a spiracular tube of bony-fish type was present (Miles, 1964; 1965: 238). There is no good evidence of the existence of an external spiracular opening in acanthodians, and it is possible that the dorsal end of the spiracular tube in *Acanthodes* ended blindly at the smooth concavity on the deeply undercut surface of the brain-case immediately below the otic condyle (position marked by arrow in Fig. 3 A), which Watson identifies as the "articulation of the otic process" in pl. 13, fig. 3 of his paper of 1937.

The small gill-rakers on the hyomandibula in *Acanthodes* presumably projected into the pharynx when they were erected as the upper jaw was abducted, and their presence might well be correlated with the existence of a large hyoidean hemibranch (Miles, 1964).

Neither the Acanthodii nor the Placodermi (see e.g. Stensiö, 1963) appear to be aphetohyoid, and there are no strong reasons for placing these fishes in the same class. The similarities between *Acanthodes* and some elasmobranchs in the jaw apparatus appear to me to be without phylogenetic significance. It seems probable that the Acanthodii have broad phylogenetic relations with bony fishes rather than with elasmobranchs (Miles, 1965).

References

Allis, E. P. Jr. (1923). The cranial anatomy of *Chlamydoselachus anguineus*. *Acta zool. Stockh.*, **4**: 123–221.
Bertmar, G. (1959). On the ontogeny of the chondral skull in Characidae, with a discussion on the chondrocranial base and the visceral chondrocranium in fishes. *Acta zool. Stockh.*, **40**: 203–364.
Daniel, J. F. (1934). *The elasmobranch fishes*. 3rd Ed. Berkeley: University of California Press.
Dean, B. (1907). Notes on acanthodian sharks. *Am. J. Anat.*, **7**: 209–222.
de Beer, G. R. (1937). *The development of the vertebrate skull*. Oxford: Oxford University Press.

Eaton, T. H. Jr. (1939). Suggestions on the evolution of the operculum of fishes. *Copeia*, **1939**: 42–46.

Edgeworth, F. H. (1935). *The cranial muscles of vertebrates*. Cambridge: Cambridge University Press.

Gadow, H. (1888). On the modifications of the first and second visceral arches, with especial reference to the homologies of the auditory ossicles. *Phil. Trans. R. Soc.*, (B) **179**: 451–485.

Goodey, T. (1910). A contribution to the skeletal anatomy of the frilled shark, *Chlamydoselachus anguineus* Gar. *Proc. zool. Soc. Lond.*, **1910**: 540–571.

Goodrich, E. S. (1909). Cyclostomes and fishes. In *A Treatise on Zoology*, ed. Lankester, E. R., vol. **9**: Vertebrate Craniata, fasc. 1. London: A. & C. Black.

Gregory, W. K. (1904). The relations of the anterior visceral arches to the chondrocranium. *Biol. Bull. Woods Hole*, **7**: 55–69.

Holmgren, N. (1942). Studies on the head of fishes. An embryological, morphological and phylogenetical study. 3. The phylogeny of elasmobranch fishes. *Acta zool. Stockh.*, **23**: 129–261.

Holmgren, N. (1943). Studies on the head of fishes. An embryological, morphological, and phylogenetical study. 4. General morphology of the head in fish. *Acta zool. Stockh.*, **24**: 1–188.

Hotton, N. 3rd. (1952). Jaws and teeth of American xenacanth sharks. *J. Paleont.*, **26**: 489–500.

Huxley, T. H. (1876). Contributions to morphology. Ichthyopsida. No. 1. On *Ceratodus forsteri*, with observations on the classification of fishes. *Proc. zool. Soc. Lond.*, **1876**: 24–59.

Jaekel, O. (1899). Über die Zusammensetzung des Kiefers und Schultergürtels von *Acanthodes*. *Z. dt. geol. Ges.*, **51**: 56–60 (Verh.).

Jaekel, O. (1906). Über die Mundbildung der Wirbeltiere. *Sber. Ges. naturf. Fr. Berl.*, **1906**: 7–32.

Jaekel, O. (1925). Das Mundskelett der Wirbeltiere. *Morph. Jb.*, **55**: 402–484.

Jarvik, E. (1954). On the visceral skeleton in *Eusthenopteron* with a discussion of the parasphenoid and palatoquadrate in fishes. *K. svenska VetenskAkad. Handl.*,(4) **5**: 1–104.

Jollie, M. (1962). *Chordate morphology*. New York: Reinhold Publ. Corp.

Kner, R. (1868). Über *Conchopoma gadiforme* nov. gen. et sp. und *Acanthodes* aus dem Rothliegenden (der unteren Dyas) von Lebach bei Saarbrücken in Rheinpreussen. *Sber. Akad. Wiss. Wien*, **52**: 278–305.

Luther, A. (1909). Untersuchungen über die von N. trigeminus innervierte Muskulatur der Selachier (Haie und Rochen) unter Berücksichtigung ihrer Beziehungen zu benachbarten Organen. *Acta Soc. Sci. fenn.*, **36**: 1–176.

Miles, R. S. (1964). A reinterpretation of the visceral skeleton of *Acanthodes*. *Nature, Lond.*, **204**: 457–459.

Miles, R. S. (1965). Some features in the cranial morphology of acanthodians and the relationships of the Acanthodii. *Acta zool. Stockh.*, **46**: 233–255.

Miles, R. S. (1966). The acanthodian fishes of the Devonian Plattenkalk of the Paffrath Trough in the Rhineland, with an appendix containing a classification of the Acanthodii and a revision of the genus *Homalacanthus*. *Ark. Zool.*, (2) **18**: 147–194.

Nelson, G. J. (1966). Gill arches of teleostean fishes of the order Anguilliformes. *Pacific Sci.*, **20**: 391–408.

Nielsen, E. (1942). Studies on Triassic fishes from East Greenland 1. *Glaucolepis* and *Boreosomus*. *Palaeozool. Groenland.*, **1**: 1–403.

Nielsen, E. (1949). Studies on Triassic fishes from East Greenland 2. *Australosomus* and *Birgeria*. *Palaeozool. Groenland.*, **3**: 1–309.

Reis, O. M. (1895). Illustrationen zur Kenntnis des Skeletts von *Acanthodes Bronni* Agassiz. *Abh. senckenb. naturf. Ges.*, **49**: 48–64.
Reis, O. M. (1896). Ueber *Acanthodes Bronni* Agassiz. *Morph. Arb.*, **6**: 143–220.
Schaeffer, B. (1967). Comments on elasmobranch evolution. In *Sharks, Skates and Rays*, ed. Gilbert, P. W., Mathewson, R. F. & Rall, D. P., :3–35. Baltimore: John Hopkins Press.
Schaeffer, B. & Rosen, D. E. (1961). Major adaptive levels in the evolution of the actinopterygian feeding mechanism. *Am. Zool.*, **1**: 187–204.
Smith, B. G. (1937). The anatomy of the Frilled Shark *Chlamydoselachus anguineus* Garman. *Am. Mus. nat. Hist., B. Dean Mem. Vol.*, **6**: 335–505.
Stensiö, E. A. (1932). Triassic fishes from East Greenland collected by the Danish Expeditions in 1929–1931. *Medd. Grønland*, **83**: 1–305.
Stensiö, E. A. (1936). Anatomical studies on the arthrodiran head. 1. Preface, geological and geographical distribution, the organisation of the arthrodires, the anatomy of the head in the Dolichothoraci, Coccosteomorphi and Pachyosteomorphi. Taxonomic appendix. *K. svenska VetenskAkad. Handl.*, (4) **9**: 1–419.
Thomson, K. S. (1967). Mechanisms of intercranial kinetics in fossil rhipidistian fishes (Crossopterygii) and their relatives. *J. Linn. Soc. (Zool)*, **46**: 223–253.
Watson, D. M. S. (1937). The acanthodian fishes. *Phil. Trans. R. Soc.*, (B) **228**: 49–146.

Gill-arch structure in *Acanthodes*

By Gareth J. Nelson

Department of Ichthyology, The American Museum of Natural History,
New York, N.Y. USA

Structure of the gill arches of acanthodians really is well known only in the Permian genus *Acanthodes*. Nevertheless, its gill arches have been variously interpreted by Reis (1890; 1894; 1895; 1896), Dean (1907), Watson (1937), and Miles (1964; 1965), and used to support various theories of acanthodian relationships: according to Reis, gill-arch structure indicates a relationship with elasmobranchs, according to Miles with teleostomes. Dean and Watson differently interpreted gill-arch structure in *Acanthodes*, but neither assessed its phyletic significance. Dean, however, referred to the acanthodians as "sharks," but Watson (1937: 144; 1959: 25), regarded them as vertebrates so primitive in organization that they cannot reasonably be called even fishes. The present paper contains a new interpretation of the gill arches of *Acanthodes bronni* and an evaluation of their systematic significance in the light of comparisons with gill arches of fossil and Recent fishes.

General features of the branchial skeleton in fishes

The branchial skeleton in fishes is generally composed of both endoskeletal and dermal components: the endoskeleton is formed by the gill arches—the bones and cartilages supporting the pharynx and the gills. Normally in gnathostome fishes there is a series of five arches, each one composed of several bones or cartilages, both paired and unpaired, arranged in a ring-like manner around the lumen of the pharynx (Fig. 1). They conveniently can be divided into ventral and dorsal parts, the ventral parts including a medial basibranchial, and paired hypo- and ceratobranchials; the dorsal parts normally including only paired elements, the epi- and pharyngobranchials. In some crossopterygian and primitive actinopterygian fishes there are two types of pharyngobranchials: infra- and supra-.

Superficial to the endoskeleton may occur dermal elements. Generally these are in the form of placoid scales (in elasmobranchs) or of tooth plates and gill-rakers (in teleostomes). Among Recent teleostomes, the dermal elements vary a great deal between species. The most structurally stable seem to be those

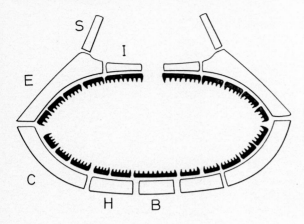

Fig. 1. Diagram of cross-section of a gill arch of a primitive actinopterygian. Dermal elements black.
B, basibranchial; *C*, ceratobranchial; *E*, epibranchial; *H*, hypobranchial; *I*, infrapharyngobranchial; *S*, suprapharyngobranchial.

tooth plates supported by the basibranchials, the fifth ceratobranchials and the infrapharyngobranchials.

The dermal elements primitively are toothed and separate from the endoskeleton, but in an advanced condition in actinopterygians sometimes are toothless. Whether toothed or not, they often are fused with their endoskeletal supports. Among Recent actinopterygians, tooth plates are known to be fused with each of the endoskeletal elements of a typical arch, with the exception of the suprapharyngobranchial.

Detailed accounts of gill-arch structure in fishes occur in Gaupp (1905), Corsy (1933), Holmgren & Stensiö (1936), Holmgren (1940; 1942; 1943), and Devillers (1958). For elasmobranchs, several types are illustrated by Fürbringer (1903) and Garman (1913). For various Recent teleostomes, Gregory (1933), Harrington (1955), Millot & Anthony (1958), Harder (1964), Daget (1964), Fox (1965), and Nelson (1966; 1967a; 1967b; 1968) either discuss gill arches in detail or cite pertinent literature. Among fossil teleostomes gill arches are well known only in a few forms (Nielsen, 1942; 1949; Lehman, 1949; Jarvik, 1954) and among fossil elasmobranchiomorphs in none.

Previous interpretations of gill arches of Acanthodes

According to Reis the arches are very similar to those of Recent elasmobranchs in having both hypo- and pharyngobranchials posteriorly directed (Figs. 2 A, B). According to Dean's figure, the arches are very similar to those of Recent teleostomes in having both hypo- and pharyngobranchials apparently anteriorly directed (Fig. 2 C). According to Watson, both hypo- and pharyngobranchials are anteriorly directed (Fig. 2 E). According to Miles, the arches are as shown by Watson, with the addition of a small suprapharyngobranchial (Fig. 2 D).

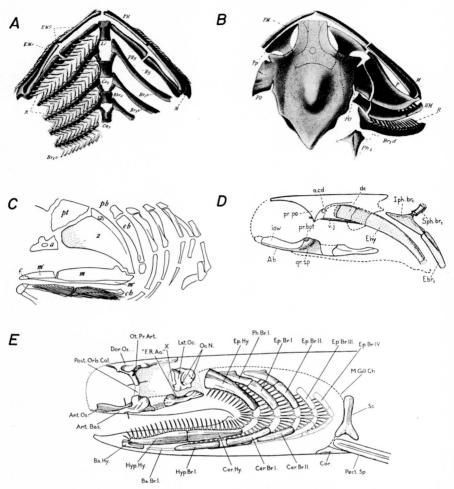

Fig. 2. Previous interpretations of the gill-arch skeleton of *Acanthodes*. *A*, ventral parts, dorsal view, from Reis (1896, pl. 6, fig. 2); *B*, anterodorsal parts, including cranium, dorsal view, from Reis (1896, pl. 6, fig. 3); *C*, lateral view of left side, from Dean (1907, fig. 12); *D*, lateral view of anterodorsal elements of left side, from Miles (1964, fig. 1 B); *E*, lateral view of left side, from Watson (1937, fig. 18 A).

Gill-arch structure according to the present study

Restorations of the gill-arch skeleton of *Acanthodes bronni* are shown in Fig. 3 A–C. The gill arches apparently number five (see also Reis, 1896: 159) and include basi-, hypo-, cerato-, epi-, and pharyngobranchials. The orientation of the hypobranchials is anterior, as in many teleostomes, but the orientation of the pharyngobranchials apparently is posterior, as in all elasmobranchs in which it is known.

Reis apparently erred in figuring posteriorly directed hypobranchials, for they seem to be in line with the ceratobranchials as Dean and Watson figured

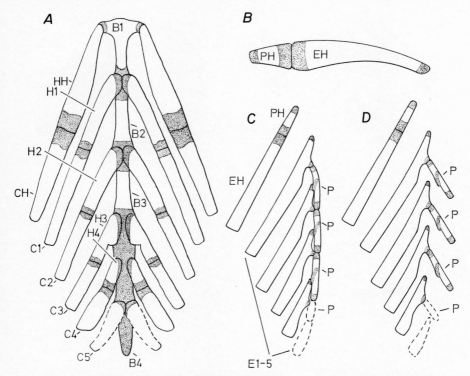

Fig. 3. *Acanthodes bronni*, reconstruction of gill-arch skeleton. *A*, ventral parts, ventral view; *B*, dorsal parts of hyoid arch, lateral view of left side; *C*, hyoid and gill arches, dorsal parts of left side, dorsal view, one possible reconstruction; *D*, as in *C*, but an alternative reconstruction. Cartilage stippled, bone clear. Cartilaginous articular surfaces at posterior end of cerato- and epibranchials not shown.

B 1–4, basibranchials 1–4; *C 1–5*, ceratobranchials 1–5; *CH*, ceratohyal; *E 1–5*, epibranchials 1–5; *EH*, epihyal; *H 1–4*, hypobranchials 1–4; *HH*, hypohyal; *P*, pharyngobranchial. *PH*, pharyngohyal.

them. Both Reis and Dean apparently correctly interpreted the dorsal parts, figuring a long epibranchial and a single pharyngobranchial. Watson and Miles apparently erred in showing the epibranchial divided in two pieces, and in terming one of these either a pharyngobranchial (Watson) or an infrapharyngobranchial (Miles). Finally, the suprapharyngobranchial of Miles appears to be part (the dorsal process, as described below) of the same pharyngobranchial figured by Reis and possibly also by Dean.

In *Acanthodes*, the basi-, hypo-, cerato-, and epibranchials are without features of importance in the present discussion. The pharyngobranchial, however, is of a complex form, with anterior, dorsal and posterior processes, each in life apparently cartilage-tipped (Fig. 4 A). The anterior process is equipped with two articular surfaces, these fitting against articular areas of the epibranchial in front. Whether the tips of the dorsal and posterior processes articulated with any structure is unknown.

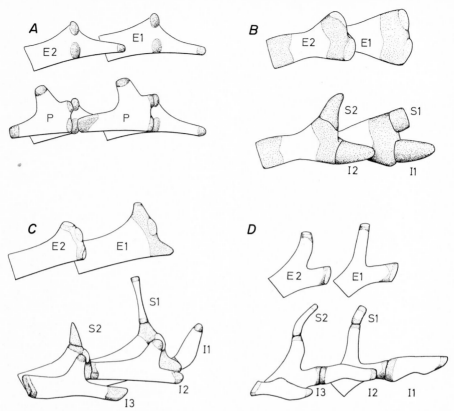

Fig. 4. Gill arches, medial view of anterodorsal parts of left side. *A, Acanthodes bronni; B, Acipenser sturio; C, Elops hawaiiensis; D, Australosomus kochi* (drawn from a wax model made by Nielsen, 1949). Above, anterior ends of epibranchials 1–2 (*E 1–2*), showing articular areas for pharyngobranchials (*P*); infrapharyngobranchials 1–3 (*I 1–3*), and suprapharyngobranchials 1–2 (*S 1–2*) added below.

One possibility is that the pharyngobranchial, like that of teleostomes, contacted the epibranchial behind, as suggested in Figs. 3 C, 4 A. An alternative possibility, is that the pharyngobranchial, like that of elasmobranchs, projected posteromedially, without contacting the epibranchial behind (Fig. 3 D). The present evidence is inadequate to determine which of these conditions obtained. Without this information it is impossible with certainty to determine to which arch a given pharyngobranchial can be attributed, in short, whether the pharyngobranchial is anteriorly or posteriorly directed. But in any case, the *Acanthodes* pharyngobranchial bears some resemblance to the infrapharyngobranchial of some primitive actinopterygians (Figs. 4 C, 5 C, 6 A–D). Among Recent forms this element typically has a broad articulation with the epibranchial behind, anteriorly it forks, forming two processes, a lateral and a medial. Only the lateral contacts the epibranchial in front, the medial extends toward the midline and supports the pharyngeal roof. In some primitive teleostomes (Figs.

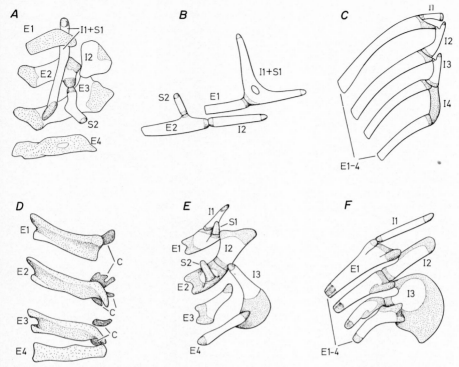

Fig. 5. *A*, *Latimeria chalumnae*, gill-arch skeleton, dorsal view of dorsal parts of left side, modified from Millot & Anthony (1958, fig. 14). *E1–4*, epibranchials 1–4; *I1 + S1*, first infra- and suprapharyngobranchials probably fused; *I2*, infrapharyngobranchial 2; *S2*, suprapharyngobranchial 2. *B*, *Eusthenopteron foordi*, gill-arch skeleton, medial view of anterodorsal parts of left side (drawn from a wax model described by Jarvik, 1954). *E1–2*, medial parts of epibranchials 1–2. Other symbols as in *A*. *C*, *Pachycormus macropterus*, gill-arch skeleton, ventral view of dorsal parts of right side, modified from Lehman (1949, fig. 9); symbols as in *A*. *D*, *Neoceratodus forsteri*, gill-arch skeleton, dorsal view of dorsal parts of left side. *C*, cartilaginous nodules (5) underlying medial ends of epibranchial 1–4 (*E1–4*). *E*, *Lepisosteus osseus*, as in *A*. *F*, *Amia calva*, as in *A*; from Allis, (1897, fig. 49).

4 B, D, 5 B), the pharyngobranchial is of a simpler form, seeming to lack the typical fork and contact with the epibranchial in front. Others (Fig. 5 E, F) seem intermediate in tending to have an anterior (second) pharyngobranchial more nearly forked than a posterior (third). Most of those forms having simple (unforked) pharyngobranchials have them contacting the cranium (*Eusthenopteron:* Jarvik, 1954: 35; sturgeons: Van Wijhe, 1882: 226; Holmgren, 1943: 45), whereas a forked pharyngobranchial rarely contacts the cranium (as apparently secondarily e.g., in the teleost *Osteoglossum*; Nelson, 1968). Possibly contact either with the epibranchial in front or with the cranium is an alternative source of mechanical support for the dorsal gill-arch elements. But it is not clear which of these should be considered the more primitive. Furthermore, the primitive nature of the relation between the axial and visceral endoskeletons of vertebrates

never has been clarified (Gaupp, 1905: 848): whether the visceral arches originally segmented from, or alternatively grew up toward, the cranium remain equally plausible but mutually contradictory hypotheses (see e.g., de Beer, 1937: 408; and Jarvik's 1954: 89–95 comparison of *Eusthenopteron* and a cephalaspid).

Among actinopterygian fishes, however, it does seem that the tendency has been toward reduction and loss of suprapharyngobranchials, and loss of contacts between the gill-arch skeleton and cranium. These possibly were necessary stages in the development of movable upper pharyngeal bones (see e.g., Nelson, 1967c). Thus within teleostomes it might be argued that suprapharyngobranchials and their contact with the cranium are relatively primitive characters. Possibly primitive also are the simple infrapharyngobranchials and their contact with the cranium. If so, the pharyngobranchials of *Acanthodes* would be of a more complex structure than those of primitive teleostomes and their various processes could not be directly compared with those of the infrapharyngobranchial of e.g., *Elops* (Fig. 4 C). If, as suggested by both Reis (1896: 159) and Stensiö (1963: 410), the *Acanthodes* pharyngobranchial is posteriorly directed, it would appear to be comparable with that of Recent elasmobranchiomorphs. The hypothesis of a posterior orientation here is supported by three observations: (1) The apparent absence of a pharyngobranchial associated with, and located anterior to, the first epibranchial. (2) The double articulation of the pharyngobranchial with the epibranchial in front. (3) The apparent absence of an articulation with the epibranchial behind.

Thus, being assignable to the arch in front, the dorsal process possibly could correspond to the suprapharyngobranchial of crossopterygians and primitive actinopterygians, as Miles (1964; 1965) thought it was. In these fishes the suprapharyngobranchial occurs sometimes in association with a dorsally directed process of the epibranchial. However, in the first arch but not the second, of *Eusthenopteron* (Fig. 5 B), and also in *Latimeria* (Fig. 5 A), the supra- appears to be fused with the infrapharyngobranchial. Because in fishes the structurally primitive condition in general seems to be better preserved in the anterior rather than the posterior arches, one might suspect that this compound pharyngobranchial might be primitive relative to the usually separate condition of supra- and infrapharyngobranchials. If so, *Acanthodes*, in having these elements united in at least arches one-three, would exhibit the most primitive condition yet known among gnathostomes. If not homologous to the suprapharyngobranchial, the dorsal process of *Acanthodes*, would appear to have no equivalent among teleostomes.

Suprapharyngobranchials among Recent fishes

The interpretation of the *Acanthodes* pharyngobranchial is complicated by the theory of Allis (1925), Holmgren (1942: 143; 1943: 45), and others that the pharyngobranchials of Recent elasmobranchs are homologous only with the suprapharyngobranchials of teleostomes, and that supra- and infrapharyngobranchials in some way may be exceedingly and equally primitive elements of the vertebrate visceral endoskeleton (see also Schmalhausen, 1923: 539; 1950: 437–440; Jarvik, 1954; 1960: 78–88).

It was generally assumed by early anatomists that the elements of a single arch most primitively were aligned in a single plane. Gegenbaur (1872: 153), for example, believed that the elasmobranch hypo- and pharyngobranchials secondarily became posteriorly directed as a result of mechanical forces generated during the process of swallowing. Allis (1925: 76) and later Holmgren (1940; 1942; 1943), rejected the idea of a secondary change in orientation of the pharyngobranchial but not of the hypobranchial (Holmgren, 1942: 143). Allis considered the condition in elasmobranchs to be the more primitive, and the infrapharyngobranchial of teleostomes to have arisen by segmentation from the medial end of the epibranchial. Holmgren (1943: 47) also believed the infrapharyngobranchial to be represented in elasmobranchs, either by the medial end of the epibranchial or by the interarcual ligament, which according to him sometimes chondrifies.

That the interarcual cartilages of Holmgren (see also Allis, 1915: 577–580), probably are secondary structures is suggested by the apparent ease by which such rudimentary cartilages are generated in Recent fishes. In addition to those discussed below, a good example is the secondary segmentation of the third infrapharyngobranchial in some teleostean fishes of the family Osteoglossidae (Nelson, 1968; figs. 5–6). A similar condition occurs in the second or third arch of some salmonids (e.g. *Osmerus* and *Salmo*, personal observations). Finally may be mentioned the peculiar cartilages associated with the medial ends of the epibranchials of *Neoceratodus* (Fig. 5 D). These to the present author seem anatomically to be independent of the branchial endoskeleton, for they underlie the medial ends of the epibranchials. Sometimes, however, they have been termed (infra-) pharyngobranchials (Van Wijhe, 1882: 296; Fürbringer, 1904: 488, fig. 29).

The arches of *Neoceratodus*, it may be added, are peculiar in having the epibranchials posteriorly directed. This orientation convinced Allis (1915: 602–603) that they "conclusively ... are pharyngobranchials, and that the epibranchials have not yet segmented off from the dorsal ends of the ceratobranchials." Although the arches of *Neoceratodus* secondarily are reduced relative to those usually present in fishes, and those of other lungfishes even more so (see e.g., Kisselewa, 1929; figs. 5, 15, 21), there is no reason to doubt that the elements in *Neoceratodus* simply are posteriorly directed epibranchials, and that orientation of gill-arch elements sometimes varies among Recent fishes (see below).

The present author has been unable to find any evidence except that of orientation to support the view that the elasmobranch pharyngobranchial is homologous only with the teleostome suprapharyngobranchial. And even the orientation of the elasmobranch pharyngobranchial does not entirely agree with that of the suprapharyngobranchial. In adults, the former is directed posteromedially, supporting the pharyngeal roof; the latter is directed dorsally, attaching to the lateral wall of the otic region of the cranium. That evidence of orientation alone is insufficient to support the view of Allis and Holmgren is evident from the variability in orientation of gill-arch elements in Recent fishes.

For the above reasons it is here suggested that the pharyngobranchials of Recent elasmobranchs and the infrapharyngobranchials of teleostomes—which support the pharyngeal roof and denticles, teeth and tooth plates when present —be considered homologous structures. Presumably, the pharyngobranchials of *Acanthodes* supported the pharyngeal roof, and should be considered also a probable homologue of the elasmobranch-teleostome pharyngobranchial.

It is further suggested that the term suprapharyngobranchial be restricted to that element which extends dorsally, generally from the epibranchial and articulates with the otic region of the endocranium. This element, not known ever to have been associated with dermal elements, seems to be restricted to rhipidistians (*Eusthenopteron*: Fig. 5 B, see also Jarvik 1954), actinistians (*Latimeria*: Fig. 5 A, see also Millot & Anthony, 1958: 53), primitive actinopterygians (*Polyodon*: Bridge, 1879: fig. 9; sturgeons: Fig. 4 B, see also Van Wijhe, 1882, pl. 15, fig. 2; *Pteronisculus*: Nielsen, 1942: 187–195; *Australosomus*: Fig. 4 D, see also Nielsen, 1949: 121–128; *Lepisosteus*: Fig. 5 E, see also Hammarberg, 1937), and teleostean fishes of the families Elopidae (Fig. 6 A), Alepocephalidae (Fig. 6 B) and possibly as a rudimentary cartilage in the first arch of the Albulidae, Pterothrissidae and Aulostomidae (personal observations).

The term "suprapharyngobranchial" erroneously has been applied to the median element in the first arch of some clupeids (by Moona, 1963), to a cartilage between the first epi- and second infrapharyngobranchial of some acanthopterygian fishes (by Allis, 1903: 124), and to the dorsally directed process ("uncinate process" of Harrington, 1954; 284) often present on actinopterygian epibranchials (Holmgren, 1943: 47, Bertmar, 1959: 305). The element of clupeids has been discussed elsewhere (Nelson, 1967a: 391), that in acanthopterygians later by Allis (1915: 580) himself, neither of these elements apparently being suprapharyngobranchials. Allis's element called by him an "interarcual cartilage," seems to have originated secondarily and probably allows greater movement between the first epibranchial and second infrapharyngobranchial. If so, it may have developed in relation with movable upper pharyngeal bones. It is common in perciform fishes, e.g. *Epinephelus* (Fig. 6 D). That the uncinate process also does not correspond to a suprapharyngobranchial is suggested by

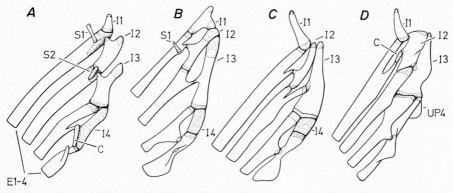

Fig. 6. A, *Elops hawaiiensis*, gill-arch skeleton, dorsal view of dorsal parts of left side. *C*, a presumably secondary "interarcual cartilage" (see text); *E1–4*, epibranchials 1–4; *I1–4*, infrapharyngobranchials 1–4; *S1–2*, suprapharyngobranchials 1–2. B, *Alepocephalus macropterus*, as in *A*. C, *Hime japonica*, as in *A*. D, *Epinephelus hexagonatus*, as in *A*. *UP4*, upper pharyngeal tooth plate 4.

two observations: firstly, in the anterior arches the uncinate process usually terminates in a cartilaginous surface articulating with the lateral process of the infrapharyngobranchial behind. This connection appears to be a primitive feature of the actinopterygian branchial apparatus, and usually occurs whether or not an uncinate process or a suprapharyngobranchial is developed. Secondly, both an uncinate process and a suprapharyngobranchial sometimes occur on the same arch, e.g., in *Alepocephalus* (Fig. 6 B). The loss of suprapharyngobranchials, and the origin of an uncinate process and Allis's element, seem to be secondary modifications of the actinopterygian branchial apparatus, and seem to be shown in the series: *Elops-Alepocephalus-Hime-Epinephelus* (Figs. 6 A–D), which in these respects the present author believes approximates an evolutionary series.

If the above interpretations are correct, suprapharyngobranchials might have arisen in the immediate ancestors of the crossopterygians and actinopterygians, possibly as secondary features associated with the crowding together of the gill arches beneath the cranium, hyoid arch and associated dermal bones. If so, the occurrence of suprapharyngobranchials would indicate a close phylogenetic relationship between these two groups. That the independent suprapharyngobranchial may have arisen by a secondary segmentation of a pharyngobranchial of the type found in *Acanthodes* has been suggested above.

Orientation of gill-arch elements

It is frequently stated that the elasmobranchs have the endoskeletal elements of a typical arch when viewed from the side arranged in a sigma (Σ) and the teleostomes in a lazy V ($>$). However, among various groups of apparently specialized teleostomes one or more of the arches are secondarily rearranged such that either hypo-, epi-, or infrapharyngobranchials are posteriorly directed (see e.g., Nelson, 1966: 398, figs. 26, 30; 1967b: 292, figs. 2–4; 1968, fig. 10).

Among elasmobranchs, the hypobranchials of the first arch are anteriorly directed according to Holmgren (1940: 106, 141, 169, 188, 195, 218, 227, figs. 138, 158; 1942: 143) and according to most authors (Gegenbaur, 1898, fig. 268; Holmgren, 1942: 207, fig. 34; Devillers, 1958, fig. 377) so they are also in most or all of the arches of chimaeras. Thus it is possible that among elasmobranchiomorphs the primitive orientation of the hypobranchials was as among teleostomes, and that the change in orientation may have followed the general rule and have proceeded from behind forwards.

Among elasmobranchiomorphs, the pharyngobranchials are posteriorly directed regardless of their relative position (some chimaeras have the gill arches beneath rather than behind the cranium). The acanthodians agree with the elasmobranchs in having the gill arches posterior in position and the pharyngobranchials posteriorly directed. If secondary, one or both of these resemblances may be phyletically significant, indicating a relationship with elasmobranchs.

Final remarks

For the reasons given above there seems to be little agreement in advanced gill-arch characters between *Acanthodes* and primitive teleostomes. Such characters as they have in common seem to be primitive ones to be expected also in primitive elasmobranchiomorphs. If so, they are of little phyletic significance. In short, gill-arch structure of *Acanthodes*, in so far as it is known, seems to lend no support to the theory of a relationship with early teleostomes. Rather the balance, small though it is, seems to favor elasmobranchiomorphs.

As already noted by Dollo (1906), *Acanthodes* seems to be without teeth on the jaws, and tooth plates on the gill arches. And in comparison with such forms as *Climatius* and *Ischnacanthus* it might well represent a line of specialization involving the adoption of microphagous habits. Gillrakers, for example, are well developed and numerous, and apparently extend far medially both on the ventral and dorsal parts of the gill arches. Thus gill-arch structure of *Acanthodes* is not necessarily the same as it was in the earliest and most primitive acanthodians. Nevertheless, so far as known, there is no character of the endoskeletal gill arches of *Acanthodes* which clearly removes the acanthodians from an early or for that matter, ancestral phyletic position relative to the elasmobranchs.

Finally, the present author admits to being impressed by the occurence of gillrakers on both the ventral (hypo- and ceratohyals) and dorsal (epihyal = hyomandibula) parts of the hyoid arch of *Acanthodes* as an indication of a primitive condition and a possible hyoidean gill slit. Despite Holmgren's statement to the contrary, the present author has been unable to find any equivalent situation in Recent fishes, including the chimaeras and catfishes. Holmgren, in stating that hyoidean gillrakers occur in these Recent fishes seems to have referred to fleshy papillae rather than ossified gillrakers. Also, it appears that the hyoidean rakers of *Acanthodes* project anteriorly as Watson figured them. If this is true they would seem to represent the anterior not the posterior set of rakers often present on teleostome gill arches, as Holmgren (1942: 145) believed. From a functional point of view, it is difficult to interpret this condition except as an indication of an at least partially open hyoidean gill slit (cf. Miles, 1964: 4–5; 1965: 238).

This conclusion is supported to some extent by the apparently primitive organization of the hyoid arch of *Acanthodes* (Fig. 3), with separate hypo-, cerato-, epi- and pharyngohyals, and possibly one other element (the "stylohyal" of Miles, 1965: 241).

In contrast, both elasmobranchs, usually with only one paired element ventrally in the hyoid arch, and teleostomes, usually with three or more, may be secondarily modified. Whether the single element of elasmobranchs represents one or both of the elements of *Acanthodes* is difficult to say. The teleostomes, however, seem to have retained both of the elements of *Acanthodes* and added either one (in chondrosteans and holosteans) or two (in teleosteans) apparently new elements, the so-called "hypohyals" (see e.g., McAllister, 1968). However, this may be, *Acanthodes*, in having the ventral part of the hyoid arch subdivided in much the same manner as the succeeding branchial arches, in this respect seems to approach a condition apparently ancestral for gnathostomes as a whole.

But however primitive the acanthodians might have been, their status as fishes seems secure. And as a group of fishes they must be related to something in the Recent fauna. If forced to choose between elasmobranchiomorphs and teleostomes, only on the basis of gill-arch structure of *Acanthodes*, the present author for the reasons mentioned above, would have to side with Reis, Dean and the elasmobranchiomorphs.

This study was done in the Section of Palaeozoology of the Swedish Museum of Natural History, Stockholm, while the author held a National Science Foundation postdoctoral fellowship. Some information on the Recent fishes is from a hitherto unpublished part of a dissertation submitted to the Graduate Division, University of Hawaii, in partial fulfillment of the requirements for the degree of Doctor of Philosophy.

References

Allis, E. P. Jr. (1897). The cranial muscles and cranial and first spinal nerves in *Amia calva*. *J. Morph.*, **12**: 487–772.
Allis, E. P. Jr. (1903). The skull, and the cranial and first spinal muscles and nerves in *Scomber scomber*. *J. Morph.*, **18**: 45–328.
Allis, E. P. Jr. (1915). The homologies of the hyomandibula of the gnathostome fishes. *J. Morph.*, **26**: 563–624.
Allis, E. P. Jr. (1925). On the origin of the V-shaped branchial arch in the Teleostomi. *Proc. zool. Soc. Lond.* **1925**: 75–77.
Bertmar, G. (1959). On the ontogeny of the chondral skull in Characidae, with a discussion on the chondrocranial base and the visceral chondrocranium in fishes. *Acta. zool. Stockh.*, **40**: 203–364.
Bridge, T. W. (1879). On the osteology of *Polyodon folium*. *Phil. Trans. R. Soc.*, **169**: 638–733.
Corsy, F. (1933). *Évolution de l'appareil hyo-branchial.* Marseille: Ciarfa.
Daget, J. (1964). Le crâne des téléostéens. *Mem. Mus. Hist. nat., Paris*, (A) **31**: 163–341.
Dean, B. (1907). Notes on acanthodian sharks. *Am. J. Anat.*, **7**: 209–222.
de Beer, G. R. (1937). *The development of the vertebrate skull.* Oxford: Oxford University Press.
Devillers, C. (1958). Le crâne des poissons. In *Traité de Zoologie*, ed. Grassé, P.-P., **13:1**: 551–687. Paris: Masson.
Dollo, L. (1906). Sur quelques points d'éthologie paleontologique relatifs aux poissons. *Bull. Soc. belg. Géol. Paléont. Hydr.*, **20**: 1–3.
Fox, H. (1965). Early development of the head and pharynx of *Neoceratodus* with a consideration of its phylogeny. *J. Zool.*, **146**: 470–554.
Fürbringer, K. (1903). Beiträge zur Kenntnis des Visceralskelets der Selachier. *Morph. Jb.*, **31**: 360–442.
Fürbringer, K. (1904). Beiträge zur Morphologie des Skelets der Dipnoer nebst Bemerkungen über Pleuracanthiden, Holocephalen und Squaliden. In Zoologische Forschungsreisen in Australien und der Malayischen Archipel, hrsg. Semon, R., Vol. **1:1**: 423–510; *Denkschr. med.-naturw. Ges. Jena*, **4**.
Garman, S. (1913). The plagiostomia. (Sharks, skates, and rays.). *Mem. Mus. comp. Zool. Harv.*, **36**: i–xiii, 1–515.
Gaupp, E. (1905). Das Hyobranchialskelet der Wirbelthiere. *Ergeb. Anat. Entw-Gesch.*, **14**: 808–1048.
Gegenbaur, C. (1872). *Untersuchungen zur vergleichenden Anatomie der Wirbelthiere. 3. Das Kopfskelet der Selachier: ein Beitrag zur Erkenntniss der Genese des Kopfskeletes der Wirbelthiere.* Leipzig: Engelmann.
Gegenbaur, C. (1898). *Vergleichende Anatomie der Wirbelthiere mit Berücksichtigung der Wirbellosen.* Vol. **1**. Leipzig: Engelmann.
Gregory, W. K. (1933). Fish skulls. A study of the evolution of natural mechanisms. *Trans. Am. phil. Soc.*, (N.S.) **23**: 75–481.
Hammarberg, F. (1937). Zur Kenntnis der ontogenetischen Entwicklung des Schädels von *Lepidosteus platystomus*. *Acta. zool. Stockh.*, **18**: 209–337.
Harder, W. (1964). Anatomie der Fische. In *Handbuch der Binnenfischerei Mitteleuropas*, hrsg. Demoll, R., Maier, H. N. & Wundsch, H. H., **2:A**: 1–308. Stuttgart: Schweizerbart.
Harrington, R. W. Jr. (1955). The osteocranium of the American cyprinoid fish *Notropis bifrenatus*, with an annotated synonymy of teleost skull bones. *Copeia*, **1955**: 267–290.
Holmgren, N. (1940). Studies on the head in fishes. Embryological, morphological

and phylogenetical researches. 1. Development of the skull in sharks and rays. *Acta zool. Stockh.*, **21**: 51–267.

Holmgren, N. (1942). Studies on the head of fishes. An embryological, morphological and phylogenetical study. 3. The phylogeny of elasmobranch fishes. *Acta zool. Stockh.*, **23**: 129–261.

Holmgren, N. (1943). Studies on the head of fishes. An embryological, morphological and phylogenetical study. 4. General morphology of the head in fish. *Acta zool. Stockh.*, **24**: 1–188.

Holmgren, N. & Stensiö, E. A. (1936). Kranium und Visceralskelett der Akranier, Cyclostomen und Fische. In *Handbuch der vergleichenden Anatomie der Wirbeltiere*, hrsg. Bolk, L., Göppert, E., Kallius, E. & Lubosch, W., **4**: 233–500. Berlin & Wien: Urban & Schwarzenberg.

Jarvik, E. (1954). On the visceral skeleton in *Eusthenopteron* with a discussion of the parasphenoid and palatoquadrate in fishes. *K. svenska VetenskAkad. Handl.*, (4) **5**: 1–104.

Jarvik, E. (1960). *Théories de l'évolution des vertébrés reconsidérées a la lumière des récentes découvertes sur les vertébrés inférieurs.* Paris: Masson.

Kisselewa, Z. N. (1929). Zur vergleichend anatomischen Kenntnis des Skelets der Dipnoi. *Trud. Inst. zool. Leningr.*, **3**: 1–44. In Russian.

Lehman, J.-P. (1949), Étude d'un *Pachycormus* du Lias de Normandie. *K. svenska VetenskAkad. Handl.*, (4) **1**: 1–44.

McAllister, D. E. (1968). The evolution of branchiostegals and associated gular, opercular and hyoid bones in teleostome fishes. *Bull. natn. Mus. Canada.* In press.

Miles, R. S. (1964). A reinterpretation of the visceral skeleton of *Acanthodes*. *Nature, Lond.*, **204**: 457–459.

Miles, R. S. (1965). Some features of the cranial morphology of acanthodians and the relationships of the Acanthodii. *Acta zool. Stockh.*, **46**: 233–255.

Millot, J. & Anthony, J. (1958). *Anatomie de Latimeria chalumnae. 1. Squelette, muscles et formation de soutien.* Paris: Cent. natn. Rech. Scient.

Moona, J. C. (1963). Studies on the cranial osteology of the Indian clupeoid fishes. 4. The skull of *Nematalosa nasus* (Bloch). *J. Morph.*, **113**: 345–358.

Nelson, G. J. (1966). Gill arches of teleostean fishes of the order Anguilliformes. *Pacific Sci.*, **20**: 391–408.

Nelson, G. J. (1967a). Gill arches of teleostean fishes of the family Clupeidae. *Copeia*, **1967**: 389–399.

Nelson, G. J. (1967b). Gill arches of some teleostean fishes of the families Girellidae, Pomacentridae, Embiotocidae, Labridae and Scaridae. *J. nat. Hist.*, **1**: 289–293.

Nelson, G. J. (1967c). Epibranchial organs in lower teleostean fishes. *J. Zool.*, **153**: 71–89.

Nelson, G. J. (1968). Gill arches of some teleostean fishes of the division Osteoglossomorpha. *J. Linn. Soc. (Zool.).* In press.

Nielsen, E. (1942). Studies on Triassic fishes from East Greenland. 1. *Glaucolepis* and *Boreosomus*. *Palaeozool. Groenland.*, **1**: 1–403.

Nielsen, E. (1949). Studies on Triassic fishes from East Greenland. 2. *Australosomus* and *Birgeria*. *Palaeozool. Groenland.*, **3**: 1–309.

Reis, O. M. (1890). Zur Kenntnis des Skelets der Acanthodinen. *Geogn. Jahresh.*, **3**: 1–43.

Reis, O. M. (1894). Ueber ein Exemplar von *Acanthodes bronni* Ag. aus der geogn. Sammlung der "Pollichia". *Mitt. Pollichia*, **53**: 316–332.

Reis, O. M. (1895). Illustrationen zur Kenntnis des Skeletts von *Acanthodes bronni* Agassiz. *Abh. senckenb. naturf. Ges.*, **19**: 49–64.

Reis, O. M. (1896). Ueber *Acanthodes bronni* Agassiz. *Morph. Arb.*, **6**: 143–220.

Schmalhausen, I. I. (1923). Das Suspensorialapparat der Fische und das Problem der Gehörknöchelchen. *Anat. Anz.*, **56**: 534–543.
Schmalhausen, I. I. (1950). O prikrieplenii vistseralnuikh dug k osevomu cherepu u ruib (On the attachment of the visceral arches to the neurocranium in fishes). *Zool. Zh.*, **29**: 435–448. In Russian.
Stensiö, E. A. (1963). Anatomical studies on the arthrodiran head. 1. Preface, geological and geographical distribution, the organization of the arthrodires, the anatomy of the head in the Dolichothoraci, Coccosteomorphi and Pachyosteomorphi. Taxonomic appendix. *K. svenska VetenskAkad. Handl.*, (4) **9**: 1–419.
Watson, D. M. S. (1937). The acanthodian fishes. *Phil. Trans. R. Soc.* (B) **228**: 49–146.
Watson, D. M. S. (1959). The myotomes of acanthodians. *Proc. R. Soc. Lond.*, (B) **151**: 23–25.
Wijhe, J. W. van (1882). Ueber das Visceralskelett und die Nerven des Kopfes der Ganoiden und von *Ceratodus*. *Niederl. Arch. Zool.*, **5**: 207–320.

The spinal plate in *Homostius* and *Dunkleosteus*

By Anatol Heintz

Palaeontological Museum, Oslo, Norway

In 1959 Stensiö introduced a new systematical division of the Euarthrodira. Instead of dividing them into the order Dolichothoraci, composing older and more "primitive" forms, and the order Brachythoraci, including younger and more "advanced" types, as approved by many other specialists, Stensiö proposed two new superorders, Aspinothoracidi and Spinothoracidi.

Stensiö based his new systematical division on a study of the structure of the pectoral girdle and the pectoral fins in different Arthrodira. According to him, the superorder Aspinothoracidi must be regarded as a more primitive group in spite of the fact that it comprises mostly the Middle and Upper Devonian forms. A characteristic feature of this superorder is *inter alia* that all forms belonging to it primarily lack the spinal plate.

All forms belonging to the second superorder—Spinothoracidi—possess a more or less well developed spinal plate. According to Stensiö this superorder embraces one group of Brachythoraci (Coccosteomorphi), all Dolichothoraci and some smaller groups of Euarthrodira. Although all these forms are known from the Lower and partly Middle Devonian, Stensiö believes that the Spinothoracidi are more advanced forms.

Stensiö thus emphasizes the presence or absence of the spinal plate and his systematical division of the Euarthrodira is to a great extent based on this character.

Earlier than 1959, the spinal plate was only mentioned in one form belonging to Stensiö's superorder Aspinothoracidi, namely in *Dunkleosteus* from the Upper Devonian of USA (Heintz, 1932). According to Stensiö (1959), however, the spinal plate in *Dunkleosteus* is not homologous with the spinal plate in the Spinothoracidi because it has another origin and lacks an inner cavity for the endoskeletal part of the shoulder girdle. Stensiö therefore called the spinal plate in *Dunkleosteus* prepectoral or pseudo-spinal.

In 1963 Mark-Kurik described two spinal plates in *Homostius* from Estonia. She compared the spinal plate of *Homostius* with that of *Plourdosteus* (Gross, 1938) and of *Dunkleosteus* (Heintz, 1932) and found that it shows great structural similarity in all three forms (Mark-Kurik, 1963, figs. 3, 4).

A reconstruction in side view of the body carapace in *Homostius* given by

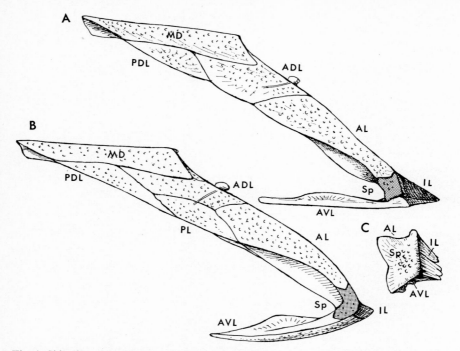

Fig. 1. Side view reconstruction of the body-carapace of *Homostius*; *A*: according to Heintz; *B*: according to Mark-Kurik (1963). *C*, spinal plate of *Homostius* (from Mark-Kurik, 1963).
ADL, antero-dorso-lateral; *AL*, antero-lateral; *AVL*, antero-ventro-lateral; *IL*, intero-lateral; *PDL*, postero-dorso-lateral; *PL*, postero-lateral; *Sp*, spinal plate.

Mark-Kurik (1963, fig. 7 A; Fig. 1 B in this paper) shows the position of the spinal plate in relation to other plates of the body-carapace. In my opinion, however, this reconstruction is not wholly successful in the following two respects:

(1) The spinal plate is depicted far too large and strongly bent as compared with the figures given elsewhere by Mark-Kurik (1963, figs. 1, 3, 4 A, B; pl. 1, figs. 1–5; Fig. 1 C in this paper). Also the AVL plate seems to be too strongly curved in the front part. In fact the AVL in *Homostius* is unusually flat (Heintz, 1933, figs. 42–45.)

(2) Mark-Kurik has depicted a rather large PL plate in her reconstruction of *Homostius* (1963, fig. 7 A; Fig. 1 B in this paper). This plate has never before been found in *Homostius* and I have reason to doubt its existence. To my knowledge no traces of overlapping-margins caused by this plate have ever been found on the adjacent plates (ADL, PDL and AL; Heintz, 1933). The AL plate in *Homostius* is somewhat modified and partly reduced, and it is therefore rather unexpected to find a large PL plate in connection with it. Furthermore, the PL plate does not exist in any of the well preserved, complete specimens of *Homostius milleri* from Scotland. It would be rather remarkable if two so closely

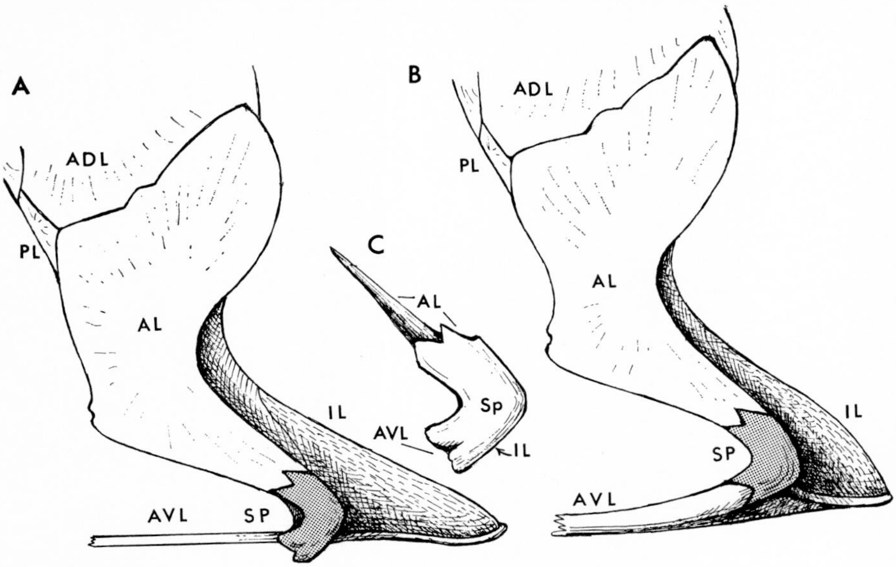

Fig. 2. Reconstruction in side view of the lower part of the body-carapace of *Dunkleosteus*; *A*: according to Heintz (1932); *B*: according to a sketch made by Heintz in the Cleveland Museum of Natural History. *C*, spinal plate of *Dunkleosteus* (from Heintz, 1932).
Abbreviations as in Fig. 1.

related *Homostius* species as those from Estonia and Scotland should differ in such an important way.

After studying casts of the spinal, antero-lateral and intero-lateral plates of *Homostius* at my disposal I have therefore arrived at a somewhat different reconstruction of the lower part of the body carapace in this form (Fig. 1 A).

After describing the spinal plate of *Homostius* and comparing it with those of *Plourdosteus* and *Dunkleosteus*, Mark-Kurik concludes, reasonably enough, that the resemblance between these plates is so great that they undoubtedly are homologous with, and of the same origin as, the spinal plates in all other Spinothoracidi. Thus they cannot be regarded as pseudo-spinals.

In the side view reconstruction of the carapace of *Dunkleosteus*, given by me (Heintz, 1932, fig. 67; Fig. 2 A in this paper), the position of the spinal plate is somewhat misleading. Its postero-ventral corner protrudes downwards and does not come in contact with the AVL plate but lies below it. It is possible that just this incorrect reconstruction has mislead Stensiö into regarding the spinal plate of *Dunkleosteus* as a "pseudo-spinal".

However, among the number of sketches of *Dunkleosteus* which I made in the Cleveland Museum of Natural History in 1930, there is a more correct drawing of the connection between the spinal plate and the AL, AVL and IL plates. It can be seen from Fig. 2 B that the postero-ventral corner of the spinal plate really comes in contact with the antero-lateral corner of the AVL—exactly as in all other Euarthrodira with well developed spinals.

In my opinion, there can no longer be any doubt than the bone named spinal in *Dunkleosteus* (Heintz, 1932) and *Homostius* (Mark-Kurik, 1963) is homologous with the spinal plate in other Euarthrodira.

Accordingly, it seems that in two very important but different forms included in Stensiö's Aspinothoracidi, the spinal plate is present. Thus the systematical division of Euarthrodira as proposed by Stensiö is questionable, and it is more reasonable to accept the more recent classification given by Obruchev (1964).

References

Gross, W. (1938). Über das Spinale und die angrenzenden Knochen der Brachythoraci. *N. Jb. Min. Geol. Paläont., Beil.-Bd.*, **79**, B: 403–418.
Heintz, A. (1932). The structure of *Dinichthys*, a contribution to our knowledge of the Arthrodira. *Am. Mus. nat. Hist., B. Dean Mem. Vol.*, **4**: 115–224.
Heintz, A. (1933). Revision of the Estonian Arthrodira. 1. Family *Homostiidae* Jaekel. *Eesti Loodustead. Arh.* (1) **10**: 1–115.
Mark-Kurik, Elga (*née* Mark). (1963). On the spinal plate of Middle Devonian arthrodire *Homostius*. *Trud. Inst. geol. Acad. Sci. Est. SSR*, **13**: 189–200. In Russian with English summary.
Obruchev, D. V. (1964). Klass Placodermi. Plastinokozhie. In *Osnovui paleontologii (Fundamentals of Palaeontology)*, ed .Orlov, I. A., **11**: 118–178. Moscow: Acad. Sci. U.S.S.R. In Russian.
Stensiö, E. A. (1959). On the pectoral fin and shoulder girdle of the Arthrodires. *K. svenska VetenskAkad. Handl.*, (4) **8**: 1–229.

Discussion

E. A. Stensiö

Our knowledge of the arthrodires still is very incomplete. Like all other classifications of the arthrodires, that proposed by myself (Stensiö, 1959; 1963; 1968a; 1968b) therefore is a temporary arrangement which should be changed on the basis of new evidence.

According to Heintz it is doubtful whether the classification of the two genera *Homostius* and *Dunkleosteus* suggested by me can be retained. In reply it should first be pointed out that the Pachyosteomorphi may turn out to be a heterogenous order and that the orders Pachyosteomorphi and Coccosteomorphi, the former of which includes all the forms so far known of the Aspinothoracidi, are not based solely on the structure of the shoulder girdle and pectoral fin but also on several other important anatomical characters in the cranium.

The most important characters of the better known pachyosteomorphs are briefly as follows (Stensiö, 1959; 1963; 1968a; 1968b). The otic region of the endocranium has undergone a considerable lengthening posterior to the glossopharyngeus canal and may be regarded as being composed of two divisions, a broad, preglossopharyngeal division which houses the labyrinth cavity and a narrower postglossopharyngeal division in the dorsal wall of which the endoskeletal part of the endolymphatic duct ran backwards. On each lateral face the latter division of the otic region exhibits a conspicuous cucullaris fossa for the insertion *inter alia* of the foremost trunk myomeres, a fossa which extends backwards to a point immediately behind the external opening of the vagus canal. The occipital region usually was shorter than the other three regions of the endocranium and at least in some forms its length may have been only about one half of that of the otic region; it possessed a powerful, paired supravagal process on the dorsal side of which the endolymphatic duct emerged from the endocranium and whence then it continued almost straight laterally through the paranuchal plate. Because of the small length of the occipital region the nuchal and paranuchal plates both are short. With rare exceptions (*Leptosteus*) the lateral wall of the exoskeletal shoulder girdle is short with a more or less conspicuous posterior emargination for the pectoral fin. The thoracic division of the pectoral fin frequently was persisting far forwards along the corresponding division of the scapulo-coracoid, and in consequence a prepectoral spine (including both endoskeleton and a spinal plate) is absent. In those forms where the thoracic division of the fin had undergone a more or less extensive reduction from in front backwards the radialia have disappeared entirely and thus have not fused with the scapulo-coracoid. There is reason to believe, however, that the original scales of the reduced foremost part of the fin in some forms may have fused to form a spinal plate without endoskeleton and that this spinal plate hence has arisen independently of that in the Spinothoracidi. Under all conditions there naturally existed in the early ancestors of the Pachyosteomorphs all the necessary prerequisites for the evolution of a prepectoral spine consisting of both endoskeleton and exoskeleton.

The Coccosteomorphi (as defined in Stensiö, 1959; 1963; 1968*a*; 1968*b*) differ considerably from the Pachyosteomorphi in the following respects. The otic region was of the ordinary broad not particularly long type in that its postglossopharyngeal division had not undergone any lengthening whatever. On the other hand, the occipital region was long in proportion to the other three regions of the endocranium; in several forms this region was even very long, its length amounting to about one half or more of the total length of the endocranium. This region possessed a long, paired cucullaris fossa, which lay entirely posterior to the external opening of the vagus canal. The supravagal process has completely disappeared and has been replaced by a paired craniospinal process which, arising from the hindmost part of the occipital region far behind the vagus canal, formed the support for the posterior part of the exoskeletal skull roof. The endolymphatic duct emerged on the dorsal side of the endocranium approximately at the transition between the otic and occipital regions and thus far anterior to the posterior end of the endocranium. Posterior to its exit from the endocranium this duct then ran obliquely backwards in the paranuchal plate dorso-lateral to the occipital region. Owing to the lengthening of the latter region the nuchal and paranuchal plates also are long; in some forms their length may amount to about 50% or even more of the total length of the exoskeletal skull roof. The lateral wall of the exoskeletal shoulder girdle generally is long or fairly long and is perforated by a pectoral fenestra through which the pectoral fin articulated against the scapulo-coracoid, but in some forms it had undergone a reduction to such an extent as to be approximately as short as in the pachyosteomorphs and, like in these, has a more or less distinct posterior emargination for the pectoral fin. The thoracic division of this fin generally was represented only by a short posterior part, whereas the longer anterior part had modified to form a prepectoral spine consisting of both endoskeleton and of a spinal plate, the latter of which was wedged in between the dorsal and ventral cleithral plates. Some forms (*Rhachiosteus*) had lost the prepectoral spine altogether and were secondarily aspiniferous.

As may readily be gathered from what has been stated concerning their anatomical differences, the orders Pachyosteomorphi and Coccosteomorphi have to be retained, but their comprehension must naturally be changed according to our increasing knowledge of the individual genera. After all this may also apply to the two superorders Aspinothoracidi and Spinothoracidi, the latter of which *inter alia* includes the Coccosteomorphi.

Heintz has not mentioned that because of the length of the nuchal and paranuchal plates and for other reasons too in 1963 (Stensiö, 1963), without knowledge of Mark-Kuriks investigations (1963) and thus independently of her, I could establish that *Homostius* is a Coccosteomorph and must have possessed a spinal plate.

Concerning *Dunkleosteus* it seems now practically certain that this genus possessed a spinal plate. Moreover, as pointed out by me in two forthcoming works (Stensiö, 1968a; 1968b) *Dunkleosteus* also is reminiscent of the Coccosteidae in the following four respects: (1) the general shape of the biting part of the jaw bones; (2) the position of the worn areas on the jaw bones; (3) the presence of a long, deeply descending bone lamina on the basal face of the exoskeletal skull roof; and (4) the fact that this lamina possesses a powerful ventral postocular process (see Stensiö, 1963; 1968a; 1968b). On the other hand, *Dunkleosteus* resembled clearly the Pachyosteomorphi in two important characters, the small length (forward extent) of the nuchal and paranuchal plates which means that the occipital region was short and in the general shape of the exoskeletal shoulder girdle which is nearest as in the Leiosteids and Trematosteids. In view of the two latter characters I have in my forthcoming publications provisionally classified *Dunkleosteus* and related genera in a family of their own, Dunkleosteidae, amongst the Pachyosteomorphi. A definitive decision of the question where the dunkleosteids should be classified will probably be arrived at on the basis of the gigantic new material from the Cleveland shales recently collected by the Cleveland Museum of Natural History.

Mark-Kurik, Elga (*née* Mark). (1963). On the spinal plate of the Middle Devonian arthrodire *Homostius*. *Trud. Inst. geol. Acad. Sci. Est. SSR*, **13**: 189–200. In Russian with English summary.

Stensiö, E. A. (1959). On the pectoral fin and shoulder girdle of the arthrodires. *K. svenska VetenskAkad. Handl.*, (4) **8**: 1–229.

Stensiö, E. A. (1963). Anatomical studies on the arthrodiran head. 1. Preface, geological and geographical distribution, the organization of the arthrodires, the anatomy of the head in the Dolichothoraci, Coccosteomorphi and Pachyosteomorphi. Taxonomic appendix. *K. svenska VetenskAkad. Handl.*, (4) **9**: 1–419.

Stensiö, E. A. (1968*a*). Les arthrodires. In *Traité de Paléontologie*, ed. Piveteau, J., **2:2**. Paris: Masson. In press.

Stensiö, E. A. (1968*b*). Les arthrodires. Taxonomie. *Annls Paléont*. In press.

The bradyodont elasmobranchs and their affinities; a discussion

By Svend Erik Bendix-Almgreen
Mineralogical Museum, Copenhagen, Denmark

The group Bradyodonti was erected in 1921 by Woodward to include the four Palaeozoic families Cochliodontidae, Psammodontidae, Copodontidae and Petalodontidae. Most of the members of these families were known only by parts of their dentition. A common feature of all forms included was thought to be the tooth histology. A characteristic feature also seemed to be that the dentition consisted of a very few teeth or tooth plates, replacing each other extremely slowly and retained within the mouth during the complete life of the animal. Because of certain resemblances between the cochliodonts and some of the early members of the chimaeroids—more exactly the shape of the tooth plates and the presence of peculiar spines on the posterior end of the lower jaw—the bradyodonts were suggested to be intermediate between primitive elasmobranchs and the chimaeroids (Woodward, 1921: 34, 37).

Nielsen (1932) proposed to include the Edestidae and the Orodontidae in the bradyodont group because of the microscopic tooth structure. He was followed in this by Moy-Thomas and by most writers who later dealt with the bradyodont question (Berg, 1958; Gregory, 1951; Obruchev, 1953; 1964; Arambourg & Bertin, 1958; Patterson, 1965; etc.). Because of the wider comprehension thus given the bradyodont group, the character concerning slow tooth replacement was no longer valid; however, for practical reasons, the name bradyodonts was retained.

New evidence gained by palaeohistological investigations (Radinsky, 1961) has shown that the microscopic tooth structure cannot be used as a systematic character in the classification of elasmobranch fishes. This radically altered view on a feature which has played such a prominent role in the discussion of the bradyodonts raises the question whether or not the bradyodonts still can be maintained as a natural group. Recent investigations (Ørvig, 1967) have also thrown doubt upon the resemblances between the microscopic tooth structure of bradyodonts and of fossil and recent chimaeroids.

The hypothesis put forward by Dean (1906) and later elaborated by i.a. Holmgren (1942: 87–92) concerning the descent of the chimaeroids from ptyctodont arthrodires has recently been revived by Ørvig (1960; 1962) who maintains that there is no evidence of a connection between the Palaeozoic bradyodonts and the chimaeroids (Ørvig, 1962: 57; see also Stahl, 1967: 192).

The present paper aims at an analysis of the problems thus posed.

Taxonomic value of the histology and morphology of the dentition

Histologically, the teeth of bradyodonts are generally assumed to consist of tubular dentine. This kind of hard tissue, occurring in the coronal part of the teeth, is composed of nearly vertically arranged denteons containing each a central canal and separated by an interstitial hard tissue referred to as enamel (Nielsen, 1932; 1952), enamel-like hard tissue (Ørvig, 1951), or interosteonal tissue (Radinsky, 1961). The denteons are penetrated by dentinal tubules, the distal parts of which penetrate into the interstitial hard tissue.

The interstitial hard tissue shows in ordinary light a relatively homogeneous structure (Nielsen, 1932, pl. 9, figs. 1–2, pl. 11, figs. 2–3; Radinsky, 1961, pl. 7, fig. 16), while in polarized light interlacing bands joining each other at about right angles are seen (Radinsky, 1961, pl. 7, fig. 17). As the location and general character of this hard tissue is very like the interstitial hard tissue present in the recent dipnoan tooth plates, the ontogenetic development of which is well known (Lison, 1941), it most probably agrees with the latter in its ectomesenchymatic origin and its mode of formation.

The formation of the interstitial hard tissue, as it is known from the tooth plates of recent dipnoans, can briefly be described as follows. In the initial stages of mineralization of the organic matrix, lime salt crystallites are laid down in alignment with the collagen fibrils. During the subsequent stages the crystallites retain the same orientation as those first formed, even when the collagen fibrils are reduced. Thus there is reason to believe that the pattern of interlacing bands, characterizing the hard tissue when observed in polarized light, reflects the original orientation of the collagen fibrils (cf. Ørvig, 1967: 91–92). The process of mineralization, as well as the subsequent hypermineralization, takes place in a basal direction from the upper surface of the tooth plate. It continues uninterruptedly even in the adult fish to form successively deeper parts of the tooth plate. These processes result in the formation of the high columns of interstitial hard tissue between the tube-like denteons, characteristic of the microstructure of the complex dipnoan tooth plate.

In the bradyodont teeth, on the other hand, the interstitial hard tissue is restricted to the crown and shows a distinct boundary to the upper surface of the dentine of the root. This suggests that the process of mineralization and subsequent hypermineralization of the interstitial hard tissue ceased when fully developed at the upper surface of the root after having penetrated the complete height of the crown. This, however, does not impair the conclusion that the interstitial hard tissue of the bradyodont teeth probably formed by the same process of hypermineralization as in dipnoans.

The absence of a layer of pallial dentine in all hitherto known bradyodont teeth (Radinsky, 1961: 80; Patterson, 1965: 108) is probably also due to the process of mineralization and subsequent hypermineralization in the most superficial parts of the tooth papilla presumably commencing very early in ontogeny. If so, the most superficial parts of the hard tissue would during its development be affected by this process and thus be developed as an integral part of the interstitial hard tissue.

The name "coronal pleromic hard tissue" has been coined by Ørvig (1967) for the interstitial hard tissue in bradyodont teeth. It represents one of several kinds of "pleromic hard tissues" of which that occurring in the dipnoans already has been mentioned, another kind is found in the tritoral columns of the Devonian ptyctodonts and the recent chimaeroids (Ørvig, 1967: 89, 92–96). There can be little doubt of its nature as an adaptation to specific mechanical demands on the dental elements.

Among the elasmobranchs the "coronal pleromic hard tissue" (+denteons ="tubular dentine") is present not only in the bradyodonts, but also in the teeth of various Mesozoic selachians as e.g. *Asteracanthus* and *Ptychodus* (Radinsky, 1961, pl. 1, figs. 3–4, pl. 2, figs. 5–6, pl. 6, figs. 14–15). The occurrence of this pleromic hard tissue in elasmobranchs, as also in such a remote group as the dipnoans, does seriously impair the significance of this hard tissue in regard to classification. Concerning the bradyodonts, therefore, all forms known only by detached teeth must be left out of consideration in the present connection. This refers to the psammodonts, the copodonts and the cochliodonts (not including the genera *Helodus*, *Deltoptychius* and *Menaspis*; see Patterson, 1965) as well as many of the genera referred to the edestids. The possession of specialized symphyseal teeth is usually accepted as a general character of edestids. However, specializations of the symphyseal portion of the dentition to form curved tooth rows or tooth spirals are met with also in other groups of lower vertebrates e.g. acanthodians (Watson, 1937: 77, pl. 9, fig. 2) and certain crossopterygians (Jarvik, 1963, fig. 15 A, B; Jessen, 1966: 388, 394–395, figs. 84, 86). Recent investigations of *Helicoprion* (Bendix-Almgreen, 1966) have shown that this form is not closely related to *Fadenia* and *Sarcoprion* which are generally regarded as typical edestids. This indicates that the development of symphyseal teeth into spirals and curved tooth rows could have taken place independently within different elasmobranch lines. The only edestids to be considered here are *Fadenia*, *Sarcoprion*, *Erikodus*, *Ornithoprion* and *Agassizodus*, all represented also by parts of their cranial and postcranial skeleton. The term edestids in the previous sense is ambiguous, but is retained for practical reasons.

For similar reasons as with regard to the edestids, the majority of the genera placed as petalodonts are left out of consideration here. In the following this group will be discussed only on the basis of the genera *Janassa* and *Ctenoptychius* in which, apart from the dentition, some features are known of the anatomy and general body form.

Anatomical features

The only bradyodonts in which something is known of the anatomy are the following: *Chondrenchelys problematica, Helodus simplex, Menaspis armata, Deltoptychius armigerus, D.moythomasi, Janassa bituminosa, Ctenoptychius apicalis, Fadenia crenulata, Sarcoprion edax, Erikodus groenlandicus, Ornithoprion hertwigi* and *Agassizodus sp.* (Jaekel, 1891; 1899; Woodward, 1921; 1932; Weigelt, 1930; Nielsen, 1932; 1952; Moy-Thomas, 1935; 1936a; 1936b; 1939; Bendix-Almgreen, 1962; 1967; Patterson, 1965; Zangerl, 1966). According to Dr. C. Patterson (personal comm.) the interpretation of *Eucentrurus paradoxus* by Moy-Thomas (1937) as a probable relative of *Chondrenchelys problematica* is extremely doubtful. This form is therefore left out of consideration in the present connection.

Neurocranium and visceral endoskeleton. The preorbital portion of the neurocranium, consisting of the rostral and the ethmoidal regions, is relatively long in *Chondrenchelys, Fadenia, Sarcoprion* and *Ornithoprion*. An exact determination of the boundary between these regions is impossible in *Chondrenchelys*. In *Fadenia, Sarcoprion* and *Ornithoprion* the olfactory capsules are known, enabling us to state that the rostrum, which is not preserved in its anterior part in *Sarcoprion*, is completely closed, strong, gently pointed, passing over proximally into the ethmoidal region without any noticeable change in external appearance. The proximity of the olfactory capsules of both sides suggests that the precerebral cavity did not enter the rostrum in *Fadenia* and *Sarcoprion*, as is possibly the case also in *Ornithoprion*. In *Helodus* the preserved parts indicate that the preorbital portion of the neurocranium is high and comparatively short and broad. The position of the olfactory cavities is not known, but both the rostral and the ethmoidal regions are presumably short. On the preorbital portion there is a pair of dorso-lateral, vertically directed laminae. Moy-Thomas (1936a) interpreted these laminae as the lateral walls of an incomplete ethmoidal canal, an interpretation which is rather unlikely (Patterson, 1965: 187).

In *Chondrenchelys, Fadenia* and *Ornithoprion* the orbito-temporal region is comparatively short and high. Dorsally, it appears to be fairly broad in *Chondrenchelys* and comparatively narrow in the two other forms. *Helodus*, on the other hand, has an extremely long and broad but relatively low orbito-temporal region. A well developed preorbital process occurs in *Chondrenchelys* and *Helodus* and most probably also in *Fadenia* but seems to be absent in *Ornithoprion*. The orbit is large in all four forms, but the actual size of the eye is known only in *Fadenia*, where a large, well developed sclerotic ring of calcified cartilage is known (Nielsen, 1932: 49, pl. 2). In *Chondrenchelys*, *Helodus* and *Fadenia* the postorbital process is continued in a ventrally directed ridge, which passes over into the subotic shelf. Anteriorly this shelf is continuous with the subocular part of the palatoquadrate. The cranial cavity appears to be comparatively wide between the orbits in all three forms. *Ornithoprion* has a less well developed postorbital process, the ventral continuation of which forms the posterior wall of the orbit. An interorbital septum, probably membraneous, may be present in this form, indicating that the brain had a ventral position.

The otic and the occipital region of the neurocranium are short in *Chondrenchelys* and in *Helodus*. Judging from the descriptions and figures (Moy-Thomas, 1935; 1936*a*; Patterson, 1965) both regions are high in *Chondrenchelys*, while in *Helodus* they are low and broad, rapidly tapering posteriorly toward the area around the foramen magnum, ventro-laterally to which two occipital condyles are situated. In *Ornithoprion* the otico-occipital portion of the neurocranium is extremely short and in lateral view tapers rapidly in posterior direction. There is no indication of the boundary between the two regions. Two processes occur on both sides of the posterior part of the otico-occipital portion. The anterior of these is directed ventrally, articulating with the posterior part of the presumed palatoquadrate. Zangerl suggests (1966: 12) that this process represents "a hyomandibular that might have become fused to the neurocranium", but this is rather unlikely. The posterior process is directed caudally and situated laterally to the foramen magnum. The exact shape of the otic and the occipital region is unknown so far in *Fadenia*, *Erikodus* and *Sarcoprion*. However, judging from unpublished material, the otic region was almost as broad as the orbital region and comparatively short with the anterior semicircular canal situated just behind the postorbital ridge. The occipital region seems to be short and tapering rapidly posteriorly.

The palatoquadrates are completely fused with the neurocranium in *Chondrenchelys*, *Helodus*, *Fadenia*, *Sarcoprion* and *Erikodus*. In *Helodus* they are comparatively short, in contrast to the condition found in the other forms. The articulation for the lower jaw is situated below the orbital region (*Helodus*, *Chondrenchelys*, *Fadenia* and probably in *Sarcoprion* and *Erikodus*) or just behind that region (*Ornithoprion*). In *Ornithoprion* (and *Agassizodus*; Zangerl, personal comm.) the palatoquadrates are presumed to be independent elements which, at any rate in *Ornithoprion*, articulate with the neurocranium at two points, viz. anteriorly below the anterior part of the orbito-temporal region, and posteriorly at the antero-ventrally directed process of the otico-occipital portion of the neurocranium. There is no symphyseal connection between the palatoquadrates; and the upper dentition is, according to Zangerl (1966), situated directly on the ventral surface of the neurocranium well forward of the anterior ends of the separate palatoquadrates. Zangerl suggests that these curious features reflect a certain stage in a progressive reduction of the palatoquadrates, which is supposed to have reached its ultimate stage in the later edestids (e.g. *Sarcoprion, Fadenia, Erikodus*) with a total reduction of the dorsal parts of the mandibular arch (Zangerl, 1966: 33–35). The process is apparently thought to start off from an amphistylic jaw suspension as represented in the pleuropterygians (Zangerl, 1966: 39). However, with regard to the mandibular skeleton and its mode of attachment to the neurocranium the pleuropterygians are, in my opinion, specialized in a way which makes it improbable that they can have given origin to such features as found in *Ornithoprion* and in the Greenlandic Permian edestids.

The fact that the broad subocular laminae in *Fadenia, Sarcoprion* and *Erikodus* are carrying the posterior part of the dentition laterally on their lower surface and are reaching without interruption to the mandibular articulation is not consistent with Zangerl's assumption of a total reduction of the palatoquadrates in these forms.

In recent selachians the palatine parts of the palatoquadrates have been shown to form by differentiation from the same mesenchymatic tissue which is giving rise to the upper dentition (Holmgren, 1940: 120, 124–125, 131–132, 139–140, 145, 154, 159; 1943: 29). This makes it conceivable that the well developed upper dentition in *Ornithoprion* as well as in *Fadenia, Sarcoprion* and *Erikodus* indicates the presence of palatine parts of the palatoquadrates incorporated in the neurocranium.

In *Deltoptychius* there is reason to suspect that the jaw suspension was of the endocranio-autostylic type (Patterson, 1965: 156).

The lower jaw is long and slender in *Chondrenchelys*. It is also long in *Fadenia, Sarcoprion* and *Erikodus*, but in these genera it is relatively strong and completely fused to its antimere at the symphysis, forming here a long and strong base for the curved row of symphyseal teeth. *Helodus* has a short but high and strong lower jaw. The lower jaw in *Ornithoprion*, composed of three parts, is extremely long (Zangerl, 1966). A long anterior element—the "mandibular rostrum"—articulates against the anterior margins of the left and the right jaw rami. The curved row of symphyseal teeth and the adjacent parasymphyseal teeth are situated proximally on the "mandibular rostrum" on a semicircular elevation. The lower jaw rami in *Deltoptychius* are short and strong and both carry a spine posteriorly; a feature thought to characterize also *Menaspis* (Moy-Thomas, 1936b: 782–784; Patterson, 1965: 156–159, 170; and in this Volume). In

Janassa the lower jaw rami are completely fused at the symphysis (Jaekel, 1899). They are remarkably high, but taper rapidly toward the articular end. Also a couple of elements, interpreted as labial cartilages, were described here. The head in *Janassa* is dorso-ventrally flattened, but cannot be as short as indicated by Jaekel's reconstruction (1899, fig. 2). That this reconstruction is incorrect is clearly shown by the specimen figured by Stromer (1912, fig. 28). Probably the general shape of the head of *Ctenoptychius* resembles that of *Janassa* (Woodward, 1932: 90–91).

The postmandibular visceral skeleton is known in some detail only in *Fadenia* and *Ornithoprion*. Traces of it are found in *Helodus*, but these traces only indicate a position below the neurocranium. Traces are known also in *Chondrenchelys*. In *Ornithoprion* the remains of the hyoid and branchial arches consist of ten rod-shaped elements and a few small pieces. The rod-shaped elements are interpreted as the ceratohyal and the ceratobranchials of six postmandibular arches (Zangerl, 1966: 19). This interpretation, however, is uncertain. The material seems to show that the greater part of the branchial basket was situated caudally to the neurocranium. In *Fadenia* the hyoid arch comprises a short, unspecialized hyomandibula and a long, slender ceratohyal, which articulates distally with a short, not particularly broad basihyal. A series of short slender branchial rays are supported by the ceratohyal. There are at least four branchial arches, each composed of a suprapharyngobranchial, an epibranchial to which the infrapharyngobranchial is completely fused, and a long, slender ceratobranchial. Possibly even a fifth arch was present. Of the basibranchial series only unrecognizable fragments are preserved. The location of the postmandibular visceral skeleton is below the neurocranium, but continues also a short distance behind it.

Brain, cranial nerves and blood vessels. Almost nothing is known of these parts in the bradyodonts. In *Fadenia* the foremost part of the cranial cavity indicates that the olfactory tracts were long and the olfactory filaments well developed. The brain was located between the orbits, and this was probably also the case in *Sarcoprion* and *Erikodus*. In *Helodus*, likewise, the brain had such a location (Patterson, 1965: 179), that is not ventrally to the orbits as suggested by Moy-Thomas (1936a), whereas in *Ornithoprion*, according to Zangerl (1966), it actually occupied a ventral position. The short occipital region, as it appears in all the better known forms, suggests a short medulla oblongata.

In *Helodus* a number of foramina for nerves and blood vessels were interpreted by Moy-Thomas on the basis of comparisons with the recent chimaeroids (Moy-Thomas, 1936a: 494–496, figs. 4, 6, 7). This interpretation must be treated with some reserve.

Axial skeleton. In the few forms where parts of the axial skeleton are known it appears that the notochord was persistent. In *Chondrenchelys* ring-like calcifications occur segmentally together with normal neural and haemal arches, the latter restricted to the posterior part of the body. In *Helodus* a synarcual is the only preserved part of the axial skeleton (Patterson, 1965: 179, fig. 38). The presence of a synarcual is also indicated in *Deltoptychius* (Patterson, 1965: 162, fig. 27). No such structure is found in *Fadenia* and it is not known in *Sarcoprion* and *Erikodus* either. These forms possess the normal neural and haemal arches, the latter without spines in the caudal region. In the most caudal portion extensive fusion has taken place between the arcuals, forming a strong, comparatively long, triangular plate which comprises both dorsal and ventral elements. This plate surrounds the posterior part of the notochord and of the spinal cord and forms a strong support for the caudal fin (Bendix-Almgreen, 1962).

In *Agassizodus* there is no such fusion (Zangerl, personal comm.). The neural arches in the foremost part of the axial skeleton in *Ornithoprion* are highly modified, forming rather large, leaf-shaped elements (Zangerl, 1966). The structure of the axial skeleton is not known in *Janassa* or *Ctenoptychius*. With respect to the axial skeleton in *Menaspis* see Patterson (in this Volume).

Median fins. These differ in structure in the forms where they are preserved. In *Chondrenchelys* the dorsal fin is continuous with the epichordal lobe of the caudal fin and its endoskeleton consists of two rows of radials; the whole structure is very like that occurring in the pleuracanthids. No traces of an anal fin or of a hypochordal lobe of the caudal fin are preserved. *Helodus* has two dorsal fins of which the foremost is supported by a well developed smooth spine and a triangular basal plate articulating with the synarcual. The second dorsal fin is long and low, but nothing is known of its endoskeleton. This applies also to the heterocercal caudal fin. No anal fin is known. In *Fadenia* the first dorsal fin consists of a large, roughly triangular basal, supporting twelve radials. There is not known any second dorsal fin or any anal fin. The caudal fin is a remarkable structure in the edestids from the Greenlandic Permian (*Fadenia, Sarcoprion, Erikodus*), resembling that of *Cladoselache* (Dean, 1909) in general shape, but being more complex in structure (Bendix-Almgreen, 1962). The same general type of caudal fin occurs in *Agassizodus*, but here the endoskeleton is less complex (Zangerl, personal comm.). This is true also of the anterior dorsal fin in this form which is supported only by a large, triangular basal (Zangerl, personal comm.). *Janassa* has two triangular dorsal fins located caudally to the pelvic region; nothing is known of their endoskeleton. Neither the caudal nor an anal fin is known, but judging from the *Squatina*-like shape of the body, the latter was probably not developed. No information exists about these fins in *Ctenoptychius*. In *Menaspis*, *Deltoptychius* and *Ornithoprion* the median fins are unknown.

Paired fins and girdles. Generally speaking four principal types of pectoral fins are represented. In *Chondrenchelys* the archipterygial type occurs. In *Helodus* the pectoral fin is of a dibasal type, including a small propterygium and a long and strong metapterygium with which a series of segmented radials articulate. Anteriorly some of the proximal segments of the radials have fused to form one large element. A similar compound element is found anteriorly in the pectoral fin in *Menaspis*, but the state of preservation of the material does not make a close comparison with *Helodus* possible. The plesodic pectoral fin characterizing *Fadenia* (Bendix-Almgreen, 1962; 1967), *Erikodus* (Bendix-Almgreen, unpubl.) and *Agassizodus* (Zangerl, personal comm.) resembles to a certain degree those of the cladoselachids (Dean, 1909). However, in *Fadenia* the presence of a number of small postmetapterygial elements suggests that fusion has taken place between the proximal parts of the posterior radials. This indicates also that the shoulder girdle and its articular crest are slightly more shortened from behind than is probably the case in the cladoselachids (cf. Jarvik, 1965: 160–165). If this is also true of *Erikodus* and *Agassizodus* is at present unknown, but somewhat similar conditions may have occurred even in these genera. Nothing is known of the pectoral fin in *Ornithoprion*. However, as seen in Zangerl's illustrations (1966, figs. 8, 10, 11, 12) of the shoulder girdle, the articulation area for the pectoral fin is much restricted. This suggests that considerable fusion has taken place between the radials in the proximal part of the fin, probably resulting in the formation of larger compound elements (cf. Jarvik, loc. cit.). The pectoral fin in *Ornithoprion* was thus probably of a much more specialized type than in the other edestids with regard to its endoskeleton. *Janassa* and *Ctenoptychius* are both characterized by large rounded flap-like fins, resembling the pectorals in *Squatina*. In *Janassa* two trapezoid endoskeletal elements are described as the left and the right propterygium (Jaekel, 1899), but this interpretation is uncertain.

The pectoral girdle is known in *Chondrenchelys*, *Helodus*, *Fadenia*, *Erikodus*, *Agassizodus* and *Ornithoprion*. In all these genera the two halves of the girdle are separate and both the scapular and the coracoid parts are well developed. A large, elongated cartilage rod, described as a "sternal element", is found in *Ornithoprion* (Zangerl, 1966: 21–22, figs. 10–12). This element articulates with the distal margins of the coracoids and extends forward below the branchial basket. In *Chondrenchelys*, *Helodus* and *Ornithoprion* the articular surface for the pectoral fin is situated on a small, backwardly directed process, while in *Fadenia*, *Erikodus* and probably also in *Agassizodus* a ventro-laterally directed articular crest supports the extremely broad-based fin.

The endoskeleton of the pelvic fin is known in *Helodus*, *Menaspis* and *Chondrenchelys*. The two former have a long, rather slender basal element supporting a series of short, unsegmented radials, while in *Chondrenchelys* the fin is of the archipterygial type with a preaxial series of jointed radials supported by a slender, segmented axis, which is presumed distally to be developed as a clasper. The pelvic fins in *Janassa* are only known by their general shape, i.e. a relatively well developed, gently rounded anterior lobe and a larger, rounded, flap-like posterior lobe. No description of the pelvic fins in *Ctenoptychius* has ever been published, but it is not inconceivable that they were of the same general shape as in *Janassa*. At present nothing is known of the pelvic fins and girdle in *Fadenia*, *Sarcoprion*, *Erikodus* and *Ornithoprion*.

The pelvic girdle is known in *Helodus*, *Chondrenchelys* and *Janassa*. In the two former the two halves are separate. This is also suggested in *Janassa* where there are also indications of a well developed iliac process. In *Ctenoptychius* the pelvic girdle is unknown, and the traces of it found in *Menaspis* give limited information (see further Patterson, in this Volume).

Dentition. The dentition is composed either of a few, large tooth plates in the upper and the lower jaw (*Chondrenchelys*, *Menaspis*, *Deltoptychius*) or of isolated teeth arranged in several rows on the jaws in a "selachian" type of dentition (*Janassa*, *Ctenoptychius*, *Fadenia*, *Erikodus*, *Sarcoprion*, *Agassizodus*). *Ornithoprion* is characterized by the latter type, although a certain reduction has taken place in the lower jaw where the lateral part of the dentition, according to Zangerl (1966: 6–8, 16), is represented only by a few tooth rows on the proximal part of the "mandibular rostrum" adjacent to the symphyseal tooth row. The dentition of *Helodus* resembles the "selachian" type but in the middle of each jaw fusion has taken place between the successive teeth in one or two rows, forming curved tooth plates. Specializations are also found in the symphyseal region in *Fadenia*, *Erikodus*, *Sarcoprion*, *Agassizodus* and *Ornithoprion* which all have a strongly developed curved tooth row at the mandibular symphysis and a similar but usually less strong or less curved row at the upper jaw symphysis.

Squamation, dermal armour, fin spines and frontal spines. In *Deltoptychius* the scales are of the cyclomorial type, while in *Helodus* only the synchronomorial type (placoid scales) is found (Patterson, 1965: 163–164, 179–182). In the edestids from the Permian of East Greenland both types of scales occur with about 50 % of each (Stensiö, 1961). The squamation of *Ornithoprion* is composed of cyclomorial scales and of what has been interpreted as single lepidomoria, occurring in great number on the ventral side of the body. The scales on both the rostrum and on the "mandibular rostrum" have fused basally to form long rods and plates. In *Agassizodus* the squamation appears to consist mainly of cyclomorial scales, while lepidomoria are said to be present in great numbers (Zangerl, 1966: 28–29, 32–33, 35–36, 40, figs. 17–19, 23–26). *Condrenchelys* is characterized by a poorly developed squamation consisting of small, conical scales (Moy-Thomas, 1935: 401). *Janassa* possesses a squamation of mushroom-shaped placoid scales (Ørvig, 1966: 32–35, fig. 6). In *Menaspis* the trunk is covered by scales with conical crowns. The two rows of scales situated along the dorsal midline are much enlarged (Patterson, in this Volume). Similar conditions occur in *Deltoptychius*, where the much enlarged scales are spine-like (Patterson, 1965: 162–164, 170–171).

A strong dermal armour covers the dorsal side of the head in *Deltoptychius* and in *Menaspis*, and a remarkable similarity exists between the shields in the two forms (Patterson, 1965).

The fin spine in *Helodus* and the mandibular spines in *Deltoptychius* and *Menaspis* have already been mentioned. *Menaspis* does possess three further pairs of slender, sickle-shaped, smooth spines, situated on the anterior part of the dermal armour. The similarity of these to the frontal clasper in the chimaeroids has been discussed by Patterson (1965: 199–200, see, however, also Patterson, in this Volume).

Discussion

(1) The most important anatomical feature shared by the genera *Chondrenchelys*, *Helodus*, *Fadenia*, *Sarcoprion* and *Erikodus* is the endocranio-autostylic jaw suspension. Further common features are (a) the large, deep orbits, indicating well developed eyes, (b) the short otic region, and (c) the short occipital region. Most other known features of the neurocranial and visceral skeleton in these genera owe their origin to the fusion between the palatoquadrates and the neurocranium. These features are (d) the location in all five genera of the upper dentition close to the ventral surface of the neurocranium, (e) the position of the mandibular articulation below the orbito-temporal region in *Chondrenchelys*, *Helodus*, *Fadenia* and probably also in *Sarcoprion* and *Erikodus*, (f) the small, unspecialized, non-suspensorial hyomandibular present in *Fadenia* as probably also in *Chondrenchelys*, *Helodus*, *Sarcoprion* and *Erikodus*, (g) the location of the branchial basket primarily below the neurocranium in *Helodus* and *Fadenia* and probably also in *Sarcoprion* and *Erikodus*, and (h) the ventrally directed continuation of the postorbital process in a ridge joining a subotic shelf in *Chondrenchelys*, *Helodus* and *Fadenia* and probably also in *Sarcoprion* and *Erikodus*; anteriorly this shelf is continuous with the subocular part of the palatoquadrate.

On the basis of these similarities it is not unreasonable to classify *Chondrenchelys*, *Helodus*, *Fadenia*, *Sarcoprion* and *Erikodus* within the same major group.

This conclusion is not, to my mind, seriously contradicted by the following diversities in the cranial structure of the five genera: (a) the differences in the proportions of the ethmoidal and orbito-temporal regions, (b) the apparent location in *Chondrenchelys* of the branchial basket behind rather than below the neurocranium, (c) the closed rostrum in *Fadenia* and *Sarcoprion*, and probably in *Erikodus*, which appears to have no equivalent in *Chondrenchelys* and *Helodus* and (d) the vertically directed laminae in the ethmoidal region of *Helodus* which do not occur in any of the other genera under consideration.

In *Ornithoprion* the general shape of the neurocranium and the position of the upper dentition are in good accordance with the condition of *Fadenia*, *Sarcoprion*, and *Erikodus*. The structure of the mandibular arch in *Ornithoprion*, as well as in *Agassizodus*, and the location of the postmandibular visceral skeleton in *Ornithoprion* do, however, make it difficult to trace any relationship between those two Carboniferous forms, on the one hand, and the Upper Permian *Fadenia*, *Sarcoprion* and *Erikodus*, on the other. Because of the specialized features in its postcranial skeleton *Ornithoprion* cannot possibly be a Carboniferous predecessor of any of the three Upper Permian forms. *Agassizodus*, on the other hand, possesses a postcranial skeleton which, although somewhat simpler, resembles that in *Fadenia* (the best known of the Permian forms). This, however, probably does not mean much in terms of relationship, since about the same kind of specialization of the postcranial skeleton as in *Agassizodus* and *Fadenia* is met with also in the cladoselachians, which, in my opinion, are not closely related to the edestids as represented by the three Upper Permian forms. In this case differences in the visceral skeleton are more likely to indicate that *Agassizodus* and *Fadenia* belong to separate evolutionary lines. In conclusion, it is hardly justified to maintain any close kinship between *Ornithoprion* and *Agassizodus*, on the one hand, and *Fadenia*, *Sarcoprion* and *Erikodus*, on the other.

In their dermal armour and dentition *Menaspis* and *Deltoptychius* bear a strong resemblance to each other and are usually regarded as closely related forms. In *Deltoptychius* there is reason to suspect a jaw suspension of the endocranio-autostylic type, and the existence of a relationship between *Deltoptychius* and *Menaspis*, on the one hand, and the genera *Chondrenchelys*, *Helodus*, *Fadenia*, *Sarcoprion* and *Erikodus*, on the other, is perhaps not inconceivable.

The few characters known at present from the petalodonts *Janassa* and *Ctenoptychius* (the *Squatina*-like body form, the shape of teeth and sequence of tooth succession, the shape of lower jaw, the squamation) provide no evidence for a close relationship with any other genera here under discussion, nor, for that matter, with any other members of the elasmobranchiomorphs.

When summarizing the above one finds: (a) Provisionally the genera *Chondrenchelys*, *Helodus*, *Deltoptychius*, *Menaspis*, *Fadenia*, *Sarcoprion* and *Eriko-*

dus may be placed within the same major group—the Bradyodonti (the old name being retained here for purely practical reasons). (b) Because of the wide range of variation in the cranial and postcranial skeleton, the dermal skeleton, as well as in the dentition, the Bradyodonti, in this sense, must be regarded as including at least four evolutionary lines,—viz. the chondrenchelyids, the helodontids, the menaspids, and the edestids, which evidently had been separate for a long time. (c) The Pennsylvanian *Ornithoprion* and *Agassizodus* cannot on present evidence be included in the edestids as represented by *Fadenia*, *Sarcoprion* and *Erikodus*. (d) On present evidence it is not warranted to place the petalodonts as represented by *Janassa* and *Ctenoptychius* in the bradyodont group; it is here proposed to classify these fishes as elasmobranchiomorphs inc. sed.

(2) At this point attention may be called to certain characteristic features of the dentition in the bradyodonts. This dentition can be divided into two types, —the "selachian" type and the "dental plate" type—which show a different arrangement and development in the upper symphyseal region. Thus a well developed dentition covering the whole symphyseal region is typical for the "selachian" type and is found in all edestids. In the genera with the "dental plate" type (*Chondrenchelys*, *Deltoptychius*, *Menaspis*) the upper dentition is clearly differentiated into left and right parts, between which a symphyseal interspace occurs, as is clearly seen in i.a. *Deltoptychius* (Patterson, 1965, fig. 31). The upper dentition in *Helodus* is also divided into left and right parts and apparently represents a slightly diverging pattern of the "dental plate" type.

If these characters are compared with the ontogenetical development of the upper tooth blastemas and the corresponding parts of the palatoquadrates in recent selachians and chimaeroids, some features concerning the anterior portion of the palatoquadrates in the fossil genera under discussion may be deduced.

In selachian embryos it has been shown that the antero-ventral part of the palatoquadrate incorporates the infrapharyngo-premandibular as a symphyseal process (Jarvik, 1954; infrapharyngo-mandibular: Holmgren, 1940; 1943: 63). These processes of the palatoquadrates of both sides meet in the median line below the ethmoidal region, as do the corresponding tooth blastemas. The formation of a well developed dentition, evenly distributed over the whole symphyseal region, takes place from these tooth blastemas.

In the recent chimaeroids, on the other hand, each infrapharyngo-premandibular may be assumed to have fused with the ventral side of the ethmoidal region (Stensiö, 1963: 41–42, 406, 407), not meeting its antimere in the midline. The dentition developing from the anterior tooth blastemas, therefore, becomes divided into a left and a right anterior tooth plate.

As may thus be gathered, the "selachian" type and the "dental plate" type of dentition in the upper symphysial region of the bradyodonts have equivalents in recent selachians and chimaeroids, respectively. This may indicate that also the ontogenetical development of this part of the dentition corresponded to that of the recent selachians in the edestids, and to that of recent chimaeroids in *Chondrenchelys*, *Helodus*, *Deltoptychius* and *Menaspis*. In the fossil forms discussed in this paper, all characterized by a holostylic jaw suspension, there is no way of telling in what manner and in what relations to each other the palatoquadrates and infrapharyngo-premandibulars formed during ontogeny and fused with the neurocranium. The arrangement of the dentition, however, gives a clue to the position occupied by the tooth blastemas belonging to the infrapharyngo-premandibulars, and hence also to the location of these particular elements when wholly integrated in the adult neurocranium. In the edestids, more precisely, the infrapharyngo-premandibulars of both sides were in contact at the symphysis, as in selachians, whereas in the chondrenchelyids, the helodontids and the menaspids they were clearly separated from each other anteriorly, presumably in the same way as in the chimaeroids. Consequently, the edestids on the one hand, and the chondrenchelyids, helodontids and menaspids on the other, in all probability represented two separate lines of specialization with regard to the modifications of the anterior parts of the visceral endoskeleton. In these circumstances the group Bradyodonti, in the comprehension it has been given here, may constitute a rather heterogeneous assemblage of late Palaeozoic elasmobranchs.

(3) It seems appropriate to end this discussion with an evaluation of the opinion that the chimaeroids are descendants of the bradyodonts.

The holostylic jaw suspension shared by the chimaeroids and the better known bradyodonts (*Chondrenchelys*, *Helodus*, *Deltoptychius*, *Menaspis*, *Fadenia*, *Sarcoprion* and *Erikodus*) has been held to show a relationship between those groups. However, this is not convincing, since holostyly clearly evolved repeatedly in lower vertebrates and therefore might just as well indicate convergent specialization in the present case.

None of the bradyodonts show the principal part of the ethmoidal region to be situated lower down than the postethmoidal part of the skull. The reverse is true of both recent and fossil chimaeroids, with perhaps *Squaloraja* as on only exception. This remarkable difference seems also to indicate that the bradyodonts and the chimaeroids differ significantly from each other with regard to the general shape and proportions of the palatoquadrates.

The complete absence of labial cartilages in all better known bradyodonts may seem to exclude them as close relatives or predecessors of the chimaeroids.

It is possible, however, that in bradyodonts these elements originally lacked cartilage calcification and thus were not preserved in the fossil state.

Among the bradyodonts the rostral region is fairly well known only in *Fadenia* and *Sarcoprion*. Here it is very different from that in the recent chimaeroids which contains three rod-like cartilage elements: one median located dorsally to the two lateral, all articulating with the anterior part of the ethmoidal region close to the internasal septum. However, the differences are less pronounced if comparison is made with some of the Jurassic chimaeroids (*Squaloraja, Metopacanthus, Acanthorhina*; Patterson, 1965) where the median rostral cartilage is a long, closed forward extension of the ethmoidal region, possibly corresponding to the rostrum in the edestids. In view of the large frontal clasper present in *Squaloraja, Metopacanthus*, and probably also in *Acanthorhina*, on the other hand, the strong rostrum probably represents a specialized feature which does not reflect primitive rostral morphology of the chimaeroids in general. Elements corresponding to the pair of lateral rostral cartilages in the chimaeroids are at present unknown in the edestids. Although the true nature of these cartilages in the recent chimaeroids is so far obscure (Holmgren, 1942: 77–78), the absence of corresponding structures in the edestids (and perhaps in all the genera here retained in the bradyodont group) probably speaks against a bradyodont-chimaeroid relationship.

The otic and the occipital regions are short in *Chondrenchelys, Helodus, Fadenia* and *Sarcoprion*, thus resembling the corresponding regions in the chimaeroids. Because of the great variation in the length of these regions within the elasmobranchs as a whole, however, these resemblances are not necessarily suggestive of close relationship.

The skulls of *Helodus, Chondrenchelys* and *Fadenia* show a well defined ridge in the anterior part of the otic region connecting the postorbital process with the subotic shelf, and a corresponding structure is met with in the chimaeroids. As this is probably a secondary character due to the fusion between the palatoquadrates and the neurocranium, it can just as well reflect convergent specializations within the two groups.

The similarities actually found in different parts of the postcranial skeleton of some bradyodonts and certain chimaeroids are: (a) The dibasal pectoral fin in *Helodus*, (b) the presence of calcified rings in the vertebral column in *Chondrenchelys*, (c) the separate halves in the pelvic girdle in *Helodus* and *Chondrenchelys*, and (d) the synarcual and the structure of the first dorsal fin in *Helodus*. None of these features, however, are sufficient to suggest a direct link between any of these bradyodonts and the chimaeroids.

The above comparisons indicate that among the better known bradyodonts, at least the genera *Chondrenchelys, Helodus, Fadenia, Sarcoprion* and *Erikodus*, are no close relatives or predecessors of the Mesozoic and later chimaeroids.

Whether or not this applies also to *Deltoptychius* and *Menaspis* is uncertain. In these forms the endocranial structure is practically unknown, but the dermal armour shows a certain similarity to that of the head in some of the early Mesozoic chimaeroids (*Myriacanthus, Metopacanthus, Chimaeropsis*; Patterson, 1965: 131, 142–143, 172–174, 201–205). If this should be taken to indicate that the ancestors of the myriacanthids are to be found among the menaspids, the latter are not to be classified as bradyodonts, with which they on present knowledge share, at best, only one important feature, the endocranio-autostylic jaw suspension. A descent of myriacanthids from menaspids, however, cannot yet be taken as anything else than a possibility, and requires more substantial evidence before it can be accepted.

This work has been supported by grants from the Copenhagen University. For permission to refer to unpublished data on *Agassizodus* the author is indebted to Dr. R. Zangerl.

References

Arambourg, C. & Bertin, L. (1958). Sous-class des bradyodontes (Bradyodonti). In *Traité de Zoologie*, ed. Grassé, P.-P., **13**:3: 2057–2067. Paris: Masson.

Bendix-Almgreen, S. E. (1962). De østgrønlandske perm-edestiders anatomi, med særligt henblik på *Fadenia crenulata. Medd. dansk geol. Foren.*, **15**: 152–153.

Bendix-Almgreen, S. E. (1966). New investigations on *Helicoprion* from the Phophoria Formation of South-east Idaho, U.S.A. *Biol. Skr. K. danske VidenskSelsk.*, **14**: 1–54.

Bendix-Almgreen, S. E. (1967). On the fin-structure of the Upper Permian edestids from East Greenland. *Medd. dansk geol. Foren.*, **17**: 147–149.

Berg, L. S. (1958). *System der rezenten und fossilen Fischartigen und Fische*. Berlin: VEB Deutscher Verl. d. Wiss.

Dean, B. (1906). Chimaeroid fishes and their development. *Publ. Carneg. Instn.*, **32**: 1–194.

Dean, B. (1909). Studies on fossil fishes (sharks, chimaeroids and arthrodires). *Mem. Am. Mus. nat. Hist.*, **9**: 211–287.

Gregory, W. K. (1951). *Evolution emerging*. New York: Macmillan.

Holmgren, N. (1940). Studies on the head in fishes. Embryological, morphological and phylogenetical researches. 1. Development of the skull in sharks and rays. *Acta zool. Stockh.*, **21**: 51–267.

Holmgren, N. (1942). Studies on the head of fishes. An embryological, morphological and phylogenetical study. 3. The phylogeny of elasmobranch fishes. *Acta zool. Stockh.*, **23**: 129–261.

Holmgren, N. (1943). Studies on the head of fishes. An embryological, morphological and phylogenetical study. 4. General morphology of the head in fish. *Acta zool. Stockh.*, **24**: 1–188.

Jaekel, O. (1891). Ueber *Menaspis* nebst allgemeinen Bemerkungen über die systematische Stellung der Elasmobranchii. *Sber. Ges. naturf. Fr. Berl.*, **1891**: 115–131.

Jaekel, O. (1899). Ueber die Organisation der Petalodonten. *Z. dt. geol. Ges.*, **51**: 258–298.

Jarvik, E. (1954). On the visceral skeleton in *Eusthenopteron* with a discussion of the

parasphenoid and palatoquadrate in fishes. *K. svenska VetenskAkad. Handl.*, (4) **5**: 1–104.
Jarvik, E. (1963). The composition of the intermandibular division of the head in fish and tetrapods and the diphyletic origin of the tetrapod tongue. *K. svenska Vetensk-Akad. Handl.*, (4) **9**: 1–74.
Jarvik, E. (1965). On the origin of girdles and paired fins. *Israel J. Zool.*, **14**: 141–172.
Jessen, H. L. (1966). Struniiformes. In *Traité de Paléontologie*, ed. Piveteau, J., **4**:3: 387–398. Paris: Masson.
Lison, L. (1941). Recherches sur la structure et l'histogenèse des dents des poissons dipneustes. *Arch. Biol. Liége*, **52**: 279–320.
Moy-Thomas, J. A. (1935). The structure and affinities of *Chondrenchelys problematica* Tr. *Proc. zool. Soc. Lond.*, **1935**:2: 391–403.
Moy-Thomas, J. A. (1936a). On the structure and affinities of the Carboniferous cochliodont *Helodus simplex*. *Geol. Mag.*, **73**: 488–501.
Moy-Thomas, J. A. (1936b). The structure and affinities of the fossil elasmobranch fishes from the Lower Carboniferous rocks of Glencartholm, Eskdale. *Proc. zool. Soc. Lond.*, **1936**:3: 761–788.
Moy-Thomas, J. A. (1937). On the Carboniferous fish *Eucentrurus paradoxus* Traquair. *Geol. Mag.*, **74**: 183–184.
Moy-Thomas, J. A. (1939). The early evolution and relationships of the elasmobranchs. *Biol. Rev.*, **14**: 1–26.
Nielsen, E. (1932). Permo-Carboniferous fishes from East Greenland. *Medd. Grønland*, **86**: 1–63.
Nielsen, E. (1952). On new or little known *Edestidae* from the Permian and Triassic of East Greenland. *Medd. Grønland*, **144**: 1–55.
Obruchev, D. V. (1953). Izuchenie edestid i rabotui A. P. Karpinskogo (Edestid researches in the works of A. P. Karpinskij). *Trud. Inst. palaeont. Acad. Sci. U.S.S.R.*, **45**: 1–85. In Russian.
Obruchev, D. V. (1964). Podklass Holocephali. Tselnogovuie, ili khimerui. In *Osnovui paleontologii (Fundamentals of Palaeontology)*, ed. Orlov, I. A., **11**: 238–266. Moscow: Acad. Sci. U.S.S.R. In Russian.
Ørvig, T. (1951). Histologic studies of placoderms and fossil elasmobranchs. 1. The endoskeleton, with remarks on the hard tissue of lower vertebrates in general. *Ark. Zool.*, (2) **2**: 321–454.
Ørvig, T. (1960). New finds of acanthodians, arthrodires, crossopterygians, ganoids and dipnoans in the Upper Middle Devonian Calcareous Flags (Oberer Plattenkalk) of the Bergisch Gladbach-Paffrath Trough. 1. *Paläont. Z.*, **34**: 295–325.
Ørvig, T. (1962). Y-a-t-il une relation directe entre les arthrodires ptyctodontides et les holocéphales? *Colloques int. Cent. natn. Rech. Scient.*, **104**: 49–61.
Ørvig, T. (1966). Histologic studies of ostracoderms, placoderms and fossil elasmobranchs. 2. On the dermal skeleton of two late Palaeozoic elasmobranchs. *Ark. Zool.*, (2) **19**: 1–39.
Ørvig, T. (1967). Phylogeny of tooth tissues: Evolution of some calcified tissues in early vertebrates. In *Structural and Chemical Organization of Teeth*, ed. Miles, A. E. W., vol. **1**: 45–110. New York & London: Academic Press.
Patterson, C. (1965). The phylogeny of the chimaeroids. *Phil. Trans. R. Soc.*, (B) **249**: 101–219.
Radinsky, L. (1961). Tooth histology as a taxonomic criterion for cartilaginous fishes. *J. Morph.*, **109**: 73–81.
Stahl, Barbara S. (1967). Morphology and relationships of the Holocephali with special reference to the venous system. *Bull. Mus. comp. Zool. Harv.*, **135**: 141–213.
Stensiö, E. A. (1961). Permian vertebrates. In *Geology of the Arctic*, ed. Raasch, G. O., vol. **1**: 231–247. Toronto: University of Toronto Press.

Stensiö, E. A. (1963). Anatomical studies on the arthrodiran head. 1. Preface, geological and geographical distribution, the organisation of the arthrodires, the anatomy of the head in the Dolichothoraci, Coccosteomorphi and Pachyosteomorphi. Taxonomic appendix. *K. svenska VetenskAkad. Handl.*, (4) **9**: 1–419.

Stromer v. Reichenbach, E. (1912). *Lehrbuch der Paläozoologie.* Vol. **2**: *Wirbeltiere*. Leipzig & Berlin: B. G. Teubner.

Watson, D. M. S. (1937). The acanthodian fishes. *Phil. Trans. R. Soc.*, (B) **228**: 49–146.

Weigelt, J. (1930). Wichtige Fischreste aus dem Mansfelder Kupferschiefer. *Leopoldina*, **6**: 601–624.

Woodward, A. S. (1921). Observations on some extinct elasmobranch fishes. *Proc. Linn. Soc. Lond.*, **133**: 29–39.

Woodward, A. S. (1932). *K. A. v. Zittel: Text-book of palaeontology.* Vol. **2**. London: Macmillan.

Zangerl, R. (1966). A new shark of the family Edestidae, *Ornithoprion hertwigi*. *Fieldiana: Geol.*, **16**: 1–43.

Menaspis and the bradyodonts

By Colin Patterson
Department of Palaeontology, British Museum (Natural History), London, England

The two most important questions concerning the various Palaeozoic elasmobranchs which have been included in the Bradyodonti are first, whether this group is a real one, that is, whether the constituent subgroups are more closely related to each other than to any other taxon, and secondly, whether there is close phylogenetic relationship between the bradyodonts and the living chimaeroids. That these questions remain unanswered almost fifty years after the Bradyodonti was erected (Woodward, 1921) is due mainly to lack of information on the constituent groups. Recently, new material from the Permian of Greenland (Bendix-Almgreen, 1962; 1967) and North America (Bendix-Almgreen, 1966) and from the Pennsylvanian of Indiana (Zangerl, 1966) has indicated that *Helicoprion* is not related to the edestids, and thrown doubt both on the unity of the remaining edestids and on the relationship between the edestids and other bradyodonts. New information on the ptyctodont arthrodires (Ørvig 1960; 1962; Miles, 1967) and the armoured cochliodonts (Patterson, 1965) has raised again the question of the relationships of the chimaeroids. Much new material from the Pennsylvanian of Indiana still awaits description (Zangerl & Richardson, 1963) and this can be expected to clarify some points, but the existing information on the psammodonts, copodonts, petalodonts, helodonts, cochliodonts and menaspoids rests on a handful of specimens. The purpose of this paper is to provide new information on the Upper Permian *Menaspis* and on some of the Lower Carboniferous bradyodonts as a basis for comment on points raised in Bendix-Almgreen's discussion (this Volume) of existing knowledge of the various bradyodont groups.

Menaspis, from the German Kupferschiefer, occupies a central position in discussions of the possible relationship between bradyodonts and chimaeroids (Jaekel, 1891; 1892; Woodward, 1891*b*; 1921; Reis, 1895*a*; 1895*b*; Dean, 1904; 1906; Moy-Thomas, 1936*a*; 1936*b*; Patterson, 1965), since both in time and in certain anatomical structures it has been held to occupy an intermediate position between Carboniferous fishes with cochliodont dentition and the earliest chimaeroids, the Lower Jurassic Myriacanthidae and *Squaloraja*. *Menaspis armata* Ewald is known only by six specimens (Weigelt, 1930), three showing little more than the dentition, one isolated head spine, one more or less complete skull (Weigelt, 1930, pl. 8, fig. 1) and an almost complete fish, first described and figured by Giebel (1856), later prepared and more completely described by Jaekel (1891), whose paper has been the basis of subsequent discussion of the animal. This complete specimen, No. 462054 in the possession of the Geologisch-Paläontologisches Institut, Martin-Luther Universität, Halle (Saale), is redescribed in the sequel. In the text, references are also made to certain bradyodont specimens belonging to the British Museum (Natural History), London (abbreviated BMNH).

The Halle specimen of Menaspis

Jaekel prepared this specimen with a wire brush: this has abraded off the high points of the exoskeleton so that many of the tubercles and scale crowns are worn down to the pulp cavity and have a smooth, polished surface. Since Jaekel removed almost all the overlying matrix the only additional preparation undertaken has been cleaning and slight etching of parts of the head with an S. S. White Industrial Airbrasive machine, using the softest abrasive, sodium bicarbonate. The specimen has been studied in conjunction with a series of stereoscopic radiographs. In radiographs of the whole specimen (Fig. 2) the stereoscopic effect was exaggerated to compensate for crushing of the animal by tilting the specimen 9° left and right of the vertical (target distance 38 cm) in the two exposures. Enlarged details of the head (Figs. 6, 9) were taken on an X-ray microscope, with the tilt angles increased to 12° left and right (film to target distance 37 cm, specimen to target distance 16 cm). Positive prints of the radiographs were made on an EMI Logetronic printer.

Cleaning and radiography of the specimen showed that almost 10 mm was missing from the tip of the longest of the three head spines on the left side, being modelled in plaster: this was presumably removed by Jaekel, who commented on the histology of these spines. Since part of the spine was already missing, a further 2 mm was removed and thin sections prepared to settle the vexed question of the composition of these spines.

General features. Jaekel has given a good description of the main features of the fossil (Fig. 1). The fish is preserved dorsal side up anteriorly, lying more on its left side posteriorly. It shows the armoured dorsal side of the head, fragments of the dentition and sensory canals, the anterior part of the trunk and the right pectoral and pelvic fins. The caudal end of the fish is missing: as preserved the specimen is 170 mm long. The anterior part of the animal is covered by a more or less complete armour of enlarged scales and tuberculated plates, ending above the middle part of the pectoral fin in a pair of posteriorly directed conical spines. Jaekel interpreted this armoured portion as comprising both the head and the anterior division of the trunk, but features of the endoskeleton described

Fig. 1. *Menaspis armata* Ewald. The Halle specimen photographed under alcohol, nat. size.

below and comparison with the Lower Carboniferous *Deltoptychius armigerus*, in which the head shield is fused into a single plate (Fig. 4), confirm Dean's (1904: 50) interpretation of the armoured region as consisting of the head only. On the anterior part of the head there are four pairs of spines. Three pairs of slender, smooth, sickle-shaped spines, the second the largest and the third the smallest, articulate with the dorso-lateral surface of the skull by expanded bases: these will be referred to as "frontal spines" (*fr.sp.*, Fig. 3). Below the frontal spines there is a pair of broader, tuberculated spines which insert on the ventro-lateral surface of the skull: these will be referred to as "mandibular spines" (*md.sp.*, Fig. 3). The conical, tuberculated spines which terminate the skull roof behind will be termed "dorsal occipital spines" (*d.occ.sp.*, Fig. 3). Leading forwards from the lateral margin of the dorsal occipital spine there is a tuberculated ridge, the supraorbital ridge (*spo.r.*, Fig. 3). On the lateral margin of the skull there is a pair of smaller tuberculated spines or projections, that on the right side lying opposite the tip of the second frontal spine: these will be referred to as "lateral occipital spines" (*l.occ.sp.*, Fig. 3). In the centre of the skull roof there is a moderately large, median plate which will be referred to as the "central plate" (*cen.*, Fig. 3): this term is purely topographic and is not intended to imply homology with the central plate of arthrodires. To the left of the animal, just behind the dorsal occipital spine, there is a group of six moderately large scales or tesserae, each with a central tubercle (Fig. 1). Dean (1904: 50) interpreted these as lying in their natural position, indicating that the trunk was broader than it appears to be, but Jaekel's original interpretation, that they are detached from the animal, is certainly correct.

In addition to the exposed parts of the fossil, the radiographs show both the upper and lower dentition, but apart from a pair of structures below the posterior part of the skull (see below) they show that there are no other sizeable calcified elements concealed within the matrix.

The armour of the head. Jaekel (1891: 120) described the dorsal surface of the head (head and foremost division of the trunk in his terminology) as being completely armoured with "thick, scale-like dentine structures", writing "some parts show only a strong shagreen, in others the granules of this shagreen are enlarged into knob- or thorn-like tubercles." In fact, the cover of scales and plates on the skull roof is far from complete: what Jaekel referred to as "a strong shagreen" is the dorsal surface of the cartilaginous neurocranium, showing the prismatic calcifications characteristic of elasmobranchs. This calcified cartilage is exposed over large areas (Fig. 3), especially in a broad zone on each side of the mid-line. Sparsely scattered over these areas of cartilage are dermal structures ranging from small scale-like structures, resembling the smallest trunk scales, to moderately large, tuberculated plates. Common to all these

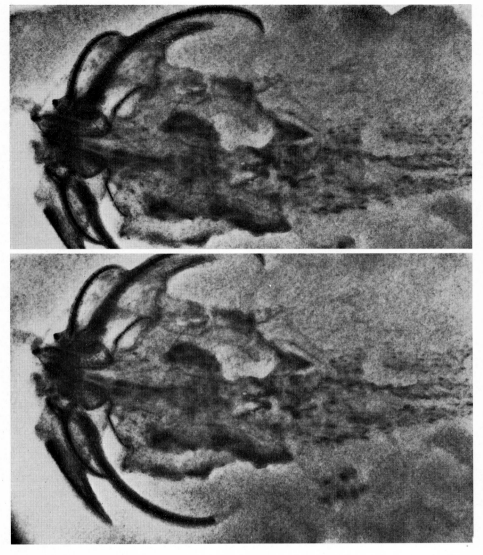

Fig. 2. *Menaspis armata*, same specimen as Fig. 1. Positive prints from a stereoscopic pair of radiographs, nat. size.

structures, whatever their size, is a broad, flat basal plate, showing a radial fibrous structure under xylene and often with the margins notched by radial furrows, and a central longitudinal keel, highest posteriorly. On some of these structures, in the vicinity of areas covered by the large plates described below, the central keel forms a single tubercle with a core of trabecular tissue, but in the majority the keel contains a longitudinal series of three to five separate pulp cavities, each cavity culminating in a low cusp (Figs. 10, 11 B), as in the trunk scales (see below). Anteriorly, between the frontal spines, these keeled scales are replaced by simple, spine-like scales, each with a single pulp cavity (Fig. 10), resembling certain scales of *Helodus* (Patterson, 1965, fig. 41 G) and the scales on the snout of chimaeriforms. On the posterior margin of the neurocranium, opposite the paired unarmoured zones, there is a pair of moderately large tuberculated plates (Fig. 3) with two or three smaller plates in a diminishing series in front of each.

In the mid-line, the posterior part of the neurocranium is unarmoured, but from the level of the lateral occipital spines forwards to the snout the medial part of the neurocranium is almost covered by a mosaic of plates. These larger plates resemble the small scale-like structures in having a broad basal plate with a radial structure, becoming thin and notched marginally, but instead of a central longitudinal crest they have a central elevation (always abraded), both this elevation and the more marginal parts of the plate bearing rounded tubercles. Each tubercle has a core of trabecular tissue rather than a single pulp cavity. These tuberculated plates are in contact marginally and occasionally fuse, lines of fusion being recognizable under xylene by changes in the orientation of the radially arranged fibres. On this central part of the skull roof there is a series of median plates of which the foremost, the "central plate" (*cen.*, Fig. 3) is the largest, preceded by a series of five or six paired plates which decrease in size and are well separated anteriorly. Surrounding the central plate and the median plates behind it are several paired plates. These are bilaterally symmetrical, but the symmetry is not perfect, as is shown by the plates lying postero-lateral to the central plate: on the right side there is a single large plate, on the left side two smaller plates, the foremost partially fused to the central plate.

Turning to the lateral part of the skull roof, the paired dorsal occipital spines (*d.occ.sp.*, Fig. 3) resemble those of *Deltoptychius* (Fig. 4) very closely; radiographs show that, as in *Deltoptychius*, their ventral surface is complete almost to the base, and they have the same flattened, triangular section, deepest laterally (Patterson, 1965, fig. 34 D), and similar ornament of rounded tubercles which tend to lie in longitudinal rows. But in contrast to *Deltoptychius*, the dorsal occipital spine is not an integral part of the head shield, being separated from the plates in front of it by a transverse suture at the base. Leading forwards from the antero-lateral corner of the dorsal occipital spine is a series of large, angulate, tuberculated plates forming the supraorbital ridge, the lateral border of the skull roof. The supraorbital ridge (*spo.r.*, Fig. 3) is complete only on the left side, where there are three large, angulate plates, the first and last showing evidence of being compound in origin, with a series of smaller plates along their medial border. On the right side only the foremost of the three large plates is preserved, and is again evidently of compound origin. Anteriorly, the supra-

Menaspis 177

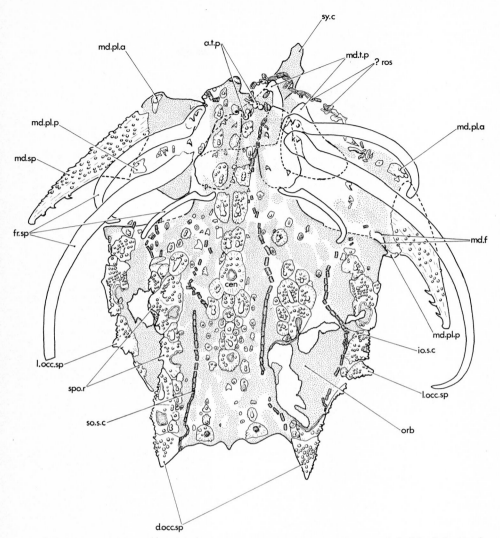

Fig. 3. *Menaspis armata*, same specimen as Fig. 1, the skull; approx. × 1.5. Stippled areas indicate calcified cartilage. The outlines of concealed parts of the tooth plates, mandibular spines and preorbital margin of neurocranium (broken lines) added from radiographs.

a.t.p, possibly anterior upper tooth plates; *cen*, "central" plate; *d.occ.sp*, dorsal occipital spines; *fr.sp*, frontal spines; *io.s.c*, infraorbital sensory canal ossicles; *l.occ.sp*, lateral occipital spines; *md.f*, facets for articulation between mandible and neurocranium; *md.pl.a, md.pl.p.*, small plates at anterior and posterior margins of mandibular spines; *md.sp*, mandibular spine; *md.t.p*, mandibular tooth plates; *orb*, floor of orbit; *?ros*, possibly an uncalcified rostrum, carrying the supraorbital sensory canals; *so.s.c*, supraorbital sensory canal ossicles; *spo.r*, supraorbital ridge; *sy.c*, symphysial cartilage of mandible.

orbital ridge grades into small, scattered, scale-like plates which trend anteromedially, as does the anterior part of the supraorbital ridge in *Deltoptychius* (Fig. 4).

On the right side of the specimen, where the posterior part of the supraorbital

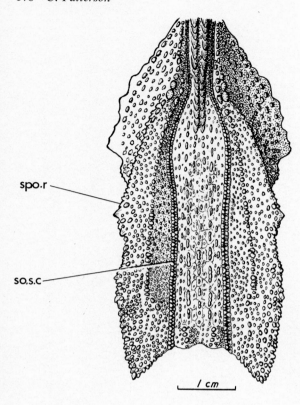

Fig. 4. *Deltoptychius armigerus* (Traquair), Lower Carboniferous, Scotland. Restoration of the head-shield in dorsal view, from Patterson (1965).
so.s.c, course of supraorbital sensory canal; *spo.r*, supraorbital ridge.

ridge is broken away, a vertical tuberculated plate, exposed edge-on, connects the antero-lateral corner of the dorsal occipital spine with the lateral occipital spine (*l.occ.sp.*, Fig. 3), a short, blunt spine projecting postero-laterally, which is borne on a small, conical tuberculated plate. At the medial edge of this plate there is a small plate and in front of it there are two large tuberculated plates, resembling the large plates on the supraorbital ridge in shape and size, with a few smaller plates in front of them. On the left side of the fish the vertical plate linking the lateral occipital spine with the base of the dorsal occipital spine is partially covered by the supraorbital ridge, and the plate bearing the lateral occipital spine is larger than on the right side, either incorporating or extending forwards to occupy the position of the plate in front of the right lateral occipital spine. In front of the left lateral occipital spine there is thus only one large plate, with three or four small plates in front of it.

Although no thin sections of the dermal plates on the skull have been prepared, enough of their microstructure can be seen on abraded surfaces to show that they must have resembled the head-shield of *Deltoptychius* (Patterson, 1965, fig. 35) in having a thick basal layer of dense lamellar tissue and an upper layer of osteodentine, which also forms the surface tubercles.

The three pairs of frontal spines have been described by Jaekel and can be seen in Figs. 1–3 (*fr.sp.*). The surface of the distal parts of these spines is quite smooth, marked only by irregularities caused by the openings of scattered vascular canals. Proximally, the spines turn forwards and expand into broad, spoon-shaped bases, concave dorsally. There are two or three tubercles within this concavity on the two large anterior pairs of spines. Radiographs (Figs. 2, 6) show that the free, distal parts of the spines have a central, smooth-walled pulp cavity, occupying almost half the diameter of the spine. Basally, the pulp cavity breaks up into ramifying vascular canals in the spongy tissue of the expanded base. Jaekel originally described these spines as consisting of vasodentine (osteodentine in modern terminology), but did not figure or describe thin sections. Reis (1895*a*; 1895*b*), although he did not study the specimen, found it more probable that the spines consist of calcified fibro-cartilage, like the frontal claspers of chimaeroids, pointing out that Jaekel thought the frontal clasper to be a dermal structure and might have mistaken calcified fibro-cartilage for osteodentine because of the abundant vascular canals. Reis's interpretation has been accepted by later authors (Dean, 1904: 52; 1906: 139; Patterson, 1965: 172), but a thin section (Fig. 5) from the tip of the second spine on the left side shows that Jaekel's original interpretation of the spines as dermal structures is correct. The spine consists of a dense, acellular tissue, opaque and completely inactive in polarized light. Surrounding the pulp cavity (*p.c*) and occupying about 15% of the thickness of the wall there is a zone of lamellar tissue (*l.t.*), laid down in concentric layers around the pulp cavity and containing no vascular canals in the sectioned material. The outer part of the spine consists of homogeneous tissue, presumably osteodentine, without lamellae and containing sparsely distributed vascular canals (*v.can.*). Except in the outermost part of the spine (seen in a fragment sectioned tangentially) the vascular canals are completely occluded by hard tissue (*occ.v.can.*). These occluded vascular canals are surrounded by narrow concentric zones which resemble denteons but are evidently due to infilling of the lumen of the canal, since the dentine tubules emanating from the canals originate at the outer limit of the zones. The tangential section shows that in the outer part of the spine, where the vascular canals are open, there is no sign of denteons around them. Passing out from the vascular canals there are sparse dentine tubules (*d.t.*), sinuous and branched. Similar but longer tubules (*b.t.*) penetrate the basal layers of lamellar tissue and pass up into the vascular tissue. In structure the frontal spines resemble the dermal armour of *Deltoptychius* (Patterson, 1965, figs. 29, 35), which also has a basal layer of lamellar tissue penetrated by vertical, branching tubules, and an outer layer of osteodentine with abundant interosteonal tissue and few tubules given off from the vascular canals. They are even more like the dermal armour of the Lower Jurassic chimaeriform *Myriacanthus* (Patterson, 1965, fig. 15), which shows the same occlusion of the vascular canals, absence of

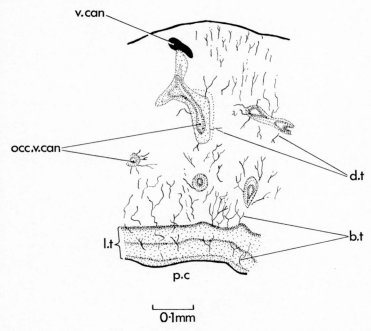

Fig. 5. *Menaspis armata*, same specimen as Fig. 1. Transverse section through part of the wall of the distal part of the second left frontal spine.
b.t, tubules penetrating basal lamellar tissue; *d.t*, dentine tubules; *l.t*, basal lamellar tissue; *occ.v.can*, occluded vascular canals; *p.c*, pulp cavity; *v.can*, vascular canal.

denteons around many of the vascular canals, abundant interosteonal tissue, etc., but has the lamellar tissue much thicker.

The frontal spines are therefore dermal structures and cannot be compared with the frontal claspers of chimaeriforms: Woodward's (1891b: 424) comparison of them with the supraorbital spines of male hybodont sharks is more apt, for these are osteodentine structures which appear to be greatly enlarged scales. The composition of the frontal spines of *Menaspis* and the presence of tubercles on the expanded bases of the two large anterior pairs of spines suggest that they too are hypertrophied scale-like plates, with the free part of the spine perhaps representing a single tubercle or scale cusp.

The mandibular spines (*md.sp.*, Fig. 3) have been described by Jaekel and are shown in Figs. 1–3, 6. Spaced along the hind edge of the distal part of the spine there are four recurved denticles, that nearest the tip almost abraded away on both sides. As Jaekel noted, the collapse of the central part of the spines shows that they had a wide pulp cavity, as in *Deltoptychius* (Patterson, 1965, fig. 29), and this is confirmed by radiographs (Figs. 2, 6). Proximally the spine broadens rapidly, principally by increased curvature of the posterior margin, and the anterior margin extends further medially than the posterior, as in *Deltoptychius* (Patterson, 1965, fig. 28 C). At the proximal end of both the anterior and posterior margins of the spine there is a separate tessera or tuberculated plate, (*md.pl.a*, *md.pl.p.*, Fig. 3). Similar tesserae occur at the base of the anterior margin of the spine provisionally assigned to the mandible of *Deltoptychius moythomasi* (Patterson, 1965: 168, pl. 28, fig. 69). Nothing is known of the microstructure of the mandibular spine, but presumably it is similar to that of the plates on the skull roof, as it is in *Deltoptychius*.

The neurocranium. As noted above, considerable areas of the prismatic calcified cartilage of the neurocranium are exposed between the dermal plates on the skull (Fig. 3). The exposed parts of the roof of the neurocranium show a continuous, uninterrupted surface, with no discernible foramina, fontanelles or fossae. The posterior margin of the neurocranial roof is clearly marked between the bases of the dorsal occipital spines, showing a pointed median protuberance and a smaller bulge behind the pair of plates on this margin. On the right side of the specimen, where the supraorbital ridge and the dorsal wall of the neurocranium are broken away, another flat surface of calcified cartilage is exposed (interpreted by Jaekel as the squamation of the ventral surface of the fish). This is not merely the ventral wall of the neurocranium (though it is probably continuous with that medially) since anteriorly, between the foremost plates of the supraorbital ridge and the third frontal spine, it is continuous with the roof of the neurocranium on both sides of the specimen, and since a sensory canal (*io.s.c.*, Fig. 3) lies on its exposed upper surface. Because of these facts, this cartilage, lying medial to the lateral occipital spine and below the supraorbital ridge, can only be interpreted as the floor of a large orbit (*orb.*, Fig. 3). This confirms the interpretation of the supraorbital ridge both in *Menaspis* and in *Deltoptychius* (Patterson, 1965: 160) and shows that the otic and occipital regions of the neurocranium must have been short, as in other bradyodonts and chimaeroids (Bendix-Almgreen, in this Volume). The vertical plate running from the antero-lateral corner of the dorsal occipital spine to the lateral occipital spine must have covered a ridge connecting the postorbital process and subotic shelf, as in other bradyodonts and chimaeroids (Bendix-Almgreen, in this Volume), and the lateral occipital spine and the plates in front of it must have covered the margin of the subotic and subocular shelf. Anteriorly, at the level of the tips of the third frontal spines and the foremost plates on the subocular shelf, the radiographs (Figs. 2, 6) show that the neurocranium was sharply constricted, the margin turning in almost at right angles and leading to a short and narrow preorbital region, confined between the bases of the two anterior frontal spines. There is no trace of a calcified rostrum. On the anterior surface of the transverse, preorbital neurocranial margin, both the radiographs (Figs. 2, 6) and the right side of the specimen show a thickened, concave area (*md.f.*, Fig. 3) which must mark the articular surface for the mandible. The similarity

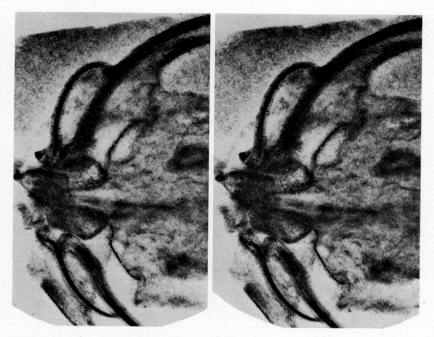

Fig. 6. *Menaspis armata*, same specimen as Fig. 1. Positive prints from a stereoscopic pair of radiographs of the anterior part of the skull, × 1.2.

between the neurocranial structures of *Menaspis* just described and those of *Helodus* (Moy-Thomas, 1936b), *Squaloraja* and chimaeroids (de Beer & Moy-Thomas, 1935) are obvious. Although it is not yet possible to investigate the jaw suspension in *Menaspis*, the similarity in shape between the neurocrania of *Menaspis*, *Helodus* and chimaeroids, which are known to be holostylic, and the presence of the articular surface for the mandible on what appears to be an integral part of the neurocranium indicate that *Menaspis* was also holostylic. The occipital condyles of the neurocranium are discussed below.

The dentition. The dentition of *Menaspis*, exposed in von Ewald's specimen and in several others (Weigelt, 1930), appears to consist of only a single pair of tooth plates in each jaw. In the Halle specimen Jaekel observed only the tips of two tooth plates, which he was unable to interpret. Radiographs (Figs. 2, 6) show that the dentition is completely preserved below the armour of the snout, the two plates seen by Jaekel being the tips of the right and left mandibular tooth plates (*md.t.p.*, Fig. 3), which are displaced forwards and a little to the right. The upper tooth plates lie in their natural position, below and between the bases of the frontal spines, and in fact considerable areas of both plates, especially their thick, inrolled lateral margins (Fig. 7), are exposed in the fossil. In shape and orientation the tooth plates closely resemble the large tooth plates

Fig. 7. *Menaspis armata*, same specimen as Fig. 1. Detail of the anterior part of the skull, photographed under alcohol, showing at top right the tips of the mandibular tooth plates and the base of the symphysial cartilage and "rostrum" (cf. Fig. 3), on the right the bases of the two anterior frontal spines partially overlying the outer margin of the right upper tooth plate, and (centre) the structures tentatively identified as anterior upper tooth plates (*a.t.p*); × 6.

of *Deltoptychius* and *Myriacanthus* (Patterson, 1965, figs. 16, 31), and the radiographs show the typical reticulate pattern of "tubular dentine" over the whole surface of the crown (Fig. 6).

Lying between the tips of the upper tooth plates in the fossil there are two elongate structures (*a.t.p.*, Figs. 3, 7), the more anterior lying at an angle to the posterior and more completely exposed, which consist of dense, glossy hard tissue, with no sign of vascular canals, and which are divided by constrictions into a series of three or four bead-like swellings. These structures lie below the cartilage of the roof of the neurocranium, which has been abraded away above them, and must be interpreted as having lain in the roof of the mouth. The radiographs do not help in interpreting these structures since they lie over the tooth plates in an area of dense shadow. Investigation of other specimens of *Menaspis* is necessary before any conclusions can be formed about these structures, but they lie in the position of the small anterior tooth plates in the upper jaw of *Deltoptychius* and the myriacanthoids. The bead-like swellings on the anterior upper tooth plates of *Myriacanthus* (Woodward, 1891a, pl. 2, fig. 1) should also be mentioned.

The visceral skeleton. Radiographs (Figs. 2, 6) show that the lower dentition is displaced forwards and to the right relative to the upper, the left lower tooth plate lying slightly anterior to the right. Relative to the frontal spines, which appear undisturbed, the left mandibular spine has moved forwards and to the right, the right mandibular spine back and to the left. The bases of the mandibular spines surround a mass of calcified cartilage which disappears beneath the frontal spines on the left side, but on the right side can be seen to pass *below* the margin of the mandibular tooth plate. The radiographs show that the mass of cartilage emerging from the left mandibular spine has moved forwards with the spine and is separated from the neurocranium by a gap. A pair of thickened facets has been described on the anterior surface of the preorbital part of the neurocranium, and on the right side of the fossil a corresponding concave, thickened area (*md.f.*, Fig. 3) can be seen on the cartilage embraced by the mandibular spine. Each of these facts indicates that the cartilage surrounded by the mandibular spines is Meckel's cartilage: taken together they prove that the mandibular spines were borne on the cartilage of the mandible, as they were thought to be both in *Menaspis* and *Deltoptychius* by Moy-Thomas (1936a: 784). Since they are not attached to the pectoral girdle they are not homologous with the spinal plates of arthrodires, as Stensiö (1925: 188, footnote) suggested they might be.

Cleaning of the specimen has exposed a tongue-like strip of calcified cartilage (*sy.c.*, Fig. 3) projecting from the tip of the snout, between the mandibular tooth plates. This strip of cartilage is bilaterally symmetrical, and emerges from *below* the mandibular tooth plates: it is therefore not a rostrum but a submandibular structure, and since it appears to be median, not paired, it can most reasonably be compared with the symphysial cartilage of *Myriacanthus* (Patterson, 1965, fig. 13) and embryonic *Callorhynchus* (Holmgren, 1942, fig. 25), rather than with the premandibular cartilages of chimaeriforms, which are primarily paired structures (Holmgren, 1942: 241, figs. 49–53).

No trace of the branchial skeleton is visible, either on the surface of the fossil or on the radiographs, but the position of the pectoral girdle and fin, which lie close behind or even below the occipital part of the skull, indicates that the branchial arches must have lain below the neurocranium, as in *Helodus*, edestids and chimaeriforms.

Postcranial skeleton. Of the postcranial skeleton, Jaekel mentioned only the pectoral and pelvic fins and the pelvic girdle, writing (1891: 122) "of the vertebral column nothing at all is preserved, it was therefore uncalcified". In fact, a great deal of calcified cartilage is preserved below the squamation of the trunk (Fig. 8).

From the level of the hind edge of the pectoral fin to the posterior end of the specimen there is clear evidence of a series of postero-dorsally directed rods of cartilage (*n.sp.*, Fig. 8). These are interpreted as neural arches and spines, not dorsal fin radials, since they are broader proximally than distally and since they are in series anteriorly with the synarcual (see below). To the right of the neural arches in front of the pelvic girdle there are traces of a series of shorter, broader rods of calcified cartilage (*bv.c.*, Fig. 8), the first few inclined antero-ventrally, the remainder postero-ventrally: these must be basiventral cartilages or short ribs, and the space between them and the neural arches must have been occupied by the uncalcified notochord. The series of basiventrals or ribs can be traced forwards to the level of the pectoral girdle, but the neural arches and spines end farther back, at the level of the hind edge of the pectoral fin: anterior to this there is an elongate mass of cartilage whose postero-dorsal margin seems to be split into three finger-like processes, resembling the succeeding neural spines and evidently serially homologous with them (*syn.*, Fig. 8). This mass of cartilage thus forms the anterior part of the vertebral column and can only be a long synarcual, similar to those of chimaeriforms, batoids and some arthrodires.

Immediately behind the neurocranium there are two saucer-like structures of calcified cartilage. The larger one (*occ.syn.*, Fig. 8) lies behind the median protuberance on the hind edge of the neurocranium and as preserved its concavity faces forwards and to the left. The smaller one (*?l.glen.*, Fig. 8) lies medial to the base of the right dorsal occipital spine and as preserved faces upwards. In radiographs (Figs. 2, 9) the large, median structure throws a dense, crescentic shadow and there is a second similar shadow to the right of it; the smaller structure on the right throws a weaker, circular shadow. Directly anterior to the two median crescentic shadows there are two dense shadows below the posterior part of the neurocranium, V-shaped with thickened bases. Stereoscopic radiographs (Fig. 9) show that these structures project postero-ventrally below the neurocranium: they appear to be cartilaginous, not dermal, although it is impossible to be certain of this. The most reasonable interpretation of these two pairs of structures, a pair of cartilaginous knobs projecting postero-ventrally from the rear of the braincase with a pair of condyle-like depressions behind them, is that they represent the occipital joint between the braincase and the synarcual. The paired postero-ventral projections carrying the occipital condyles of the neurocranium seem to agree with *Helodus* (Moy-Thomas, 1936*b*, figs 4, 6, 7; BMNH P. 8212) in which there is a pair of heavily calcified ridges, projecting well below the flat ventral surface of the neurocranium and growing

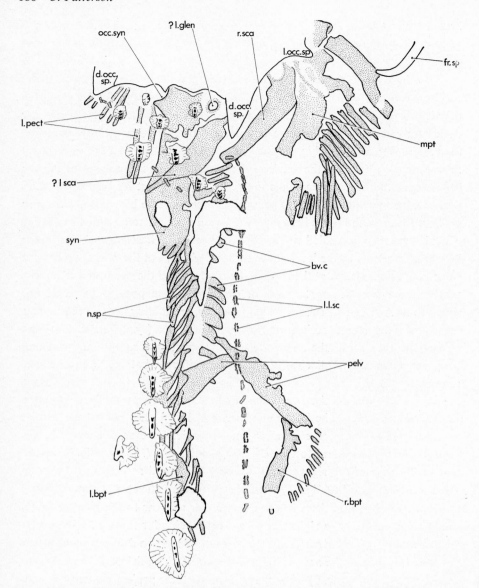

Fig. 8. *Menaspis armata*, same specimen as Fig. 1. Postcranial skeleton; approx. ×1.5. The drawing has been simplified by the omission of all but the largest scales. Stippled areas indicate calcified cartilage. The rear of the skull is shown in outline with the dorsal occipital spines (*d.occ.sp*), right lateral occipital spine (*l.occ.sp*) and the tip of the second right frontal spine (*fr.sp*).

bv.c, basiventral cartilages or ribs; *l.bpt*, left pelvic basipterygium; *? l.glen*, possibly glenoid facet of left pectoral girdle; *l.l.sc*, modified scales carrying the lateral line; *l.pect*, left pectoral fin; *?l.sca*, possibly left scapula; *mpt*, metapterygium of right pectoral fin; *n.sp*, neural spines; *occ.syn*, occipital condyle of synarcual; *pelv*, left and right halves of pelvic girdle; *r.bpt*, right pelvic basipterygium; *r.sca*, right scapula; *syn*, synarcual.

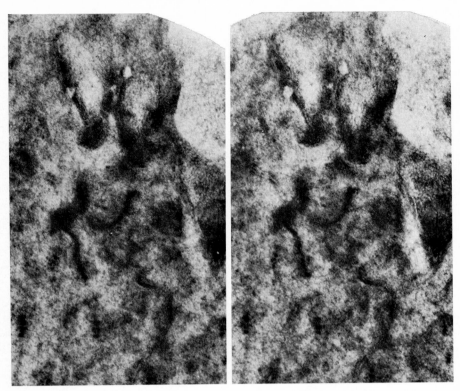

Fig. 9. *Menaspis armata*, same specimen as Fig. 1. Positive prints from a stereoscopic pair of radiographs of the occipital region of the skull and vertebral column, ×3.

deeper and narrower posteriorly, enclosing the aortal groove between them and bearing the occipital condyles behind.

Jaekel described the right pectoral fin as originating below the lateral occipital spine ("der rechten Seitenecke") and containing a broad anterior radial and about 19 more slender radials: I can find traces of only 17 radials behind the broad anterior one (Fig. 8). The first seven or eight radials are parallel, the posterior ones coming to lie antero-posteriorly. The last radial appears thicker than its predecessors, and is probably compound. The radials are unsegmented so far as they are preserved. Jaekel did not mention the pectoral girdle or the basals of the fin. The greater part of the broad right metapterygium (*mpt.*, Fig. 8) is preserved, but the lateral occipital spine of the skull covers the anterior part of the basals and it is impossible to be certain whether the fin was dibasal, as in *Helodus* and chimaeriforms, although the visible part of the metapterygium suggests that it was. An oblique line across the metapterygium opposite the fourth radial is not a line of junction, but appears to be due to crushing of the cartilage over part of the girdle. The coracoid part of the right pectoral girdle and the glenoid facet are concealed by the lateral occipital spine, but a shadow on the radiographs (Fig. 2) suggests that the broad coracoid may have curved medially below the plates covering the anterior part of the supraorbital ridge of the skull. A long, slender, postero-dorsally directed scapula of typical elasmobranch type (*r.sca.*, Fig. 8) extends back behind the right dorsal occipital spine: the scapula is evidently somewhat displaced since near its tip it lies external to a large scale.

Jaekel figured traces of the left pectoral fin projecting from the margin of the trunk, writing that these traces were destroyed during preparation. Nothing remains in this position, but fragments of the radials are visible behind the left dorsal occipital spine and below the scales behind this (*l.pect.*, Fig. 8). The position of the left pectoral fin suggests that the girdle must be crushed below the skull, but it cannot be seen in the radiographs. Emerging from below the right dorsal occipital spine, just in front of the tip of the right scapula, and passing medially over the exposed surface of the synarcual, there is strip of calcified cartilage (*?l.sca.*, Fig. 8) which resembles the right scapula in shape and size: possibly this is the displaced left scapula. If so, the saucer-like structure just in front of it (*?l.glen.*, Fig. 8) may be the glenoid facet of the left pectoral girdle. This interpretation of the structure is speculative, but no other can be offered at present.

The two separate halves of the pelvic girdle (*pelv.*, Fig. 8) are well preserved. Running back from the tip of the right half of the girdle there is an elongate basipterygium (*r.bpt.*, Fig. 8) and there are traces of ten radials, the first probably articulating directly with the girdle, not with the basipterygium. Two or three more radials may be concealed beneath the scales behind the basipterygium. There is no trace of pre-pelvic tenacula or of claspers, but the absence of claspers (if the specimen was not a female) may only be due to incomplete preservation. Traces of the left pelvic basipterygium can be seen beneath the enlarged scales (*l.bpt.*, Fig. 8). Lying along the anterior edge of the distal part of the right pelvic girdle there is an elongate, spine-like structure (Fig. 1), but this consists only of pyrite with a calcite infilling and appears to be an accident of preservation, not part of the fish.

The squamation of the trunk. Jaekel thought that the scaling of the trunk was complete, but the "ganz dicht gedrängtes Chagrin" which he describes between the enlarged scales is the calcified cartilage of the axial skeleton and fin girdles. In fact, the scales were scattered quite sparsely in the skin (Fig. 11 A). The scales are all of the same type, differing only in size. They have a broad, flat basal plate, radially striated and notched marginally, especially in the larger scales (Fig. 8), where the plate is broader than long. In the centre of the basal plate there is a single, longitudinal keel, rising at the rear into a posteriorly directed spine. This keel contains a row of three (in the smallest scales) to five

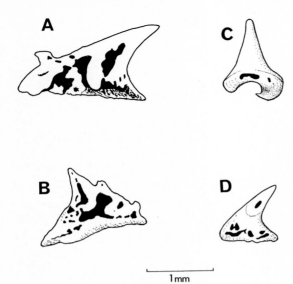

Fig. 10. *Menaspis armata*, scales from the same specimen as Fig. 1. *A*, abraded scale from near the dorsal mid-line of the trunk, anterior to the left; *B*, similar scale from the skull roof, near the "central" plate, anterior to the right; *C*, *D*, simple, spine-like scales from the snout, between the bases of the anterior frontal spines.

(in the largest) separate pulp cavities: in unabraded scales each pulp cavity culminated in a low, partially distinct cusp (Fig. 10). In every detail, these scales resemble the trunk scales of *Deltoptychius* (Patterson 1965, fig. 32).

On the flank, near the lateral line (Fig. 11 A), the scales are fairly dense and of moderate size, 1–2.5 mm long, the largest lying anteriorly. Close behind the skull roof there are five enlarged scales (Fig. 8), about 4 mm long, one median and two pairs, with smaller scales between and around them. Behind these there are small scales closely packed along the dorsal mid-line. From the pelvic girdle back to the edge of the block there is a series of six enlarged scales, growing larger posteriorly, the largest with a basal plate about 10 mm wide and a central keel about 7 mm long. Jaekel thought that these enlarged scales are median, but Patterson (1965: 170) suggested that they are paired: detailed examination of the specimen indicates that the latter interpretation is correct, although all but one of the left-hand series are missing (together with most of the squamation of the left side). As preserved, the belly and paired fins appear to be without scales, but the skin may be missing from these areas.

No explanation can be offered for the patch of six large tesserae or scales lying behind the left dorsal occipital spine (Fig. 1): they have a single central spine or tubercle and therefore resemble neither the tuberculated plates on the skull nor the keeled scales of the trunk. These plates indicate only that the skin of the animal disintegrated partially before fossilization and that there were armoured areas in *Menaspis* of which we still know nothing.

Sensory canals of the head and trunk. The series of modified scales carrying the lateral line of the right side is almost completely preserved, running from the

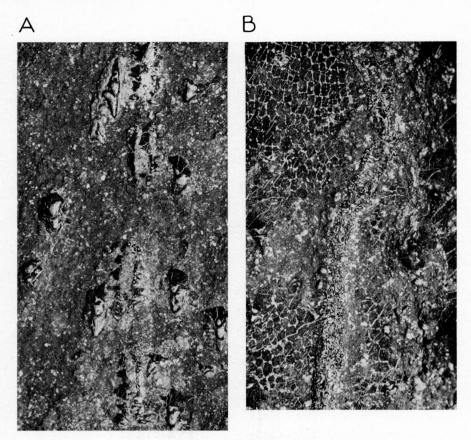

Fig. 11. *Menaspis armata*, same specimen as Fig. 1. Details of the head and trunk, photographed under alcohol, to show the modified scales surrounding the sensory canals; × 10.7. *A*, region to the left of the right pelvic basipterygium (cf. Fig. 8) showing four lateral line ossicles and a few trunk scales; *B*, region between the "central" plate and the right supraorbital ridge (cf. Fig. 3) showing a series of supraorbital sensory canal ossicles overlying the calcified cartilage of the neurocranium.

tip of the scapula back to the edge of the block (*l.l.sc.*, Fig. 8). These structures (Fig. 11 A) are about 2 mm long and consist of a narrow, elongate basal plate carrying four or five pairs of leaf-like denticles arching over the longitudinal groove which contained the sensory canal. Each of these leaf-like denticles contains a row of four or five minute pulp cavities and has the free margin notched or serrated. 29 of these structures are preserved along the right lateral line: a gap in the series would have contained three more. There are also three isolated ossicles lying on the synarcual and a series of three near the tip of the left dorsal occipital spine (Fig. 8), probably representing the left lateral line and apparently emerging from *below* the spine.

On the head Jaekel recognized three sensory canals, two symmetrically placed near the mid-line on the skull roof and one on the right side behind the tip of the third frontal spine: Patterson (1965: 171) interpreted these as the supraorbital and infraorbital canals respectively. The course of these canals is marked by series of incomplete cylinders, directly overlying the cartilage of the neurocranium (Fig. 3), which resemble the modified scales surrounding the lateral line on the trunk except that they are more closely packed, less regular in size, a little broader and appear to have no basal plate (fig. 11 B). These chains of ossicles are clearly visible in radiographs (Fig. 2); no other similar shadows, indicating sensory canals concealed within the matrix, can be identified.

The right supraorbital canal can be traced from the base of the dorsal occipital spine forwards almost to the tip of the snout, the left supraorbital canal is less complete (*so.s.c*, Fig. 3). As in *Deltoptychius*, where the supraorbital canals are similar in structure and position (Fig. 4; Patterson 1965, figs. 34, 64–66), there is no sign of branching or anastomosis along the course of the canal. Anteriorly, there is a series of dermal structures running from the tip of the snout towards the mid-point of the first right frontal spine (*?ros.*, Fig. 3). These structures resemble the ossicles surrounding the supraorbital sensory canals, but are thicker and a little larger. The preservation is not sufficiently good to determine with certainty whether these are a continuation of the supraorbital sensory canals, but this appears probable, suggesting that *Menaspis* may have had a slender, uncalcified rostrum, possibly flexible in life.

The right infraorbital sensory canal can be traced from the tip of the third frontal spine back to the antero-lateral corner of the dorsal occipital spine. On the left side the infraorbital canal is concealed by the supraorbital ridge, but a few ossicles can be seen near the tip of the third frontal spine (*io.s.c.*, Fig. 3).

It is impossible to be certain whether the lateral line of the trunk anastomosed with the supraorbital or infraorbital canal, but the absence of any direct connection with the supraorbital canal, although the dorsal side of the fish is well preserved, the fact that the left lateral line appears to emerge from *below* the dorsal occipital spine, and comparison with *Deltoptychius moythomasi* (Patterson 1965: 168), in which the supraorbital canal runs out to the tip of the projecting dorsal occipital spine and can hardly have met the lateral line, all suggest that the lateral line was continuous with the infraorbital canal.

Discussion. The limited information previously available on the structure of *Menaspis* indicated that its closest known relative is the Lower Carboniferous *Deltoptychius* (Moy-Thomas 1936*a*: 783; Patterson 1965: 171), which resembles *Menaspis* in the form of the head-shield and tooth plates, the presence of paired, spine-bearing plates on the ventro-lateral surface of the neurocranium (the lateral occipital spines of this paper), of paired spines on the mandible and of paired rows of enlarged scales along the back. New information on the head-shield of *Menaspis* confirms the resemblance to *Deltoptychius* in the dorsal occipital spines, supraorbital ridge, narrow preorbital region, etc., indicating that the neurocranium of *Deltoptychius* was very like that of *Menaspis*; this together with the similarity in the shape and position of the upper tooth plates, confirms that *Deltoptychius* must have been holostylic, like *Menaspis*. Additional evidence of relationship between *Menaspis* and *Deltoptychius* is provided by the presence in *Menaspis* of a synarcual, a structure tentatively identified in *Deltoptychius* (Patterson 1965: 162), and by the structure of the scales in *Menaspis*, which are exactly as in *Deltoptychius*. *Deltoptychius* is so incompletely known that no other useful comparisons can be made.

Menaspis was found to differ from *Deltoptychius* (Patterson 1965: 171) principally in the fragmentation of the head shield, the apparent absence of anterior tooth plates in the upper jaw, and in the large frontal spines. Certainly the armour of the skull roof in *Menaspis* (Fig. 3) is greatly reduced in comparison with *Deltoptychius* (Fig. 4), and also the lateral occipital spine is borne on one of a series of plates rather than on a single elongate plate. But the Halle specimen of *Menaspis* is one of the smallest known (Weigelt, 1930), and in the holotype (Weigelt, 1930, pl. 8, fig. 1) a cast (BMNH P. 15346) suggests that the plates which surround the lateral occipital spine in the Halle specimen are fused into a single plate, as in *Deltoptychius*. In the Halle specimen a few of the plates on the skull roof are fused, and presumably fusion would increase as the animal grew. The microstructure of the frontal spines of *Menaspis* (Fig. 5) proves that they are dermal structures, not cartilaginous. This and the presence of tubercles on the expanded bases of the spines indicates that they are probably hypertrophied dermal tubercles or scale cusps, and the spines are therefore no obstacle to close relationship with *Deltoptychius*: possibly they are homologous with enlarged tubercles lateral to the supraorbital sensory canal in front of the supraorbital ridge in *Deltoptychius* (Fig. 4). Finally, there is a possibility that *Menaspis* possessed small anterior upper tooth plates (Fig. 7) like *Deltoptychius*, but the evidence for this must be regarded as a subject for investigation rather than a basis for discussion. In summary, the information on *Menaspis* now available shows that it is closer to *Deltoptychius* than was previously supposed, and it is even doubtful whether familial separation of the two (Patterson, 1965) can be justified.

Turning to a comparison of *Menaspis* and *Deltoptychius* with other bradyodonts (Bendix-Almgreen, in this Volume), they agree with the better known forms, *Helodus*, *Chondrenchelys* and the edestids (*Ornithoprion* excluded), in all the characters common to the latter: the holostylic suspension, large orbits, short otic and occipital regions, position of the branchial skeleton below the braincase (inferred in *Menaspis* from the position of the pectoral girdle and fin) and the high, slender scapula. They differ from all in the extensive armour on the head, though the fusion of scales into splints sheathing the rostrum of

Ornithoprion (Zangerl, 1966: 33) may be comparable in principle. In almost every other feature, the affinities of *Menaspis* and *Deltoptychius* are with *Helodus*, not with *Chondrenchelys* or the edestids. *Menaspis* resembles *Helodus* and differs from *Chondrenchelys* and the edestids in the short preorbital region and long, broad orbito-temporal region of the braincase, in the position of the mandibular articulation below the anterior part of the orbit, the short palatoquadrate and short, broad mandible, the presence of a synarcual, the structure of the pectoral fins, with a broad anterior radial and a long metapterygium, and the structure of the pelvic fins (otherwise known only in *Chondrenchelys*). This is an impressive list of similarities. In many of these features, especially those of the neurocranium, jaws and vertebral column, *Chondrenchelys* resembles the edestids, but in the pectoral fins *Chondrenchelys* and the edestids differ from each other as much as they do from *Helodus* and *Menaspis*. In features of the dentition to which Bendix-Almgreen (in this Volume) draws attention, the presence of a "selachian" type of dentition without a symphysial interspace in the upper dentition, or of tooth plates with such an interspace, *Chondrenchelys* falls with *Helodus* and *Menaspis* in opposition to the edestids.

If these comparisons have any meaning, it is that *Menaspis* and *Deltoptychius* are more closely related to *Helodus* than to any other well known bradyodont, these three genera forming a group which is closer to *Chondrenchelys* than the edestids in the dentition. The differences between *Menaspis* and *Deltoptychius* on the one hand and *Helodus* on the other include the presence of a dorsal fin spine and the absence of armour on the head in *Helodus*, and the few large tooth plates and cyclomorial scales of *Menaspis* and *Deltoptychius* contrasted with the many small teeth and synchronomorial scales (Patterson, 1965, fig. 41) of *Helodus*. These are not insuperable obstacles to close relationship: in the fin spine and armour there is an analogous difference among Lower Jurassic chimaeriforms between *Squaloraja* and the myriacanthoids, while features of the dentition discussed below seem to confirm that *Helodus*, *Deltoptychius* and *Menaspis* are related.

The jaws of Cochliodus

In the bradyodont family Cochliodontidae, a great many species have been erected on the basis of isolated tooth plates, but in one or two species parts of the jaws are also preserved.

The holotype of the Lower Carboniferous *Cochliodus contortus* Agassiz (1838: 115, pl. 19, fig. 14; BMNH P. 2424), the type-species of the genus, later figured by Davis (1883, pl. 52, fig. 3), Woodward (1889, fig. 8; 1932, fig. 140) and others, has been removed from the rock with acetic acid. This reveals (Fig. 12) that the calcified cartilage of the fused rami of the jaw is almost completely

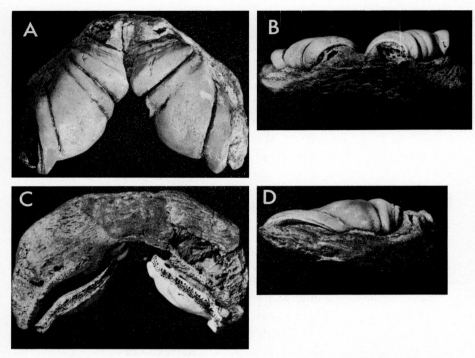

Fig. 12. *Cochliodus contortus* Agassiz. Mandible (the holotype, BMNH P. 2424) from the Carboniferous Limestone of Armagh, Ireland, in dorsal (*A*), anterior (*B*), ventral (*C*) and right lateral (*D*) views, nat. size.

preserved below the tooth plates. The question whether this specimen is the upper jaw or the lower (as has been assumed by several writers) can be answered only by analogy with other fishes. Tooth plates similar to those of *Cochliodus* have been found *in situ* in the skull only in *Deltoptychius* and *Menaspis*: in the form of the large, lateral tooth plates and in the angle of divergence between the rami of the jaw the specimen of *Cochliodus* resembles the mandibles of *Deltoptychius* and *Menaspis*, not the upper jaw. Also, in *Menaspis*, *Deltoptychius*, *Helodus* and edestids the upper dentition is borne directly on the neurocranio-palatoquadrate complex, not on a bar of cartilage. In *Helodus*, edestids (excepting *Ornithoprion*), myriacanthids and chimaeroids the rami of the lower jaw are completely fused at the symphysis, as in the specimen of *Cochliodus*. The only living elasmobranchs in which both Meckel's cartilages and the palatoquadrates are completely fused at the symphysis are the most specialized batoids, myliobatids such as *Aetobatus* (even in *Rhinoptera*, where the dentition is highly specialized, the rami of the jaws are partially distinct). In *Aetobatus* the fused palatoquadrates have a strongly concave upper surface, but the undersurface of the mandible resembles the specimen of *Cochliodus* closely. Also, it is difficult to imagine how a smooth, convex aboral surface, as in the jaw of *Cochlio-*

dus, could articulate with the braincase as a palatoquadrate. All these points confirm that the specimen is the lower jaw and its dentition.

The ventral surface of the mandible (Fig. 12) is smooth and convex in the longitudinal plane (a depression on the right side is due to crushing), and consists of dense, fibrous calcified cartilage, the fibres running transversely. The antero-lateral surface of each ramus slopes outwards away from the bases of the tooth plates and meets the ventral surface in a projecting ledge: the mandible of *Helodus simplex* (BMNH P. 8212) shows a similar ledge. Posterolaterally, the rami of the jaw taper to blunt points, with no sign of articular facets, but the cartilage is abraded ventrally and evidently the posterior parts of both rami and the articular surfaces are missing. In a specimen figured by Owen (1867, pl. 4, figs. 2–5, as *"Tomodus convexus"*, see also Davis, 1883, pl. 52, fig. 6), although the tooth plates are broken and distorted the tip of the right ramus appears to be complete, showing that the mandible of *Cochliodus* was short, broad and deep, as in *Helodus*. The smooth, tapering postero-lateral parts of the jaw rami and the absence of associated dermal plates in this or any other specimen show that in *Cochliodus*, in contrast to *Deltoptychius* and *Menaspis*, there were no mandibular spines or plates.

The two pairs of tooth plates are well known, and need not be described except to mention that there is a wear surface on the broad, central swelling of both posterior tooth plates. Anteriorly, the bolster-like cartilage which supports the tooth plates continues beyond the anterior tooth plate, suggesting that a third pair of tooth plates was present in life, as Agassiz surmised.

If this specimen, other less complete examples of the same species (Owen 1867, pl. 3, figs. 1–5, pl. 4, figs. 2–5; Davis 1883, pl. 52, figs. 1, 2, 6) and similar American specimens (St. John & Worthen, 1883, *Cochliodus vanhornii*, pl. 7, fig. 7; *Poecilodus sanctiludovici*, pl. 8, fig. 8) are mandibles, the question arises of the upper dentition in *Cochliodus*. On the basis of dissociated tooth plates which appear to belong to a single individual, Newberry & Worthen (1866: 89) suggested that the upper jaw of *Cochliodus* contained tooth plates of the type named *Streblodus* (see also Jaekel, 1926, fig. 36; Obruchev, 1964: 239; etc), and in particular, that the principal upper tooth plates of *Cochliodus contortus* are those named *Streblodus oblongus* (Portlock), which occur in the same beds and are about as abundant as the mandibular tooth plates. That this is correct is indicated by the discovery of the complete dentition of *Deltoptychius armigerus* (Moy-Thomas, 1936*a*, fig. 17; Patterson 1965: 161, 168, fig. 31), in which the principal upper tooth plates are very like *Streblodus colei*.

One specimen of *Streblodus oblongus* shows the tooth plates of both rami in position. This specimen (BMNH P. 2414; Davis 1883, pl. 53, fig. 1) has now been partially removed from the rock with acetic acid (Fig. 13). Davis (1883: 424) and Woodward (1889: 210) thought that the two rami of the jaw are crushed

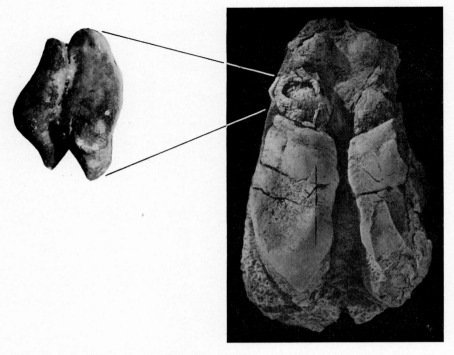

Fig. 13. "*Streblodus oblongus* (Portlock)", inferred to be the upper dentition of *Cochliodus contortus*. Incomplete upper jaws (BMNH P. 2414), Carboniferous Limestone, Armagh, Ireland, in ventral view, nat. size. On the left is an enlarged (× 5) lateral view of two *Helodus*-like tooth crowns from the anterior tooth plate on the left.

together in this specimen, having been more widely divergent in life, but this is not so since the two rami are joined throughout their length by a flat, almost undisturbed sheet of calcified cartilage. The rami of the upper jaw in *Cochliodus* evidently diverged much less than those of the lower jaw, just as in *Deltoptychius* and *Menaspis*. The sheet of calcified cartilage connecting the two rami is thin and is entirely different from the bulky cartilage of the lower jaw (Fig. 12). This sheet of cartilage must have formed the roof of the mouth and can only be the floor of the neurocranium. Since it is fused with the cartilages carrying the upper dentition, *Cochliodus* must have been holostylic, like *Helodus*, *Deltoptychius* and *Menaspis*. The specimen shows the two large posterior tooth plates ("*Streblodus oblongus*"), that of the left side (right in the figure) being much damaged, that of the right side bearing a deep wear facet. In front of the right principal tooth plate there is a fragmentary tooth plate preceded by a hump of cartilage; in front of the left tooth plate only two humps of cartilage, both anterior tooth plates having been lost. The fragmentary anterior tooth plate on the right side, although all the coronal surface is lost, confirms Woodward's observation (1889: 210) that it is not simply a rounded plate, as are the anterior tooth plates in the lower jaw, but consists of a transverse series of small *Helodus*-like teeth (Fig. 13) borne on a common base of lamellar tissue (see below). Presumably this is also true of the foremost tooth plates, missing on both sides. Newberry & Worthen (1866: 89, pls. 6, 7), on the basis of the dispersed dentition of a single individual, concluded that in *Cochliodus latus* there were two pairs of partially fused whorls of *Helodus*-like teeth in front of the principal upper tooth plates.

From these specimens the following conclusions can be drawn:

(1) The jaw suspension in *Cochliodus* was holostylic, as in other bradyodonts.

(2) As in more completely known fishes with cochliodont tooth plates (*Deltoptychius*, *Menaspis*) and as in myriacanthid chimaeriforms (Patterson 1965, fig. 16), the rami of the lower dentition diverged much more strongly than the upper.

(3) *Cochliodus* differs from *Deltoptychius* and *Menaspis* in lacking dermal spines on the mandible, in having at least two, probably three pairs of mandibular tooth plates, and in the structure of the two anterior pairs of upper tooth plates, which are whorls of small, *Helodus*-like teeth.

(4) In the shape of the mandible and the absence of dermal armour *Cochliodus* resembles *Helodus*, and the anterior upper teeth of *Cochliodus* form a link between the tooth plates of *Deltoptychius* and the whorls of small teeth in *Helodus*.

Tooth structure in bradyodonts

Bendix-Almgreen (in this Volume) discusses the significance of the histological structure of the bradyodont tooth crown in the light of recent work (Radinsky, 1961; Ørvig, 1967). He concludes that the presence of coronal pleromic hard tissue (Ørvig, 1967: 104) cannot be considered evidence of relationship and that in consequence, all groups known only by isolated tooth plates, such as the psammodonts and copodonts, must be considered *incertae sedis*. This is a sound conclusion so far as the crown of the tooth is concerned (although the absence of pallial dentine in all known bradyodonts is possibly significant). But it should be pointed out that in well preserved tooth plates of copodonts (*Copodus*), psammodonts (*Psammodus*), cochliodonts (*Cochliodus*, *Sandalodus*, etc.), menaspoids (*Deltoptychius*) and helodonts (*Helodus*, *Psephodus*) the base of the root is closed by a layer of dense tissue laid down in parallel lamellae (Fig. 14). In *Copodus* (Fig. 14 A) this layer is very thin and almost structureless, in *Psammodus* (Fig. 14 B) it is thicker, and in cochliodonts (Fig. 14 E), menaspoids (Fig. 14 F) and helodonts (Fig. 14 C, D) it may make up almost half the thickness of the root. This lamellar tissue is acellular and is penetrated from the base by long, sinuous, branching tubules resembling those in the basal lamellar tissue of the dermal armour of *Deltoptychius* and *Myriacanthus* (Patterson,

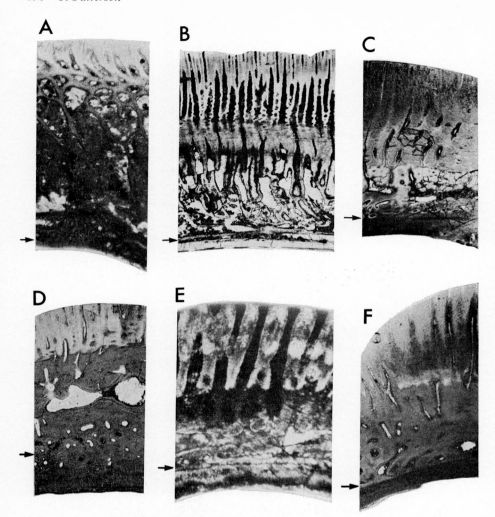

Fig. 14. Thin sections of Carboniferous bradyodont teeth. A, *Copodus spatulatus* Davis, BMNH P. 5363, Carboniferous Limestone, Ticknall, Derbyshire; approx. ×20; B, *Psammodus rugosus* Agassiz, BMNH P. 2578, Carboniferous Limestone, Armagh, Ireland; approx. ×7; C, *Psephodus magnus* (M'Coy), BMNH 46814, Carboniferous Limestone, Derbyshire; approx. ×10; D, *Helodus simplex* Agassiz, BMNH P. 8426, Upper Coal Measures, Collyhurst, near Manchester, Lancashire; approx. ×10; E, "*Streblodus oblongus* (Portlock)" (upper tooth plate of *Cochliodus contortus*), BMNH P. 1406, Carboniferous Limestone, Armagh, Ireland; approx. ×25; F, *Deltoptychius acutus* (M'Coy), BMNH P. 5352, Carboniferous Limestone, Ticknall, Derbyshire; approx. ×20. The arrow on the right of each photograph indicates the upper limit of the basal lamellar tissue of the root.

1965, figs. 35, 15) and in the similar tissue lining the pulp cavity of the fin spines of chimaeriforms and many sharks (Patterson, 1965: 113, 195; Ørvig, 1967: 102, footnote). In selachians, although "tubular dentine" like that of bradyodonts develops in *Asteracanthus*, *Ptychodus* and rays, the root is always trabecular throughout its depth and the basal surface is porous, not smooth and dense as

Fig. 15. *Metopacanthus granulatus* (Agassiz). Thin section of postero-lateral part of left mandibular tooth plate from Geological Survey TN 8005 (Dean 1906, fig. 141), Lower Lias, Lyme Regis, Dorset; × 13. The anterior border of the tooth plate is to the left, the oral surface upwards. The pleromic hard tissue in the tritoral area (the central hump on the oral surface) is decalcified in this specimen, as it is in several myriacanthid specimens from the Lias clay.

in the bradyodonts mentioned above (see figures in Agassiz, 1833–44; Casier, 1947; 1953; etc.). The presence of a basal layer of lamellar tissue in the root of the tooth suggests that *Psammodus* is related to the helodonts, cohliodonts and menaspoids. In *Copodus* the basal layer of lamellar tissue is very thin, but it may indicate similar relationships. The great thickness of the lamellar tissue in helodonts, cochliodonts and menaspoids is further evidence that these groups are closely related, supporting the conclusion drawn above from the structure of *Menaspis* and from the jaws of *Cochliodus*. In *Chondrenchelys*, lack of material has prevented me from preparing thin sections, but broken tooth plates suggest that there was a thick basal layer of lamellar tissue, indicating relationship with the groups mentioned above. There seems to be no basal lamellar tissue in the teeth of petalodonts, nor in *Orodus* or the edestids (Nielsen, 1932, pls. 4–7, 10; Zangerl, 1966, figs. 16, 21).

Finally, in the earliest and most primitive chimaeriforms, the Lower Jurassic myriacanthids, the basal surface of the tooth plate is not open and trabecular as it is in ptyctodont arthrodires and chimaeroids (Ørvig 1967: 92–95, fig. 44), but is closed by a layer of lamellar tissue (Fig. 15) resembling that in Carboniferous bradyodonts in every way except that it is even thicker and its basal part is vascular, containing vascular canals from which some of the tubules penetrating the upper, non-vascular lamellae arise. This basal lamellar tissue in myriacanthid tooth plates, together with the shallowness of both the tritors and the whole tooth plate, indicates that in these chimaeriforms the pleromic hard tissue in the tritors can hardly have grown continuously, as it appears to do in chimaeroids and ptyctodonts (columnar pleromic hard tissue, Ørvig, 1967: 104), but was closer to the coronal pleromic hard tissue of bradyodonts. Thus in the coronal and basal tissue of the tooth plates, the earliest chimaeriforms seem to resemble the cochliodonts and their relatives rather than the later chimaeroids or the ptyctodonts.

Conclusions

Interrelationships of bradyodonts. The Bradyodonti has been held to comprise the cochliodonts (*Cochliodus, Deltodus, Sandalodus*, etc.), helodonts (*Helodus, Psephodus*), menaspoids (*Deltoptychius, Menaspis*), psammodonts, copodonts, chondrenchelyids (*Chondrenchelys* only), edestids and petalodonts. I agree with Bendix-Almgreen (in this Volume) that there is no good evidence for relating the petalodonts to the other groups. The psammodonts and copodonts are known only by tooth plates, but the combination of a crown consisting of coronal pleromic hard tissue without pallial dentine and a root closed basally by lamellar tissue (very thin in copodonts) seems to me sufficient evidence for relating at least the psammodonts to the other Carboniferous groups sharing these characters, cochliodonts, helodonts, menaspoids and (probably) *Chondrenchelys*.

Bendix-Almgreen (in this Volume) finds that the more completely known bradyodonts, *Helodus*, the menaspoids, *Chondrenchelys* and the edestids, share a number of characters indicating that they are related. Further evidence presented above on the structure of *Menaspis* and the jaws of *Cochliodus* indicates that these fishes fall into three major groups, the edestids, the chondrenchelyids, and a group comprising the cochliodonts, helodonts and menaspoids. Helodonts and menaspoids resemble each other and differ from *Chondrenchelys* and the edestids (so far as these are known) in:

(1) The neurocranium with a short, narrow, preorbital region widening abruptly into a long, broad orbito-temporal region.

(2) Short, broad upper and lower jaws, with the mandibular articulation below the anterior part of the orbit.

(3) Tooth plates or teeth with the root closed basally by a thick layer of acellular lamellar tissue penetrated by branching, sinuous tubules (this tissue is also probably present in *Chondrenchelys*).

(4) Dibasal pectoral fins with a long metapterygium and a broad anterior radial.

(5) Pelvic fins with a long basipterygium and a series of parallel radials.

The cochliodonts (*Cochliodus*) are in some ways intermediate between helodonts and menaspoids, resembling menaspoids in the form of the upper and lower dentition, helodonts in the absence of mandibular spines, and having more tooth plates than the menaspoids with whorls of '*Helodus*' teeth in the upper jaw.

In the dentition, *Chondrenchelys* resembles the helodont-cochliodont-menaspoid assemblage in having large tooth plates, a symphysial interspace in the upper jaw, and probably a basal layer of lamellar tissue in the root, suggesting that it is closer to this assemblage than are the edestids.

How these conclusions are to be expressed in taxonomy depends on the nature of the relationship, if any, between these bradyodonts and the Chimaeriformes.

Relationships between bradyodonts and chimaeriforms. Previously (Patterson, 1965: 205) I reached the conclusion that the only bradyodonts showing evidence of close relationship with chimaeriforms are the menaspoids, *Menaspis* and *Deltoptychius*. This evidence comprised resemblances between the dentitions of menaspoids and myriacanthids (in number, structure and arrangement of the tooth plates), similarities in the dermal armour on the head of menaspoids and myriacanthoids (tuberculated plates or spines on the skull roof, the posteroventral part of the neurocranium and the mandible, consisting of basal acellular lamellar tissue and superficial osteodentine), the presence of modified scales around the sensory canals of the head in menaspoids and chimaeriforms, the similar structure of the paired fins in *Menaspis* and chimaeriforms, and the possible homology between the median frontal clasper of chimaeriforms and the paired frontal spines of *Menaspis* (Patterson, 1965: 175).

New information on *Menaspis* shows that one of these resemblances, the supposed homology of the chimaeriform frontal clasper and the frontal spines of *Menaspis*, is mistaken. The frontal spines of *Menaspis* are dermal structures, probably hypertrophied tubercles or scale cusps, and bear no relationship to the frontal clasper, a cartilaginous structure. The frontal spines must be considered a specialization of *Menaspis* which surely exclude this fish from the direct ancestry of any known form. Nevertheless, in the Lower Carboniferous there existed animals bearing paired spines (the unfortunately named *Menaspacanthus* Patterson) which resemble the frontal spines of *Menaspis* but have an entirely different microstructure (cf. Fig. 5 and Patterson 1965, fig. 37), similar to the calcified fibro-cartilage of chimaeriform frontal claspers.

The remaining similarities between menaspoids and chimaeriforms still stand, and are strengthened by the following:

(1) The neurocranium of *Menaspis* is now known to have been similar to those of *Squaloraja* and the chimaeroids in shape, with a narrow preorbital region, a wide and long orbito-temporal region, short otic and occipital regions and the mandibular articulation below the anterior part of the orbit.

(2) It is now confirmed that the mandibular spines of *Menaspis* and *Deltoptychius* were attached to Meckel's cartilage. The only other elasmobranchiomorphs bearing such structures are the myriacanthoids.

(3) The tooth plates of myriacanthids (Fig. 15) resemble those of menaspoids

(Fig. 14) in the thin tritors and the thick layer of lamellar tissue which closes the root basally and suggests that growth of the pleromic hard tissue was not continuous as it is in chimaeroids.

(4) *Menaspis* had a large median symphysial cartilage in front of the mandible: a similar cartilage is well developed in myriacanthids (Patterson, 1965, fig. 13) and is present in embryonic *Callorhynchus*.

(5) *Menaspis* had a long synarcual, articulating with the braincase by two well developed condyles, as in chimaeriforms.

(6) *Menaspis* had a high, slender scapula, lying close behind the skull, as in chimaeriforms, and the pectoral fin was probably dibasal.

(7) The halves of the pelvic girdle were separate in *Menaspis*, as they are in chimaeriforms.

(8) The only other fishes known to have the cephalic sensory canals surrounded by calcifications like those in *Menaspis* (incomplete cylinders consisting of a series of fused incomplete rings, Fig. 11 B) are the myriacanthids *Myriacanthus* and *Metopacanthus* (Patterson, 1965: 136, 141, fig. 19, pl. 24, fig. 57). Similar calcifications surround the lateral line on the trunk of *Menaspis*, as they do in chimaeriforms.

(9) There is a possibility that *Menaspis* had small anterior upper tooth plates resembling those of myriacanthids (Fig. 7), and a slender, uncalcified rostrum which carried the supraorbital sensory canals (Fig. 3), as in *Squalaroja* and myriacanthids.

These resemblances are strong evidence of relationship. This is especially true of the peculiar characters (tooth structure and arrangement, dermal armour, large symphysial cartilage, sensory canal ossicles) in which menaspoids resemble the myriacanthoids in particular (none of these characters is found in ptyctodonts). While there are still considerable differences between *Menaspis* and the earliest myriacanthoids (absence of symphysial tooth plate, frontal clasper and dorsal fin spine, presence of frontal spines and the peculiar course of the supraorbital sensory canal in *Menaspis*, etc.), I do not believe that a relationship between menaspoids and myriacanthoids can now be doubted. On the other hand, in many of the characters listed above (those of the neurocranium, synarcual, paired fins and girdles) *Helodus* also resembles the chimaeriforms, and as has been shown, *Helodus* is linked with the menaspoids not only by these characters but by features of the jaws and dentition, with the poorly known *Cochliodus* lying somewhere between the two. The characters in which *Helodus* resembles menaspoids and chimaeriforms are just those which distinguish helodonts and menaspoids from other bradyodont groups, the chondrenchelyids and edestids.

In summary, the evidence presented here points to two conclusions. (i) The better known bradyodonts fall into three groups, the edestids, the chondrenchelyids and the helodonts+cochliodonts+menaspoids. (ii) The helodont, cochliodont, menaspoid assemblage is linked with the chimaeriforms by a number of general resemblances in the skull, vertebral column and paired fins, and by special similarities between the menaspoids and the most primitive chimaeriforms, the myriacanthoids.

My previous attempt at a classification of these fishes (Patterson, 1965: 106) was as follows:

Class Holocephali
    Order Chimaeriformes
        Suborder Chimaeroidei
        Suborder Squalorajoidei
        Suborder Myriacanthoidei
        Suborder Menaspoidei
            incertae sedis Family Cochliodontidae
    Order Helodontiformes
    Order Chondrenchelyiformes
    Order Edestiformes

I still believe that the term Bradyodonti serves no useful purpose in formal taxonomy, unless used to dignify a group of families *incertae sedis* (Romer, 1966: 351). If forced to express the conclusions reached in this paper in taxonomic form, I would modify my earlier classification as follows:

Class Holocephali
    Order Chimaeriformes
        Suborder Chimaeroidei
        Suborder Squalorajoidei
        Suborder Myriacanthoidei
        Suborder Menaspoidei
        Suborder Cochliodontoidei
        Suborder Helodontoidei
    Order Chondrenchelyiformes
    Order Edestiformes

I realise that the order Chimaeriformes so constituted is far too comprehensive in comparison with orders in other vertebrate groups. But although tempted to divide it at the convenient break in the record in the Triassic, I believe that the evidence of relationship between the menaspoids and myriacanthoids does not justify this at present.

For the loan of the Halle specimen of *Menaspis* the writer is indebted to Prof. H. W. Matthes. The radiographs of that specimen were taken by Mr. M. Hobdell, King's College, London, and printed by Mr. S. Waterman, University College, London.

References

Agassiz, J. L. R. (1833–43). *Recherches sur les Poissons fossiles.* 5 vols. Neuchâtel.
Bendix-Almgreen, S. E. (1962). De østgrønlandske perm-edestiders anatomi med særligt henblik på *Fadenia crenulata. Medd. dansk geol. Foren.*, **15**: 152–153.
Bendix-Almgreen, S. E. (1966). New investigations on *Helicoprion* from the Phosphoria Formation of South-east Idaho, U.S.A. *Biol. Skr. K. danske VidenskSelsk.*, **14**: 1–54.
Bendix-Almgreen, S. E. (1967). On the fin-structure of the Upper Permian edestids from East Greenland. *Medd. dansk geol. Foren.*, **17**: 147–149.
Casier, E. (1947). Constitution et évolution de la racine dentaire des Euselachii. 2. Étude comparative des types. *Bull. Mus. R. Hist. nat. Belg.*, **23**: 1–32.
Casier, E. (1953). Origine des Ptychodontes. *Mém. Inst. R. Sci. nat. Belg.*, (2) **49**: 1–51.
Davis, J. W. (1883). On the fossil fishes of the Carboniferous Limestone Series of Great Britain. *Scient. Trans. R. Dubl. Soc.* (2) **1**: 327–600.
Dean, B. (1904). In the matter of the Permian fish *Menaspis. Am. Geol.*, **34**: 49–53.
Dean, B. (1906). Chimaeroid fishes and their development. *Publ. Carnegie Instn*, **32**: 1–194.
de Beer, G. R. & Moy-Thomas, J. A. (1935). On the skull of Holocephali. *Phil. Trans. R. Soc.* (B) **224**: 287–312.
Giebel, C. G. A. (1856). Räthselhafter Fisch aus dem Mansfelder Kupferschiefer. *Z. ges. Naturw., Berl.*, **7**: 367–372.
Holmgren, N. (1942). Studies on the head of fishes. An embryological, morphological and phylogenetical study. 3. The phylogeny of elasmobranch fishes. *Acta zool. Stockh.*, **23**: 129–261.
Jaekel, O. (1891). Ueber *Menaspis* nebst allgemeinen Bemerkungen über die systematische Stellung der Elasmobranchii. *Sber. Ges. naturf. Fr. Berl.*, **1891**: 115–131.
Jaekel, O. (1892). Ueber *Dichelodus* Gieb. und einige Ichthyodorulithen, eine Entgegung an Herrn A. Smith Woodward. *N. Jb. Min. Geol. Paläont.*, **1892**:1: 145–151.
Jaekel, O. (1926). Zur Morphogenie der Gebisse und Zähne. 3. Die Arten der Bezahnung. *Vjschr. Zahnheilk.*, **1926**: 217–242.
Miles, R. S. (1967). Observations on the ptyctodont fish, *Rhamphodopsis* Watson. In *Fossil Vertebrates*, ed. Patterson, C. & Greenwood, P. H., *J. Linn. Soc. (Zool.)* **47**: 99–120.
Moy-Thomas, J. A. (1936*a*). The structure and affinities of the fossil elasmobranch fishes from the Lower Carboniferous Rocks of Glencartholm, Eskdale. *Proc. zool. Soc. Lond.*, **1936**: 761–788.
Moy-Thomas, J. A. (1936*b*). On the structure and affinities of the Carboniferous Cochliodont *Helodus simplex. Geol. Mag.*, **73**: 488–503.
Moy-Thomas, J. A. (1939). The early evolution and relationships of the elasmobranchs. *Biol. Rev.*, **14**: 1–26.
Newberry, J. S. & Worthen, A. H. (1866). Descriptions of new species of vertebrates, mainly from the sub-Carboniferous Limestone and Coal Measures of Illinois. *Geol. Surv. Illinois, Palaeont.*, **2**: 9–134.
Nielsen, E. (1932). Permo-Carboniferous fishes from East Greenland. *Medd. Grønland*, **86**: 1–63.
Obruchev, D. V. (1964). Podklass Holocephali. Tselnogolovuie, ili khimerui. In *Osnovui paleontologii (Fundamentals of Palaeontology)*, ed. Orlov, I. A., **11**: 238–266. Moscow: Acad. Sci. U.S.S.R. In Russian.
Ørvig, T. (1960). New finds of acanthodians, arthrodires, crossopterygians, ganoids and dipnoans in the Upper Middle Devonian Calcareous Flags (Oberer Plattenkalk) of the Bergisch Gladbach-Paffrath Trough. 1. *Paläont. Z.*, **34**: 295–335.

Ørvig, T. (1962). Y a-t-il une relation directe entre les arthrodires ptyctodontides et les holocéphales? *Colloques int. Cent. natn. Rech. Scient.*, **104**: 49–61.

Ørvig, T. (1967). Phylogeny of tooth tissues: Evolution of some calcified tissues in early vertebrates. In *Structural and Chemical Organization of Teeth*, ed. Miles, A.E.W., Vol. **1**: 45–110. New York & London: Academic Press.

Owen, R. (1867). On the mandible and mandibular teeth of cochliodonts. *Geol. Mag.*, **4**: 59–63.

Patterson, C. (1965). The phylogeny of the chimaeroids. *Phil. Trans. R. Soc.* (B) **249**: 101–219.

Radinsky, L. (1961). Tooth histology as a taxonomic criterion for cartilaginous fishes. *J. Morph.*, **109**: 73–92.

Reis, O. M. (1895a). *Ueber die Kopfstacheln bei Menaspis armata Ewald*. 13 pp. München: H. Kutzner.

Reis, O. M. (1895b). On the structure of the frontal spine and the rostro-labial cartilages of *Squaloraja* and *Chimaera*. *Geol. Mag.* (4) **2**: 385–391.

Romer, A. S. (1966). *Vertebrate Paleontology*. 3rd Ed. Chicago: University of Chicago Press.

St. John, O. & Worthen, A. H. (1883). Descriptions of fossil fishes. *Geol. Surv. Illinois, Palaeont.*, **7**: 55–264.

Stensiö, E. A. (1925). On the head of the macropetalichthyids with certain remarks on the head of other arthrodires. *Publ. Field Mus. nat. Hist.: Geol.*, **4**: 87–198.

Weigelt, J. (1930). Wichtige Fischreste aus dem Mansfelder Kupferschiefer. *Leopoldina*, **6**: 601–624.

Woodward, A. S. (1889). *Catalogue of the fossil fishes in the British Museum (Natural History)*. Vol. **1**. London: Br. Mus. (nat. Hist.).

Woodward, A. S. (1891a). *Catalogue of the fossil fishes in the British Museum (Natural History)*. Vol. **2**. London: Br. Mus. (nat. Hist.).

Woodward, A. S. (1891b). Armoured Palaeozoic sharks. *Geol. Mag.* (3) **8**: 422–425.

Woodward, A. S. (1921). Observations on some extinct elasmobranch fishes. *Proc. Linn. Soc. Lond.*, **133**: 29–39.

Woodward, A. S. (1932). *K. A. v. Zittel: Text-book of Palaeontology*. 2nd Ed. Vol. **2**. London: Macmillan.

Zangerl, R. (1966). A new shark of the family Edestidae, *Ornithoprion hertwigi*, from the Pennsylvanian Mecca and Logan Quarry Shales of Indiana. *Fieldiana: Geol.*, **16**: 1–43.

Zangerl, R. & Richardson, E. S. (1963). The paleoecological history of two Pennsylvanian black shales. *Fieldiana: Geol. Mem.*, **4**: i–xii, 1–352.

The origin and basic radiation of the Osteichthyes

By Bobb Schaeffer

Dept. of Vertebrate Paleontology, The American Museum of Natural History,
New York, N.Y. USA

The origin and relationships of the three major groups of higher bony fishes (Actinopterygii, Crossopterygii and Dipnoi) have been disputed for more than half a century. As each group was differentiated by the time of its first known appearance in the fossil record the problem of osteichthyan ancestry remains unsolved, although there is evidence that the acanthodians were involved (Miles, 1965). In regard to intergroup relationships, Jarvik (1944; 1960), in particular, has favored a closer affinity between the Actinopterygii and the Crossopterygii than between either of these and the Dipnoi. By contrast, Romer (1966) has grouped the Crossopterygii and Dipnoi in the category Sarcopterygii, with a taxonomic rank equal to that of the Actinopterygii.

The purpose of this paper is, first, to consider certain character complexes possessed in common by the early actinopterygians, crossopterygians and dipnoans and, second, to attempt some understanding of the basic osteichthyan radiation.

In the course of this work I have had the opportunity to examine various specimens, several of which are unique, in Copenhagen and Stockholm, as well as wax plate reconstructions of *Pteronisculus* (Nielsen, 1942), *Eusthenopteron* (Jarvik, 1954), and *Diplocercides* (Stensiö, 1932).

The problem of ancestry

The characters or character complexes considered below are common to all three major groups of osteichthyans. They are, of course, modified in a somewhat different way in each group (which is generally true for shared characters), but a common design is evident and can be used to assess the probability of common ancestry. The significance of the common characters is obviously made more meaningful by comparisons with the other classes of gnathostome fishes. This is done briefly, but I hope adequately, for present purposes.

(1) Notochord in adult basicranium. A notochordal canal is present in the basal plate of the palaeoniscoids, crossopterygians and Devonian dipnoans (Fig. 3), and the notochord in the adult terminated immediately behind the prootic bridge. Although the notochord was also persistent in the adult placoderm basicranium, it did not extend forward to the hypophysial region (Stensiö, 1963). A strand of connective tissue is all that remains of this structure in the basal plate of primitive and later adult elasmobranchs and holocephalans. The relationship of the notochord to the basal plate is unknown in the acanthodians.

(2) Dermal ossification on palate and mandible. In all the major groups of gnathostome fishes at least some representatives exhibit partial calcification or ossification of the palatoquadrate and the primary lower jaws. Dermal supra- and infragnathal elements are also present in the placoderms, and, of course, teeth or tooth plates in the elasmobranchs, holocephalans and in some acanthodians. The osteichthyans are the only fishes, however, that have extensive dermal ossification covering (or partly replacing) the palatoquadrate and surrounding the primary mandible. Also the roof of the mouth in these fishes is covered, in part or entirely, by the dermal parasphenoid and the vomers. A dermal element resembling the parasphenoid is present in the arthrodires (Stensiö, 1963), but there is no evidence that it is homologous with the osteichthyan parasphenoid. Dermal bones on the oral roof are absent in the elasmobranchs, holocephalans and acanthodians. The last group has a parasphenoid-like basicranial ossification, which according to Miles (1965) is endochondral (see also Miles, in this Volume). It is probable that the dermal elements associated with the osteichthyan primary upper and lower jaws represent fusions or expansions of much smaller dermal tooth plates primitively situated along the jaw margins and on the inner sides of the visceral arches.

(3) Opercular chamber covered by a mobile series of dermal bones. The acanthodians and osteichthyans are unique in having the opercular chamber covered by a mobile complex of dermal ossifications that extend to the mandibular symphysis. There is no direct evidence regarding the origin of the osteichthyan series; it could have arisen from hyoidan branchiostegals, such as those that occur in acanthodians. The submandibular elements situated along the ventral margin of the lower jaw in the rhipidistians are, in Jarvik's (1963) opinion, remnants of the mandibular gill cover (see also Jessen, in this Volume). Certainly in the osteichthyans the opercular complex, with its plate-like opercular and subopercular, is functionally related to the highly efficient pressure-suction pump mode of gill ventilation described by Hughes (1960; 1963) and Hughes & Shelton (1958). The rigid dermal gill cover in most placoderms and the opercular skin flap in the holocephalans indicate a somewhat different mechanism involving the opercular chamber.

(4) Hyoid arch components. Several additional elements are present in the osteichthyan hyoid arch that are not represented in the other major groups of

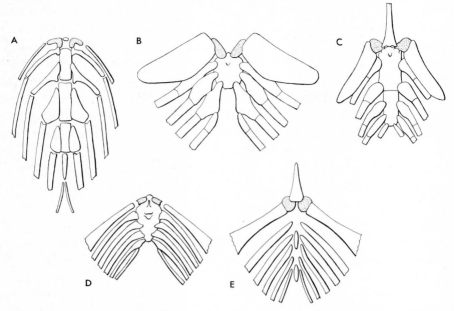

Fig. 1. Basibranchial skeleton in ventral view. *A, Pteronisculus* (after Nielsen, 1942); *B, Glyptolepis* (after Jarvik, 1963); *C, Eusthenopteron* (after Jarvik, 1954); *D, Latimeria* (after Millot & Anthony, 1958); *E, Neoceratodus*. Hypohyals stippled.

fishes. These include the symplectic, interhyal (stylohyal) and the hypohyal. There is still some question about the presence of both a symplectic and an interhyal in the palaeonisciforms and the rhipidistians. Present evidence indicates that there is a single accessory element in contact with the distal end of the hyomandibular and also with the quadrate, and in the palaeonisciforms also with the articular. It may well be that this element is, in fact, the symplectic, and that the interhyal developed only in those groups (later actinopterygians and coelacanths) that have a highly mobile ceratohyal. The situation in *Neoceratodus*, which is as close as we can come at present to the primitive dipnoan condition, is still unsettled, but there is a small element associated with the posterior end of the ceratohyal that probably represents the symplectic.

Although always relatively small and seemingly of little functional significance, the hypohyals are found only in the osteichthyans (Fig. 1). Their consistent presence, however, may, in fact, be related to the important role of the ceratohyal in the lateral expansion of the orobranchial chamber.

(5) Dermal shoulder girdle. As emphasized by Jarvik (1944), the dermal shoulder girdle exhibits the same pattern in the three major osteichthyan groups, and it is probable that the suprascapular (posttemporal), supracleithrum, cleithrum and clavicle are homologous. The anocleithrum of the rhipidistians, Jarvik notes, may be the same as the actinopterygian postcleithrum. This element has not been reported in the Devonian and later dipnoans. The coelacanth postcleithrum is probably the homologue of the anocleithrum. An overlap articulation between the dermal shoulder girdle and the dermal skull roof occurs in all the osteichthyans. In addition to the marked resemblance in the composition and relationships of the dermal shoulder girdle, the scapulo-coracoid in the three groups is relatively small and is attached to the median side of the cleithrum. The dermal shoulder girdle of the placoderms must have evolved independently of the osteichthyan one. The fact that the pectoral girdle in the acanthodians, elasmobranchs and holocephalans is composed only of the scapulo-coracoid, which is a single inflexible unit, may mean that it plays little or no functional role in the expansion of the orobranchial chamber. Increasing participation of the pectoral girdle in the movements of the feeding—respiratory mechanism in the ancestors of the osteichthyans perhaps involved modification of one or more anterior vertical scale rows into the dermal pectoral girdle and reduction of the scapulo-coracoid.

(6) Pleural ribs. Confined to the Osteichthyes, pleural ribs probably represent an expression of increased locomotor efficiency. In adult palaeoniscoids they were not always ossified, but their presence has been noted in a number of genera. Short, robust pleural ribs occur in the rhipidistians (*Eusthenopteron*), and delicate ossified ones are present in some later coelacanths, e.g., *Coelacanthus*, *Diplurus* and *Chinlea*. Among the Devonian dipnoans, *Fleurantia* (Graham-Smith & Westoll, 1937) has long bony ribs, but they were probably cartilaginous in most forms. From these data it may be assumed that a capability for pleural rib chondrification or ossification has existed in the osteichthyans from the time of their origin.

(7) Lepidotrichia. In the Osteichthyes the fin rays are composed of bone primitively covered by dentine and enamel. Palaeoniscoids and less advanced rhipidistians (porolepids and osteolepids) possess the two outer layers. These layers were lost in the rhizodontid and holoptychid rhipidistians, in the coelacanths and in the later dipnoans (Jarvik, 1959). Jointing of the lepidotrichia is the rule rather than the exception, although there are palaeoniscoids with partly fused fin rays (e.g. the haplolepids). The ceratotrichia of the placoderms, sharks and holocephalans may be calcified in varying degrees, but they are not ossified. The unsegmented "ceratotrichia" of some acanthodians are partly ossified (Watson, 1937), but there is no enamel or dentine.

Lepidotrichia are dermal structures long recognized as being derived from scales. In the early osteichthyans they must have developed in much the same way as the scales or teeth, with a dental organ responsible for the deposition of enamel and dentine. As discussed recently by Jarvik (1959), the ectomesenchyme in the fin folds, or the corium that differentiates from it, produces successive generations of cell laminae that give rise to the skeletogenous tissues in the successively produced generations of lepidotrichia. If the fin rays remain in a superficial position, dental organs develop and the rays are covered with enamel or dentine. If the rays sink into the corium early in ontogeny, bone alone differentiates. As neural crest cells play an important role in skeletal morphogenesis, it is reasonable to conclude that the partly ossified acanthodian fin rays could evolve into the primitive osteichthyan type without any fundamental change in the developmental mechanism.

(8) Scale structure. The obvious differences in scale structure between the actinopterygians on the one hand and certain primitive rhipidistians and dipnoans on the other were presumably related to the mode of scale growth. As is well known, the presence of numerous layers of enamel and the frequent presence of several generations of dentine tubercles, indicates that there is no appreciable resorption in palaeonisciform scales, and that these scales grew by the addition of dental tissues and bone to the outer scale surface. It is of interest, therefore, that Gross (1956) in *Porolepis* and Denison (in this Volume) in a Lower Devonian dipnoan have found several tubercle generations in the body scales. This indicates that there was little or no resorption during scale growth, and it suggests that the characteristic cosmoid scale type of the more primitive rhipidistians and dipnoans may have evolved more or less independently in each group in relation to a resorption—redeposition mechanism for scale growth. These remarks also pertain to the dermal skull elements where the resorption—redeposition phenomenon is frequently obvious in the rhipidistians.

(9) Endochondral ossification. The differences in the expression of this character among the early osteichthyans make it difficult to postulate the ancestral condition. Devonian coelacanths have extensive perichondral but little endochondral bone (Stensiö, 1937). The same is true for the dipnoans, although Devonian representatives such as *Chirodipterus* (Säve-Söderbergh, 1952) have more endochondral ossification in the basicranial and occipital regions than younger members of this group. Both palaeoniscoids and rhipidistians have well-ossified endoskeletons. In view of this evidence, it might be argued that the condition in the coelacanths and dipnoans is secondary. There is, however, no proof for this, and I suspect that the degree and location of endochondral ossification, even in the earliest actinopterygians, was adaptive in relation to a complicated interplay between genotypic, kinetic, buoyancy, balancing (around center gravity) and protective factors (Schaeffer, 1961).

Remarks. Any attempt to describe the "first" osteichthyans or proto-osteichthyans is, of course, highly conjectural and essentially typological. Nevertheless, by taking into consideration the common characters discussed above, along with some reasonable suppositions, it is possible to postulate certain attributes of this unknown ancestral stock. These are as follows:

Neurocranium ossified perichondrally and also endochondrally in areas of greatest stress (e.g. around basipterygoid processes). Prechordal and chordal components of basicranium separated, probably in relation to cranial kinesis. Skull moderately platybasic. Pila antotica and prefacial commissure present. Lateral commissures and myodomes probably absent; articulation of hyomandibular dorsal to lateral head vein. Basal plate with notochordal canal; notochord in adult extending forward to prootic bridge. Visceral skeleton mostly under basicranium; numerous dermal tooth plates embedded in oral mucosa of primary jaws, hyoid and branchial arches. Palatoquadrate moveably articulated with braincase in ethmoid region, at basipterygoid process and through the hyomandibular. Dermal skull, except opercular series, composed of small ossifications extending ventrally to upper jaw margin and in part covering primary lower jaw. Parasphenoid covering only prechordal part of basicranium. Dermal elements covering common opercular gill chamber including two relatively large dorsal plates, overlapping branchiostegals and ventral gulars. Typical osteichthyan gill structure and pressure—suction pump gill ventilation mechanism. Notochord unconstricted; neural and haemal elements ossified; centra absent; pleural ribs cartilaginous or ossified. Shoulder girdle composed of reduced scapulo-coracoid and a vertical series of dermal bones in articulation with skull roof. Paired fins composed of radials in parallel series and segmented lepidotrichia. Caudal fin heterocercal. Scales rhomboidal, with layers of enamel, dentine, vascular bone and laminated bone.

On the basis of this provisional characterization and what we know about acanthodian organization (Miles, 1965), there is a greater probability that the osteichthyans arose from some Silurian acanthodian stock than from any other known group of gnathostome fishes. As no other class of fishes apparently possessed the requisite combination of characters to give rise separately to one or more of these major osteichthyan groups, it may be presumed that one or more acanthodian lineages evolved into a primitive osteichthyan stock. It is also probable that this osteichthyan stock had differentiated into actinopterygian, crossopterygian and dipnoan lineages by the Late Silurian.

Basic radiation in the neurocranium

Differences in the palaeonisciform, crossopterygian and dipnoan feeding and locomotor mechanisms indicate significant differences in the way of life. The palaeonisciforms were large-eyed active swimmers and included nibblers, grazers and carnivores. The carnivorous crossopterygians presumably stalked prey by moving along the bottom with their lobed fins and then swimming swiftly toward it once it was sighted. The early dipnoans, as their living representatives indicate, have probably always used their paired fins much like the crossopterygians. They swam sluggishly in search of food that consisted mostly of mollusks and soft-bodied invertebrates.

Although much has been written about the morphology of the head in the three major osteichthyan groups, there has been little interest until recently in the mechanics of the feeding mechanism in the earliest representatives (Schaeffer & Rosen, 1961; Thomson, 1967). One important aspect of this problem, particularly considered by Thomson for the crossopterygians, is the role played by the braincase during the opening and closing of the mouth. Analysis of the adult braincase in terms of its embryonic components demonstrates that these components have been combined in a different way in each group, and it is evident that each of these combinations is related to a basically distinctive feeding mechanism.

Palaeonisciforms. The adult palaeonisciform braincase is usually composed of two distinct ossifications (Nielsen, 1942; Rayner, 1951). The larger one extends from the ethmoid region posteriorly to the lateral occipital fissure, which includes the exit for the vagus nerve (Figs. 2 A, 3 A). The smaller occipital ossification is united with the basicranial plate that contains the notochordal canal and extends forward to the prootic bridge. In terms of the embryonic components (Fig. 4 A), the basal portion of the anterior moiety must represent the trabeculae and the polar cartilages, as it includes the basipterygoid processes. Dorsally the orbital and auditory cartilages were joined through the taenia marginalis, but the auditory cartilage remained separate from the parachordal basal plate and from the occipital arch by the persistent basicapsular and occipital fissures. There is reason to believe, however, that both the anterior and posterior basicapsular commissures were partially developed. The anterior one contributed to the wall behind the posterior exit of the jugular canal and the posterior one separated the glossopharyngeal and vagus nerves. The large vestibular fontanelle that opens into the saccular cavity must represent a persistent part of the basicapsular fissure.

Another feature of the palaeonisciform neurocranium is a connection between the auditory capsule and the polar cartilages. As this connection is lateral to the incipient trigeminofacialis chamber and head vein, it must be the lateral commissure. At this level in actinopterygian evolution, however, there is no connection between the lateral commissure and the postpalatine process of the parachordals (the trabecular and parachordal divisions probably fused

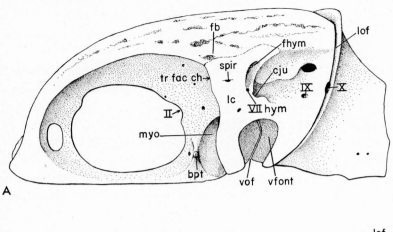
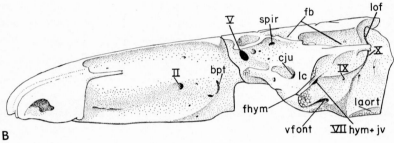
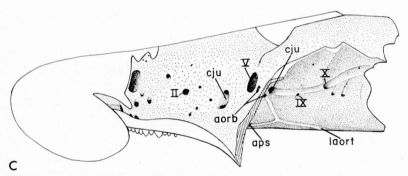

Fig. 2. Neurocrania in lateral view. *A*, *Kentuckia*, a Mississippian palaeonisciform (after Rayner, 1951); *B*, *Eusthenopteron*, an Upper Devonian rhipidistian (after Jarvik, 1954); *C*, *Chirodipterus*, an Upper Devonian dipnoan (after Säve-Söderbergh, 1952).

aps, efferent pseudobranchial artery; *aorb*, orbital artery; *bpt*, basipterygoid process; *cju*, jugular canal; *fb*, fossa bridgei; *fhym*, hyomandibular facet; *laort*, lateral aorta; *lc*, lateral commissure; *lof*, lateral occipital fissure; *myo*, myodome; *spir*, spiracle; *trfacch*, trigemino-facialis chamber; *vfont*, vestibular fontanelle; *vof*, ventral otic fissure; *II*, foramen for optic nerve; *V*, foramen for trigeminal nerve; *VII hym*, foramen for hyomandibular branch of facial nerve; *VII hym + jv*, foramen for hyomandibular branch of facial nerve and jugular vein; *IX*, foramen for glossopharyngeal nerve; *X*, foramen for vagus nerve.

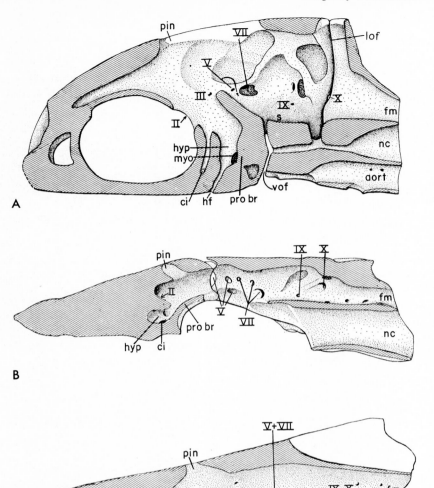

Fig. 3. Neurocrania in mid-saggital section. *A, Kentuckia* (after Rayner, 1951); *B, Ectosteorhachis* (after Romer, 1937); *C, Chirodipterus* (after Säve-Söderbergh, 1952).

aort, orbital artery; *ci*, internal carotid artery; *fm*, foramen magnum; *hf*, hypophysial foramen; *hyp*, hypophysial recess; *lof*, lateral occipital fissure; *myo*, myodome; *nc*, notochordal canal; *pin*, pineal opening; *probr*, prootic bridge; *s*, sacculus; *vof*, ventral otic fissure; *II*, foramen for optic nerve; *III*, foramen for oculomotor nerve; *V*, foramen for trigeminal nerve; *VII*, foramen for facial nerve; *IX*, foramen for glossopharyngeal nerve; *X*, foramen for vagus nerve.

numerous times independently, and when this happened the lateral commissure was joined to the postpalatal process). As the posterior myodome and the trigeminofacialis chamber of *Kentuckia* (Rayner, 1951) are separated by a ridge, and the exits for the main branches of the trigeminal and facial nerves are distinct, the presence of a prefacial commissure in the early palaeonisciforms is implied. Nielsen (1942) has also postulated an incipient pila lateralis in *Pteronisculus*, and Rayner (1951) has suggested that a dorsal remnant of the pila antotica existed between the exits for the oculomotor and profundus nerves.

The persistence of the cranial fissure and the separation of the basal plate from the prootic bridge are difficult to explain unless we assume the presence of neurokinesis. This may have involved a slight elevation and depression of the large anterior ossification in conjunction with the elevation and depression of the palate during opening and closing of the mouth.

Crossopterygians. The two moieties of the braincase undoubtedly represent the trabecular and parachordal divisions of the embryonic basicranium (Fig. 4 B; Romer, 1937). Jarvik (1954) also has observed a more or less independent occipital ossification in *Eusthenopteron* (Fig. 2 B) that is joined ventrally to the basal plate. This inconsistent subdivision suggests that some rhipidistians primitively had an occipital-parachordal segment similar to that of the palaeonisciforms.

The ethmosphenoid ossification includes a pila antotica extending from the posterior part of the orbital cartilage to the polar cartilage and the basitrabecular process. The eyes are relatively much smaller than in the palaeonisciforms, the interorbital region is correspondingly much broader, and myodomes are absent. As in the palaeonisciforms, the prootic bridge is separated from the more posterior part of the basicranium, but in the crossopterygians the division at this level is complete dorsally to form a functional intracranial joint. The joint surface on the anterior moiety thus involves the upper part of the prootic bridge and the posterior portion of the orbital cartilages.

The otico-occipital moiety includes the exits for the branches of the trigeminal nerve, which is thus far removed from the orbit. These differences from the palaeonisciform condition are probably related to the formation of the intercranial joint. In fact, this portion of the crossopterygian neurocranium can best be compared to the actinopterygian by assuming, figuratively, that the palaeoniscoid postorbital wall has swung laterally from the level of the lateral commissure to about that of the optic foramen. Actually, the crossopterygian condition could have arisen embryonically through greater lateral separation of the pila metoptica, pila antotica and the several prefacial commissures. In addition to these "bridges", there was apparently a pretrigeminal commissure between the intercranial joint and the main trigeminal foramen.

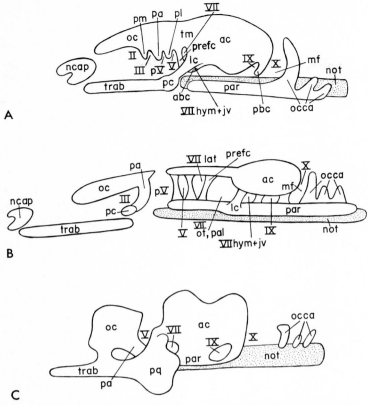

Fig. 4. Components of the embryonic neurocranium. *A*, Palaeonisciform (based on *Kentuckia*, Rayner, 1951); *B*, Crossopterygian (partly after Romer, 1937); *C*, Dipnoan (*Neoceratodus*, partly after de Beer, 1937).

abc, anterior basicapsular commissure; *ac*, auditory capsule; *lc*, lateral commissure; *mf*, metotic fissure; *ncap*, nasal capsule; *not*, notochord; *occa*, occipital arch; *oc*, orbital cartilage; *pV*, profundus branch of trigeminal nerve; *pa*, pila antotica; *par*, parachordal; *pbc*, posterior basicapsular commissure; *pc*, polar cartilage; *pl*, pila lateralis; *pm*, pila metoptica; *pq*, palatoquadrate; *prefc*, prefacial commissure; *tm*, taenia marginalis; *trab*, trabecula; *II*, foramen for optic nerve; *III*, foramen for oculomotor nerve; *V*, foramen for trigeminal nerve; *VII hym+jv*, foramen for hyomandibular branch of facial nerve and jugular vein; *VII lat*, foramen for lateralis branch of facial nerve; *VII ot, pal*, foramen for otic and palatine branches of facial nerve; *IX*, foramen for glossopharyngeal nerve; *X*, foramen for vagus nerve.

The preotic portion of the otico-occipital moiety is variable in length among the different rhipidistians. The trigeminal and facial nerve foramina also alter their relative positions (Thomson, 1967). In the coelacanths the trigeminal has its exit at the intracranial joint. Behind the lateral commissure the posterior moiety is more conservative and can be more readily compared with the same area in the palaeonisciform neurocranium. It includes such features as the spiracular canal, the crista parotica, the fossa bridgei, the lateral commissure, the essentially vertical hyomandibular articulation situated partly on the lateral commissure, and finally the vestibular fontanelle. The characteristic absence of the basal plate in the crossopterygians around the anterior part of the notochord presumably has some functional significance, probably related to cranial kinesis.

In regard to function, it is probable that the kinetic neurocranium of the crossopterygians is related in some way to the feeding mechanism. Even the limited motion of the anterior moiety in the rhipidistians could be explained in this way. Present evidence indicates, however, that the kinetic neurocranium does not function to increase the force of the bite (the forces of the upper and lower jaws, as in the akinetic dipnoan skull, must be equal and opposite). Until further analysis is carried out, the meaning of the kinetism in both the crossopterygian and the palaeonisciform skull remains speculative.

Dipnoans. Most students of the dipnoan skull have been impressed with the close resemblance of the Late Devonian *Chirodipterus* (Säve-Söderbergh, 1952) neurocranium (Fig. 2 C) to that of *Neoceratodus*. Clearly, the development of the akinetic braincase must have been very similar in these two genera (Fig. 4 C). The endochondral ossification of the *Chirodipterus* braincase is not as extensive as in the palaeonisciform or rhipidistian neurocranium, but it is probably concentrated in areas subjected to greatest stress. An important aspect of the dipnoan neurocranium is the fusion of the reduced palatoquadrate to the otic region; correlated with this the upper part of the hyoid arch is greatly reduced. It is also important to note that the massive levator or adductor mandibulae musculature had extended its origin to the posterodorsal surface of the neurocranium beneath the dermal bone cover. The akinetic neurocranium, the autostylic jaw suspension and the extensive origin of the adductor jaw musculature are related to the manner in which the mandible operates for chewing and crushing. Also, the autostylic condition must have reduced the expansion capacity of the orobranchial chamber.

As in the palaeonisciforms and crossopterygians, the notochord of the adult *Chirodipterus* probably terminated just behind the low prootic bridge (Fig. 3 C). As discussed earlier, certain features of the skull, including the opercular series and the pressure—suction pump respiratory mechanism indicate common ancestry with the other major osteichthyan groups in spite of the obvious and ancient dipnoan specializations.

Remarks. The embryonic components of the neurocranium were combined differently in the three major osteichthyan groups, and these combinations were correlated with three independently evolved feeding mechanisms. We have long inferred that the primitive dermal skull of the osteichthyans, like that of the acanthodians, was composed of a mosaic of small units. Sometime before the separation into the three main osteichthyan groups, the suction—

pressure pump respiratory mechanism, including the dermal opercular-branchiostegal series, must have evolved. As the three types of feeding mechanisms were perfected, the dermal bone cover of each was subjected to somewhat different (and somewhat similar) forces, both during feeding and during gill ventilation. It is my belief, albeit unproveable, that the dermal bone cover responded to these forces in a somewhat different way in each group. One major exception to this thesis involves the opercular-branchiostegal series, which is basically the same in all three. The parasphenoid may be another exception.

In the actinopterygians and crossopterygians the forces transmitted to the dermal skull during feeding and respiration, and perhaps even during swimming, favored rather extensive fusion or elimination of the primitive dermal units. These forces were enough alike to bring about somewhat similar patterns, but they were different enough to result in rather distinctive patterns. Because the dipnoan skull is relatively akinetic, the early dipnoans more or less retained the ancestral mosaic of small dermal units. For these reasons, I believe it is practically impossible to homologize dermal elements between the actinopterygians, crossopterygians and dipnoans on a one-to-one basis. In fact, the only reasonably certain homology involves the parasphenoid and the elements of the opercular—branchiostegal series, but even here a one-to-one homology is not probable. It is probable that most of the resemblances in dermal bone patterns among the major osteichthyan groups are primarily related to similar forces produced by the feeding and respiratory mechanisms. The skull roof of the long-snouted Devonian dipnoan *Soederberghia* (Lehman, 1959) graphically demonstrates the independent acquisition of a teleostome-like pattern.

Summary

The characters shared in common by the three major groups of osteichthyans are such that it is reasonable to assume the existence of an ancestral osteichthyan stock that evolved from some unknown acanthodian lineage or lineages sometime during the Silurian. Points of conflict with an acanthodian origin (e.g. posterior inclination of the pharyngobranchials and the scapulo-coracoid shoulder girdle) are not regarded as serious impediments to this hypothesis. If we regard the Acanthodii as a class, the origin would be a monophyletic one in the sense of Simpson (1961), but it may well have involved experimentation with various combinations of ancestral and descendant characters in several lineages that attained the osteichthyan level (Schaeffer, 1965).

It is not difficult to imagine further experimentation leading to the invasion of new environments and exposures to new selection pressures. Certainly many or all of the common characters were preadapted to the changing conditions.

First through modifications in behavior, then through morphological and physiological change, three major sorts of organization, representing the actinopterygian, crossopterygian and dipnoan, came into existence. The first two groups retained important aspects of the ancestral feeding mechanism, including the relationship of the hyomandibular to the palatoquadrate, and the articulation of the latter with the basipterygoid process. Also in these groups, the adductor mandibulae musculature continued to arise only from the lateral surface of the palatoquadrate. The braincase in both was kinetic, but the kineticism evolved in a different way in each through a distinctive union of the embryonic neurocranial components. Presumably the forces produced by these different feeding mechanisms influenced the evolution of the dermal skull, finally resulting in the characteristic patterns of each group. Aside from the opercular-branchiostegal series, the parasphenoid and possibly some of the dermal palatal elements, dermal bone homologies between the two groups must be viewed with suspicion (see discussion of bone fragmentation versus fusion in Lehman, 1966).

The dipnoans obviously deviated more radically from the ancestral pattern than either the actinopterygians or the crossopterygians. Because of the early development of the akinetic neurocranium and the autostylic jaw suspension (which was accompanied by a very different arrangement of the adductor mandibulae musculature), it may be presumed that the forces transmitted to the dermal skull during feeding were less effective in influencing modification of the ancestral dermal bone pattern. The dipnoan branchial skeleton was modified in relation to the autostylic jaw suspension and perhaps in relation to an increase in air breathing. In particular, the dorsal segments of hyoid and branchial arches were affected. Either the hyomandibular was greatly reduced or indistinguishably fused with the neurocranium and palatoquadrate. The pharyngobranchials must have been reduced and at least partly lost early in dipnoan history.

There is ample evidence that ossified centra appeared independently in all the osteichthyan groups except the coelacanths (Schaeffer, 1967b). Rhipidistians evolved several different types that surrounded the notochord while several Devonian dipnoans replaced the notochord with solidly ossified centra —only to lose all central ossification in their subsequent history.

Evidence of experimentation in the structure of the paired fin skeleton is apparent in the cladodont sharks (Schaeffer, 1967a), with one group evolving the archipterygial type. Also the archipterygial pectoral fins must have developed independently in the holocephalan *Chondrenchelys* (Patterson, 1965). A well-known figure of White's (1939, fig. 12 F, G, H, I) indicates considerable diversity in the structure of the actinopterygian pectoral fin skeleton, including modifications that suggest how an archipterygial fin might have arisen in this group (although it actually never happened). In view of this evidence, it is at least possible that the archipterygial fins of crossopterygians and dipnoans arose separately.

On the basis of dermal bone and scale structure it would appear, however, that the rhipidistians and dipnoans are closely related. As noted above, there are only two ways that these structures can grow—either by the addition of calcified tissue around the periphery or by resorption and subsequent deposition of a greater amount of tissue than was present previously. Rhipidistians and dipnoans must have increased the dimensions of the basal bone layer in the same way as the palaeonisciforms, but there is evidence (see Jarvik, 1948: 21–33) that the dentine and even part of the vascular bone layer were periodically resorbed and redeposited in greater thickness. The condition in *Porolepis* and certain Devonian dipnoans (Denison, in this Volume) could mean that the earliest crossopterygians and dipnoans had the same mode of scale growth as the palaeonisciforms and acanthodians. It also suggests that there was experimentation with the two types of scale and bone growth at the time the crossopterygians and dipnoans arose. It is therefore possible that the resorption-redeposition mechanism as well as the related histological structure evolved independently in the two groups.

In conclusion, I can find little unquestionable evidence for believing that any two of the major groups of osteichthyans are more closely related to each other than either is to the third group. At present I favor assigning equal rank to the Actinopterygii, Crossopterygii and Dipnoi within the Class Osteichthyes.

References

Fox, H. (1965). Early development of the head and pharynx of *Neoceratodus* with a consideration of its phylogeny. *J. Zool.*, **146**: 470–554.

Graham-Smith, W., & Westoll, T. S. (1937). On a new long-headed dipnoan fish from the Upper Devonian of Scaumenac Bay, P. Q., Canada. *Trans. R. Soc. Edinb.*, **59**: 241–266.

Gross, W. (1956). Über Crossopterygier und Dipnoer aus dem baltischen Oberdevon im Zusammenhang einer vergleichenden Untersuchung des Porenkanalsystems paläozoischer Agnathen und Fische. *K. svenska VetenskAkad. Handl.*, (4) **5**: 1–140.

Hughes, G. M. (1960). A comparative study of gill ventilation in marine teleosts. *J. exp. Biol.*, **37**: 28-45.

Hughes, G. M. (1963). *Comparative physiology of vertebrate respiration*. Cambridge: Harvard University Press.

Hughes, G. M., & Shelton, G. (1958). The mechanism of gill ventilation in three freshwater teleosts. *J. exp. Biol.*, **35**: 807–823.

Jarvik, E. (1944). On the exoskeletal shoulder-girdle of teleostomian fishes, with special reference to *Eusthenopteron foordi* Whiteaves. *K. svenska VetenskAkad. Handl.*, (3) **21**: 1–32.

Jarvik, E. (1948). On the morphology and taxonomy of the Middle Devonian osteolepid fishes of Scotland. *K. svenska VetenskAkad. Handl.*, (3) **25**: 1–301.

Jarvik, E. (1954). On the visceral skeleton in *Eusthenopteron* with a discussion of the parasphenoid and palatoquadrate in fishes. *K. svenska VetenskAkad. Handl.*, (4) **5**: 1–104.

Jarvik, E. (1959). Dermal fin-rays and Holmgren's principle of delamination. *K. svenska VetenskAkad. Handl.*, (4) **6**: 1–51.

Jarvik, E. (1960). *Théories de l'évolution des vertébrés reconsidérées a la lumière des récentes découvertes sur les vertébrés inférieurs.* Paris: Masson.

Jarvik, E. (1963). The composition of the intermandibular division of the head in fish and tetrapods and the diphyletic origin of the tetrapod tongue. *K. svenska Vetensk Akad. Handl.*, (4) **9**: 1–74.

Lehman, J.-P. (1959). Les dipneustes du Dévonien supérieur du Groenland. *Medd. Grønland*, **160**: 1–58.

Lehman, J.-P. (1966). Actinopterygii. In *Traité de Paléontologie.* ed. Piveteau, J., **4**:3 1–242. Paris: Masson.

Miles, R. S. (1965). Some features in the cranial morphology of acanthodians and the relationships of the Acanthodii. *Acta zool. Stockh.*, **46**: 233–255.

Nielsen, E. (1942). Studies on Triassic fishes from East Greenland. 1. *Glaucolepis* and *Boreosomus. Palaeozool. Groenland.*, **1**: 1–403.

Patterson, C. (1965). The phylogeny of the chimaeroids. *Phil. Trans. R. Soc.*, (B) **249**: 101–219.

Rayner, Dorothy H. (1951). On the cranial structure of an early palaeoniscid, *Kentuckia*, gen. nov. *Trans. R. Soc. Edinb.*, **62**: 53–83.

Romer, A. S. (1937). The braincase of the Carboniferous crossopterygian *Megalichthys nitidus. Bull. Mus. comp. Zool. Harv.* **82**: 1–73.

Romer, A. S. (1966). *Vertebrate paleontology.* 3rd Ed., Chicago: University of Chicago Press.

Säve-Söderbergh, G. (1952). On the skull of *Chirodipterus wildungensis* Gross, an Upper Devonian dipnoan from Wildungen. *K. svenska VetenskAkad. Handl.*, (4) **3**: 1–28.

Schaeffer, B. (1961). Differential ossification in the fishes. *Trans. N. Y. Acad. Sci.*, (2) **23**: 501–505.

Schaeffer, B. (1965). The role of experimentation in the origin of higher levels of organization. *Syst. Zool.*, **14**: 318–336.

Schaeffer, B. (1967a). Comments on elasmobranch evolution. In *Sharks, Skates and Rays*, ed. Gilbert, P. W., Mathewson, R. F. & Rall, D. P.,: 3–35. Baltimore: John Hopkins Press.

Schaeffer, B. (1967b). Osteichthyan vertebrae. In *Fossil Vertebrates*, ed. Patterson, C. & Greenwood, P. H., *J. Linn. Soc. (Zool.)*, **47**: 185–195.

Schaeffer, B. & Rosen, D. E. (1961). Major adaptive levels in the evolution of the actinopterygian feeding mechanism. *Am. Zool.*, **1**: 187–204.

Simpson, G. G. (1961). *Principles of animal taxonomy.* New York: Columbia University Press.

Stensiö, E. A. (1932). Triassic fishes from East Greenland collected by the Danish Expeditions in 1929-1931. *Medd. Grønland.*, **83**: 1–305.

Stensiö, E. A. (1937). On the Devonian coelacanthids of Germany with special reference to the dermal skeleton. *K. svenska VetenskAkad. Handl.*, (3) **16**: 1–56.

Stensiö, E. A. (1963). Anatomical studies on the arthrodiran head. 1. Preface, geological and geographical distribution, the organisation of the arthrodires, the anatomy of the head in the Dolichothoraci, Coccosteomorphi and Pachyosteomorphi. Taxonomic appendix. *K. Svenska VetenskAkad. Handl.*, (4) **9**: 1–419.

Thomson, K. S. (1967). Mechanisms of intercranial kinetics in fossil rhipidistian fishes (Crossopterygii) and their relatives. *J. Linn. Soc. (Zool.)*, **46**: 223–253.

Watson, D. M. S. (1937). The acanthodian fishes. *Phil. Trans. R. Soc.* (B) **228**: 49–146.

White, E. I. (1939). A new type of palaeoniscoid fish, with remarks on the evolution of the actinopterygian pectoral fins. *Proc. zool. Soc. Lond.*, **109**: 41–61.

The systematic position of the Dipnoi

By Erik Jarvik

Section of Palaeozoology, Swedish Museum of Natural History, Stockholm, Sweden

During the period of time when elasmobranchs were thought to be the most primitive gnathostomes it was taken for granted that dipnoans are descendants of primitive shark-like fishes. Then it was natural to compare dipnoans with elasmobranchs, and many similarities between them were recorded. Soon, however, interest declined in a possible relationship between these two groups and has up to the present time been virtually nonexistent. The main reason for this was the growing interest in dipnoans as possible ancestors of tetrapods; and many seemingly important resemblances with amphibians (presence of autostyly, lungs, choana and multicellular glands, similarities in the structure of the heart, the vascular and nervous system, etc.) were established by the early anatomists (see Semon, 1901). Later, similarities in the heart were further emphasized (Goodrich, 1909; 1930) and resemblances in histologic structure (Kerr, 1932), musculature (Edgeworth, 1935; Kesteven, 1942–1945) and pituitary (Wingstrand, 1956) were recorded. Moreover, Holmgren and students accepting his view about the diphyletic origin of tetrapods tried to show that dipnoans agree with urodeles but differ from anurans and amniotes in limb and skull structure (Holmgren, 1933; 1949), excretory system (Kindahl, 1938), chromosomes (Wickbom, 1944), forebrain (Rudebeck, 1945) and musculature (Säve-Söderbergh, 1945). Finally, Fox (1965) partly on the basis of his own detailed ontogenetical studies presented lists of similarities in support of the view that dipnoans and amphibians are related.

It might seem these many resemblances would have been sufficient to demonstrate a close relationship between lungfishes and amphibians, at any rate the urodeles, but this was not so. As early as the eighties was advanced the view that the similarities may be due to convergence, and in the years before the turn of the century several prominent anatomists (Pollard; Kingsley; Dollo; Gegenbaur; Klaatsch; and others), as later did Sewertzoff (1926), adopted this view and claimed that the tetrapods are closer to the crossopterygians, by which was meant mainly *Polypterus*.

Important differences between dipnoans and amphibians, notably in the dentition, were demonstrated by these writers, and Goodrich (1930), who thought dipnoans and amphibians are related, had to admit the possibility of convergence with regard also to certain similarities in heart structure. Foxon (1955: 222; 1964: 192) shares this opinion but claims that dipnoans in the heart agree better with reptiles than with amphibians. According to Wingstrand (1956) and Bertmar (in this Volume) resemblances in the pituitary also are probably due to convergence. Jarvik (1942) was unable to find any significant common characters in the structure of the snout and demonstrated that the so-called choana in the dipnoans is no true choana. According to Brien & Bouillon (1959; Brien, 1962; 1963) there are important differences in the position, internal subdivision, and nerve and vascular supply of the lungs and they conclude that these organs have arisen independently in dipnoans, brachiopterygians and tetrapods. Bertmar (1966; and in this Volume), finally, claims that dipnoans are most closely related to actinopterygians (including brachiopterygians) and concludes (1966: 147) that they "have probably departed from the actinopterygian stem very early in evolution, and independently achieved certain urodelan-like structures. These are accordingly examples of convergence in evolution".

Because dipnoans lack choana, because their lungs probably have arisen independently and because their similarities with tetrapods in the heart and other organs are probably due to convergence, the opinion of a close relationship between dipnoans and tetrapods certainly rests on a weak basis. This opinion is unsupported also by the studies of the early fossil fishes.

Ever since Cope and Baur before the turn of the century made the first hints of a possible relationship between tetrapods and Devonian rhipidistids, it has become increasingly clear that tetrapods are descendants of this group of fishes and this view is now generally accepted, in the first hand thanks to the pioneer works by Goodrich, Gregory, Watson and Stensiö (for bibliography see Stensiö, 1921; Schaeffer, 1965) and the discovery of the ichthyostegids (Säve-Söderbergh, 1932). However, rhipidistids include two groups, Porolepiformes and Osteolepiformes (Jarvik, 1942), which although specialized in different ways no doubt are more closely related to each other than to other fishes, and which probably evolved from a not too remote pre-Devonian common ancestor. If so dipnoans cannot be derived from osteolepiforms independent of porolepiforms, which is the view held by many students (see Westoll, 1943; 1949); nor can they be more closely related to porolepiforms than to osteolepiforms, as suggested by Säve-Söderbergh (1945: 39). If dipnoans are in some way related to rhipidistids, as most students apparently influenced by the similarities between dipnoans and tetrapods seem to believe, they can share with them at most a common ancestor. However, it is to be taken into consideration also that rhipidistids are generally grouped together with coelacanthiforms into a common systematic unit, the Crossopterygii. If correct this means that coelacanthiforms are more closely related to rhipidistids than to dipnoans. Accordingly, if dipnoans are allied to rhipidistids and tetrapods it must be a relationship via the common ancestor of dipnoans and *all* crossopterygians. This raises the questions which are the common features of crossopterygians (including the new group Struniiformes), and on the basis of our present knowledge whether it is justified to keep the Crossopterygii as a systematic unit?

Discussion of the Crossopterygii. The main common character of fishes classified as crossopterygians is that the skull is divided into two moieties, the divisio cranialis anterior and the divisio cranialis posterior, connected by a so-called intracranial joint. In crossopterygians the notochord is wide and extends unconstricted forward in the basis cranii to close behind the hypophysis. Because in *Eusthenopteron* the posterior end of the ethmosphenoid housing the anterior tip of the notochord is suggestive of the posterior end of the otico-occipital and of the adjoining vertebrae I assumed (1960) that the vertebral column has a long cranial division and that arcual elements behind the hypophysis form an essential part of the basis cranii. With the idea in mind that the intracranial joint may be a persisting vertebral joint I suggested (1960: 39) that the presence of this joint may be a primitive feature and therefore not necessarily one indicating relationship.

The investigations of *Latimeria* (Millot & Anthony, 1958b) have shown that the intracranial joint is more complex than earlier there was reason to assume. Of great importance is the discovery of (1) independent cartilaginous or partly ossified elements, no doubt representing independent arcualia, in the basis cranii (*Da.f*, *Va.f*, *Va.gl*, Stensiö, 1963b, fig. 44), (2) the fact that the anterodorsal of these elements (*Da.f*) shares in the formation of the joint, and (3) a strong paired muscle (muscle sous-cranien, Millot & Anthony, 1965: 5; basicranial muscle, Bjerring, 1967), spanning the joint ventrally.

Starting from the conditions in *Latimeria* Bjerring (1967) has described the complex joint, by him called the intracranial juncture apparatus, in several fossil coelacanthiforms, porolepiforms and osteolepiforms. He has observed independent arcual elements (called zygal plates) in early members (*Diplocercides*, *Nesides*, *Porolepis*, *Glyptolepis*, *Eusthenopteron*) of the three groups, established that a basicranial muscle was present and made out its extent in a great number of forms.

In the juncture apparatus Bjerring distinguishes ventral or infracerebral and dorsal or supracerebral divisions. According to him, the infracerebral division is situated in the somitic (axial) part of the skull (formed by the cranial division of the vertebral column) and represents the original intrasegmental vertebral joint of the mandibular (second) metamere. Disregarding a secondary lengthening of the basicranial muscle in the coelacanthiforms the infracerebral division is very similar in coelacanthiforms, porolepiforms and osteolepiforms and here we may be concerned with very primitive conditions, more primitive probably than in any other vertebrates, the cyclostomes included. In the supracerebral division, on the other hand, there are considerable differences between the three groups. This is quite natural in view of the fact that this division lies in the endo- and exoskeletal cranial roof, that is in the ectomesenchymatic part of the skull. It certainly appears to be a secondary formation having arisen independently in the three groups as a consequence of the persistence of the intrasegmental joint in the axial part. As is well known the supracerebral division of the joint has a different position in the three groups, being situated farther forwards in coelacanthiforms than in osteolepiforms and porolepiforms (see also Jarvik, 1954, figs. 1, 4; 1967b, fig. 9).

If the similarities in the intracranial juncture apparatus and associated structures (basicranial muscle, notochord, parasphenoid) are due to the persistence of primitive characters, these similarities cannot be used to support the theory of a close relationship between coelacanthiforms and rhipidistids. It may therefore be asked if there are any other significant resemblances justifying the retention of the Crossopterygii as a systematic unit.

This is hardly the case. It may be true that coelacanthiforms, porolepiforms and osteolepiforms agree in the general patterns of the sensory lines and dermal bones, but these resemblances are, partly at least, shared with brachiopterygians (Jarvik, 1947) and probably are of little importance. Coelacanthiforms lack a choana and in this and other respects they differ strongly from porolepiforms and osteolepiforms. As evidenced by the detailed information now available on the anatomy of *Latimeria* (Millot & Anthony, 1958a; 1958b; 1965) coelacanthiforms have several structures (rostral organ; muscular suprabranchial organs, Anthony & Robineau, 1967) not found in any other vertebrates, whereas they in certain regards (heart structure) seem to be primitive. In some structures (e.g. the eye) they agree with elasmobranchs, in others with actinopterygians (the forebrain, Nieuwenhuys & Hickey, 1965; see also Jarvik, 1964: 83), but most surprisingly they seem to have no significant characters in common with tetrapods. The absence of tetrapod characters in coelacanthiforms is really remarkable, because many such characters have been recorded in both osteolepiforms and porolepiforms, each of which independently has given rise to tetrapods. Under these circumstances it is questionable whether coelacanthiforms can be classified together with porolepiforms and osteolepiforms. This classification gives the false impression that coelacanthiforms are some kind of cousins to tetrapods. Because positive evidence is lacking I prefer to reject the term Crossopterygii as was already done by Stensiö (1963b; see also Goodrich, 1909).

Dipnoans and coelacanthiforms. Millot & Anthony studying *Latimeria* have not recorded any resemblances of importance between these two groups; nor does the palaeontological evidence support a close relationship. Dipnoans and coelacanthiforms were highly specialized in different ways when in the Devonian they first appeared in the fossil record. Because they have no significant specializations in common they cannot be classified together and the term Sarcopterygii is to be rejected (see also Jarvik, 1964: 40).

Dipnoans and rhipidistids. Because rhipidistids in most respects differ fundamentally from coelacanthiforms and cannot be grouped together with them there is a possibility that Rhipidistia and Dipnoi are closely related and if so should be placed in a common systematic unit. Evidence for such a relationship has been given by several writers (Goodrich, 1909; Watson & Gill, 1923; Säve-Söderbergh, 1934; Holmgren & Stensiö, 1936; Westoll, 1949; and others). Let us now critically examine the alleged resemblances between the two groups.

(1) Structure of dermal bones and scales.—"The most primitive members of both groups have very similar histology of dermal bones and scales, with cosmine" (Westoll, 1949: 122).

Because Westoll states categorically that it "is significant that these points of resemblance are in sharp contrast with all other contemporary and earlier fishes" it is to be emphasized that a very similar histology of the exoskeletal elements is found in cephalaspids which also have a cosmine-like layer (Gross, 1956: 136–137; cf. Goodrich, 1909, fig. 190 and Denison, 1947, fig. 1).

Nor in certain early dipnoans (*Uranolophus*, Denison, in this Volume) can the presence of rhombic scales with an articular ridge on the inner side be attributed any great importance. Such an articular ridge is developed e.g. in anaspids (Kiær, 1924, fig. 29) and actinopterygians (see Gross, 1966; Schultze, 1966), and transformation of rhombic scales into cycloid has happened independently in several evolutionary lines both in rhipidistids and actinopterygians.

(2) Westoll-lines.—"In both the peculiar markings termed 'Westoll-Linien' ... may be developed on the cosmine-surfaces" (Westoll, 1949: 122).

Such lines have never been observed in Devonian porolepids and osteolepids (Jarvik, 1950; Gross, 1956: 8) and are consequently characteristic only of Dipnoi. The nature of the lines in the cosmine layer of the maxillary in *Megalichthys* (Jarvik, 1966, pl. 4, fig. 3) is uncertain.

(3) Body form and median fins.—Resemblances between Middle Devonian types in "the general body form, possession of two separate dorsal and an anal fin, and a heterocercal tail with an epichordal lobe" (Watson & Gill, 1923: 210; see also Westoll).

In Dipnoi, in contrast to porolepiforms and osteolepiforms, the head is generally small and triangular in shape as seen in lateral view. In *Scaumenacia* the body has a narrow flattened ventral side, and transverse sections through its anterior part are triangular, not oval as in rhipidistids (see Jarvik, 1948, fig. 3).

Two dorsal fins, an anal fin and a heterocercal caudal with an epichordal lobe occur also in many elasmobranchiomorphs and in them, too, the supporting endoskeletal elements may fuse into more or less complex units. If the long, continuous dorsal fin found in certain Devonian and in post-Devonian dipnoans is primitive, or if short, separate fins with complex endoskeletal supports have become secondarily lengthened as held by Dollo is still uncertain (see Jarvik, 1952: 24–25). In favour of the former alternative speak embryological data and the facts that the dorsal median fin in certain Devonian forms (*Scaumenacia*, *Phaneropleuron*) extends forwards almost to the head and that in *Neoceratodus* it apparently has undergone a reduction from in front (Jarvik, 1960: 34; see also Goodrich, 1930: 103; Lehman, 1959: 5; 1966: 278, 293).

(4) Paired fins.—One of the main reasons for the assumption that dipnoans are closely related to rhipidistids is that they "have fleshy lobate fins covered with scales and with a well developed internal skeleton of archipterygial type" (Romer, 1955; see also Westoll, 1949: 122).

Fleshy lobate fins covered with scales occur in many elasmobranchiomorphs and are characteristic also of brachiopterygians and many actinopterygians. Also, a biserial archipterygium more like that of *Neoceratodus* than the dichotomously branched fin-skeleton of osteolepiforms is found in *Pleuracanthus* and other elasmobranchiomorphs.

Another common characteristic feature of dipnoans and elasmobranchiomorphs is that the fin-scales cover almost the whole fin (Jarvik, 1959).

Because dipnoans show in addition a tendency to form horny fin-rays similar to the ceratotrichia of elasmobranchs and furthermore the articular knob of the shoulder joint in elasmobranchs and dipnoans, in sharp contrast to osteolepiforms and tetrapods, is on the shoulder girdle, the structure of the paired fins suggests a relationship with elasmobranchiomorphs rather than with rhipidistids.

(5) Pattern of dermal bones.—Similarities in "the number and relations of the dermal bones of the top of the head" (Watson & Gill, 1923: 210; Säve-Söderbergh, 1934: 7).

The pattern of the dermal bones of the cranial roof (and cheek) in dipnoans is so different from that in other groups of fishes that most modern writers (see White, 1965) have found it necessary to use a special terminology.

(6) Opercular and gular bones.—Similarities in "the possession of a very elaborate opercular apparatus, including opercular and sub-opercular, a series of lateral gulars, a pair of principal gulars, and a median gular" (Watson & Gill, 1923: 210).

In early dipnoans and in rhipidistids occur a submandibular series and an operculo-gular series (Jarvik, 1963; 1967a, fig. 5). However, these series are differently developed in the two groups and moreover both series are probably present in certain elasmobranchiomorphs (Jarvik, 1963, fig. 12) and actinopterygians (Jessen, in this Volume).

(7) Exoskeletal shoulder girdle.—With reference to my paper of 1944 Westoll (1949: 122) claims that the "shoulder-girdles are rather closely similar".

Studies of *Scaumenacia* (unpublished) have revealed that the exoskeletal shoulder girdle, at any rate of this Devonian form, is of the same type as that of *Neoceratodus* and differs more distinctly from that of porolepiforms and osteolepiforms than in 1944 there was reason to assume. Most probably the exoskeletal shoulder girdle, which is of a different type in each of the crossopterygian groups (as regards the struniiforms, see Jessen, 1966a; 1966b; 1967), has arisen more or less independently and very likely the general similarities in structure between dipnoans, various teleostomes and arthrodires are due mainly to similar functions (Jarvik, 1965b: 153–155).

(8) Sensory lines.—"The latero-sensory system is very similar in general plan, and in particular the jugal line is developed as an enclosed canal" (Westoll, 1949: 122; see also Säve-Söderbergh, 1934: 5; Holmgren & Stensiö, 1936: 386).

With regard to the sensory canals of the head it is difficult to find any specializations shared only with rhipidistids. An enclosed "jugal canal" is present also in holocephalians (preopercular canal, Stensiö, 1947, figs. 28 C, 29 C; as regards other similarities with holocephalians see below) and moreover there is a "well-developed oral line which recalls conditions met with chiefly in the *Chondrichthyes*" (Pehrson, 1949: 153; see also Jarvik, 1967a).

In the skull-roof of dipnoans there are most often three pit-lines, generally homologized with the so-called anterior, middle and posterior pit-lines of rhipidistids and other fishes.

Pehrson's studies (1949) on the ontogeny of the pit-lines of the skull-roof in *Protopterus* resulted in the following partly new interpretations (Fig. 1 A–D):

(a) The pineal line. This pit-line (*pin*), present only in larvae, is a posteromedial branch of the supraorbital canal in the area between the eyes.

(b) The anterior pit-line. This line (*apl*) arises as a small outgrowth from the point where the supraorbital canal joins the infraorbital canal. In advanced stages it becomes T-shaped with anterior and posterior shanks. The posterior shank becomes reduced in *P.annectens*, but is retained in *P.aethiopicus* (Fig. 1 G).

(c) The middle pit-line. This line (*mpl*) arises above the area between the primordia of the otic and postotic canals. In the adult it forms a somewhat transverse line.

(d) The posterior pit-line. This line consists of two portions, anterior (*pla*) and posterior (*plp*), arising from separate primordia. The posterior portion

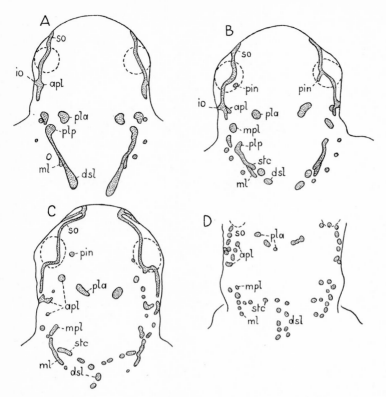

Fig. 1. *A–D, Protopterus annectens*. Four stages (12.4, 14.0, 17.5 and 19.0 mm) in the development of the sensory lines of the dorsal side of the head; from Pehrson, 1949. *E, Rhinodipterus ulrichi*. Part of the skull-roof in dorsal aspect; from Ørvig, 1961. *F, Dipterus valenciennesi*. Part of the skull-roof in dorsal aspect; after Specimen No. 53371, Geol. Surv., London; original. *G, Protopterus aethiopicus*. Head in dorsal view; original. *H, Scaumenacia curta*. Attempted restoration of the head in dorsal aspect; original.

B, bone B of skull-roof; *apl*, anterior pit-line; *dsl*, dorsal sensory line of body; *io*, infraorbital sensory canal; *ml*, main sensory line of body; *mpl*, middle pit-line; *pin*, pineal pit-line; *pla*, *plp*, anterior and posterior portions of posterior pit-line; *so*, supraorbital sensory canal; *stc*, supratemporal commissural canal.

becomes reduced. The anterior portion migrates forwards in a remarkable way and forms in the adult (Fig. 1 G) a long, oblique pit-line, which although innervated by lateralis fibres associated with the n.vagus, extends forwards in front of the anterior pit-line.

These data suggest the following interpretations of the pit-lines of the skull-roof in Devonian dipnoans (Fig. 1 E, F, H).

The pineal pit-line is probably developed in certain forms (*Rhinodipterus*, Fig. 1 E, *pin*) as indicated by the presence of numerous pores in the area between the supraorbital canals postero-medial to the eyes.

The anterior pit-line in Devonian forms (*Scaumenacia*, Fig. 1 H, *apl*; Stensiö,

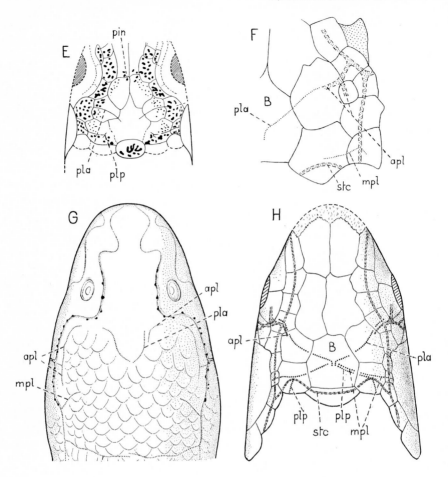

1947, *ap*, fig. 32; *Dipterus*, White, 1965, *soc*, figs. 18, 29; *Soederberghia*, Lehman, 1959, *ap.c*, fig. 2) is a short canal given off close to the junction between the supraorbital and infraorbital canals. In one specimen of *Dipterus* (Fig. 1 F) in which it has been dissected it is branched, with anterior and posterior shanks somewhat as in *Protopterus*.

The middle pit-line (*mpl*) is a more or less transverse line sometimes extending medially into bone B.

The posterior pit-line consists of two portions, as in *Protopterus*. The anterior portion (*pla*), hitherto generally assumed to be the anterior pit-line (central line, Stensiö, 1947) in its antero-lateral part is situated outside and possibly slightly in front of the canal which is the anterior pit-line. It runs in a posteromedial direction to the central part of bone B, where it, close to the medial end of the middle pit-line, generally bends backwards and sometimes slightly laterally and may be continuous with the posterior portion. The latter (*plp*) which in contrast to that of *Protopterus* is retained in the adult is the pit-line which hitherto has been regarded as the posterior pit-line.

In the development of the pit-lines of the cranial roof there is thus a close agreement between the Devonian forms and *Protopterus* and obviously the peculiar migration forwards of the anterior portion of the posterior pit-line has taken place early in the phylogeny of the Dipnoi.

(9) Internal nostril.—The presence of a nostril in the roof of the oral cavity, assumed to be homologous to the choana of tetrapods, has been an important argument for the view that dipnoans are related to rhipidistids (Säve-Söderbergh, 1934). However, as is now well established (Jarvik, 1942; Bertmar, 1966; and in this Volume; Panchen, 1967) the presumed choana in dipnoans is equivalent to the posterior (excurrent) nostril in other fishes and its position in the roof of the mouth cavity is secondary.

In sharks the position of the nostrils varies and the excurrent aperture may be situated more or less close to the margin of the mouth and sometimes even within the mouth cavity. Oral to the nostrils in sharks runs the infraorbital sensory canal accompanied by the r.maxillaris V, the r.buccalis VII and vessels (see Jarvik, 1942, figs. 2, 4). If, as is reasonable to assume, the change in position is due to migration towards the margin of the mouth (cf. Allis, 1919: 153) this of course will affect the course of the infraorbital canal and accompanying nerves and vessels, and in *Heterodontus* in which the excurrent nostril "is enclosed within the buccal cavity when the mouth is closed" the infraorbital canal "encircles its oral margin" (Allis, 1919: 159). Assuming that this migration proceeds a little farther in an oral direction so that the excurrent nostril becomes situated well within the mouth cavity it follows that this ultimately will lead to a reduction of the part of the infraorbital sensory canal oral to it. This apparently is what has happened both in dipnoans and in holocephalians (Jarvik, 1964, fig. 12) in which the excurrent nostril has become displaced into the roof of the mouth cavity forming a pseudo-choana. In both these groups the infraorbital sensory canal, in contrast to other fishes, passes antero-dorsal (aboral) to the incurrent (anterior) nostril and it may join the supraorbital canal at the tip of the snout (*Chimaera*, Devillers, 1958, figs. 688, 689; *Neoceratodus*, Pehrson, 1949, fig. 20; Jarvik, 1967a, fig. 3). That we are concerned with a migration is evidenced by the fact that the r.maxillaris V in all modern dipnoans (Jarvik, 1942; Bertmar, 1966: 147) runs forward in the roof of the mouth postero-medial to the excurrent nostril, that is it has retained its original position in relation to that nostril.

(10) Lower jaw.—"The lateral-line bones of the mandible in these early forms are easily comparable" (Westoll, 1949: 122; see also Watson & Gill, 1923; Säve-Söderbergh, 1934; Holmgren & Stensiö, 1936).

Fig. 2. *Protopterus dolloi*. Photographs of living specimen in the Zoological Institute of the University of Stockholm taken in April 1963 by Mr. U. Samuelson. The specimen shows well the transverse sensory lines of the body which previously appear to have been unknown in modern dipnoans.

As was recently demonstrated (Jarvik, 1967 a) dipnoans, with regard not only to the dermal bones and sensory canals of the lower jaw, but also to other structures (endoskeleton, dentition, etc.), differ fundamentally from rhipidistids and it seems impossible to find a single character in the lower jaw indicating relationship.

(11) *Evolution from a postulated osteolepid-like ancestor.*—Starting from the hypothesis that dipnoans are descendants of some *Osteolepis*-like ancestors Westoll made an attempt (1949: 163) "to investigate the rate of evolution of important characters in the Dipnoi" and in tables he presented data showing the step-by-step evolution of 21 "characters" in 15 fossil and modern dipnoans from the presumed ancestor to *Lepidosiren*.

As demonstrated by Lehman (1956; 1966: 293) and will now be further elucidated many of the evolutionary trends proposed by Westoll have no support in available facts and several of the data included in his tables and used for the construction of graphs illustrating "the longevity of genera" and "rate of loss of characters" are unreliable.

Concerning the paired fins Westoll (1949: 170, table 2) stated that they have increased in length in phylogeny being "moderately elongated" in *Scaumenacia* and "very elongated" in *Neoceratodus*. The fact is that in *Scaumenacia* (Hussakof, 1912, fig. 3, pls. 2, 3; Jarvik, 1960, fig. 9) they are very long, much longer proportionately than in *Neoceratodus*.

The pectoral fins are said (Westoll, 1949: 170, table 2) to be carried very low down in early dipnoans (*Dipterus, Scaumenacia*) as in rhipidistids, whereas they "arise very high up flanks" in *Neoceratodus*. As far as I can find there is no difference in this respect between *Scaumenacia* and *Neoceratodus* (Jarvik, 1960, fig. 9). The pectorals in both these forms seem to arise a little higher up than in the osteolepiforms and approach the conditions in the porolepiforms (Jarvik, 1960, fig. 12). However, a most remarkable difference is that whereas in the porolepiforms (see also Jarvik, 1944, fig. 8 E) it is the ventral part of the cleithrum that has become elongated, the elongation in the dipnoans has affected the clavicle.

An evolutionary change attributed great importance by Westoll (1949: 163, table 1; cf. Jarvik, 1967 b) is the "posterior shortening of otic region" and "reduction of otic cavities". According to him it is "a significant fact that nearly all primitive fishes ... have very elongated otic regions, with presumably large otic structures and large brain-nuclei supplying important latero-sensory systems. The change of proportions in later members may be due in part to rapid loss of the relative importance of the acoustico-lateralis system".

Because in 1949 the otic region of the neural endocranium was practically unknown in fossil dipnoans Westoll's statement about the evolution of that region was premature. The first real description of the braincase in a Devonian

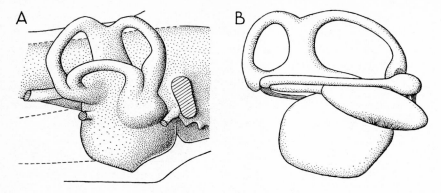

Fig. 3. *A*, *Chirodipterus wildungensis*. Cast of the right labyrinth cavity and adjoining parts of the cranial cavity in lateral view; from Säve-Söderbergh, 1952. *B*, *Neoceratodus forsteri*. Membraneous labyrinth of right side in lateral view; simplified from Retzius, 1881; after Säve-Söderbergh, 1952.

dipnoan (*Chirodipterus*) was given by Säve-Söderbergh in 1952 (as to *Dipterus*, see Säve-Söderbergh, 1952: 22; White, 1965) and according to him (pp. 15, 28), the otic region in *Chirodipterus* (Fig. 3 A) is "extremely short, high and broad" and the labyrinth organ "was apparently relatively higher and shorter" than in *Neoceratodus* (Fig. 3 B). The available facts thus suggest a slight increase in size of the otic cavities and not a reduction, and that it has hardly been a rapid loss of the acoustico-lateralis system is illustrated by the very strong development of that system in *Protopterus* and *Lepidosiren*. As is well shown in particular in specimens of *P.dolloi* (Figs. 2, 4 B) there are on the body, besides longitudinal lines (*l.sl*), also numerous transverse lines (*tr.sl*). Such transverse body-lines which previously seem to have been unknown in modern dipnoans have been observed in *Scaumenacia* and certain other fossil fishes (e.g. acanthodians, Troschel, 1857; osteolepids, Jarvik, 1948: 136) but generally not so complete as in *Protopterus*. In *Neoceratodus* I have been able to find only short transverse lines on a few of the sensory canal scales, but in certain other recent fishes, notably in holocephalians (Fig. 4 A; Ruud, 1917) and blennoid teleosts (Fig. 4 C; Makushok, 1961) there are numerous such lines suggestive of those in *Protopterus*.

With regard also to the alleged shortening of the cheek in the phylogeny of dipnoans (Westoll, 1949: 163, table 1), available facts suggest an evolutionary change in the reverse direction. Westoll claims that the cheek is fairly long in *Dipterus*, "further shortened" in *Scaumenacia* and still more shortened in *Neoceratodus*. The fact is that the cheek is shorter in *Dipterus* (Fig. 5 A) than in *Scaumenacia* (Fig. 5 B; Stensiö, 1947, figs. 24, 32). Nor by available facts is the statement born out that "the attitude of the quadrate" is "somewhat forwardly directed in *Dipterus*, and very forwardly directed in *Scaumenacia* and *Epicerato*-

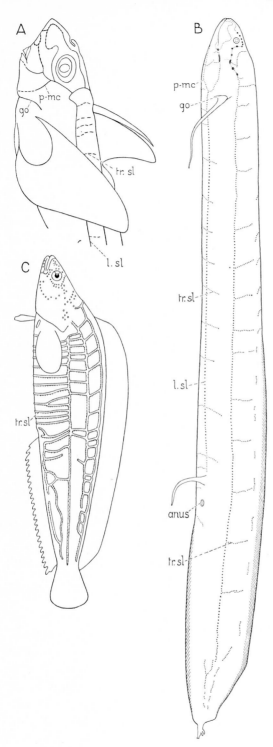

Fig. 4. Transverse sensory lines of the body in A, *Chimaera monstrosa*; from Ruud, 1917; B, *Protopterus dolloi*, after the specimen shown in Fig. 2; and, C, *Stichaeopsis nana*; from Makushok, 1961. *go*, gill opening; *l.sl*, longitudinal sensory lines of body; *p-mc*, preoperculo-mandibular sensory canal; *tr.sl*, transverse sensory lines of body.

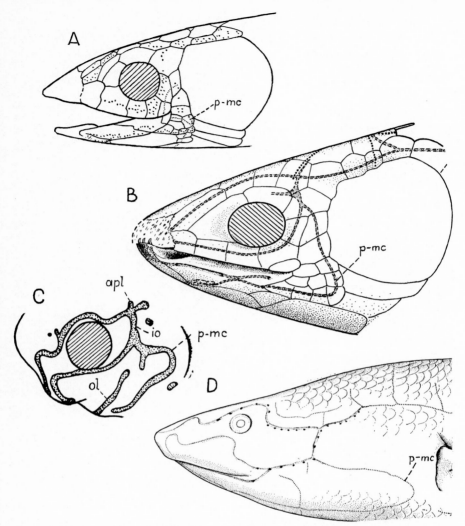

Fig. 5. *A*, *Dipterus valenciennesi*. Restoration of the skull in lateral view; from Westoll, 1949. *B*, *Scaumenacia curta*. Attempted restoration of the skull in lateral view; original. *C*, *Protopterus annectens*. Sensory lines of the cheek in larva (13.6 mm); from Pehrson 1949. *D*, *Protopterus aethiopicus*. Head in lateral view; original.

apl, anterior pit-line; *io*, infraorbital sensory canal; *ol*, oral sensory canal; *p-mc*, preoperculomandibular sensory canal.

dus, starting in each case from the otic process". Judging from the figure recently published by White (1965, fig. 43) the quadrate in *Dipterus* is situated farther forward in relation to the otic process than in *Chirodipterus* (Säve-Söderbergh, 1952, fig. 5) and *Neoceratodus* (Greil, 1913, pl. 16, figs. 21–23; Holmgren & Stensiö, 1936, fig. 231; Fox, 1965). In *Neoceratodus* the attitude of the quadrate is nearly vertical, a condition Westoll ascribed to *Dipnorhynchus*, in which the palatoquadrate (like that of *Scaumenacia*) is unknown.

Of interest in this connection is the remarkable lengthening backwards of the loop of the preoperculo-mandibular canal in the ontogeny of *Protopterus* (Fig. 5 C, D; Pehrson, 1949). In the adult this canal extends backwards on the gill-cover almost to the gill-opening, that is somewhat as in holocephalians (Fig. 4 A).

As now demonstrated all the similarities previously offered as proof of a close relationship between dipnoans and rhipidistids are either nonexistent or insignificant, and in certain characters in which similarities have been claimed (e.g. in the structure of the lower jaw) profound differences have been established. Moreover, the phyletic changes in dipnoans have not proceeded in the way that has been supposed on the assumption that dipnoans are osteolepid derivatives, but with regard to important characters in exactly the opposite direction. Because, in addition, dipnoans in numerous other regards (vertebral column, ribs, neural endocranium, visceral endoskeleton, anatomy of snout, etc., Jarvik, 1942; 1952; 1954; 1955; 1960; 1964) differ fundamentally from rhipidistids the idea of a close relationship between these two groups lacks foundation and is accordingly untenable. As a matter of fact it still is difficult to find a single specialization characteristic only of these two groups, which as far as may be judged at present must be far apart (see also White, 1965; 1966).

Dipnoi and Struniiformes. In the new group, Struniiformes, recently described by Jessen (1966a; 1966b; 1967), the cranial roof is divided into fronto-ethmoidal and parietal shields indicating the presence of an intracranial juncture apparatus of the same type as that of "crossopterygians". Struniiforms are in certain regards (lower jaw, operculo-gular series, etc.) very specialized but seem to be more closely related to rhipidistids than to other fishes. Dipnoans differ strongly from struniiforms in all known regards and cannot on present evidence be closely allied to them either.

Dipnoans compared with brachiopterygians and actinopterygians. Bertmar recently (1966; and in this Volume) suggested that dipnoans are closer to *Polypterus* and actinopterygians than to other fishes.

The systematic position of *Polypterus* has been much debated. It has been regarded as a crossopterygian (Woodward, 1891; and others), it has been thought to be close to the ancestry of the tetrapods (see above), it has (together with *Calamoichthys*) been referred to a group of its own (Polypterini, Goodrich, 1909; Brachiopterygii, Stensiö, 1921), and it has been referred to the Actinopterygii and regarded as a palaeoniscid derivative (Goodrich, 1928; and others). As is now evident *Polypterus* has very little in common with tetrapods and obviously it cannot be as closely related to coelacanthiforms as there was some reason assume (Jarvik, 1942; 1947) before the anatomy of *Latimeria* was known.

The opinion that it is an actinopterygian rests almost exclusively on a single character: the similarity with palaeoniscids in the histological structure of the scales. However, the significance of this similarity is doubtful (see Ørvig, 1967: 84–89) and because *Polypterus* in numerous regards (structure of median and paired fins, pattern of dermal bones, structure of nasal sac and the snout as a whole, absence of myodome, presence of lungs instead of air-bladder, etc.) differs considerably from palaeoniscids and other actinopterygians it is hardly so closely allied to them as has become customary to assume after the appearance of Goodrich's paper (1928). Since 1942 I therefore have preferred to group *Polypterus* and *Calamoichthys* as Brachiopterygii. This view has now been supported by the anatomical and embryological investigations carried out by Daget, Bauchot, Bauchot & Arnoult (1964) who conclude that the brachiopterygians have evolved independently from a preteleostomian stage.

Among the many differences between brachiopterygians and actinopterygians one deserves special attention: the former have lungs, the latter an air-bladder.

The origin and development of the lungs have been much studied during the last few years (*Protopterus*, urodeles, Brien & Bouillon, 1959; Brien, 1962; 1963; Poll, 1962; *Polypterus*, de Smet, 1966; and others) and as now seems to be well established (see also Goodrich, 1930; Bertin, 1958 *b*) the lungs in dipnoans, brachiopterygians and tetrapods arise as a median outpocketing in the floor of the pharynx about at level with, but independent of, the hindmost pair of branchial pouches. This outpocketing in early stages (Fig. 6 A) is suggestive of the thyroid evagination (Fig. 6 B) which arises farther forwards in the midventral wall of the pharynx (about at a level with the second pair of branchial pouches) and like that it grows backwards and becomes bilobate or remains single. In *Latimeria* (Millot & Anthony, 1958 *a*: 2584) the lung, which in that form is a single cylindrical body filled with adipose tissue, is also connected to the midventral area of the pharynx. Because of the presence of a true choana and numerous other tetrapod characters it may be reasonable to assume that true lungs were developed in the porolepiforms and the osteolepiforms, too.

In recent actinopterygians, on the other hand, there is an air-bladder which arises as a median outgrowth of the dorsal wall of the pharynx (Goodrich, 1930; Rauther, 1937). It is true that the air-bladder is situated in about the same part of the pharynx as the lungs and that like them it sometimes (*Amia*) receives its blood from the 6th pair of aortic arches. But all attempts to account for the fact that the lungs are a *ventral* and the air-bladder is a *dorsal* outgrowth of the pharynx (e.g. by migration of the pneumatic duct) have failed (see also Bertin, 1958 *b*: 1383–1384) and as far as may be judged at present we may be concerned with two different, non-homologous organs.

In any event there thus seems to be an important difference between actinopterygians, on the one hand, and dipnoans, brachiopterygians, "crossoptery-

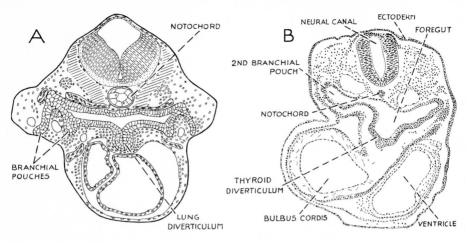

Fig. 6. Two transverse sections to illustrate the similarities in the early development of the lungs and the thyroid gland. *A*, axolotl, embryo; from Brien, 1962. *B*, human embryo, 14-somite stage; after Heuser; from Hamilton, Boyd & Mossman, 1952.

gians" and tetrapods, on the other. If all these groups are of a common origin we may have to assume that both lungs and air-bladder were lacking in their ancestor. It may be true that the air-bladder is a significant actinopterygian character. The lungs, on the other hand, have according to Brien & Bouillon (1959; Brien, 1962; 1963) arisen independently in dipnoans, brachiopterygians and tetrapods and if so, the dipnoans may therefore have evolved separately from forms which like recent elasmobranchs lacked lungs.

Bertmar's opinion concerning a close relationship between Dipnoi and *Polypterus* and Actinopterygii rests mainly on the assumption that the discrete elements and centres of chondrification (Bertmar, 1966, figs. 32–34; and in this Volume) which arise in the ectomesenchymatic skeletogeneous tissue of the anterior part of the head may be homologized and are of phyletic importance. It seems possible to identify elements of visceral arches (see Jarvik, 1954) and no doubt also arcual elements (cf. above) that enter into the neural endocranium but the existence of discrete elements formed in the ectomesenchymatic tissues and inherited independently in various groups from the Preteleostomi or even the Preichthyes is hard to imagine and has no support in palaeontological evidence. As is well known the walls of the nasal capsules and the neural endocranium as a whole are comparatively complete in the early fossil forms and they undergo a retrogressive development in phylogeny. Because the roof and sidewalls of the nasal capsule are complete in early Dipnoi (see e.g. Gross, 1965) it is difficult to imagine that the bars which constitute the fenestrated nasal capsule in the adult and which apparently have arisen independently in dipnoans in connection with partial reduction of the nasal walls may be homologized with similarly situated bars in other groups. Nor is it possible to accept the view (Bertmar, 1966: 119–120) that the so-called trabecular horn (which likely has nothing to do with the trabecula or infrapharyngomandibular, Jarvik, 1954) of dipnoans, situated dorsal (aboral) to the incurrent nostril, is homologous to the "trabecular horn" of urodeles which lies ventral (oral) to that nostril (see Jarvik, 1942, fig. 13). Bertmar's statement containing the main argument for the proposed homology, namely that "each trabecular horn is formed from the antero-lateral part of the planum ethmoidale in exactly the same way" is rather hard to understand because the "trabecular horn" of urodeles seems to be present before the planum ethmoidale (trabecular or internasal plate) is formed (see e.g. Fox, 1954, fig. 10; 1959, fig. 3; as regards dipnoans see also Jarvik, 1942: 276).

Disregarding the alleged resemblances between chondrification centres neither Bertmar nor anybody else has presented any convincing evidence in favour of the view that the dipnoans are more closely allied to brachiopterygians and actinopterygians than to other fishes, and because of the many obvious differences it is difficult to believe that such a close relationship exists.

Dipnoans and elasmobranchiomorphs. As recently demonstrated (Jarvik, 1964; 1967a) dipnoans differ fundamentally from "crossopterygians", other teleostomes and lower tetrapods but agree in a remarkable way with holocephalians in the structure of the upper and lower jaws. In dipnoans as in holocephalians there is no outer dental arcade, corresponding to that in teleostomes and tetrapods, formed in the upper jaw by the maxillary and premaxillary and in the lower jaw by the dentary (the elements identified as maxillary and premaxillary in certain Devonian dipnoans are doubtful). Moreover, the three coronoids of the lower jaw and the dermopalatine and the ectopterygoid of the upper jaw have no equivalents in dipnoans and holocephalians either, and that the so-called vomerine plate is homologous to the vomer of the teleostomes remains doubtful (see Jarvik, 1967a: 175). The main biting elements in dipnoans as in holocephalians are the mandibular and palatine tooth-plates. The teeth in both groups are similar in histology, being made up of columnar pleromic hard tissue and they grow in a basal direction (Ørvig, 1967: 89–96, 104–105). In front of the mandibular tooth-plates in dipnoans there is a median tooth-bearing element (MdY, Jarvik, 1967a) suggestive of the median tooth-plate in the corresponding position in certain fossil holocephalians. Furthermore the meckelian cartilages have fused anteriorly forming a symphysial plate and in front of that there are in both groups so-called labial cartilages.

Among the many other similarities between dipnoans and elasmobranchiomorphs notably holocephalians, the following may be mentioned:

Important resemblances in the cranial muscles were demonstrated by Edgeworth who on the basis of his detailed studies of the ontogenetic development and anatomy of both muscles and endoskeletal elements concluded (1935: 230) that the "Holocephali have sprung from primitive Dipnoi". Similarities in these regards were emphasized also by Kesteven (1942–1945) and the fact that the operculo-gular membranes of both sides have fused in the area between the jaws (Jarvik, 1967 a: 176) may perhaps also be of some importance. "The kidneys, the gonads, and their ducts differ but little in *Ceratodus* from those of other primitive fish, such as the Elasmobranchs" (Goodrich, 1909: 253). As in holocephalians the enteric canal is straight, stomach is lacking (Pernkopf & Lehner, 1937; Bertin, 1958 a: 1264) and there are spiral valves (developed much as in *Scyllium*, Parker, 1892: 141). As in holocephalians, too, the palatoquadrate has fused with the neural endocranium and in both it extends forwards to the ethmoidal region although the pars palatina of dipnoans is represented only by a ligament (Bertmar, 1966: 89, 123). Moreover, a spiracle is lacking, there is an hyoidean gill-cover, stiffened by hyoid rays (Fürbringer, 1904, pl. 50, figs. 40–42; Goodrich, 1909, fig. 134; Ørvig, 1960, fig. 4) and the preoperculo-mandibular canal extends far backwards in the skin of the gill-cover (see above; cf. Stensiö, 1963 a: 381). As in elasmobranchs there are two efferent arteries in the gill arches (Goodrich, 1909: 247, 250, fig. 57) and dorsal anastomoses between the efferent branchial vessels (Goodrich, 1930: 514, figs. 531, 534). As in elasmobranchs in general (cf. p. 228) the fin-scales extend almost to the margin of the fleshy lobate fins, there is a tendency to form ceratotrichia and the articular fossa of the shoulder joint is on the fin. A biserial archipterygium suggestive of that of *Neoceratodus* may occur in elasmobranchs. The vertebrae of early dipnoans are of the same type as those of selachians (Jarvik, 1952: 46–47). "In its general form the olfactory organ of *Protopterus* comes nearest to that of Elasmobranchs with which it has many points of similarity", and "the position of the nostrils is essentially selachian" (Parker, 1892: 125; also Huxley, 1876). In dipnoans as in holocephalians the posterior (excurrent) nostril has moved into the mouth cavity forming a pseudo-choana and the infraorbital sensory canal has been reduced orally to it (see above). The labyrinth organ is similar to that in holocephalians (Retzius, 1881: 146; Parker, 1892: 129) and there is a close agreement with selachians and holocephalians in the structure of the forebrain (Rudebeck, 1945: 138, 140).

Concluding remarks. It is not easy to form an opinion of the value of these many similarities between dipnoans and holocephalians or to what extent they, like the resemblances between dipnoans and urodeles, may be due to convergence. Several of them are shared with other groups or are for other reasons insignificant. Considering the many obvious differences it is evident that there cannot be a very close relationship, in particular not if it is true that the holocephalians are closely allied to the Devonian ptyctodontid arthrodires (Ørvig, 1960; 1962;

Stensiö, 1968) which certainly differ greatly from the contemporaneous dipnoans. As we have seen it is difficult to find significant resemblances indicating close relationship to any of the teleostome groups. Dipnoans differ fundamentally from all these groups in the structure of the lower jaw and the dentition, but on the other hand, they differ from holocephalians and other elasmobranchs e.g. in the presence of lungs and the absence of claspers. It is therefore difficult to decide if dipnoans are teleostomes (cf. Schaeffer, in this Volume) or if they are elasmobranchiomorphs. We are here facing the same problem as that of the Acanthodii and the decision is not made easier by the fact that acanthodians and dipnoans differ widely and as far as can be seen cannot possibly be closely allied. This raises the question if it is justified to distinguish between teleostomes and elasmobranchiomorphs and in order to be able to discuss this and other problems touched upon in this article we have to consider the phylogeny of the Vertebrata as a whole (Jarvik, in this Volume).

References

(For further references see Jarvik, 1967 a)

Allis, E. P. Jr. (1919). The lips and the nasal apertures in the gnathostome fishes. *J. Morph.*, **32**: 145–197.
Anthony, J. & Robineau, D. (1967). Le cercle céphalique de *Latimeria* (poisson coelacanthidé). *C. R. Acad. Sci. Paris*, **265**: 343–346.
Bertin, L. (1958a). Appareil digestif. In *Traité de Zoologie*, ed. Grassé, P.-P., **13:2**: 1248–1302. Paris: Masson.
Bertin, L. (1958b). Vessie gazeuse. Organes de la respiration aérienne. In *Traité de Zoologie*, ed. Grassé, P.-P., **13:2**: 1342–1398. Paris: Masson.
Bjerring, H. C. (1967). Does a homology exist between the basicranial muscle and the polar cartilage? *Colloques int. Cent. natn. Rech. Scient.*, **163**: 223–267.
Brien, P. (1962). Formation du cloaque urinaire et origine des sacs pulmonaires chez *Protopterus*. *Annls Mus. R. Afr. Centr.*, (8) **108**: 1–51.
Brien, P. & Bouillon, J. (1959). Ethologie des larves de *Protopterus dolloi* Blgr. et étude de leurs organes respiratoires. *Annls Mus. R. Congo belg.* (8) **71**: 23–74.
Daget, J., Bauchot, M.-L. & R., & Arnoult, J. (1964). Développement du chondrocrâne et des arcs aortiques chez *Polypterus senegalus* Cuvier. *Acta zool. Stockh.*, **46**: 201–244.
Denison, R. H. (1947). The exoskeleton of *Tremataspis*. *Am. J. Sci.*, **245**: 337–365.
de Smet, W. (1966). Le développement des sacs aériens des Polyptères. *Acta zool. Stockh.*, **47**: 151–183.
Devillers, C. (1958). Le système latéral. In *Traité de Zoologie*, ed. Grassé, P.-P., **13:2**: 940–1032. Paris: Masson.
Fox, H. (1954). Development of the skull and associated structures in the Amphibia with special reference to the urodeles. *Trans. zool. Soc. Lond.*, **28**: 241–295.
Fox, H. (1959). A study of the development of the head and pharynx of the larval urodele *Hynobius* and its bearing on the evolution of the vertebrate head. *Phil. Trans. R. Soc.*, (B) **242**: 151–205.
Foxon, G. E. H. (1955). Problems of the double circulation in vertebrates. *Biol. Rev.*, **30**: 196–228.

Foxon, G. E. H. (1964). Blood and respiration. In *Physiology of the Amphibia*, ed. Moore, J. A.: 151–209. New York & London: Academic Press.

Goodrich, E. S. (1928). *Polypterus* a palaeoniscid? *Palaeobiologica*, **1**: 87–92.

Gross, W. (1965). Über den Vorderschädel von *Ganorhynchus splendens* Gross (Dipnoi, Mitteldevon). *Paläont. Z.*, **39**: 113–133.

Gross, W. (1966). Kleine Schuppenkunde. *N. Jb. Geol. Paläont. Abh.*, **125**: 29–48.

Hamilton, W. J., Boyd, J. D. & Mossman, H. W. (1952). *Human embryology.* 2nd. Ed. Cambridge: Heffer & Sons.

Holmgren, N. (1933). On the origin of the tetrapod limb. *Acta zool. Stockh.*, **14**: 185–295.

Hussakof, L. (1912). Notes on Devonic fishes from Scaumenac Bay, Quebec. *Bull. N.Y. St. Mus.*, **158**: 127–139.

Jarvik, E. (1944). On the exoskeletal shoulder-girdle of teleostomian fishes, with special reference to *Eusthenopteron foordi* Whiteaves. *K. svenska VetenskAkad. Handl.*, (3) **21**: 1–32.

Jarvik, E. (1947). Notes on the pit-lines and dermal bones of the head in *Polypterus*. *Zool. Bidr. Uppsala*, **25**: 60–78.

Jarvik, E. (1948). On the morphology and taxomony of the Middle Devonian osteolepid fishes of Scotland. *K. svenska VetenskAkad. Handl.*, (3) **25**: 1–301.

Jarvik, E. (1950). Middle Devonian vertebrates from Canning Land and Wegeners Halvø (East Greenland). 2. Crossopterygii. *Medd. Grønland*, **99**: 1–132.

Jarvik, E. (1965*a*). Die Raspelzunge der Cyclostomen und die pentadactyle Extremität der Tetrapoden als Beweise für monophyletische Herkunft. *Zool. Anz.*, **175**: 101–143.

Jarvik, E. (1965*b*). On the origin of girdles and paired fins. *Israel J. Zool.*, **14**: 141–172.

Jarvik, E. (1966). Remarks on the structure of the snout in *Megalichthys* and certain other rhipidistid crossopterygians. *Ark. Zool.*, (2) **19**: 41–98.

Jarvik, E. (1967*a*). On the structure of the lower jaw in dipnoans: with a description of an early Devonian dipnoan from Canada, *Melanognathus canadensis* gen. et sp. nov. In *Fossil Vertebrates*, ed. Patterson, C. & Greenwood, P. H., *J. Linn. Soc. (zool.)*, **47**: 155–183.

Jarvik, E. (1967*b*). The homologies of frontal and parietal bones in fishes and tetrapods. *Colloques int. Cent. natn. Rech. Scient.*, **163**: 181–213.

Jessen, H. (1966*a*). Struniiformes. In *Traité de Paléontologie*, ed. Piveteau, J., **4**:3: 387–398.

Jessen, H. (1966*b*). Die Crossopterygier des Oberen Plattenkalkes (Devon) der Bergisch-Gladbach-Paffrather Mulde (Rheinisches Schiefergebirge) under Berücksichtigung von amerikanischem und europäischem *Onychodus*-Material. *Ark. Zool.*, (2) **18**: 305–389.

Jessen, H. (1967). The position of the Struniiformes (*Strunius* and *Onychodus*) among crossopterygians. *Colloques int. Cent. natn. Rech. Scient.*, **163**: 173–180.

Kerr, J. G. (1932). Archaic fishes—*Lepidosiren, Protopterus, Polypterus*—and their bearing upon problems of vertebrate morphology. *Jena. Z. Naturw.*, **67**: 419–433.

Kesteven, H. L. (1942–1945). The evolution of the skull and the cephalic muscles; a comparative study of their development and adult morphology. *Mem. Aust. Mus. Sydney*, **8**: 1–316.

Kiaer, J. (1924). The Downtonian fauna of Norway. 1. Anaspida. *Skr. norske Vidensk-Selsk. Oslo, Mat.-naturv. Kl.*, **1924**: 1–139.

Kindahl, M. (1938). Zur Entwicklung der Exkretionsorgane von Dipnoern und Amphibien, mit Anmerkungen bezüglich Ganoiden und Teleostier. *Acta zool. Stockh.*, **19**: 1–190.

Lehman, J.-P. (1956). L'évolution des dipneustes et l'origine des urodèles. *Colloques int. Cent. natn. Rech. Scient.*, **60**: 69–75.

Lehman, J.-P. (1966). Actinopterygii. In *Traité de Paléontologie*, ed. Piveteau, J., **4:3**: 1–242.
Makushok, V. M. (1961). Some peculiarities in the structure of the seismosensory system in northern blenniid fish (Stichaeoidae, Blennioidei, Pisces). *Trud. Inst. Oceanol. Acad. Sci. U.S.S.R.*, **43**: 225–281. In Russian.
Millot, J. & Anthony, J. (1958*a*). Crossoptérygiens actuels. In *Traité de Zoologie*, ed. Grassé, P.-P., **13:3**: 2553–2597.
Millot, J. & Anthony, J. (1958*b*). *Anatomie de Latimeria chalumnae.* **1.** *Squelette, muscles et formations de soutien.* Paris: Cent. natn. Rech. Scient.
Millot, J. & Anthony, J. (1965). *Anatomie de Latimeria chalumnae.* **2.** *Système nerveux et organes des sens.* Paris: Cent. natn. Rech. Scient.
Nieuwenhuys, R. & Hickey, M. (1965). A survey of the forebrain of the Australian lungfish *Neoceratodus forsteri. J. Hirnforsch.*, **7**: 433–452.
Ørvig, T. (1960). New finds of acanthodians, arthrodires, crossopterygians, ganoids and dipnoans in the Upper Middle Devonian calcareous flags (Oberer Plattenkalk) of the Bergisch Gladbach-Paffrath trough. 1. *Paläont. Z.*, **34**: 295–335.
Ørvig, T. (1962). Y a-t-il une relation directe entre les arthrodires ptyctodontides et les holocéphales? *Colloques int. Cent. natn. Rech. Scient.*, **104**: 49–60.
Panchen, A. L. (1967). The nostrils of choanate fishes and early tetrapods. *Biol. Rev.*, **42**: 374–420.
Pernkopf, E. & Lehner, J. (1937). Vorderdarm. In *Handbuch der vergleichenden Anatomie der Wirbeltiere.* hrsg. Bolk, L., Göppert, E. Kallius, E. & Lubosch, W., **3**: 349–562. Berlin & Wien: Urban & Schwarzenberg.
Poll, M. (1962). Étude sur la structure adulte et la formation des sacs pulmonaires des Protoptères. *Annls Mus. R. Afr. Centr.*, (8) **108**: 129–172.
Rauther, M. (1937). Die Schwimmblase. In *Handbuch der vergleichenden Anatomie der Wirbeltiere.* hrsg. Bolk, L., Göppert, E., Kallius, E. & Lubosch, W., **3**: 883–908. Berlin & Wien: Urban & Schwarzenberg.
Retzius, G. (1881). *Das Gehörorgan der Wirbelthiere.* 1. *Das Gehörorgan der Fische und Amphibien.* Stockholm: Samson & Wallin.
Romer, A. S. (1955). Herpetichthyes, Amphibioidei, Choanichthyes or Sarcopterygii? *Nature, Lond.*, **176**: 126–127.
Rudebeck, B. (1945). Contributions to forebrain morphology in Dipnoi. *Acta zool. Stockh.*, **26**: 9–156.
Ruud, G. (1917). Sinneslinien und freie Nervenhügel bei *Chimaera monstrosa. Zool. Jb. (Anat.)*, **40**: 421–439.
Säve-Söderbergh, G. (1932). Preliminary note on Devonian stegocephalians from East Greenland. *Medd. Grønland*, **94**: 1–107.
Säve-Söderbergh, G. (1934). Some points of view concerning the evolution of the vertebrates and the classification of this group. *Ark. Zool.*, **26A**: 1–20.
Säve-Söderbergh, G. (1945). Notes on the trigeminal musculature in non-mammalian tetrapods. *Nova Acta Soc. Sci. upsal.*, (4) **13**: 1–59.
Schaeffer, B. (1965). The evolution of concepts related to the origin of the Amphibia. *Syst. Zool.*, **14**: 115–118.
Schultze, H.-P. (1966). Morphologische und histologische Untersuchungen an Schuppen mesozoischer Actinopterygier (Übergang von Ganoid- zu Rundschuppen). *N. Jb. Geol. Paläont. Abh.*, **126**: 232–314.
Semon, R. (1901). Über das Verwandtschaftsverhältnis der Dipnoer und Amphibien. *Zool. Anz.*, **24**: 180–188.
Sewertzoff, A. N. (1926). Der Ursprung der Quadrupeda. *Paläont. Z.*, **8**: 75–95.
Stensiö, E. A. (1921). *Triassic fishes from Spitzbergen.* 1. Vienna.
Stensiö, E. A. (1963*a*). Anatomical studies on the arthrodiran head. 1. Preface, geological and geographical distribution, the organisation of the arthrodires, the

anatomy of the head in the Dolichothoraci, Coccosteomorphi and Pachyosteomorphi. Taxonomic appendix. *K. svenska VetenskAkad. Handl.*, (4) **9**: 1–419.

Stensiö, E. A. (1963*b*). The brain and the cranial nerves in fossil, lower craniate vertebrates. *Skr. norske VidenskAkad. Oslo., Mat.-naturv. Kl.*, **1963**: 1–120.

Stensiö, E. A. (1968). Les arthrodires. In *Traité de Paléontologie*, ed. Piveteau, J., **4:2**. Paris: Masson. In press.

Troschel, F. H. (1857). Beobachtungen über die Fische in den Eisennieren des Saarbrücker Steinkohlengebirges. *Verh. naturh. Ver. preuss. Rheinl.*, **14**: 1–19.

Westoll, T. S. (1943). The origin of the tetrapods. *Biol. Rev.*, **18**: 78–98.

Wickbom, T. (1944). Cytological studies on Dipnoi, Urodela, Anura, and *Emys*. *Hereditas*, **31**: 241–346.

Wingstrand, K. G. (1956). The structure of the pituitary in the African lungfish, *Protopterus annectens* (Owen). *Vidensk. Medd. dansk naturh. Foren.*, **118**: 193–210.

The evolutionary significance of the earliest known lungfish, *Uranolophus*

By Robert H. Denison

Dept. of Geology, Field Museum of Natural History, Chicago, Illinois, USA

The Field Museum quarry in the Beartooth Butte formation of the Bighorn Mountains of Wyoming has yielded plants, eurypterids, scorpions, as well as numerous vertebrates, including lungfish. The fauna, and the similar one at Beartooth Butte itself, has been correlated (Denison, 1958: 500), largely on the evidence of pteraspids and arctolepids, with the Lower, or possibly Middle Siegenian. If this is correct, its lungfish is the earliest one known, antedating *Dipnorhynchus lehmanni* from the Upper Siegenian Hunsrückschiefer of Germany and *D. sussmilchi* from the Emsian or Upper Siegenian of Australia. The lungfish shows resemblances to *Dipnorhynchus*, but also differences that are considered important enough to warrant the erection of a new genus. It has, therefore, been named *Uranolophus wyomingensis*. *Uranolophus* is being formally described in detail elsewhere (Denison, 1968). My purpose here is to consider such of its characters as have a bearing on the evolution of lungfishes and their relations to other fishes, especially to crossopterygians.

The most important specimen of *Uranolophus wyomingensis* is the largely articulated but incomplete type specimen (Fig. 1). Unfortunately, the tail and much of the belly are missing, but if instead, the head had been lost, it is not impossible that the rest would have been identified as a primitive crossopterygian. The scales, the dorsal fins, the shoulder girdle and to a less extent the vertebrae, have many similarities to those of osteolepids and porolepids.

In referring to specimens in the sequel, Field Museum of Natural History has been abbreviated FMNH.

Scales. A scale of the anterior or middle flank (Fig. 2) is rhombic, with narrow overlapped areas and a well developed dorsal articular process. Its inner surface shows a rather weak ridge extending from the articular process and a well marked depression for the articular process of the adjacent scale. The exposed part commonly has a smooth, cosmine-covered surface, often with Westoll lines, bounded anteriorly by a band with dentine-capped denticles which was overlapped by the scales in front. Some scales are denticulate over their entire exposed area, and thin sections reveal that beneath a cosmine coating there

Fig. 1. Largely articulated type specimen of *Uranolophus wyomingensis* Denison, FMNH PF 3874; about × $\frac{1}{2}$.

Fig. 2. Anterior flank scales of *Uranolophus wyomingensis* Denison; ×2. *A*, outer side, FMNH PF 5546; *B*, inner side, PF 5547.

may be as many as three or four generations of buried denticles, representing the surfaces of younger stages. These scales differ from typical dipnoan scales which are cycloid and have wide overlapped areas; they closely resemble scales of osteolepids and porolepids. There can be little question that the scales of *Uranolophus* are of a type that was primitive in lungfishes and crossopterygians (Ørvig, 1957: 410–413).

Dorsal fins. The only fins preserved in the articulated specimen of *Uranolophus* are the first dorsal and part of the second dorsal (Fig. 3). In general they resemble the dorsal fins of *Dipterus*, but differ in the following respects: (1) the first dorsal is larger relative to body size and is probably not much smaller than the second dorsal; (2) the lobe at the base of the first dorsal is covered with small, cycloid, cosmine-covered scales, and is small and sharply delimited; (3) the lepidotrichia are all cosmine-coated, and only the base of the proximal row of elongate elements is covered by scales. These fins are comparable to the dorsal fins of certain early crossopterygians. In their position they resemble *Thursius* and *Gyroptychius*; in their subequal size they resemble *Osteolepis* and *Thursius*; in their cosmine-covered lepidotrichia and small size of the scale-covered lobe they resemble *Porolepis* and *Gyroptychius*. I conclude that the dorsal fins of *Uranolophus* are of a type that is primitive in both crossopterygians and lungfishes, except for the cycloid shape of the scales of the basal lobe.

Fig. 3. First dorsal fin and base of second dorsal fin of *Uranolophus wyomingensis* Denison, FMNH PF 3874; ×3/2.

The advocates of the fin-fold theory might take issue with this conclusion, for the long, continuous median fin of many lungfishes conforms closely to their conception of the ancestral median fin. The facts are, however, that separate dorsal, caudal and anal fins occur in most Devonian lungfishes and only in Devonian genera, and that a continuous median fin is known only in post-Devonian forms. The Upper Devonian *Phaneropleuron* is intermediate in possessing a separate anal, but continuous dorsal and caudal fins. We must remember in drawing evolutionary conclusions from these facts that the fin-fold theory is only a theory, and that in this case, as in some other groups of fishes, it is not supported by the paleontological evidence.

Shoulder girdle. This is inadequately known in other Devonian lungfishes, so detailed comparisons are not possible. Most lungfishes have a cartilaginous scapulo-coracoid, but in one specimen of *Uranolophus* it is ossified, surely a primitive feature. Dorsal elements connecting the cleithrum to the skull roof have not been identified, possibly because they had not yet evolved into large, easily recognizable bones. The cleithrum (*CLM*, Fig. 4), has a nearly flat external face, as in *Eusthenopteron* (Jarvik, 1944, figs. 3–7); in most lungfishes, probably including *Dipterus*, the operculum overlies a depressed anterior part of the cleithrum, separated by a ridge from the posterior part of the outer face. Some *Uranolophus* cleithra have a strong antero-ventral process with a well developed inner lamina (*il*, Fig. 4) for the reception of the clavicle, but this is

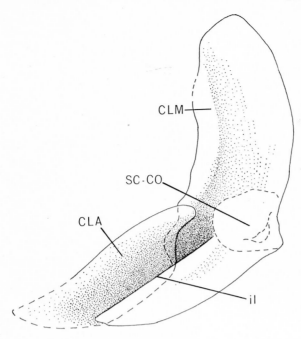

Fig. 4. Restoration of inner side of right shoulder girdle of *Uranolophus wyomingensis* Denison.
CLA, clavicle; *CLM*, cleithrum; *il*, inner lamina of ventral part of cleithrum; *SC-CO*, approximate position of scapulo-coracoid.

variable, and in one specimen it is slightly developed as in *Eusthenopteron*. A rather large, symmetrical median bone is identified provisionally as an interclavicle. A small interclavicle may occur in *Dipterus*, but in later lungfishes it has become cartilaginous. Interclavicles occur in crossopterygians, where they may be large and closely comparable to that of *Uranolophus*.

Vertebral column. The few vertebrae that have been exposed in the type specimen of *Uranolophus wyomingensis* (Fig. 5) each consists of a neural spine fused to a neural arch, and a ventral element that is probably an intercentrum (*ic*, Fig. 5). No separate pleurocentrum has been found. The structure is comparable to that of *Neoceratodus*, though in the latter the neural arches and central elements are cartilaginous, and a small pleurocentrum occurs anteriorly. There is nothing resembling the ossified amphicoelous centra described by Jarvik (1952: 40–47, figs. 16–19) in the Upper Devonian dipnoans, *Soederberghia* and *Jarvikia*, and so one may inquire into the significance of the latter. Similar spool-shaped centra may also occur in the Scottish Upper Devonian lungfish, *Rhynchodipterus* (Säve-Söderbergh, 1937: 8), though these have not been adequately described. These three genera are all long-snouted, probably toothless forms, and may belong to a specialized side branch; if so, their spool-shaped centra may also be a specialization. Jarvik has claimed (1952: 47–48, fig. 21) that spool-shaped centra may occur in the more primitive Devonian lungfish, *Dipterus*. In the specimens he figures, however, the vertebrae are covered by

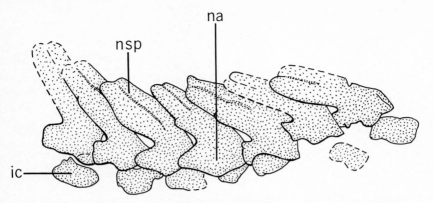

Fig. 5. Seven vertebrae of *Uranolophus wyomingensis* Denison, FMNH PF 3874; × 2. *ic*, intercentrum; *na*, neural arch; *nsp*, neural spine.

scales, and it is known that centra are absent in most *Dipterus*. Traquair (1878: 10–11) concluded that the raised, beaded line sometimes seen in *Dipterus* might be the enlarged bases of neural arches.

Skull. While the postcranial body of *Uranolophus* might well be mistaken for that of a crossopterygian, the same is not true of the skull, for it already possesses most of the typical features of a lungfish. It does differ, however, from *Dipterus* and later lungfishes in certain respects, which I believe are significant in an evolutionary sense.

The pattern of the bones of the skull roof, as I interpret it, is very similar in *Uranolophus* (Fig. 6 A) and *Dipterus* (Fig. 6 B), the only important difference being that bone B in *Dipterus* has expanded posteriorly to separate the pair of I-bones, and to reach the posterior edge of the cranial roof. It is of interest that in certain specimens of *Uranolophus* and *Dipnorhynchus* the snout region is subdivided into a mosaic of small dermal bones. This is excellently shown in the specimen of *D. sussmilchi* described by Campbell (1965) and in one incomplete cranial roof of *Uranolophus* (FMNH PF 3805). This condition may be interpreted as a relic of an ancestral mosaic of dermal bones (see also Schaeffer, in this Volume). A pair of lower jaws of *Uranolophus* (conceivably belonging to the same individual as PF 3805) shows apparent subdivision of the bones of the ventral symphysial region, also suggestive of an ancestral mosaic.

Some of the most obvious differences in the *Uranolophus* skull as compared with other dipnoans are those of proportions. The prepineal length is relatively much shorter in *Uranolophus* than in *Dipterus*, while *Dipnorhynchus* is intermediate. The prepineal lengthening in *Dipterus* may have been related to an increased development of the olfactory sense, for the space available for nasal

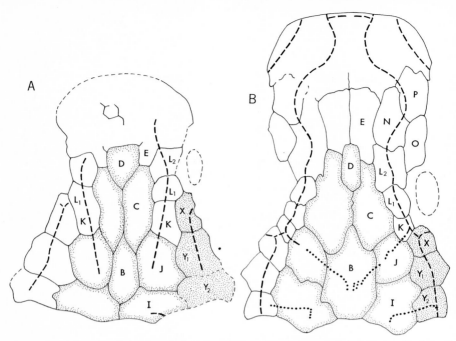

Fig. 6. Cranial roofs of A, *Uranolophus wyomingensis* Denison and B, *Dipterus valenciennesi* Sedgwick & Murchison, both enlarged so that the distance from the center of bone D to the posterior edge of the cranial roof is the same. Dashed lines are lateral line canals, dotted lines are pit lines. (B modified from Jarvik, 1950).

capsules is relatively larger in *Dipterus* than in *Uranolophus*. The supratemporal part of the skull roof (bones X-Y_1-Y_2) is long in *Uranolophus* and much shortened in *Dipterus*, while again *Dipnorhynchus* is intermediate. This change is related to the shorting of the cheek in *Dipterus*, and together with the lengthening of the snout, has resulted in a more posterior position for the orbits. It is also evident that the articulation of the jaw was much farther back in *Uranolophus* than in *Dipterus*, and again *Dipnorhynchus* appears to be intermediate; this signifies a relatively longer jaw, wider gape and weaker bite in *Uranolophus*. Presumably the ancestral dipnoan had relatively long jaws, and retention of such in *Uranolophus*, and perhaps in certain Upper Devonian genera, may be related to the weak dentition and the absence of crushing tooth plates. The short snout, long supratemporal and cheek regions, and posterior jaw articulation are all characters in which *Uranolophus* approaches closer to crossopterygians than do other lungfishes.

Uranolophus probably had two large subopercular bones (SOP_{1-2}, Fig. 7). *Dipterus* and *Scaumenacia* have two small suboperculars, but where known in other lungfishes, there is only a single, relatively small subopercular. In this feature, *Uranolophus* approaches the condition of *Osteolepis* (Jarvik, 1948, fig. 32 A).

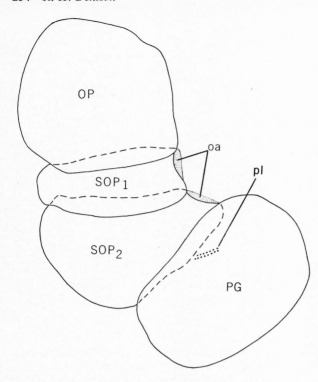

Fig. 7. Restoration of right operculars and principal gular of *Uranolophus wyomingensis* Denison.
oa, overlap areas; *OP*, opercular; *PG*, principal gular; *pl*, pit line; SOP_1, first subopercular; SOP_2, second subopercular.

The palate of *Uranolophus* (Fig. 8) shows some interesting features. The paired vomers are small, but still relatively large compared to those of other lungfishes. Elongate pterygoids bound the long, slender parasphenoid, which clearly lacks the posterior stem that is characteristic of other Dipnoi. The posterior margin of the parasphenoid and adjacent parts of the pterygoids are transverse, smooth and untuberculated. Behind them, one skull (FMNH PF 3792) shows, albeit imperfectly, what appears to be a pair of plates underlying the posterior part of the endocranium; these presumably are parotic plates such as are known to occur in crossopterygians (Jarvik, 1954, figs. 22, 25), and such as were predicted in *Dipnorhynchus* by Westoll (Lehmann & Westoll, 1952: 415). The palate of *D. lehmanni* is probably very similar to that of *Uranolophus*.

The palate and lower jaws of *Uranolophus* completely lack the tooth plates with radiating ridges or rows of denticles that characterize most lungfishes. The ventral surface of the palate and the dorsal surfaces of the prearticulars are covered with small, conical denticles composed of dentine. Instead of teeth, there occur what I have termed "tooth ridges" (Fig. 9) on the lateral margins of the pterygoids, on the margins of the "upper lip", and on the dorsal edges of the prearticulars and dentaries. These ridges, though variable in development, are typically continuous, but with numerous side ridges or projec-

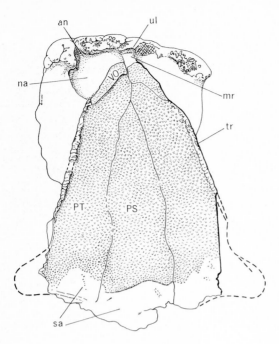

Fig. 8. Palate and "upper lip" of *Uranolophus wyomingensis* Denison, FMNH PF 3805; × 3/4.

an, anterior naris; *mr*, median ridge; *na*, position of nasal capsule; *PS*, parasphenoid; *PT*, pterygoid; *sa*, smooth untuberculated posterior margins of pterygoids and parasphenoid; *tr*, tooth ridge; *ul*, "upper lip", *VO*, vomer.

tions on each side, and are composed of a "tubular" dentine similar to that of typical lungfish teeth.

There are a number of genera of lungfishes in which tooth plates are absent, and it has been suggested by Watson & Gill (1923: 214) and Lehman (1959: 34) that this condition was secondary. To me, it seems much more likely that the ancestors of these genera never had tooth plates. In the early evolution of the group, as the marginal jaw bones were being reduced and lost, there was probably experimentation with a variety of substitute biting mechanisms. Some genera evolved tooth plates, but others, such as *Uranolophus*, did not. *Soederberghia* and *Conchopoma* retained only the primitive denticulation, while *Griphognathus*, *Ganorhynchus*, *Uronemus*, *Holodipterus* and *Dipnorhynchus lehmanni* developed enlarged denticles on the margins of the palate and prearticulars to form one or more rows of teeth. *Dipnorhynchus sussmilchi* developed whole areas of the palate and lower jaws as bulbous crushing surfaces. According to this view, genera without tooth plates were not derived from nor closely related to those that did possess them.

Two other features that have evolutionary significance should be mentioned briefly. The lateral-line canals are similar to those of *Dipnorhynchus sussmilchi* and are primitive in having the supraorbital canal separate from the infraorbital, and continuous onto bone J as a canal rather than a pit line. The dermal skeleton and endoskeleton are well ossified. The endocranium consists of la-

Fig. 9. Left pterygoid tooth ridge of *Uranolophus wyomingensis* Denison, FMNH PF 3805; ×10.

minar bone surrounding spongy bone but is badly crushed in all known specimens. Branchial arches may be ossified, as are the scapulo-coracoid and vertebral elements; ribs have not been found and may not have been ossified.

I have come to the following conclusions regarding the evolutionary position of *Uranolophus*: This genus is primitive in all known post-cranial characters, in its scales, dorsal fins, shoulder girdle, and probably in its vertebral column. In these characters it resembles primitive crossopterygians. On the other hand it is definitely dipnoan in many significant cranial characters: in the loss or reduction of marginal jaw bones, in the autostylic jaw suspension, in the probable fusion of endocranial sections, and in the pattern of cranial roofing bones. However, in its dipnoan skull adaptions it is more primitive than other lungfishes in its skull proportions, in its long lower jaws with posterior articulation, in the occasional anterior mosaic of cranial roofing bones, in the short parasphenoid and separate parotic plates, and in the absence of tooth plates. In these characters it approaches Rhipidistia.

An opinion regarding the closeness of the relationship between Dipnoi and Rhipidistia must necessarily be influenced by a subjective weighting of characters. If we wish to emphasize the distance between the two groups, we may consider post-cranial characters as "superficial" or "unessential", and thus of little importance systematically; however, there is no logical basis for such a procedure. The features that particularly characterize early lungfishes are all related to an adaptation to a new method of feeding, and resulted probably from a relatively rapid evolutionary shift. But if we consider the characters that were not involved in this adaptive shift, we find so many that approach the Rhipidistia that a close ancestral relationship seems to be clearly indicated.

References

Campbell, K. S. W. (1965). An almost complete skull roof and palate of the dipnoan *Dipnorhynchus sussmilchi* (Etheridge). *Palaeontology*, **8**: 634–637.

Denison, R. H. (1958). Early Devonian fishes from Utah. 3. Arthrodira. *Fieldiana: Geol.*, **11**: 461–551.

Denison, R. H. (1968). Early Devonian lungfishes from Wyoming, Utah and Idaho. *Fieldiana: Geol.*, **17**. :353–413.

Jarvik, E. (1944). On the exoskeletal shoulder-girdle of teleostomian fishes, with special reference to *Eusthenopteron foordi* Whiteaves. *K. svenska VetenskAkad. Handl.*, (3) **21**: 1–32.

Jarvik, E. (1948). On the morphology and taxonomy of the Middle Devonian osteolepid fishes of Scotland. *K. svenska VetenskAkad. Handl.*, (3) **25**: 1–301.

Jarvik, E. (1952). On the fish-like tail in the ichthyostegid stegocephalians with descriptions of a new stegocephalian and a new crossopterygian from the Upper Devonian of East Greenland. *Medd. Grønland*, **114**: 1–90.

Jarvik, E. (1954). On the visceral skeleton in *Eusthenopteron* with a discussion of the parasphenoid and palatoquadrate in fishes. *K. svenska VetenskAkad. Handl.*, (4) **5**: 1–104.

Lehman, J.-P. (1959). Les dipneustes du Dévonian supérieur du Groenland. *Medd. Grønland*, **160**: 1–58.

Lehmann, W. & Westoll, T. S. (1952). A primitive dipnoan fish from the Lower Devonian of Germany. *Proc. R. Soc. Lond.*, (B) **140**: 403–421.

Ørvig, T. (1957). Remarks on the vertebrate fauna of the lower Upper Devonian of Escuminac Bay, P. Q., Canada, with special reference to the porolepiform crossopterygians. *Ark. Zool.*, (2) **10**: 367–426.

Säve-Söderbergh, G. (1937). On *Rhynchodipterus elginensis* n.g., n.sp., representing a new group of dipnoan-like Choanata from the Upper Devonian of East Greenland and Scotland. *Ark. Zool.*, **29 B**: 1–8.

Traquair, R. H. (1878). On the genera *Dipterus*, Sedgw. & Murch., *Palaedaphus*, Van Beneden and De Koninck, *Holodus* Pander, and *Cheirodus*, M'Coy. *Annls Mag. nat. Hist.*, (5) **2**: 1–17.

Watson, D. M. S. & Gill, E. L. (1923). The structure of certain Palaeozoic Dipnoi. *J. Linn. Soc. (Zool.)*, **35**: 163–216.

Lungfish phylogeny

By Gunnar Bertmar

Section of Ecological Zoology, Dept. of Biology, Umeå University, Umeå, Sweden

Today most embryologists and palaeozoologists seem to agree that it is often possible to obtain an approximate picture of adult as well as embryonic stages of the fossil ancestors through comparative embryological studies. These matters were briefly discussed in 1962 by Szarski. He claimed that embryology will always supply zoology with very important information about the structure and mode of life of the ancestors of recent animals. This information, however, will be sound and reliable only when it is based on a thorough knowledge of *homology*. The concept of homology was introduced by Owen 1843, and was under discussion for a long time. It has recently been discussed by Remane (1952; 1964).

Lungfishes have for a long time been an interesting group to investigate. Their ethology is fairly unknown and their ecology and morphology are rather unusual. It is only natural that their almost amphibian behaviour and appearance made earlier zoologists believe that they dealt with real amphibians. Krefft (1870) for example, who was the first to describe *Neoceratodus forsteri*, thought that this fish is a "gigantic Amphibian allied to *Lepidosiren*"; the latter fish had already been described as an amphibian related to the urodele *Amphiuma*.

Since the beginning of this century we know that the dipnoans are fishes and not amphibians, but not so long ago facts seemed to justify the hypothesis that the urodeles have developed from dipnoan ancestors, and the other two amphibian groups from the crossopterygians (Holmgren, 1933; 1949; Säve-Söderbergh, 1934). The Amphibia accordingly should be diphyletically evolved. But both lines were said to have one essential character in common, the choana, and therefore Säve-Söderbergh (1934) classified the Dipnoi, the Crossopterygii and the Amphibia in one common group, the Choanata.

However, from the evidence available at the time Jarvik (1942) agreed with the hypothesis of Greil (1913), Allis (1919; 1932) and de Beer (1937) that the dipnoan excurrent (posterior internal) nostril is not a choana but merely the modified external excurrent nostril of the non-choanate fishes. This hypothesis, together with the increasing knowledge of the fossil crossopterygians, established with reasonable certainty that the Rhipidistia were the only immediate ancestors of the tetrapods (Jarvik, 1942; 1960; Romer, 1962; Szarski, 1962;

Schaeffer, 1965; and others). The Dipnoi were relegated to a position as primitive relatives of the Rhipidistia and Amphibia, or were regarded as being quite unrelated to them (Jarvik, 1960; 1964).

With this background it is not surprising that the development of the head of recent lungfishes has attracted much attention during the present decade. Fox worked on 8 series of *Neoceratodus* but made his approach from amphibians, whereas Thomson, studying about 20 series of *Protopterus*, made his from fossil fishes. The present writer who approached from recent fishes as well as amphibians, worked on 39 series of *Neoceratodus forsteri* ranging from egg to 52 mm, 97 series of *Protopterus annectens* (10–55 mm) and 4 series of *Lepidosiren paradoxa* (33–55 mm). The differences in collections and approach may explain disagreements in results and terminology (Fox, 1960; 1961; 1962; 1963*a*; 1963*b*; 1963*c*; 1965; Thomson, 1965; Bertmar, 1959; 1962*a*; 1962*b*; 1963*a*; 1963*b*; 1966*a*; 1966*b*; 1968*a*; 1968*b*).

Fox showed that a great many characters should be considered when larvae of *Neoceratodus*, Urodela and Anura are compared. He concluded with certainty that these larvae are fish-like and similar, but also that each comparable system in the three groups has evolved independently. He therefore suggested that "a common pro-Dipnoan-Amphibian stock originated from rhipidistian crossopterygians in the lower Devonian, and soon separated into two streams. The first has practically disappeared and is merely represented by the three genera of living lungfishes. The second stream radiated into various Amphibia throughout the Devonian and Carboniferous" (Fox, 1965: 542).

The essential problem for Thomson and me was to see if the excurrent nostril of dipnoans is homologous with the choana of tetrapods, because this would be an important piece of evidence suggesting a close relationship between the Dipnoi and the Amphibia (Tetrapoda). But in order to elucidate this problem thoroughly it proved necessary to compare the morphogenesis of all structures of the snout in these groups (the olfactory organ, nasal capsule, nerves, vascular system etc.), and to compare with the corresponding structures in fossil genera. For my part, this investigation included the skeleton of almost the whole of the head.

Structure of the head

Olfactory organ. The development of the dipnoan olfactory organ has recently been thoroughly investigated (Bertmar, 1965*b*). The following results are relevant for our discussion.

The nasal placode is of the general, ectodermal type found in fishes and amphibians (Fig. 1 A, B). As in actinopterygians and amphibians it is primarily situated on the antero-ventral side of the snout anterior to the stomodeum.

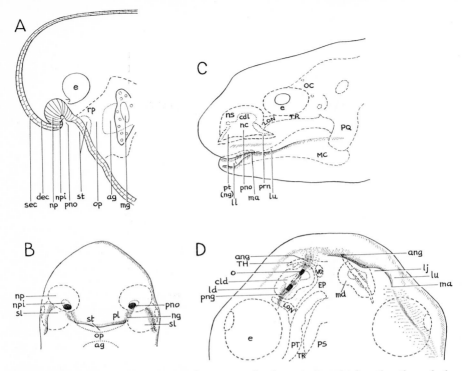

Fig. 1. Reconstructions of the snout of *Neoceratodus forsteri*. *A*, sagittal section through the left nasal pit, 8.6 mm stage; *B*, ventral view of the head, 8.6 mm stage; *C, D*, lateral and ventral view of the head, 14.5 mm stage; broken and dotted lines indicate structures and cavities of another plane. From Bertmar 1965*b*.

ag, anterior gut; *ang*, anterior naso-buccal groove; *cdl*, cavity of diverticulum laterale; *dec*, deep ectodermal layer; *e*, eye; *ld*, lateral diverticle of nasal sac; *lj*, lower jaw; *ll*, lower lip; *lu*, lateral secondary upper lip; *ma*, mouth angle; *md*, medial diverticle of nasal sac; *mg*, mandibular gill pouch; *nc*, nasal cavity; *ng*, naso-buccal groove; *np*, nasal placode; *npi*, nasal pit; *ns*, nasal sac; *o*, part of primary nasal opening seen from the exterior; *op*, oral plate; *pl*, primary upper lip; *png*, posterior naso-buccal groove; *pno*, primary nasal opening; *prn*, primary nostril; *pt*, primary nasal tube; *rp*, Rathke's pouch; *sec*, superficial ectodermal layer; *sl*, secondary upper lip; *st*, stomodeum; *EP*, ethmoidal plate; *LON*, proc. ectethmoideus; *MC*, meckelian cartilage; *OC*, orbital cartilage; *PQ*, palatoquadrate; *PS*, parasphenoid; *PT*, pterygo-palatine; *TH*, trabecular horn; *TR*, trabecula; *VO*, vomer.

But contrary to what is found in actinopterygians, it remains there when differentiating into the nasal sac. The dipnoan nasal placode differs from that of amphibians and most fishes in that the two ectodermal layers are quite distinct. This makes it possible to determine from which embryonic layer the adult components of the olfactory organs are developed.

The nasal pit is formed by invagination of the nasal placode in *Neoceratodus* as well as in Selachii, *Salmo* and Amphibia (Fig. 1 A, B). In *Protopterus* and *Lepidosiren* as well as in actinopterygians other than *Salmo* it is said to be formed by cytolysis between the nasal placode and the superficial ectodermal layer, or in the placode itself. If this is correct, the formation of the nasal pit is

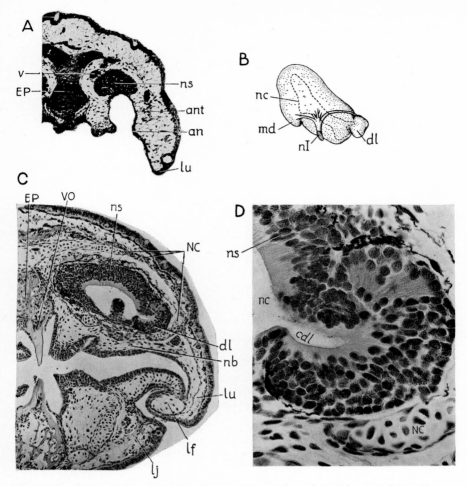

Fig. 2. The dipnoan olfactory organ. *A*, transverse section of the left part of the olfactory organ and "vestigial organ" of *Neoceratodus forsteri*, 16 mm; × 50; modified from Broman, 1939. *B*, ventral reconstruction of the right part of the olfactory organ and the olfactory nerve of *Protopterus annectens*, 20.2 mm; × 34; modified from Rudebeck, 1944. *C*, transverse section of the left side of the snout of *Protopterus annectens*, 55 mm, at the level of the diverticulum laterale; × 38. *D*, the same diverticulum laterale as in *C*, but of higher magnification; × 168. *C* and *D* from Bertmar 1965*b*.

an, anterior nostril; *ant*, anterior nasal tube; *cdl*, cavity of diverticulum laterale; *dl*, diverticulum laterale; *lf*, labial fold; *lj*, lower jaw; *lu*, lateral secondary upper lip; *md*, medial diverticle of nasal sac; *nb*, nasal bridge; *nc*, nasal cavity; *nI*, nervus olfactorius; *ns*, nasal sac; *v*, vestigial organ; *EP*, ethmoidal plate; *NC*, nasal capsule; *VO*, vomer.

accordingly of little value for general phylogenetic considerations (see Bertmar 1965*b*: 20–22).

The nasal sac has a structure which is generally comparable to that of other recent fish groups. A large diverticulum laterale is certainly formed in all three dipnoan genera (Figs. 1 C, D, 2 B, C, D) but it develops into an olfactory fold as do the other lateral diverticles of the nasal sac, and therefore it does not seem to be homologous with the vomeronasal organ (Jacobson organ) in tetrapods. The "vestigial organ" in *Neoceratodus* that Broman (1939) incorrectly described (Fig. 2 A) and homologized with a "rudimentary nasal gland" in tetrapods is obviously the ganglion terminale. There are thus no relevant specific resemblances between the nasal sac of the Dipnoi and that of the Amphibia.

A naso-buccal groove is formed in Dipnoi as well as in Holocephali and some other elasmobranchs such as *Scyllium, Raja* and *Heterodontus*, in Apoda and Amniota. It is, however, of different types and has probably evolved independently (Bertmar, 1967 b). The Holocephali, *Scyllium* and *Raja* have a posterior naso-buccal groove and a posterior nostril (Fig. 3 C). The lungfishes have a long naso-buccal groove, which primarily lies posterior to the primary nasal opening (Fig. 1 B) and then extends also anterior to the opening (Fig. 1 D). Simultaneously the naso-buccal groove deepens and its middle part, representing the nasal groove of the Actinopterygii, becomes the primary nasal tube facing to the exterior through the primary nostril (Fig. 1 C, D). Very soon the folds of the primary nasal tube (naso-buccal groove) fuse in the middle part of its length. As a result of this growth the anterior and posterior nasal tubes and nostrils are formed (Fig. 3 D). The remaining anterior and posterior parts of the naso-buccal groove then atrophy. There is accordingly no bridging of this groove posterior to the excurrent nostril and thus no choanal bridge, choanal tube or choana are formed in the Dipnoi.

The choana of the Apoda and Amniota (Fig. 3 H) is formed from the posterior end of the naso-buccal groove, whereas in dipnoans, as in other fishes (Fig. 3 A, B, C), the excurrent nostril is formed by division of the primary nostril. In the urodelan family Hynobiidae there is a short postnasal groove, which represents the anterior portion of an ancient naso-buccal groove (Bertmar, 1966 a). As far as known the other urodeles and all anurans have no such rudiment of a superficial naso-buccal groove, the choana being formed in the posterior part of the gut process (Fig. 3 G). This process later joins the choanal process of the nasal sac, and the two processes together represent a transformed naso-buccal groove (Bertmar, 1966 a).

From the morphogenesis of the olfactory organ may be concluded (1) that the posterior nostril in Dipnoi, Holocephali and some elasmobranchs is not homologous with the choana in tetrapods; (2) that it is instead equivalent to the external excurrent nostril in other non-choanate fishes; (3) that the part of the naso-buccal groove from which the anterior and posterior nasal tubes and nostrils are developed in Dipnoi, Holocephali and some elasmobranchs is homologous with the nasal groove of the Elasmobranchii and Actinopterygii; and (4) that the bridge that is formed by fusion of the edges of this part of the naso-buccal groove corresponds to the nasal bridge in the Elasmobranchii and Actinopterygii.

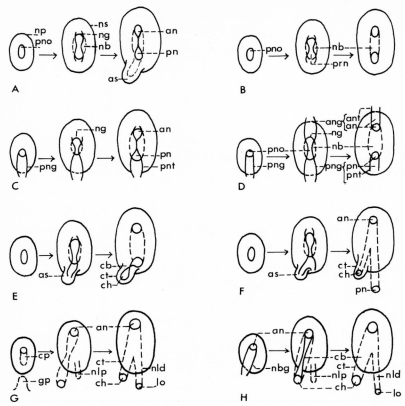

Fig. 3. Diagram of different types of embryologic development of the external nostrils and choana in Gnathostomata; ventral view; modified from Bertmar, 1966a. *A*, Preteleostomi; *B*, Coelacanthiformes, Struniiformes, Actinopterygii, most Selachii and Arthrodira; *C*, Holocephali, Ptyctodontida, *Scyllium*, *Raja*; *D*, Dipnoi; *E*, Porolepiformes; *F*, Osteolepiformes; *G*, Anura, Urodela; *H*, Apoda, Amniota. The types of the fossil groups are deduced from the anatomy of their olfactory organ and comparisons with that of recent fish. The evolution of these types is described in Bertmar, 1968b. Anterior nostril: open circle; posterior nostril: circle with vertical lines; choana: circle with horizontal lines; choanal tube: broken lines; nasolacrimal duct: dotted lines.

an, anterior (incurrent) nostril; *ang*, anterior naso-buccal groove; *ant*, anterior nasal tube; *as*, accessory nasal sac; *cb*, choanal bridge; *ch*, choana; *cp*, choanal process; *ct*, choanal tube; *gp*, gut process; *lo*, opening of naso-lacrimal duct; *nb*, nasal bridge; *ng*, naso-buccal groove; *nld*, naso-lacrimal duct; *nlp*, process of naso-lacrimal duct; *ns*, nasal sac; *pn*, posterior (excurrent) nostril; *pno*, primary nasal opening; *png*, posterior naso-buccal groove; *pnt*, posterior nasal tube; *prn*, primary nasal tube.

Endocranium. In order to elucidate the problem of the origin of the dipnoan nostrils, we must also consider their relation to the skeleton of the snout. Since we know the relationships of the choana to the nasal capsule in the Rhipidistia and Tetrapoda, we can attempt to find signs of the same morphological relationships in the Dipnoi.

As a result of his studies on *Protopterus*, Thomson suggested that the lamina nariochoanalis forms the lateral rim of the nasal capsule in the Tetrapoda and

Rhipidistia. "In the Dipnoi the lateral rim of the capsule is formed by the union of the cornu trabeculae with the posterior part of the tectum nasi ... Therefore, the lateral rim of the dipnoan capsule may not be considered to be the homologue of the lamina nariochoanalis of tetrapods ... The most likely conclusion to be drawn from the present study is that the posterior naris of Dipnoi is directly homologous with the posterior external naris of fishes, and is not a choana" (Thomson, 1965: 233–235).

Thomson added that the Dipnoi and the Crossopterygii probably "had a not too distantly removed common ancestor". He offered, however, no evidence to support this assumption.

The important question of the origin of the excurrent nostril of the Dipnoi was not discussed by Fox (1965). Apparently, however, he considered it homologous with the tetrapodean choana because in his list of main characters in common in larvae of *Neoceratodus*, Anura and Urodela, he mentioned as point 5: "Internal and (external) nares".

My own research on the development of the endocranium in fishes (Bertmar, 1959; 1962*b*; 1963; 1966*b*) showed that the dipnoan cartilaginous *nasal capsule* is not unique. It is certainly large and highly fenestrated but nevertheless its ontogeny reveals that it is composed of cartilage structures and condrification centres most of which have equivalents in other fish groups, especially in the Actinopterygii, and some also in the Amphibia (Fig. 4). In the anterior median line, the ethmoidal (or trabecular) plate (1) is joined by (2) the nasal septum, and dorsally and posteriorly to this a single bar called (3) the taenia tecti medialis anterior is formed. The nasal septum is antero-ventro-laterally connected with (4) the trabecular horn, and postero-dorso-laterally with (5) the taenia marginalis ethmoidalis; and the latter element is joined posteriorly to a separate chondrification centre, (6) the cartilago supraentethmoidalis. This cartilage constitutes the dorsal limit of the ethmoidal and orbito-temporal regions, and the ventral limit is formed by an antero-ventrally directed process from the trabecula cranii, called (7) the processus ectethmoideus.

In addition to these seven similarities of the nasal capsule there are also general similarities of the *orbito-temporal* region of the endocranium of Dipnoi, many other fish groups and Amphibia. Ventrally is (1) the trabecula cranii, and dorsally lies (2) the posterior supraorbital cartilage. These are interconnected by two or three perpendicular bars, the so-called (3) pila prooptica, (4) pila metoptica and (5) pila antotica. Each posterior supraorbital cartilage is also connected with the auditory capsule by a short bar, (6) the taenia marginalis posterior. In *Neoceratodus* is found (7) an anterior supra-orbital cartilage and this is connected to the nasal capsule by (8) the taenia marginalis anterior.

As in other fish groups and in Amphibia the dipnoan trabeculae are posteriorly joined to (1) the basal plate, (2) the auditory cartilages of the *otic* region

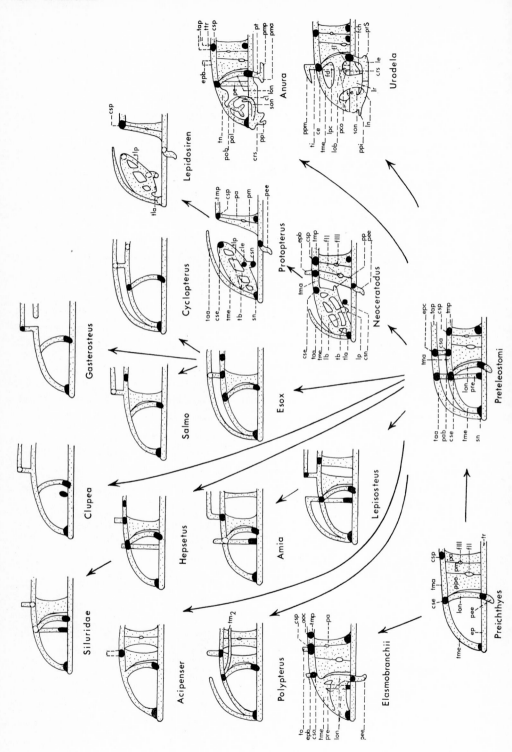

and (3) the pila occipitalia of the *occipital* region of the endocranium. The basal plate is of the primitive, distinct type also present in the Chondrostei and Selachii, i.e. on each side of the notochord it consists of (4) a parachordal (primarily segmented in the latter groups) and (5) a primarily segmented lamina basiotica (Bertmar, 1959). The Tetrapoda have a more unitary basal plate. Its composition is still rather unknown but probably it generally consists of unsegmented parachordals and basiotic laminae. (6) The lateral commissure at the anterior margin of the auditory cartilage is of the most primitive type found in recent fishes (Bertmar, 1959; 1963).

All these 21 structures which make up the dipnoan cartilaginous skull are not found solely in the Dipnoi, but some of them, it is true, are *specialized* in a characteristic way (Fig. 5). The different elements of the orbito-temporal region very early became intimately fused to form an inter-orbital wall, (1) the so-called orbital cartilage. And in the ethmoidal region the nasal sacs are supported and covered by basket-like nasal capsules. (2) The ethmoidal plate is comparatively long and narrow, (3) the nasal septum is short and high, and (4) the postnasal wall between the cartilago supraentethmoidalis and the processus ectethmoideus has degenerated. This gives the impression that the dipnoan nasal capsules are relatively "open" ventrally and posteriorly. Dorsally and laterally, however, the dipnoan nasal capsule seems comparatively "closed", partly because of some skeletal elements which have no equivalents in other fishes or in amphibians: (5) some special labial cartilages; (6) certain dorso-

---

Fig. 4. Diagram of the basic composition and evolution of the ethmoidal and orbito-temporal regions of the cartilaginous skull in Preichthyes and most recent fish and amphibian stocks. The *Hepsetus*- and dipnoan types are founded on earlier comparative embryological studies by the present writer (Bertmar, 1959; 1963; 1966 *b*); the other skull types are based on the literature. Chondrification centres: black; hypothetic or irregularly ocurring structures: broken lines.

aoc, antotic cartilage; *ce*, cartilago (columna) ethmoidalis; *cl*, cartilago lateralis; *cle*, cartilago lateroethmoidalis; *crs*, crista rostro-caudalis; *csa*, cartilago supraorbitalis anterior; *cse*, cartilago supraentethmoidalis; *csn*, cartilago subnasalis; *csp*, cartilago supraorbitalis posterior; *ep*, ethmoidal plate; *epb*, epiphyseal bridge; *epc*, epiphyseal cartilage; *fch*, fenestra endochoanalis; *fd*, fenestra dorsalis; *fe*, fenestra endonarina; *fl*, fenestra lateralis; *fII*, optic foramen; *fIII*, metoptic foramen; *lb*, longitudinal bar of nasal capsule; *le*, lamina ectochoanalis; *ln*, lamina nariochoanalis; *lob*, lamina obliqua; *lon*, lamina orbitonasalis; *lp*, lateral process of taenia lateroethmoidalis; *lpc*, lamina precerebralis; *lr*, lamina retronarina; *pa*, pila antotica; *pab*, paraphyseal bridge; *pal*, processus alaris; *pco*, planum conchale; *pe*, pila ethmoidalis; *pee*, processus ectethmoideus; *pl*, pila lateralis; *pm*, pila metoptica; *pma*, processus maxillaris anterior; *pmp*, processus maxillaris posterior; *pob*, planum obliqua; *pp*, processus posterior of taenia lateroethmoidalis; *ppi*, processus prenasalis inferior; *ppm*, processus prenasalis superior medialis; *ppo*, pila prooptica; *pre*, processus entethmoidalis; *prS*, Seydel's palatal process; *pt*, pars triangularis; *sn*, septum nasi; *son*, solum nasi; *taa*, taenia tecti medialis anterior; *tap*, taenia tecti medialis posterior; *tb*, transverse bar of nasal capsule; *ti*, tectum internasale; *tla*, taenia lateroethmoidalis anterior; *tlp*, taenia lateroethmoidalis posterior; *tm$_2$*, secondary taenia marginalis; *tma*, taenia marginalis anterior; *tme*, taenia marginalis ethmoidalis; *tmp*, taenia marginalis posterior; *tn*, tectum nasi; *to*, tectum orbitale; *tr*, trabecula cranii; *ttr*, taenia tecti ransversalis.

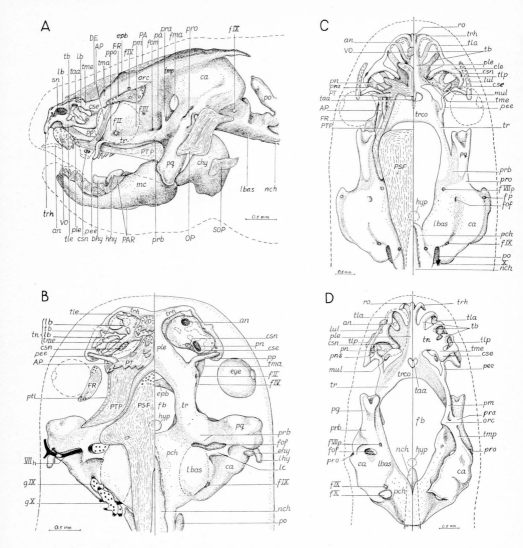

Fig. 5. Reconstructions of the cartilaginous skeleton and parts of the exoskeleton of *A, B, Neoceratodus forsteri*, 28 mm, lateral and ventral views; *C, Protopterus annectens*, 55 mm, ventral view; *D, Lepidosiren paradoxa*, 55 mm, left side in ventral view, right side in dorsal view. From Bertmar 1966b.

an, anterior (incurrent) nostril; *bhy*, basihyal; *ca*, auditory cartilage; *chy*, ceratohyal; *cle*, cartilago lateroethmoidalis; *cse*, cartilago supraentethmoidalis; *csn*, cartilago subnasalis; *ehy*, epihyal; *epb*, epiphyseal bridge; *fb*, fenestra basicranialis; *fma*, foramen sphenoticum majus; *fof*, foramen for nervus facialis; *fom*, foramen for v.ophthalmica magna; *fp*, foramen for Pinkus organ; *fII*, optic foramen; *fIII*, metoptic foramen; *fIV*, foramen for n.trochlearis; $fVII_p$, foramen for r.palatinus VII; *fIX, gIX*, foramen and ganglion of n.glossopharyngeus; *fX, gX*, foramen and ganglion of n.vagus; *hhy*, hypohyal; *hyp*, hypophysis; *lb*, longitudinal bar of nasal capsule; *lbas*, lamina basiotica; *lc*, lateral commissure; *lhy*, laterohyal; *lul*, lateral upper labial cartilage; *mul*, medial upper labial cartilage; *nch*, notochord; *orc*, orbital cartilage; *pa*, pila antotica; *pch*, parachordal; *pee*, processus ectethmoideus; *ple*, planum ethmoidale; *pm*, pila metoptica; *pn*, posterior (excurrent) nostril; *pns*, processus nasalis; *po*, pila occipitale; *pp*, processus posterior of taenia lateroethmoidalis; *ppo*, pila prooptica; *pq*, palatoquadrate;

lateral bars of the capsule; (7) the taenia lateroethmoidalis posterior constituting the posterior half of the ventro-lateral margin of the nasal capsule; and (8) the subnasal cartilage, a structure in the floor of the nasal capsule separating the anterior and posterior nostrils.

There are certainly also specializations of the endocranium within the Dipnoi: 11 differences are to be found between *Neoceratodus*, on one hand, and *Protopterus* and *Lepidosiren* on the other; and 4 differences between the latter two genera (cf. Fig. 5; and Bertmar, 1966 b: 98–104).

This survey of the morphogenesis of the dipnoan endocranium has shown some general similarities and differences between the dipnoan endocranium and that of other fishes and amphibians. A fairly close comparison of the nasal capsules in Dipnoi, Elasmobranchii, Actinopterygii, Anura and Urodela was recently presented (Bertmar, 1966 b: 120–133). The presence or absence of homologous elements was shown in four tables, the position of these elements was illustrated diagrammatically, and points of special importance for the phylogenetic discussion were commented upon.

If the Dipnoi are compared with the Elasmobranchii there are apparently certain similar structures formed in the ethmoidal and orbito-temporal regions of the skull (Fig. 4), but these are of a general type and present also in teleostomes and tetrapods. There are also similarities especially with the Holocephali, for example the autostyly, the presence of an internal excurrent nostril (pseudochoana), the absence of maxillary and premaxillary bones, and the possession of tooth-plates with similar location, shape and histological structure (Ørvig, 1951). Unfortunately we still know little of the embryology of the Holocephali. It seems, however, probable that most of these similarities with the Dipnoi are structural adaptations to the same type of feeding habits and of minor phylogenetic significance.

Comparisons of the anterior half of the cartilaginous endocranium reveals more structural similarities between the Dipnoi and Actinopterygii than between the Dipnoi and Elasmobranchii or between the Dipnoi and Amphibia (Fig. 4). It is thus reasonable also from this point of view to include the dipnoans and actinopterygians into one main group, the Teleostomi, the third member

---

pra, processus ascendens of palatoquadrate; *prb*, processus basalis of palatoquadrate; *pro*, processus oticus of palatoquadrate; *ptl*, pterygoid ligament; *ro*, rostral processus of planum ethmoidale; *sn*, septum nasi; *taa*, taenia tecti medialis anterior; *tb*, transverse bar of nasal capsule; *tla*, taenia lateroethmoidalis anterior; *tle*, taenia lateroethmoidalis, *tlp*, taenia lateroethmoidalis posterior; *tma*, taenia marginalis anterior; *tme*, taenia marginalis ethmoidalis; *tmp*, taenia marginalis posterior; *tn*, tectum nasi; *tr*, trabecula cranii; *trco*, trabecula communis; *trh*, trabecular horn; *AP*, ascending process of pterygopalatine; *DE*, dermethmoid; *FR*, frontal; *OP*, operculum; *PA*, parietal; *PAR*, prearticular (coronoid); *PSF*, parasphenoid; *PT*, toothplate of pterygo-palatine; *PTP*, pterygo-palatine; *SOP*, subopercular; *VO*, vomer; VII_h, tr.hyomandibularis VII; *X*, n.vagus.

of which is the Crossopterygii. Unfortunately no embryonic or larval material of *Latimeria chalumnae* has yet been found. But with regard to the adult forms, the Actinistia and Rhipidistia seem to differ considerably from the dipnoans, especially with respect to the structure of the snout (Jarvik, 1942: 391–392, 495, 620–621).

Where the Amphibia are concerned there are certainly some general similarities (cf. above) but also pronounced differences in the composition of the anuran and dipnoan nasal capsules (Fig. 4). It is evident that the embryonic and larval endoskeleton of the snout are more similar in the Urodela and Dipnoi than in the Anura and Dipnoi. These resemblances between urodeles and dipnoans are to be found especially in the lateral and ventro-lateral parts of the nasal capsule, as well as in the autostylic jaw suspension, and the relationships of the blood vessels and the trigeminal and facial nerves to the palatoquadrate processes. The latter resemblances Fox (1965) considered to be evidence of a common relationship of the Dipnoi and Urodela. It has to be remembered, however, that in general the autostyly is due to a similar type of feeding habits, an adaptation type which needs a firm support of the upper jaw. Furthermore, this type does not form in the actinopterygians but still three of their palatoquadrate processes—proc. basalis, proc. oticus internus and proc. pterygoideus (cf. Bertmar, 1959; 1963)—have principally the same position to the trigeminal and facial nerves and blood vessels as they have in the Dipnoi. This means that these processes are homologous with those of the Dipnoi, and therefore it seems reasonable to believe that if any actinopterygian fish group will ever attain an autostylic jaw suspension, these palatoquadrate processes will have the same relationships of nerves and vessels as they have in the Dipnoi. For this reason the autostyly and relationships of nerves and vessels to the palatoquadrate have minor relevance for a consideration of the relationships of the Dipnoi and Urodela.

Fox (1965) also considered the platytraby in this connection. This character, however, is present not only in the Amphibia and Dipnoi but in all sub-classes of fish (cf. Stensiö, 1963: 10–12). This character has thus evolved independently in all groups and it is therefore of minor importance for phylogenetical considerations.

Furthermore, it has also to be remembered that the endoskeleton of the urodelan and dipnoan snout also differ in many respects, for example with regard to the structure of the postnasal wall and the anterior part of the ethmoidal plate. And if the postero-lateral floor of the nasal sacs and capsules, the nasobuccal grooves and the internal nostrils are also considered (see above) we can reasonably draw the conclusion that from these points of view the dipnoans and urodeles do not have any close affinities.

We have thus found that chondrification centres and certain endocranial elements are very conservative structures and that they therefore are relevant for consideration of phylogenetical relationships.

Exocranium. It has been shown that the shape and composition of the dipnoan endocranium is both primitive and specialized. It can be added that it is also specialized in so far as it does not ossify. The exocranium, however, seems to be even more specialized. This conclusion was reached already by Goodrich (1909: 238) who suggested that "so far have these Dipnoi departed from the normal type, that the homology of the cranial bones cannot yet be determined with certainty".

The terminology of the dipnoan skull roof was discussed also 40 years later (Westoll, 1949: 125–127) mainly because of the difficulty in homologizing the many small roofing plates in early dipnoans. According to White (1965) the Dipnoi separated from the Preteleostomi when the skull roof was in a mosaic stage, and therefore he considered it misleading to use the nomenclature of compound names suggested by Holmgren & Stensiö (1936) which was based on that of the tetrapods and actinopterygians. Instead, he recommended a modified version of Forster-Cooper's (1937) alphabetical scheme, properly related to the standard sensory canal pattern. He also suggested that the pattern of the dipnoan skull roof has developed from an ancestral mosaic principally by loss-and-invasion of roofing plates, whereas fusion occurs for the most part as individual aberration.

Referring to the mosaic conditions in the skull roof of osteolepids and *Acipenser* Jarvik (1948: 81) suggested that "complex bones in different groups of fishes given the same name rarely, if ever, are strictly homologous". He therefore wanted a new terminology in vertebrates, but until this has been presented it seemed to him that we must retain the current terms as far as possible, and use them in their old sense in each group.

Regarding the nomenclature of the dipnoan exocranium two things should be kept in mind:

(1) Where the dipnoan palate is concerned, the position and ontogeny of the dermal bones are usually accepted as criteria of their homology; these criteria should therefore also be applied to the dermal bones of the rest of the skull (Fig. 5 A, B, C).

(2) Those authors adhering to an alphabetical scheme in the Dipnoi seem to accept the current terminology in the Tetrapoda (that is, single or compound names like frontal, intertemporo-supratratemporal, etc.) in spite of the fact that many of the ancestors of the Tetrapoda, i.e. the rhipidistian crossopterygians, had a partly mosaic skull roof of principally the same type as that of the dipnoans. This is inconsequent. Besides, a mosaic skull roof is also present in certain actinopterygians (Lehman, 1966) but nevertheless the current tetrapodean terminology is used in actinopterygians in general.

Consequently, if we wish to retain the current terminology on dermal bones in the Tetrapoda, Crossopterygii and Actinopterygii it should be applied also to the Dipnoi. In fact, studies on fossil and recent dipnoans reveal a variant of the same general pattern of exocranium present in the other teleostomian fishes (Fig. 5 A–C; see Holmgren & Stensiö, 1936; Bertmar, 1966b).

With regard to the structure of the dermal bones and scales, the Crossopterygii and Dipnoi have a similar cosmoid type. Westoll (1949) and others used this as an important evidence of relationships of these fishes. Cosmine is, however, present also in scales of the agnaths *Tremataspis* and *Cephalaspis* and is therefore of minor importance for such considerations. Besides, palaeohistological studies have recently convinced Gross (1966) that the dipnoan "Rundschuppen" have evolved "unabhängig von den Rhipidistia".

The development of the latero-sensory system of the dipnoan head was studied by Pehrson (1949). According to him it is similar to that of selachians and amphibians (*Necturus*).

Upper lips. In the vertebrates having a naso-buccal groove there are primarily two types of upper lips, a primary lip on the medial side and a secondary lip on the lateral side of the groove (Fig. 1 B). In dipnoan embryos the latter lip divides into a medial secondary upper lip along the lateral side of the naso-buccal groove and a lateral secondary upper lip constituting the upper edge of the mouth (Fig. 1 C, D). These lips are large and supported by labial cartilages. This is probably due to the atrophy of the premaxillary and maxillary bones in ancient Dipnoi and the subsequent adaptation to a new type of jaw suspension. According to Allis (1919) adult dipnoans also have a functional tertiary upper lip. However, when present, this is the anterior part of the lateral secondary upper lip (Bertmar, 1965b). Recently Jarvik (1964) has demonstrated the amazingly close similarities in Devonian and recent adult dipnoans as regards the lips and notches of the snout as well as the lips and sensory lines of the lower jaw. The notches probably lodged the anterior nasal tubes, and behind each notch there was a subnasal ridge containing the subnasal cartilage. These structures are very similar to those in recent dipnoans.

Vascular system. Information on the *arterial system* is relatively scarce, at least in fish. With regard to the development of the anterior part of the visceral arterial system the Dipnoi appear to be of a fairly primitive type which may have separately evolved from the basic type of the Preteleostomi (Fig. 6). The orbital artery has a dorsal branch, the supraorbital artery, which is the only artery of the ethmoidal region. The orbital artery is formed from the hyoid aorta and Fox (1965) homologized it with the tetrapodean stapedial artery. It is, however, also present in many teleosts (Bertmar, 1962a). With regard to the snout the

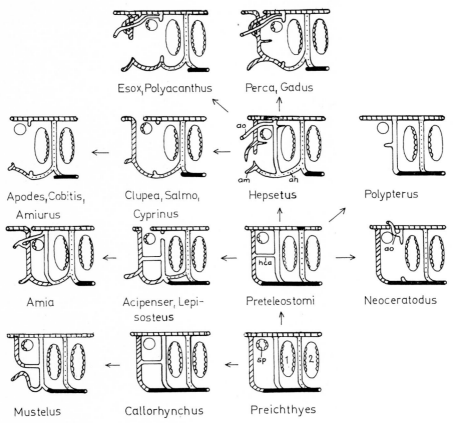

Fig. 6. Diagram of the evolution of anterior visceral artery systems in fishes. The *Hepsetus*- and *Neoceratodus*-types are founded on comparative embryological studies by the writer (Bertmar 1962 a; 1966 b) and the other types are based on the literature (cf. Bertmar, 1962 a, fig. 12 and references). Ventral aorta: black; dorsal aorta: transversely striped; mandibular aorta: obliquely striped; hyoid aorta: unmarked; 1st branchial aorta: dotted.

ah, arteria hypobranchialis; *am*, a. mandibularis; *ao*, a. supraorbitalis; *hca*, hyomandibular commissural artery; *sp*, *1*, *2*, spiraculum, 1st and 2nd gill opening.

dipnoans have quite another type of arterial system than the urodeles (Bertmar, 1966 b: 136).

It has also been shown (Bertmar, 1965 a: 1966 b) that the dipnoan *vein system* of the head is both primitive (presence of an anterior connection between the supra- and infraorbital veins) and specialized: veins being very branched; sinuses formed both in the ethmoid and orbito-temporal regions; ethmoidal region drained only from the infraorbital vein.

Where the jugular and cerebral vein systems are concerned, however, the Dipnoi seem to be of a type intermediate between that of the Elasmobranchii and the Actinopterygii (Fig. 7; see Bertmar, 1965 a).

The jugular vein has evolved from the vv. capitis medialis and lateralis. The

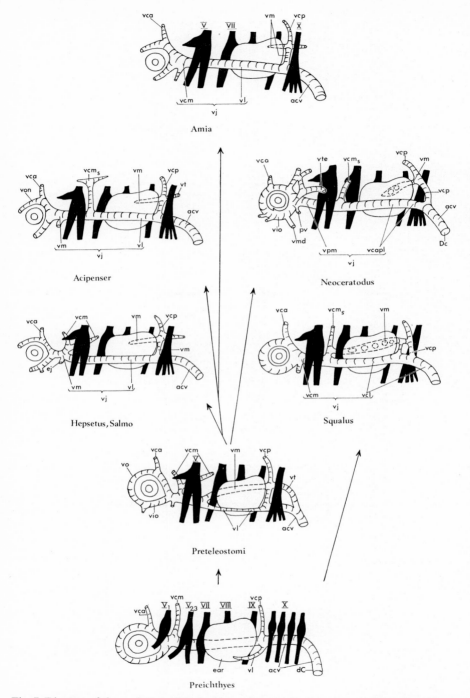

Fig. 7. Diagram of the evolution of the jugular and cerebral vein systems in fishes; founded on comparative embryological studies. Modified from Bertmar 1965a.

acv, anterior cardinal vein; *dC*, ductus Cuvieri; *ej*, external jugular vein; *pv*, pituitary vein; *vca*, vena cerebralis anterior; *vcm$_s$*, v. cerebralis media secundaria; *vcp*, v. cerebralis posterior;

vena capitis medialis was probably the dominant vein of the Preichthyes and Preteleostomi. In recent fishes it is generally formed in the embryos and retained in the adults as three portions, one constituting the anterior part of the jugular vein (except in most elasmobranchs), one draining the ventro-medial side of the ear, and one forming the terminal vein which drains the tissue along the accessory nerve. The posterior part of the v. capitis medialis is generally continuous with the anterior cardinal vein but in lungfishes this part atrophies and the v. capitis medialis instead opens into the v. cerebralis posterior. This is thus a specialized condition in the Dipnoi.

The v. capitis lateralis forms the middle and posterior part of the jugular vein of the adults. In elasmobranchs and dipnoans the posterior part of the jugular vein runs lateral to the vagus and—in dipnoans— joins the anterior cardinal vein behind the vagus ganglion. This is a specialized condition, but even more specialized are the elasmobranchs (*Squalus* at least) because in these fishes the entire anterior cardinal vein becomes atrophied (Bertmar, 1965a: 100–101). In this respect the actinopterygians are more primitive, as they have a long anterior cardinal vein running medial to the vagus and up to the ear where it divides into the vv. jugularis and capitis medialis. As there is no part of the jugular vein lying lateral to the vagus the v. jugularis is thus shorter in actinopterygians.

The so-called "secondary vein" is not any separate vein. In actinopterygians (in dipnoans it is not described) it is usually represented by the anterior portion of the v. cardinalis anterior and all the v. capitis lateralis, and in elasmobranchs it is represented only by the former portion. The term "secondary vein" should therefore not be used, at least in recent fishes (Bertmar, 1965a).

Of the cerebral veins the dipnoan v. cerebralis posterior is already mentioned. In dipnoans it opens into the v. cardinalis anterior, in elasmobranchs into the posterior end of the jugular vein, and in actinopterygians into the v. capitis medialis.

The v. cerebralis media primitively runs anterior to the trigeminus or—if this was originally divided into two ganglia—between the ganglion of the ramus ophthalmicus profundus and that of r. maxillaris and r. mandibularis. In the teleosts *Hepsetus* and *Salmo* the primitive type is developed but in *Acipenser* (*Amia* has no such vein at all), *Neoceratodus* and *Squalus* there is a secondary v. cerebralis media running between the trigeminus and facialis ganglia. This is in contrast to van Gelderen's (1933) classification.

The v. cerebralis anterior is generally formed in fish but it can be rather

---

vl, vena capitis lateralis; *vm*, v. capitis medialis; *vmd*, v. mandibularis; *vmp*, v. capitalis medialis proper; *vio*, vena infraorbitalis; *vj*, v. jugularis; *von*, v. orbito-nasalis; *vt*, v. terminalis; *vte*, v. temporalis; V_1, nervus ophthalmicus profundus; $V_{2, 3}$, r. maxillaris and r. mandibularis trigeminus; *VII*, n.facialis; *VIII*, n.acusticus; *IX*, n.glossopharyngeus; *X*, n. vagus.

vestigial. It opens into the v. supraorbitalis, not the jugular vein as van Gelderen maintained, and generally it migrates caudad during ontogeny. In Dipnoi it becomes a relatively large vessel.

The Dipnoi are definitely primitive in the respect that they have no *lymphatic system* (Bertmar, 1965a; 1966b).

Nervous pattern, brain and pituitary. The morphogenesis of the nervous system of the dipnoan snout was recently compared with that of the Amphibia (Bertmar, 1966b). There are some similarities in the proximal run of the nerves but distally there are differences, e.g. in the r. maxillaris and r. ophthalmicus superficialis. The r. buccalis lateralis is not associated with the r. maxillaris as in other fish and Urodela.

As far as the brain is concerned Stensiö (1963: 116) suggested that "the brain is of a greater importance in judging the phylogenetic affinities of the major stocks of lower vertebrates than what has as yet generally been realized ... there is reason to believe that the general shape and structure of the brain indicates a much closer affinity of the Dipnoans to the Actinistians, Porolepiforms and Osteolepiforms than to the two other stocks of lower Gnathostomes, viz. the Elasmobranchiomorphs and Actinopterygians". He added, however, that this view "is contradicted by the fact that especially in the hard and soft anatomical characters of the head the Dipnoans are very different from the Actinistians, Porolepiforms and Osteolepiforms".

The structure of the pituitary of *Protopterus* has been studied by e.g. Kerr (1949), Wingstrand (1956; 1959; 1966) and Kerr & van Oordt (1966). By histochemical technique the latter authors could compare in detail the cell types of *Protopterus* and amphibians. They found that the elaborate intermingling of intermediate- and neural-lobe tissue of the dipnoan pituitary seems to be quite comparable to the condition found in elasmobranchs and lower actinopterygians. Also the pattern of tubules penetrating the nervous tissue from the infundibular cavity is strongly reminiscent of the arrangement seen in the holosteans. Possibly the whole complex represents some very ancient vertebrate organ which loses its importance, and hence its identity, in the land animals. The distal lobe instead resembles that of an amphibian, at least in general structure.

Evolution of the nostrils

If we now return to the problem of the evolution of the dipnoan excurrent nostril (cf. above) it should be observed, in the first place, that the incurrent nostril is secondarily located at the anterior rim of the mouth cavity—its posterior half is even situated inside this cavity. In vertebrates this position is

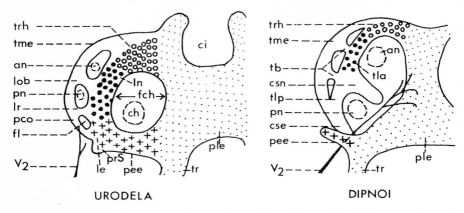

Fig. 8. Diagram of the position of the external nostrils and choana in relation to certain parts of the nasal capsule and r. maxillaris V in Urodela and Dipnoi; ventro-lateral aspect. Homologous structures are indicated in the same way. Modified from Bertmar 1966b (cf. also Jarvik, 1942, fig. 13).

an, anterior (incurrent) nostril; *ch*, choana; *ci*, crista intermedia; *cse*, cartilago supraentethmoidalis; *csn*, cartilago subnasalis; *fch*, fenestra endochoanalis; *fl*, fenestra lateralis; *le*, lamina ectochoanalis; *ln*, lamina nariochoanalis; *lob*, lamina obliqua; *lr*, lamina retronarina; *pco*, planum conchale; *pee*, processus ectethmoideus; *ple*, planum ethmoidale; *pn*, posterior (excurrent) nostril; *prS*, Seydel's palatal process; *tb*, transverse bar; *tla*, taenia lateroethmoidalis anterior; *tlp*, taenia lateroethmoidalis posterior; *tme*, taenia marginalis ethmoidalis; *tr*, trabecula cranii; *trh*, trabecular horn; V_2, r.maxillaris V.

unique and specialized (providing of course that the external position of the incurrent nostril is primitive). Consequently, it is only natural that the incurrent nostril is also secondarily related to the trabecular horn and the posterior continuation of the latter element, which I have called the taenia lateroethmoidalis anterior (Fig. 8).

Secondly, if we compare the Dipnoi with the Tetrapoda it seems reasonable to suggest (from its connection with the trabecular horn) that the dipnoan taenia lateroethmoidalis anterior is homologous with the anterior part of the tetrapodean lamina nariochoanalis, whereas there is no dipnoan equivalent to the lamina nariochoanalis posterior. This fact provides further evidence for the theory set forth on basis of the development of the olfactory organ that the excurrent nostril of Dipnoi is not homologous with the tetrapodean choana (Bertmar, 1965b: 31–34). There is also other support for this theory, for example the blood supply and the position of the maxillary branch of the trigeminal nerve (Bertmar, 1966b: 139–140). An analysis of these facts also shows that the dipnoan excurrent nostril is the external excurrent nostril which has migrated ventrally simultaneously with atrophy of the maxillary and premaxillary bones and the anterior part of the infraorbital sensory canal, and formation of new structures in the lateral part (taenia lateroethmoidalis) and ventral part (cartilago subnasalis) of the nasal capsule. A simultaneous lengthening forward of the lower jaw then contributed to make it possible for the posterior half of the incurrent nostril and the whole excurrent nostril to open into the mouth cavity.

Panchen (1967: 379) has objected that "it is very unlikely that there was ever in the ontogeny or phylogeny of the Dipnoi an actual physical ventral migration of the nares. Rather there was a failure of the dorsal migration of the nasal placode normal in the ontogeny of other bony fish. This conclusion is in fact implied by Bertmar elsewhere".

However, when the main gnathostomian groups are compared it can be deduced that the dipnoan ancestors, the Preteleostomi, probably had an olfactory organ which lay on the antero-lateral side of the snout and had external incurrent and excurrent nostrils (but no choana). Furthermore, the recent dipnoan nasal placode has certainly a ventral position but still it lies anterior to the mouth. That there is no ventral migration in the ontogeny of recent dipnoans was earlier shown (Bertmar, 1965b) but whether it has ever taken place in fossil dipnoans is more difficult to suggest. As there have existed genera without any dermal bones of the anterior part of the upper jaw (see above) it seems possible, however, that some of their species had embryos or larvae with initially external nostrils which then partly, during their ontogeny, migrated ventrally into the mouth cavity. As a parallel may be mentioned that there are amniote embryos (Bertmar, 1966a, fig. 8; 1968b, fig. 3) in which the choana primarily lies outside (anterior to) the mouth cavity.

Summary and conclusions

The present account began with a reference to Szarski (1962) who advocated basic homologies when discussing the origin of land vertebrates, and warned against an enumeration of small similarities and differences. In my opinion the nostrils and the other parts of the head discussed above—the olfactory organ, endocranium, exocranium, lips, vascular system, nervous pattern, brain and pituitary—are truly basic and can be used as criteria of phylogenetic relationships. An analysis of these criteria shows that the major stocks of the fishes may have the relationships presented diagrammatically in Fig. 9.

It was shown that the dipnoan olfactory organ has no rudiment of the vomeronasal organ, or any "rudimentary nasal gland", or other tetrapodean affinities. The nasal pit seems to have little value for general phylogenetic considerations. The lungfish embryos certainly have a naso-buccal groove but it is of a specific type, and the excurrent nostril is not formed from the posterior end of it—like a choana—but divides from the primary nostril. Further evidence of the theory that the dipnoan internal excurrent nostril is homologous with a ventrally migrated external excurrent nostril, not a choana, is supplied by studies on the olfactory organ and skull.

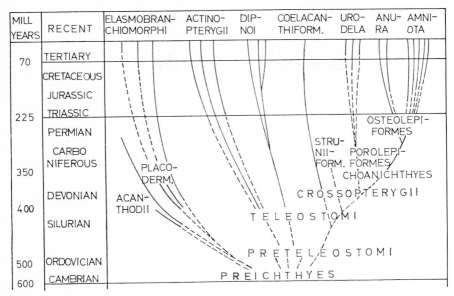

Fig. 9. Diagram of the phylogeny of the major stocks of the Gnathostomata. In the Dipnoi three lines are shown representing the evolution of the recent genera; *Neoceratodus* to the left, *Lepidosiren* to the right (Bertmar, 1968a). The phylogeny of the Elasmobranchiomorphi is according to Stensiö, 1963, that of the Tetrapoda according to Jarvik, 1964, and the position of the Struniiformes after Jessen, 1966.

Comparisons of the endocranium of other fishes and amphibians show that the chondrification centres and certain other endocranial structures are conservative and of great phylogenetic relevance. They also show that in general the dipnoan endocranium is not as specialized as was earlier believed. Its morphogenesis reveals that it is composed of structures many of which have equivalents mainly in the Actinopterygii. The specializations are to be found mainly in the ventro-lateral part of the basket-like nasal capsules, the labial cartilages and orbital cartilage.

The dipnoan exocranium is much specialized but studies on fossil and recent dipnoans reveal a possible variant of the same general pattern present in other teleostomian fishes, and therefore the current terminology on dermal bones should be applied also to the Dipnoi.

As a consequence of the atrophy of the premaxillary and maxillary bones and the subsequent adaptation to a new type of jaw suspension the upper lips are relatively large and supported by labial cartilages. The lips are ancient and conservative structures in Dipnoi.

The dipnoan type of visceral artery system is probably separately envolved from that of the Preteleostomi. The vein system of the head is both primitive and specialized. No lymphatic system is formed.

There are some similarities in the proximal portion of the nervous pattern of the Dipnoi and Amphibia. The general shape and structure of the brain indicates affinity to the Crossopterygii. And the pituitary is in part comparable both to that of elasmobranchs, lower actinopterygians and amphibians.

In the Cambrian the hypothetical basic stock Preichthyes Cambrian probably evolved into the Elasmobranchiomorphi and another hypothetical group, the Preteleostomi. But relatively early the latter stock divided into three stems, the Actinopterygii, Dipnoi and Crossopterygii. Finally the latter stem gave rise to the Amphibia and other tetrapods. According to this theory the lungfishes thus have not evolved from the crossopterygians as was suggested by Westoll (1949), Romer (1955), Fox (1965) and others, or from a stock quite unrelated to that of the Crossopterygii as Jarvik (1942; 1964) implied. The dipnoans instead have a closer affinity to the actinopterygians than was earlier believed. Furthermore, as the dipnoan excurrent nostril is not a choana but a secondarily located external excurrent nostril, the lungfishes cannot be regarded as choanate fishes or as being evolved from such fishes. Again the similarities between the Dipnoi and Urodela can be shown to be evolved by parallelism.

This study was supported by grants No. 2389-4 and 2389-8 from the Swedish Natural Science Research Council (Statens naturvetenskapliga forskningsråd), Stockholm.

References

Allis, E. P. Jr. (1919). The lips and the nasal apertures in the gnathostome fishes. *J. Morph.*, **32**: 145–197.
Allis, E. P. Jr. (1932). Concerning the nasal apertures, the lachrymal canal and the bucco-pharyngeal upper lip. *J. Anat.*, **66**: 650–658.
Atz, J. W. (1952*a*). Internal nares in the teleost, *Astroscopus*. *Anat. Rec.*, **113**: 105–115.
Atz, J. W. (1952*b*) Narial breathing in fishes and the evolution of internal nares. *Q. Rev. Biol.*, **27**: 366–377.
Bertmar, G. (1959). On the ontogeny of the chondral skull in Characidae, with a discussion on the chondrocranial base and the visceral chondrocranium in fishes. *Acta zool. Stockh.*, **40**: 203–364.
Bertmar, G. (1962*a*). On the ontogeny and evolution of the arterial vascular system in the head of the African characidean fish *Hepsetus odoë*. *Acta zool. Stockh.*, **43**: 255–295.
Bertmar, G. (1962*b*). Homology of ear ossicles. *Nature, Lond.*, **193**: 393–394.
Bertmar, G. (1963). The trigemino-facialis chamber, the cavum epiptericum and the cavum orbitonasale, three serially homologous extracranial spaces in fishes. *Acta zool. Stockh.*, **44**: 329–344.
Bertmar, G. (1965*a*). On the development of the jugular and cerebral veins in fishes. *Proc. zool. Soc. Lond.*, **144**: 87–130.
Bertmar, G. (1965*b*). The olfactory organ and upper lips in Dipnoi, an embryological study. *Acta zool. Stockh.*, **46**: 1–40.
Bertmar, G. (1966*a*). On the ontogeny and homology of the choanal tubes and choanae in Urodela. *Acta zool. Stockh.*, **47**: 43–59.
Bertmar, G. (1966*b*). The development of skeleton, blood-vessels and nerves in the

dipnoan snout, with a discussion on the homology of the dipnoan posterior nostrils. *Acta zool. Stockh.*, **47**: 81–150.
Bertmar, G. (1968a). Phylogeny and evolution in lungfishes. *Acta zool. Stockh.* In Press.
Bertmar, G. (1968b). The vertebrate nose, remarks on its structural and functional adaptation and evolution. Manuscr.
Brien, P. (1964). Les sacs gazeux des vertébrés primitifs. *Annls Soc. belg. Méd. trop.*, **44**: 385–400.
Broman, I. (1939). Über die Entwicklung der Geruchsorgane bei den Lungenfischen. *Morph. Jb.*, **83**: 85–106.
de Beer, G. R. (1937). *The development of the vertebrate skull*. Oxford: Oxford University Press.
Forster-Cooper, C. (1937). The Middle Devonian fish fauna of Achanarras. *Trans. R. Soc. Edinb.*, **59**: 223–239.
Fox, H. (1960). Early pronephric growth in *Neoceratodus* larvae. *Proc. zool. Soc. Lond.*, **134**: 659–663.
Fox, H. (1961). The segmentation of components in the hind region of the head of *Neoceratodus* and their relation to the pronephric tubulus. *Acta anat.*, **47**: 156–163.
Fox, H. (1962). A study of the evolution of the amphibian and dipnoan pronephros by an analysis of its relationship with the anterior spinal nerves. *Proc. zool. Soc. Lond.*, **138**: 225–226.
Fox, H. (1963a). Prootic anatomy of the *Neoceratodus* larva. *Acta anat.*, **52**: 126–129.
Fox, H. (1963b). Prootic arteries of larvae of Dipnoi and Amphibia. *Acta zool. Stockh.*, **44**: 345–360.
Fox, H. (1963c). The hyoid of *Neoceratodus* and a consideration of its homology in urodele Amphibia. *Proc. zool. Soc. Lond.*, **141**: 803–810.
Fox, H. (1965). Early development of the head and pharynx of *Neoceratodus* with a consideration of its phylogeny. *J. Zool.*, **146**: 470–554.
Gelderen, C. van (1933). Gefässsystem. In *Handbuch der vergleichenden Anatomie der Wirbeltiere*, hrsg. Bolk, L., Göppert, E., Kallius, E. & Lubosch, W., **6**: 685–744. Berlin & Wien: Urban & Schwarzenberg.
Goodrich, E. S. (1909). Cyclostomes and fishes. In *A Treatise on Zoology*, ed. Lankester, E. R., vol. **9**: Vertebrata Craniata, fasc. 1. London: A. & C. Black.
Greil, A. (1913). Entwicklungsgeschichte des Kopfes und des Blutgefässsystemes von *Ceratodus forsteri*. *Denkschr. med. naturw. Ges. Jena*, **4**: 935–1492.
Gross, W. (1956). Über Crossopterygier und Dipnoer aus dem baltischen Oberdevon im Zusammenhang einer vergleichenden Untersuchung des Porenkanalsystems paläozoischer Agnathen und Fische. *K. svenska VetenskAkad. Handl.*, (4) **5**: 1–140.
Gross, W. (1966). Kleine Schuppenkunde. *N. Jb. Geol. Paläont. Abh.*, **125**: 29–48.
Holmgren, N. (1933). On the origin of the tetrapod limb. *Acta zool. Stockh.*, **14**: 185–295.
Holmgren, N. (1943). Studies on the head of fishes. An embryological, morphological and phylogenetical study. 4. General morphology of the head in fish. *Acta zool. Stockh.*, **24**: 1–188.
Holmgren, N. (1949). Contributions to the question of the origin of tetrapods. *Acta zool. Stockh.*, **30**: 459–508.
Holmgren, N. & Pehrson, T. (1949). The sensory lines in fishes and amphibians. *Acta zool. Stockh.*, **30**: 249–314.
Holmgren, N. & Stensiö, E. A. (1963). Kranium und Visceralskelett der Akranier, Cyclostomes und Fische. In *Handbuch der vergleichenden Anatomie der Wirbeltiere*, hrsg. Bolk. L., Göppert, E, Kallius, E. & Lubosch, W., **4**: 233–500. Berlin & Wien: Urban & Schwarzenberg.
Jarvik, E. (1942). On the structure of the snout of crossopterygians and lower gnathostomes in general. *Zool. Bidr. Uppsala*, **21**: 235–675.

Jarvik, E. (1948). On the morphology and taxonomy of the Middle Devonian osteolepid fishes of Scotland. *K. svenska VetenskAkad. Handl.*, (3) **25**: 1–301.
Jarvik, E. (1960). *Théories de l'évolution des vertébrés, réconsidérées a la lumière des récentes découvertes sur les vertébrés inférieurs.* Paris: Masson.
Jarvik, E. (1964). Specializations in early vertebrates. *Annls Soc. R. zool. Belg.*, **94**: 11–95.
Jessen, H. (1966). Struniiformes. In *Traité de Paléontologie*, ed. Piveteau, J., **4:3**: 387–398. Paris: Masson.
Kerr, T. (1965). Histology of the distal lobe of the pituitary of *Xenopus leavis* Daudin. *Gen. comp. Endocrinol.*, **5**: 232–240.
Kerr, T. & van Oordt, P. G. W. J. (1966). The pituitary of the African lungfish *Protopterus* sp. *Gen. comp. Endocrinol.*, **7**: 549–558.
Krefft, G. (1870). Description of a gigantic amphibian allied to the genus *Lepidosiren* from the Wide Bay district, Queensland. *Proc. zool. Soc. Lond.*, **1870**: 221–224.
Kulczycki, J. (1960). *Porolepis* (Crossopterygii) from the Lower Devonian of the Holy Cross Mountains. *Acta palaeont. polon.*, **5**: 65–106.
Lehmann, J.-P. (1966). Actinoptergyii. In *Traité de Paléontologie*, ed. Piveteau, J., **4:3**: 1–242. Paris: Masson.
Medvedeva, Irene M. (1965). O lokalizatsii materiala v oboniatelnoi plakodie khvostatuikh amfibii (On the material localization of the nasal placode in urodeles). *Dokl. Acad. Sci. U.S.S.R.*, **162**: 709–712. In Russian.
Ørvig, T. (1951). Histologic studies of placoderms and fossil elasmobranchs. 1. The endoskeleton, with remarks on the hard tissues of lower vertebrates in general. *Ark. Zool.* (2) **2**: 321–454.
Ørvig, T. (1957). Remarks on the vertebrate fauna of the Lower Upper Devonian of Escuminac Bay, P. Q., Canada, with special reference to the porolepiform crossopterygians. *Ark. Zool.*, (2) **10**: 367–426.
Pehrson, T. (1949). The ontogeny of the lateral line system in the head of dipnoans. *Acta zool. Stockh.*, **30**: 153–182.
Remane, A. (1952). *Die Grundlagen des natürlichen Systems der vergleichenden Anatomie und der Phylogenetik. Theoretische Morphologie und Systematik.* 2 Aufl. Leipzig: Akad. Verlagsges.
Remane, A. (1964). Das Problem Monophylie-Polyphylie mit besonderer Berücksichtung der Phylogenie der Tetrapoden. *Zool. Anz.*, **173**: 22–49.
Romer, A. S. (1955). Herpetichthyes, Amphibioidea, Choanichthyes or Sarcopterygii? *Nature, Lond.*, **176**: 126–127.
Romer, A. S. (1962). Vertebrate evolution. Reviews and comments. *Copeia*, **1962**: 223–227.
Rudebeck, B. (1944). Does an accessory olfactory bulb exist in Dipnoi? *Acta zool. Stockh.*, **25**: 89–96.
Säve-Söderbergh, G. (1934). Some points of view concerning the evolution of the vertebrates and the classification of this group. *Ark. Zool.*, **26A**: 1–20.
Schaeffer, B. (1965). The rhipidistian—amphibian transition. *Am. Zool.*, **5**: 267–276.
Schmalhausen, I. I. (1957). On the seismosensory system of urodeles in connection with the problem of the origin of the tetrapods. *Zool. Zh.*, **36**: 100–112. In Russian with English summary.
Schmalhausen, I. I. (1958). Nostrils of fishes and their fate in terrestrial vertebrates. *Zool. Zh.*, **37**: 1710–1718. In Russian with English summary.
Sewertzoff, A. N. (1962). Der Ursprung der Quadrupeda. *Paläont. Z.*, **8**: 75–95.
Stensiö, E. A. (1963). The brain and the cranial nerves in fossil, lower craniate vertebrates. *Skr. norske VidenskAkad. Oslo., Mat.-naturv. Kl.*, **1963**: 1–120.
Szarski, H. (1962). The origin of the Amphibia. *Q. Rev. Biol.*, **37**: 189–241.

Thomson, K. S. (1965). The nasal apparatus in Dipnoi, with special reference to *Protopterus*. *Proc. zool. Soc. Lond.*, **145**: 207–238.
Westoll, T. S. (1949). On the evolution of the Dipnoi. In *Genetics, Paleontology and Evolution*, ed. Jepsen, G. L., Mayr, E. & Simpson, G. G.: 121–184. Princeton: Princeton University Press.
White, E. I. (1965). The head of *Dipterus valenciennesi* Sedgwick & Murchison. *Bull. Br. Mus. nat. Hist.: Geol.*, **11**: 1–45.
White, E. I. (1966). Presidential address: A little on lung-fishes. *Proc. Linn. Soc. Lond.*, **177**: 1–10.
Wingstrand, K. G. (1956). The structure of the pituitary of the African lungfish *Protopterus annectens* (Owen). *Vidensk. Medd. dansk naturh. Foren.*, **118**: 193–210.
Wingstrand, K. G. (1959). Attempts at a comparison between the neurohypophysial region in fishes and tetrapods, with particular regard to amphibians. In *Symposium Comparative Endocrinology*, ed. Gorbman, A.,: 393–403. New York: Wiley & Sons.
Wingstrand, K. G. (1966). Comparative anatomy and evolution of the hypophysis. In *The Pituitary Gland*, ed. Harris, G. W. and Donovan, B. T., vol. **1**: 58–126. London: Buttersworths.

A critical review of the diphyletic theory of rhipidistian-amphibian relationships

By Keith Stewart Thomson

Division of Vertebrate Zoology, Peabody Museum of Natural History,
Yale University, New Haven, Conn. USA

The subject of the relationships of the Rhipidistia, both to each other and to the Amphibia, is one that in recent years has developed enormously but, unfortunately, has become the center of a certain amount of controversy. It is probably inevitable that this short communication will add to the controversy. However, it is necessary to summarize some recent developments in studies of the rhipidistian fishes and as we shall see, this will lead to certain criticisms of existing theories of rhipidistian-amphibian relationships.

I wish to state at the outset that I cannot present a complete picture of the phylogeny of the modern Amphibia. I do not know whether the modern Amphibia evolved monophyletically or diphyletically. What I can state is that, in the context of the results of my own researches on the comparative anatomy of Rhipidistia, recent expositions in favor of each of these types of ancestry seem to me to be far from unequivocal. I conclude, therefore, that we have simply to admit that our data on this subject are presently incomplete and that any conclusions from them can only be tentative. This is frequently true of palaeozoological investigations, but it is particularly true in the present instance, when there is such a long, unbridged gap in the fossil record between the Rhipidistia and the first modern amphibians.

The controversy concerns two closely related, but strictly separate, questions: (a) are the Rhipidistia diphyletic, and (b) have the modern Amphibia evolved diphyletically from two separate rhipidistian stocks? The exclusively diphyletic view has been ably and extensively propounded by Jarvik, starting in 1942 with his fundamental study of the snout in lower gnathostomes. His position, put simply, is that known rhipidistian fishes may be divided into two basically separate groups (separate Classes of the Chordata, according to Stensiö, 1963) that differ from each other anatomically in essentially the same characteristic ways as the two principal groups of modern Amphibia—Urodela and Anura—differ from each other (see also Jarvik, in this Volume). However, recent studies (Thomson, 1962; 1964a; 1964b; 1964c; 1967) have led to the conclusion that the Rhipidistia do not differ from each other in such a fundamental manner. Particularly, the results of comparative study of certain key structural characteristics may be interpreted as offering little evidence of diphyly in the rhipidistian fishes. I have not been concerned with the Amphibia in these arguments, and in fact it is extremely important that the Amphibia be considered only *after* the evidence concerning the Rhipidistia has been interpreted, in order to avoid circular arguments and misinterpretation of rhipidistian structures.

Classification of the Rhipidistia

Various classifications of the Rhipidistia have been proposed, stressing to a greater or lesser extent the differences or resemblances between the families. For reasons that have been expounded elsewhere more fully and will be apparent from the following discussion, I find a simple division of the Rhipidistia into widely separate "Porolepiformes" and "Osteolepiformes" unsatisfactory. More in accord with the present state of our knowledge of the group as a whole is a classification into three distinct *grades* of evolution (see Thomson, 1967: 667–669). In this manner the Rhipidistia may be considered as forming three superfamilies of a single Order, as follows:

Order Rhipidistia
   Superfamily Holoptychoidea
     Family Porolepidae
     Family Holoptychidae
   Superfamily Osteolepoidea
     Family Osteolepidae
   Superfamily Rhizodontoidea
     Family Tristicopteridae
     Family Rhizodontidae
     Family Rhizodopsidae

I do not feel that our present knowledge of the Rhipidistia is adequate to allow of a fully cladal classification at the present time. The above is presented as a working system and will certainly need modification, particularly as we come to know more about the holoptychoid rhipidistians. It has the advantage of stressing the close relationships of the known forms without obscuring obvious differences, for example in features of the dermal bone patterns of the skull, that show the clear gradal situation.

Comparative study of the Rhipidistia

In the following paragraphs I wish to draw attention to some new developments in the problem of rhipidistian relationships, developments that necessitate changes of even the most recent review of the situation (Thomson, 1967). The subjects to be covered are: the arrangement of the nerves and vessels in the snout of Rhipidistia, the shape of the brain cavity in Rhipidistia, the structure and functional significance of the anterior palatal recesses in Rhipidistia and the relationship of these structures to the mandibular dentition and the nature of the vomerine bone, together with a consideration of the evolution of the tetrapod and rhipidistian paired limbs in the light of recent discussions.

The distribution of nerves and vessels in the rhipidistian snout. The interpretation of the passage of nerves and vessels in the snout region of a rhipidistian or any other fossil vertebrate is an extremely difficult matter. In certain cases a single canal, apparently for a nerve or vessel, may be traced through continuous bony material for a considerable distance. In the case of the nasal cavity itself, in most instances all we have to work from is a series of foramina in the postnasal wall and a somewhat larger number of "exit" foramina principally in the medial, anterior and dorsal nasal walls. Except where a groove on the inner surface of the nasal capsule connects two or more foramina, the only evidence that can be used in making reconstructions of interconnections between the foramina, and therefore of the passage of particular nerves through the nasal cavity, is from the comparative study of other vertebrates in which such details are known, and this, of course, means living forms. We must therefore be very careful to distinguish between our reconstructions and our basic data.

The basic data for the interpretation of the nerves and vessels of the snout consist of the foramina, canals and grooves just mentioned. In Fig. 1 the arrangement of these in three rhipidistian fishes is illustrated. The arrangement in two further forms, *Eusthenopteron* and *Gyroptychius* not illustrated here, is essentially the same as in *Ectosteorhachis* and *Osteolepis*, respectively.

Jarvik (1942) initially contrasted the situation in *Eusthenopteron* and *Porolepis* and interpreted the greater number of foramina in the dorsal and medial walls of the latter as indicating a greater development and more diffuse branching of the profundus V nerve (more closely approximating the urodelan condition) and lesser development of the ophthalmicus superficialis VII nerve. My own studies on *Osteolepis* and *Gyroptychius* show that in these "osteolepiform" genera, the numbers of "exit" foramina in the dorsal nasal wall is much greater than in either *Eusthenopteron* or *Ectosteorhachis* and much closer to the "porolepiform" condition. Clearly, then, however one *restores* the internal connections of these foramina, there is a broad spectrum of the number of foramina to be considered, rather than a polarization into two "types" of arrangement.

In all reconstructions so far given for the significance of these nerve foramina, only two main nerves have been considered as possibly having passed within or between them—the ramus profundus V and ramus ophthalmicus super-

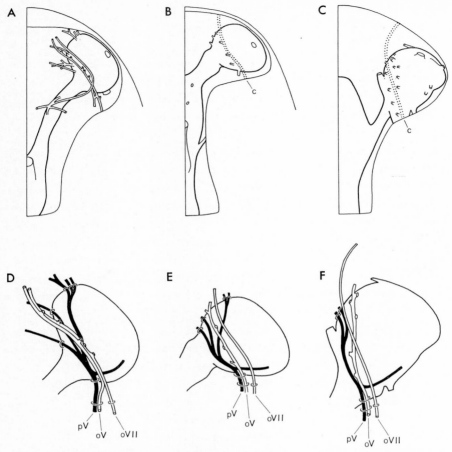

Fig. 1. Endocranial cavities of the rostral region (*A*, *B*, *C*) and restoration of the nerves (*D*, *E*, *F*) in *Ectosteorhachis* (*A*, *D*), *Osteolepis* (*B*, *E*), and *Porolepis* (*C*, *F*); data from Thomson (1964b; 1965) and Jarvik (1942; 1962).

c, position of canal for r.ophthalmicus VII; *oV*, r.ophthalmicus V; *oVII*, r.ophthalmicus VII; *pV*, r.profundus V.

ficialis VII (we are not considering palatal or buccal nerve patterns here). In previous papers I have attempted to show that, even if only these two are considered, the patterns of the nerves are not significantly different in "porolepiforms" and "osteolepiforms". However, some data suggest that we have all been wrong in restricting ourselves to these two nerves alone. Apart from the two above-mentioned nerves that are found in Amphibia (the relationships to which we are attempting to discover), the Rhipidistia, which are extremely primitive bony fishes, may in fact have possessed a separate ramus ophthalmicus V—a nerve that in other fishes such as the Dipnoi (Pinkus, 1895) is usually closely incorporated with the ramus ophthalmicus VII. Thomson (1965: 189) found evidence for such a nerve in *Osteolepis* as did Stensiö (1963, fig.

52) in *Eusthenopteron*. If we then re-examine the arrangement of the nerve foramina in the light of the fact that three cranial nerves may be present, one of them, the ramus ophthalmicus V, being known to be closely associated with the ramus ophthalmicus VII in living forms, the apparent differences between the different forms are readily explained.

In *Ectosteorhachis* (Fig. 1 A, D) the ramus ophthalmicus VII runs in a separate canal through the substance of the dorsal wall of the nasal capsule, giving off branches to the lateral line organs. Presumably the ramus ophthalmicus V passes with it, because the remaining nerve foramina seem most readily to be reconstructed as forming part of the profundus V distribution. Apparently, in *Ectosteorhachis* the ramus ophthalmicus V may pass through the orbit along with the ramus profundus V, for within the postnasal wall there is a small previously unexplained interconnection between the canals for the ramus profundus V and ramus ophthalmicus VII (see Thomson, 1964*b*: 322, fig. 3) through which presumably the ramus ophthalmicus V transferred to run alongside the latter.

In *Porolepis* and *Glyptolepis* (Fig. 1 C, F), where, according to Jarvik (1962), there is also a separate canal within the dorsal wall of the capsule presumably for the ramus ophthalmicus VII, there are also a series of foramina in the dorsal wall that follow in a curved line exactly the distribution that one would expect for a series of branches passing towards the lateral line organs. Conceivably many of these foramina carried the ramus ophthalmicus V, which entered the nasal capsule through a separate foramen in the postnasal wall. The ramus profundus V entered the nasal capsule through a separate larger foramen in the postnasal wall and was distributed into the remaining nerve foramina which are arranged in the medial and anterior walls of the capsule very much as in *Ectosteorhachis*. In view of Jarvik's (1966: 53) statement that the ethmoid region of his original material of *Porolepis* was imperfectly ossified, we should not place great emphasis upon the fact that the foramen for the ramus profundus V in the postnasal wall is extremely large. In *Osteolepis* (Fig. 1 B, E) apparently an arrangement very similar to that of *Porolepis* was found.

A blood supply of both arteries and veins also entered and left the nasal capsule, and at least some of the foramina (although it is hard to be sure which) must have contained these vessels.

It must not be overlooked that the above is only a reconstruction. However, it seems most probable that the arrangement of the nerves and vessels in the snout of Rhipidistia was based upon an essentially single ground-plan, with only minor individual modifications, principally according to whether the separate nerves may have passed together through the snout in close association one with another, or separately.

The endocranial cavity in Rhipidistia. A few remarks must be made on the subject of the shape of the endocranial cavity in Rhipidistia, specifically in the ethmoid region. One of the most convincing arguments in Jarvik's original thesis (1942) favoring the diphyletic origin of the Rhipidistia was his description in *Porolepis* of a large anterior "pars ethmoidalis cranialis" of the endocranial cavity (Fig. 1 C), situated between the olfactory tracts and apparently having contained the cerebral hemispheres of a brain arranged almost exactly like that of a urodele. Such a structure was not developed in *Eusthenopteron*. Subsequently, Kulczycki (1960) described material of *Porolepis* in which such a structure was also lacking. Although Kulczycki's results have not universally been given the attention that they deserve, it is important to note that Jarvik (1965: 112; 1966: 53, footnote) has now confirmed that the pars ethmoidalis cranialis as described originally in *Porolepis* is an artifact. In fact, the only structure in "porolepiforms" that even remotely resembles such a structure is a slight swelling of the front portion of the endocranial cavity in *Glyptolepis*. In *Porolepis* itself, there is no evidence that this slight swelling was present, and Kulczycki (1960, pl. 1, fig. A) figures a specimen in which this is clearly absent.

The anterior palatal recesses. In the front portion of the palate of Rhipidistia there is a pair of recesses of varying development in different genera, sometimes shallow, sometimes deep, that I have termed the *anterior palatal recesses*. According to Jarvik, these features are developed very differently in "porolepiforms" and "osteolepiforms" and thus offer evidence in support of the diphyletic theory. The anterior palatal recesses have previously been described in different terms. In "porolepiforms", according to Jarvik, the "cava internasalia" form a pair of deep cavities in the neural endocranium and are separated by an endocranial internasal ridge onto the posterior portion of which, as seen in Fig. 2 A, the tip of the parasphenoid extends. According to Jarvik (1966: 56), each anterior palatal recess in "osteolepiforms" comprises a "fossa apicalis" which is "an unpaired cavity formed mainly by dermal bones", an "anterior palatal fenestra" which is an "opening in the exoskeleton in the roof of the fossa apicalis" and a prenasal pit which is a "paired shallow depression in the lateral part of the fossa apicalis".

In order to examine this problem, we may note first of all the anatomy of this region, and then later consider the function of these recesses.

In my opinion, there is essentially no difference between "porolepiforms" and "osteolepiforms" in the development of the anterior palatal recesses. As shown in Fig. 2 A, D, in *Porolepis* the anterior palatal recesses are enclosed by the premaxillae and vomers of the dermal bones series (together with a small portion of the parasphenoid). In terms of the endocranium they are a pair of depressions in the endocranium and are separated by an endocranial ridge. In

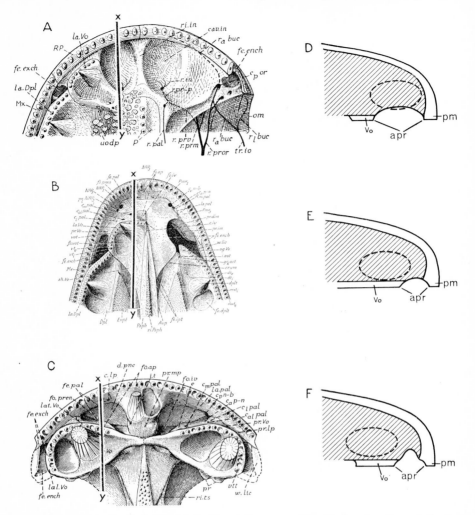

Fig. 2. The snout in palatal view (*A, B, C*) and diagrammatic vertical section through the line X–Y (*D, E, F*) in *Porolepis* (*A, D*), *Eusthenopteron* (*B, E*) and *Megalichthys* (*C, F*); modified from Jarvik (1942; 1966) and original.

Relevant structures: *apr*, anterior palatal recess; *cav.int*, cavum internasale; *fe.pal*, palatal fenestra; *pm*, premaxilla; *vo*, vomer.

Eusthenopteron (Fig. 2 B, E) the anterior palatal recesses, which are also enclosed by the premaxillae and vomers, are a pair of depressions in the front of the endocranium separated by a gentle elevation or ridge of the endocranium. In certain "osteolepiforms" such as the osteolepoid *Megalichthys* (Fig. 2 C, F) the anterior palatal recesses are slightly modified. In this case the recesses are enclosed by the premaxillae and vomers in the normal way, are roofed at least posteriorly by the endocranium, and are separated by a median process that has both an endocranial and a premaxillary portion. Without yet considering the possible use to which these recesses were put, we can, I believe, recognize the close resemblance in structure in all these Rhipidistia. The only major points of difference between these three conditions seem to lie in the relative position of the nasal organs, which are slightly more posteriorly located in "porolepiforms" and in the secondary development in certain "osteolepiforms" (see below for further discussion) of a dermal bone contribution to the ridge separating the recesses on either side. Furthermore, if we are to use separate terminologies for the exoskeletal and endoskeletal portions of the recesses in one group, we must use the same terminology for the other. However, since the exoskeletal and endoskeletal portions of the recesses can scarcely be considered separately one from another, the use of a common term, anterior palatal recess, is, in my opinion, preferable.

As we are dealing only with *recesses* in the palate, it is necessary to examine what might have fitted into these recesses before further consideration of their homology. According to Jarvik (1942), the recesses contained intermaxillary glands arranged differently in the two "lines" of Rhipidistia, each arrangement corresponding with a particular situation seen in one or other lines of modern Amphibia. However, other workers (Holmgren & Stensiö, 1936; Romer, 1937) have noted that the presence of such recesses in Rhipidistia and Stegocephalia is correlated with the development of the mandibular dentition, and that these recesses served for the reception of dentary teeth. In 1962 I was able to demonstrate this relationship in *Megalichthys*, and in the same year Jarvik demonstrated that this was also the case in *Glyptolepis* (a "porolepiform"). It is now known, due to Jarvik's researches, that the mandibles of holoptychoids bear a pair of parasymphysial tooth whorls, the tusks of which fit into the anterior palatal recesses—the arrangement of tooth whorls and recesses corresponding exactly (see Jarvik, 1962, fig. 2). Similarly, it was suggested (Romer, 1937; Thomson, 1962; 1964*a*) that the development of the anterior palatal recesses in "osteolepiforms" reflects exactly the nature of *their* mandibular dentition. This has also been demonstrated in *Megalichthys* and *Ectosteorhachis* where the anterior palatal recesses are distinctly separated one from another and are relatively deep to receive the prominent dentary tusks. The problem remains whether the shallow, only partially separated recesses in such forms as *Eusthenopteron foordi* and other Devonian genera also served for the reception of teeth. We must assume that, if teeth were received in the shallow anterior palatal fenestrae of these Devonian forms, these teeth would not be as prominently developed as in, for example, *Megalichthys*. Apart from Jarvik's description of the lower jaw in *Eusthenopteron foordi* (1944: 26), in which dentary tusks are lacking (see below), we have very little information concerning the structure of the inner surface of the lower jaw in Devonian Rhipidistia. The inner surface of the lower jaw of *Gyroptychius* sp. from the Middle Old Red

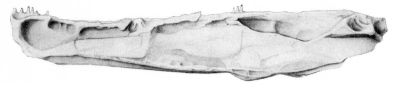

Fig. 3. *Gyroptychius* sp., Middle Old Red Sandstone, Tynet Burn, Scotland. Inner view of lower jaw; Specimen No. P 87 (DMSW-coll.), Museum of Zoology, Cambridge University; × 9/4.

Sandstone of Scotland (Fig. 3) shows clearly that the anterior portion of the dentary tooth row does not contain a single prominent tusk, but rather is slightly raised into a curved arc of small teeth that would fit exactly into the shallow anterior palatal recess of a form such as *Eusthenopteron foordi* or the specimen of *Gyroptychius milleri* figured by Jarvik (1942, fig. 62 B). According to Jarvik (1966: 57), the anterior palatal recesses in *Eusthenopteron* are always shallow and the dentary bone bears no tusk. This certainly applies to *Eusthenopteron foordi*. However, in *Eustheonpteron saevesoederberghi*, (Fig. 4) the anterior palatal recesses are separated from each other by a curious combination of *vomer* and endocranium (not by a posterior extension of the premaxillae, which seems so far to be an advanced osteolepid characteristic not seen in rhizodontids). Furthermore, the tip of each mandible evidently bore an enlarged dentary tusk that fitted into the anterior palatal recess.

As there is evidently a close correlation between the nature of the mandibular dentition and the development of the anterior palatal recesses, and as this correlated relationship may be developed differently in two species of the same genus, it may be concluded that the nature of the dentition is the main factor in determining the shape of the palatal recesses. Thus whether or not Jarvik is correct in assuming that there were glandular structures in the anterior palatal recesses (and considerable evidence can be assembled in support of such a view) we must assume that the glandular structures themselves cannot be instrumental in directing the primary arrangement of the recesses.

As the structure of the anterior palatal recesses is essentially identical in all known Rhipidistia and their arrangement is related to the pattern of the dentition, we must conclude that this system cannot be used as evidence for a division of Rhipidistia into two separate lineages, and does not suggest direct and separate relationships to the characteristic arrangements of the intermaxillary glands of modern Amphibia.

We may make one further observation concerning the arrangement of the anterior palatal recesses in Rhipidistia. There seems to be a connection between the nature of the anterior palatal recesses and the shape and position of the vomers. As previously indicated (Thomson, 1964*b*: 350) the vomers in holoptychoids are small and lateral in position, and do not meet in the midline. This

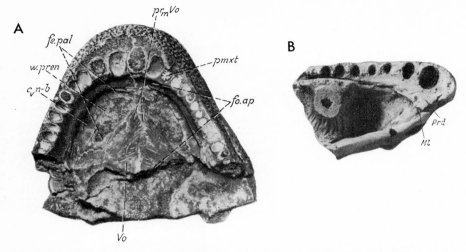

Fig. 4. Palatal view of snout (*A*) and inner view of anterior tip of mandible (*B*) of *Eusthenopteron saevesoederberghi*; from Jarvik (1942) and Vorobyeva (1962).
Relevant structures: *fe.pal*, palatal fenestra; *fo.ap*, fossa apicalis; *Vo*, vomer.

seems to be connected with the presence of the anterior palatal fenestrae between them. When we reach the osteolepoid grade, the vomers have enlarged and reach towards the midline, but it is not until the rhizodontoid grade that the vomers have an enlarged median contact and send posterior processes backwards alongside the tip of the tooth-bearing lamina of the parasphenoid. Thus, in osteolepoids the anterior palatal fenestrae, separated by an endocranial ridge, may become further separated by a posterior premaxillary process when the dentary tusk is evolved. In the rhizodontoid line, where the vomers are more strongly developed in the midline, the anterior palatal fenestrae are separated in some forms by an endocranial ridge (eg. *Eusthenopteron foordi*) which may become further reinforced by an anterior process from the vomers (*Eusthenopteron saevesoederberghi*) when the dentary tusk appears. It must be stressed that in no osteolepoid has a *posterior* vomerine process comparable to that which seems to be typical of rhizodontoids been described—the structure illustrated by Jarvik (1966, pls. 2, 3) being a broken portion of the ventral endocranium (see Thomson, 1967: 664, 673). Osteolepoids and rhizodontoids seem to have developed quite separately in this respect, even though a dentary tusk appears in both lines.

The nature and origin of the amphibian paired limbs

The last subject that I wish to treat is the oft-discussed origin of the tetrapod limb, particularly the recent discussion of this subject by Jarvik (1964; 1965) in terms of the diphyletic theory of amphibian ancestry.

Holmgren (1933, and following papers) conducted extensive researches on the structure of the tetrapod paired limbs in relation to phylogeny. He concluded that the urodele limb is significantly different from that of other tetrapods and sought to relate the limb structure of urodeles with that of dipnoans. Holmgren's description of differences between the limbs of urodeles and other tetrapods have formed the basis of later studies of the origin of paired limbs, notwithstanding the basic weakness of the plan in which urodeles and dipnoans were incorporated. What Holmgren was attempting to show, and what others have sought to show in other ways, is that there are differences between urodeles and other tetrapods that are the result of separate evolution from different ancestral stocks—a basic diphyly.

As is the case in other studies of diphyly, we must be extremely careful to separate the two discrete aspects of the study: (a) are the limbs of urodeles basically different from those of other tetrapods and (b) can we find fossil fishes that can only be considered as ancestors of one line or the other but not of both?

The urodele limb compared with that of other tetrapods. First of all, it must be noticed that in both adult structure and development, the paired limbs of the Anura, like so many aspects of their structure, are so heavily modified that, although they fall within the general plan of tetrapod structure, it is not possible to compare them meaningfully with those of other tetrapods. This being the case, we may only compare the urodelan limb with that of stegocephalian amphibians (a group that have received scant attention in most other considerations of supposed diphyly) and early reptiles.

Jarvik, in a concise discussion of the question has summarized the basic differences that, in his opinion, characterize the urodelan limb as being entirely different from that of other tetrapods. According to his account one of the main differences between urodeles and other tetrapods is that "the intermedium ray... is 'branched' and carries two digits (I and II) in the urodeles, whereas it... is unbranched and carries one digit (I) in other tetrapods (and in the osteolepiformes...). In accordance with this there is always a so-called basale commune (carpale 1+2, tarsale 1+2) in the urodelan limb. This element... certainly has a double origin... and... it can be assumed that, in its proximal parts too, the intermedium ray is of double formation..." (Jarvik, 1964: 69). In order to discuss this hypothesis, we must examine both the embryological development and the adult structure of the "intermedium ray" and the carpale (tarsale) commune for signs of a primary relationship to both the first and second digits.

If we examine the works of Sewertzoff (1908) and Schmalhausen (1910), which form the basis of Holmgren's original study, we discover that the situation does not seem in fact to offer strong support for the diphyletic view.

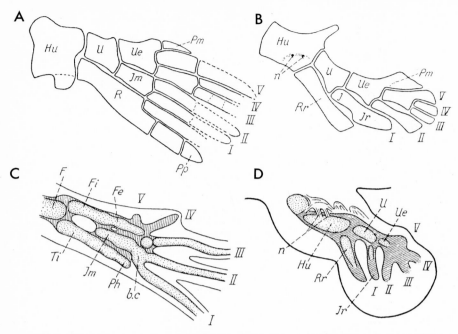

Fig. 5. Structure of the extremity of the limb in *A, Sauripteris, B, Eusthenopteron, C, Triturus* and *D, Lacerta*; from Jarvik (1964) q.v.for abbreviations.

According to Holmgren (1933: 215), who has summarized the evidence, "the carpale (and tarsale) commune" [i.e. of urodeles] "is peculiar, as two fingers (or toes) articulate with it. Its embryological development gives *no clear information*" [here italicized] "about its character of a double carpal (or tarsal). Sewertzoff, however, indicates that this may be the case, and I, myself, am inclined to accept the suggestion of Sewertzoff, (1908)". Jarvik (1964, fig. 23 B, D) produces a figure from Sewertzoff (1908, fig. 17c) of a developing limb of *Triturus* compared with that of *Lacerta* showing that, in the former, digits I and II were connected with the tarsale commune (Fig. 5 C). However, if we examine the earlier stages in the development of *Triturus* from the same work by Sewertzoff we may see that the first digit seems to arise from a separate rudiment which has, at this stage, no connection with that of digit II (Fig. 6 A). In the later stage (Fig. 6 B), however, such a connection is developed, but obviously this is secondary. In this there seems little basic difference from the situation that is typical of non-urodelan tetrapods. Schmalhausen (1910) studied the development of "*Salamandrella*" (=*Hynobius*). Again, two critical stages in the developmental series may be examined. A stage may be seen (Fig. 7 A) in which the rudiment of the first digit has no connection with that of the second digit, and apparently is a direct continuation of the blastema of the tibiale and tibia. At a slightly later stage (Fig. 7 B) the rudiment of the first

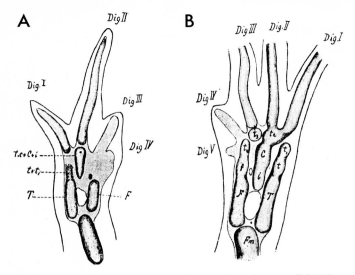

Fig. 6. Development of the foot in *Triturus*; from Sewertzoff (1908).

digit develops, at its base, a lateral connection to that of the second digit, exactly as in *Triturus*. To quote from Holmgren again (1933: 212): "The first toe arises as an independent blastema. The second toe, which at its base is prolonged by a swelling representing the rudiment of the tarsale commune, is surely not quite so independent, but its connection with the intermedio-central column is so very slight as to be hardly worth mentioning". In both cases, the carpale (or tarsale) commune develops from the blastema that gives the rudiment of the second digit and its connection with the rudiment of the first digit is, as we have seen, secondary. Thus, whether the carpale (or tarsale) commune is double or single seems to offer little evidence concerning the primary relationship of the digits. Similarly the intermedium belongs solely to the series of blastema associated with the second digit, not with that producing the first digit.

The fact that the carpale (tarsale) commune may develop embryologically from two centers of ossification (see Francis, 1934) may be interpreted, in the light of the above considerations, simply as a secondary phenomenon associated with the fact that in the adult form it comes to be an important element supporting two digits. In general there is apparently considerable variability in the number and arrangement of the carpals and tarsals in Urodela and occasionally two separate carpals (tarsals) may appear in this position (see Holmgren, 1933).

When we come to examine the adult structures of the urodelan, stegocephalian and other tetrapod limbs, we find a most remarkable situation.

Szarski (1962: 230, 231) compared the limbs of urodeles and stegocephalians

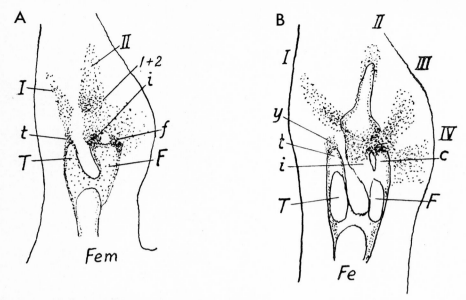

Fig. 7. The developing foot of *Hynobius*; slightly modified from Holmgren (1933) in accordance with Schmalhausen (1910).

while noting that in Urodela "the real arrangement of the skeletal elements... is only remotely similar to the schematic drawings of Holmgren". He concluded that there was no difference between urodelan and stegocephalian limbs. Jarvik (1964: 69), nothing the above remark of Szarski, used as a basic non-urodelan tetrapod limb the diagrammatic limb given by Steiner (1922). In Figs. 8 and 9, I present comparisons of the limbs of urodelan and stegocephalian amphibians, together with Steiner's reconstruction and an early reptile. As may be seen, this comparison (like those that have been made many times before) shows quite clearly the fundamental similarity in arrangement of the elements of the limbs of all these forms. This obviously argues in favor of an extremely close relationship between urodeles, stegocephalians and other tetrapods with respect

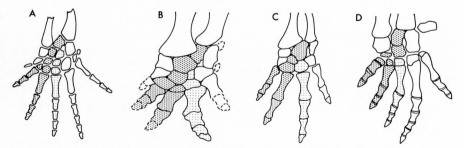

Fig. 8. Diagrammatic representation of the manus in *A* "primitive tetrapod" according to Steiner (1922), *B*, *Eryops*, *C*, *Amblystoma* and *D*, *Ophiacodon*; from Jarvik (1964) and Romer (1962).

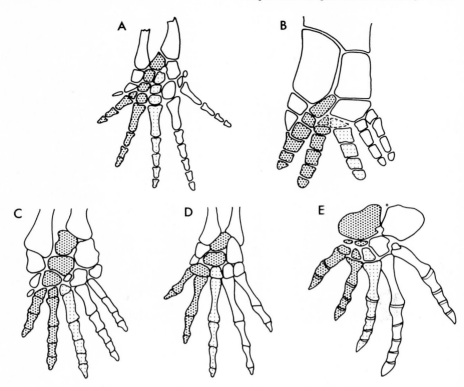

Fig. 9. The pes in *B, Ichthyostega, C, Trematops, D, Salamandra* and *E, Ophiacodon*, compared with *A*, Steiner's reconstruction of the primitive tetrapod hand; from Jarvik (1964) and Romer (1962).

to both the manus and the pes. However, it is also of great interest to note that this comparison of *adult* anatomy shows that in all of these forms, the first and second digits (together with the third) are apparently associated mechanically with the row of elements based proximally in the intermedium.

According to previous investigations (see Jarvik, 1964), the first and second digits in non-urodelan tetrapods must be considered as arising separately. The apparent mechanical association of these two digits with the intermedium ray in the adult of the forms just considered (Figs. 8, 9), must, therefore, be a secondary phenomenon. We have already seen that just this situation is shown in the embryology of the limbs in urodeles, and we may strongly suspect that the early manifestation of this phenomenon in urodeles is related to the very early development and swimming function of the paired fins.

To sum up: The adult condition of the limbs in urodeles and stegocephalians is essentially identical and accords closely with the topographic arrangement of the elements in the scheme of the basic tetrapod limb given by Steiner (1922). The connection between the first digit and the row of elements—intermedium, carpale centrale and second digit, can be continued as being embryonically

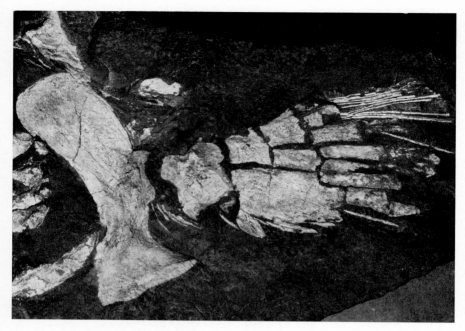

Fig. 10. *Sauripteris taylori* Hall, pectoral fin; Specimen No. 3341 belonging to the American Museum of Natural History, New York. Photograph reproduced by courtesy of that museum.

secondary in urodeles and the situation in these forms is thus not *basically* different from that of non-urodelan tetrapods.

Comparison of tetrapod and rhipidistian limbs. Jarvik, in the paper referred to previously (1964), has most ably summarized the evidence for the interpretation of the tetrapod extremity as a seven-rayed structure, and from his researches and those of Mahala Andrews (unpublished) we are able to see that basically the same arrangement is to be found in the pectoral appendage of *Eusthenopteron foordi* (although this does not apply to the pelvic appendage). It must be noted, however, that in *Eusthenopteron* (Fig. 5 B) we do not see the arrangement of the first two digits "stemming" from the intermedium as in adult stegocephalians and urodeles, but rather the arrangement seen in all *embryo* tetrapods where the first digit has a separate origin.

Jarvik (1964, fig. 23) has sought to find in the fin structure of the Upper Devonian rhipidistian fish *Sauripteris* (the familial assignation of which is completely uncertain) an antecedent stage in the development of the supposedly separate urodelan limb (Fig. 5 A). Although we have seen that the urodelan limb is probably not, in fact, distinct in structure, it is worthwhile to examine briefly the nature of the fin in *Sauripteris*, which has previously been studied by many, including Gregory & Raven (1941) and Holmgren (1933, etc.). Fig. 10

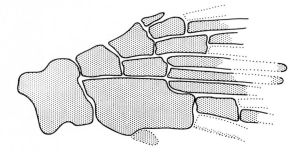

Fig. 11. Reconstruction of the pectoral fin in *Sauripteris* showing the maximum number of distal "rays" of the internal skeleton.

shows a photograph of the only known specimen of the appendage and this may be compared with the drawing given by Jarvik (1964, fig. 23; reproduced here as Fig. 5 A). As may be seen, the fin does not show a seven-rayed tetrapod structure and may, in fact, be constructed as having had as many as twelve distal "rays" (Fig. 11). Thus, far from helping clarify the general picture of the evolution of the tetrapod limb, the fin of *Sauripteris*, in the light of our knowledge of *Eusthenopteron*, raises many new problems and serves principally to illustrate how little we know about the radiation of the Rhipidistia. It certainly offers no evidence of a relationship to the urodelan tetrapods.

Conclusions

The evidence presented above seems to indicate that, with respect to the topics covered, there is no safe evidence for the view that the Rhipidistia are distinctly separated into two distant lineages or that any particular rhipidistian group is specially characterized by features indicating a unique relationship to any particular group of modern Amphibia. In previous papers I have presented evidence to show that the same conditions apply with respect to the evidence presented by other structural complexes such as the condition of the parietal organ, the palatal-ethmosphenoid articulation, the arrangement of the nares, etc.

Little has been said so far in these studies concerning the relationships of the Rhipidistia to the living Amphibia. Indeed, I have deliberately avoided reference to the latter, because, as was seen in the example of the nerve arrangement in the snout region, of the danger of qualifying the argument.

Jarvik has been able to assemble an impressive amount of information documenting the well-known fact that the Anura and Urodela are very different from each other. This body of information need not be questioned. What must be examined very carefully, however, is the question of the *overall phylogenetic significance* of the fact that these two groups of Amphibia differ from each other. A large number of scientists have studied the problem of amphibian relationships and of these many (see Thomson, 1964c; 1967) have recently concluded that the weight of the evidence supports the view that the modern groups of Amphibia have evolved from a single amphibian stock. But even all these studies, it seems to me, still cannot unequivocally resolve the problem of lissamphibian relationships—a problem that seems quite insoluble, *given present data alone*.

One important question that may, however, be included in our discussion of the diphyletic theory is: whether or not the apparent similarities seen between genera of Rhipidistia and the Anura or Urodela may be accounted for otherwise than by direct descent. Are the patterns of structure seen in modern Amphibia derivable from some common intermediate condition or conditions that could have arisen from any of the known Rhipidistia?

As an example of an alternative explanation to the diphyletic theory, we may consider the question of the branchial skeleton in Rhipidistia and Lissamphibia. Up to the present we have been unable to discuss this problem comparatively, since the branchial skeleton is only known in one holoptychoid and one rhizodontid rhipidistian.

Jarvik (1963), has shown that the branchial skeleton of *Eusthenopteron* is very similar in structure to that of non-urodelan tetrapods, while the urodeles differ from the *Eusthenopteron*-tetrapod condition in the absence of a sublingual rod and pronounced modification of the m. rectus cervicis. The branchial skeleton of *Glyptolepis* also apparently lacks a sublingual rod. Although Jarvik interprets this evidence as indicating a direct connection between *Glyptolepis* and the Urodela, separate from the remaining tetrapods, the following explanation seems to me to be equally, if not more, valid.

It may be concluded that *Eusthenopteron* shows a condition directly antecedent to that of the tetrapods in general. However, both the Anura and the Urodela show a tendency to reduction of the typical tetrapod arrangement of the branchial skeleton. In the Anura "the sublingual rod became much reduced and its anterior part supporting the m. genioglossus has always disappeared completely. The posterior part... which is associated with the m. hyoglossus, is sometimes retained and may be represented either by a small independent cartilage... or... an anterior median process... of the anterior basibranchial" (Jarvik, 1963: 65). In Urodela, reduction of the branchial skeleton has gone further and no vestige of the sublingual rod is present. This reduction in both of the principal lissamphibian groups presumably is directly connected with the other reductions in the cranial structure and with the modification of the overall proportions of the head, as compared with stegocephalian and reptilian tetrapods, particularly with foreshortening of the snout region. As for *Glyptolepis*, it is not yet known whether the structure of this form is typical of the "porolepiforms", but if it is, then the differences from the *Eusthenopteron* condition presumably indicate that these fishes were a side-branch, unconnected with the evolution of tetrapods, and that their structure is modified "convergently" with Urodela, again possibly in connection with a relative shortening of the snout region.

If any reminder were needed of the dangers of reconstruction of wide-ranging phylogenies from limited numbers of widely separated forms, one need look no further than the example of the convincing case that was formerly made for a direct relationship between the Lissamphibia and the Dipnoi. Here was an example of extremely detailed points of anatomical resemblance giving a logical though erroneous (so it seems) impression that was only altered by consideration of a wider sample of vertebrates—notably including the Rhipidistia. The point of these remarks is not to demean in any way the efforts of those who worked so well on the problem of amphibian-dipnoan relationships, but simply to explain my viewpoint that it seems, currently, impossible to make firm conclusions concerning the ancestry of the modern Amphibia when so little known about the relationships of either group to any of the known fossil amphibians which lived during the vast intervening span of time. Potentially there has been sufficient time for the type of structure seen in modern Amphibia to have developed from almost any of the known amphibian lineages or, in fact, from any of the majority of "osteolepiform" Rhipidistia, at least when only the specific structures considered in Jarvik's studies are examined.

An additional facet of the diphyletic theory is that the Urodela and Anura are assumed to have evolved directly from the Rhipidistia without any intervening stage with which known stegocephalian fossil amphibians could be associated (see Jarvik, 1960, fig. 29). This hypothesis can only be *disproved* by the discovery of an "intermediate" fossil form. *Proof* of this hypothesis also seems impossible given present data. Jarvik has proposed that a "conservatism" of structure results in the direct descent of rhipidistian characteristics into the Lissamphibia. However, it must be noted that such extreme structural "conservatism" is only known without doubt in those groups—Dipnoi or Coelacanthini—in which the rate of structural change is known to have been extremely slow between the Devonian and the present. Furthermore, the constancy of structure that is seen *within* Anura or Urodela also gives no evidence of Devonian-to-Recent conservatism, but reflects merely the constancy of the overall ground-plan of structure of each group of Lissamphibia. In view of the fact that overall evolution between Devonian and Recent in the Amphibia has been immense and that considerable doubt has to be thrown upon the notion that the structural characteristics of lissamphibian groups may be discovered exclusively in one or another rhipidistian group, we cannot invoke a concept of conservatism in this instance. In short, we come down to the circular statement that only those groups that show little structural divergence at the major group level (e.g. ordinal level) show extreme constancy of structure.

My conclusions are thus as follows:

(1) It is not possible to establish that the same structural characteristics which distinguish and characterize lissamphibian groups are developed exclusively within separate groups of Rhipidistia. Comparative study of a wide range of Rhipidistia reveals that distinctions which once held good between "porolepiforms" and "osteolepiforms" are broken down.

(2) While there can be no doubt whatsoever that the three groups of Lissamphibia—Anura, Urodela and Gymnophiona—are clearly separated from each other by a large range of structural characteristics, the fact that these structural characteristics cannot be found to occur exclusively in any one line of rhipidistian fishes or stegocephalian amphibians, leads to the following conclusions:

(a) The data are insufficient to establish any direct lineage linking rhipidistian fishes and Lisamphibia, with or without intermediate stegocephalian stages.

(b) The data are insufficient to decide with any certainty whether or not the three groups included within the Lissamphibia form a single natural assemblage.

(3) A final conclusion would be the obvious one, that what is needed most at the present time is detailed morphological information concerning as wide a range of rhipidistian fishes as possible, and the inclusion within the compass of this type of investigation of data concerning the stegocephalian Amphibia, particularly such forms as the Ichthyostegalia.

This work was supported by grant GB-4814 from the National Science Foundation.

References

Francis, E. T. (1934). *The anatomy of the salamander*. Oxford: Oxford University Press.

Gregory, W. K. & Raven, H. C. (1941). Studies on the origin and early evolution of paired fins and limbs. 1. Paired fins and girdles in ostracoderms, placoderms, and other primitive fishes. *Annls N. Y. Acad. Sci.*, **42**: 273–360.

Holmgren, N. (1933). On the origin of the tetrapod limb. *Acta zool. Stockh.*, **14**: 185–295.

Holmgren, N. & Stensiö, E. A. (1936). Kranium und Visceralskelett der Akranier, Cyclostomen und Fische. In *Handbuch der vergleichenden Anatomie der Wirbeltiere*, hrsg. Bolk, L., Göppert, E., Kallius, E. & Lubosch, W., **4**: 233–500. Berlin & Wien: Urban & Schwarzenberg.

Jarvik, E. (1942). On the structure of the snout of crossopterygians and lower gnathostomes in general. *Zool. Bidr. Uppsala*, **21**: 235–675.

Jarvik, E. (1944). On the exoskeletal shoulder-girdle of teleostomian fishes, with special reference to *Eusthenopteron foordi* Whiteaves. *K. svenska VetenskAkad. Handl.*, (3) **21**: 1–32.

Jarvik, E. (1960). *Théories de l'évolution des vertébrés, réconsidérées a la lumière des récentes découvertes sur les vertébrés inférieurs*. Paris: Masson.

Jarvik, E. (1962). Les porolepiformes et l'origine des urodeles. *Colloques int. Cent. natn. Rech. Scient.*, **104**: 87–101.
Jarvik, E. (1963). The composition of the intermandibular division of the head in fish and tetrapods and the diphyletic origin of the tetrapod tongue. *K. svenska VetenskAkad. Handl.*, (4) **9**: 1–74.
Jarvik, E. (1964). Specializations in early vertebrates. *Annls Soc. R. zool. Belg.*, **94**: 11–95.
Jarvik, E. (1965). Die Raspelzunge der Cyclostomen und die pentadactyle Extremität der Tetrapoden als Beweise für monophyletische Herkunft. *Zool Anz.*, **175**: 101–143.
Jarvik, E. (1966). Remarks on the structure of the snout in *Megalichthys* and certain other rhipidistid crossopterygians. *Ark. Zool.*, (2) **19**: 41–98.
Kulczycki, J. (1960). *Porolepis* (Crossopterygii) from the Lower Devonian of the Holy Cross Mountains. *Acta palaeont. polon.*, **5**: 65–106.
Pinkus, F. (1895). Die Hirnnerven des *Protopterus annectens*. *Morph. Arb.*, **4**: 275–346.
Romer, A. S. (1937). The braincase of the Carboniferous crossopterygian *Megalichthys nitidus*. *Bull. Mus. comp. Zool. Harv.*, **82**: 1–73.
Romer, A. S. (1962). *The vertebrate body*. 3rd Ed. Philadelphia & London: Saunders.
Schmalhausen I. I. (1910). Die Entwicklung des Extremitätenskelettes von *Salamandrella Keyserlingii*. *Anat. Anz.*, **37**: 431–466.
Sewertzoff, A. N. (1908). Studien über die Entwicklung der Muskeln, Nerven und des Skeletts der Extremitäten der niederen Tetrapoda. Beiträge zu einer Theorie des pentadactylen Extremität der Wirbeltiere. *Bull. Soc. Imp. Nat. Moscou*, (N.S.) **21**: 1–432.
Steiner, H. (1922). Die ontogenetische und phylogenetische Entwicklung des Vogelflügelskelettes. *Acta zool. Stockh.*, **3**: 307–360.
Stensiö, E. A. (1963). The brain and cranial nerves in fossil, lower craniate vertebrates. *Skr. norske VidenskAkad. Oslo. Mat-naturv. Kl.*, **1963**: 1–120.
Szarski, H. (1962). The origin of the Amphibia. *Q. Rev. Biol.*, **37**: 189–241.
Thomson, K. S. (1962). Rhipidistian classification in relation to the origin of the tetrapods. *Breviora*, **117**: 1–12.
Thomson, K. S. (1964a). Revised generic diagnoses of the fossil fishes *Megalichthys* and *Ectosteorhachis* (family Osteolepidae). *Bull. Mus. comp. Zool. Harv.*, **131**: 283–311.
Thomson, K. S. (1964b). The comparative anatomy of the snout in rhipidistian fishes. *Bull. Mus. comp. Zool. Harv.*, **131**: 313–357.
Thomson, K. S. (1964c). The ancestry of the tetrapods. *Sci. Progr.*, **52**: 451–459.
Thomson, K. S. (1965). The endocranium and associated structures in the Middle Devonian rhipidistian fish *Osteolepis*. *Proc. Linn. Soc. Lond.*, **176**: 181–195.
Thomson, K. S. (1967). Notes on the relationship of the rhipidistian fishes and the ancestry of the tetrapods. *J. Paleont.*, **41**: 660–674.

Discussion

H. Szarski

It has been demonstrated during this Symposium that there is an impressive list of dissimilarities between the two rhipidistian groups, parallelled by a corresponding list of differences between Urodela and Anura. It has been shown also that in many details Anura are similar to Osteolepiformes and Urodela to Porolepiformes. In these circumstances it may appear astonishing that there are still people who are not convinced of amphibian polyphyletism. Without discussing this extensive question in any detail, I would like to bring to mind three major motives which speak for monophyletism of all Tetrapoda.

The first is based on the general observation that every adaptive radiation took place after the acquirement of a major improvement. It is assumed that the improvement of the tetrapod ancestors was the ability to walk, respire, and gather food outside water, and that the adaptive radiation took place only after all this had been achieved.

The second refers to the impressive list of similarities connecting all Tetrapoda, and in particular Lissamphibia. Many of these similarities seem to be good homologies (cf. review in Szarski, 1962; for further facts see Parsons & Williams, 1963).

The third concerns the significance of dissimilarity. Not every dissimilarity speaks against a phylogenetic affinity. A mouse and a whale are very dissimilar; they are nevertheless both mammals. There are, however, dissimilarities which are valid arguments against affinity. The tail-fin of a whale is in a way similar to that of a fish. However, the fish fin works sidewards, whereas the whale fin moves up and down. This difference is important as it shows that the ancestors of whales must have been mammals, which held the body high above the ground and in running made galloping movements. Had these ancestors moved by side-to-side undulations, as the majority of fishes and present day reptiles do, the tail fin of whales would have developed as a vertical, not horizontal skin fold. Another example: in Rodentia the mandible is broader that the maxilla, in Lagomorpha it is narrower; the teeth are accordingly worn in a different manner, and it is improbable that the one arrangement was changed into the other. This is one of the arguments which induce us to believe that the common ancestor of Rodentia and Lagomorpha was little specialized in the tooth arrangement. Other examples could easily be found among the patterns of enamel folds in various herbivorous mammals.

A dissimilarity is a strong argument against affinity *only* if it can be demonstrated that the intermediate condition is incompatible with function or with the known mode of development, or if some other reason speaks against the possibility of a shift in structure. I do not feel that among the differences between Urodela and Anura on the one hand, and between Porolepiformes and Osteolepiformes on the other, there are such which fulfill this requirement. On the other hand, these differences exist and ought to be explained in some way. Unfortunately I am not able to formulate now a hypothesis worthy of a detailed argument. Possible causes could perhaps lie in the mode of food capture, or in the mechanisms of ontogenetic development? Further research will, I hope, suggest some explanation.

Szarski, H. (1962). The origin of the Amphibia. *Q. Rev. Biol.*, **37**: 189–241.
Parsons, T. S. & Williams, E. E. (1963). The relationships of the modern Amphibia. *Q. Rev. Biol.*, **38**: 26–53.

Remarques concernant la phylogénie des Amphibiens

Par Jean-Pierre Lehman

Institut de Paléontologie, Muséum National d'Histoire Naturelle, Paris, France

Dans divers articles récents, Williams (1959), Parsons & Williams (1963) arrivent à des conclusions phylogéniques très particulières concernant l'origine des Amphibiens actuels en se fondant sur l'anatomie des vertèbres des Tétrapodes et des Poissons. Selon Williams (1959) dont les conclusions ont été reprises par Parsons & Williams (1963), les vertèbres des Amniotes et des Amphibiens actuels sont comparables parce qu'elles sont chez ces deux groupes d'origine intersegmentaire, étant formées chacune essentiellement par une partie antérieure représentant l'arrière d'un sclérotome et une partie postérieure représentant l'avant du sclérotome suivant. D'autre part, les centra chez les Amniotes comme chez les Amphibiens actuels ont pour origine un tissu périchordal et se développent indépendamment des arcs. Ce tissu périchordal dans la région intermédiaire entre les deux vertèbres consécutives forme le disque intervertébral embryonnaire qui aurait même position que l'intercentre des Reptiles et des Mammifères quand cet élément existe. Le corps vertébral serait donc exclusivement pleurocentrique. Nous examinerons d'abord ces conceptions avant de passer à la discussion des conséquences phylogéniques que Parsons & Williams pensent pouvoir en déduire.

La première différence entre vertèbres des Tétrapodes actuels et vertèbres des Poissons selon Williams, c'est-à-dire la *resegmentation*, n'est pas une nouveauté chez les Tétrapodes; il est vrai que, chez les Poissons, la « resegmentation n'a pas ce caractère général reconnu pour les Amniotes et que souvent la vertèbre peut être intramétamérique » (Devillers); il n'en reste pas moins que la vertèbre est, semble-t-il, au moins en général, intermétamérique chez les Sélaciens sauf dans la queue (cas de diplospondylie) (Goodrich, 1930: 20, fig. 20); par contre chez *Amia*, la vertèbre est intrasegmentaire mais, chez de nombreux Poissons fossiles, elle est intersegmentaire comme le montre la présence de l'orifice de l'artère intersegmentaire (ex. *Boreosomus*: Lehman, 1952: 90, fig. 60; *Pteronisculus*: Nielsen, 1942: 215, fig. 49; *Australosomus*: Nielsen, 1949: 137, fig. 43; *Eusthenopteron*: Jarvik, 1955: 62, fig. 6). La resegmentation ne représente donc pas un caractère apparaissant chez les Tétrapodes et l'importance de ce critère ne doit pas être surestimée.

Quant à l'existence du *tissu périchordal*, elle n'est pas non plus exclusivement caractéristique des Tétrapodes; chez les Téléostéens, le corps vertébral paraît surtout être périchordal, autrement dit être un autocentre dans la terminologie de Remane; mais de plus, chez de nombreux Téléostéens, s'observe au cours du développement une ossification vertébrale chordacentrique. Même si le « chordacentre provenant de la calcification localisée de la gaine chordale » semble exister dans certains cas au cours de l'évolution des Actinoptérygiens (François, 1967), il n'en reste pas moins que l'ossification des centres vertébraux des Téléostéens et du Polyptère est essentiellement autocentrique. Le rôle du tissu périchordal semble donc tout à fait comparable chez les Téléostéens et chez les Tétrapodes.

Le corps vertébral peut-il enfin chez les Amphibiens actuels être considéré comme l'homologue d'un *pleurocentre*? Les données de l'embryologie, d'ailleurs bien exposées par Williams, nous semblent susceptibles d'une autre interprétation que celle proposée par ce savant; en effet selon Williams, chez les Anoures, de la matière du disque intervertébral embryonnaire est graduellement assimilée aux centra, ce qui revient à dire que le disque intervertébral est un intercentre — comme l'admet Williams — mais que du matériel de cet intercentre s'incorpore à chaque corps vertébral qui est donc non pas pleurocentrique mais intercentro-pleurocentrique. Cette interprétation est confirmée par la paléontologie, car il existe des cas dans lesquels la fusion d'un intercentre et d'un pleurocentre semble bien la seule explication plausible : ainsi chez *Kotlassia*, Seymouriamorphe du Permien supérieur russe, dans les vertèbres dorsales, les vertèbres 9 et 10 ont des intercentres indépendants, mais les vertèbres suivantes ont un corps vertébral unique qui pourrait bien être un intercentropleurocentre (Buistrov, 1944); de même dans une vertèbre du Trias de Madagascar j'ai cru devoir admettre la fusion intercentre-pleurocentre (parastéréospondylie, Lehman, 1961) admise d'ailleurs antérieurement par Buistrov & Efremov (1940) chez *Dvinosaurus*. C'est un point de vue comparable que défend Panchen (1967) dans un article nouveau. Panchen souligne que les homologies proposées par Williams (pleurocentre des Labyrinthodontes correspondant au centrum des Amphibiens actuels et des Amniotes; intercentre des Labyrinthodontes correspondant au disque intervertébral des autres vertébrés) sont incompatibles avec la position des myocommes chez les Stégocéphaliens.

Y a-t-il d'autre part *homologie* des éléments vertébraux appelés intercentre et pleurocentre chez tous les Vertébrés? C'est peu probable. Si l'on renonce, comme la plupart des auteurs, à la théorie de Gadow selon laquelle chez tous les Vertébrés, sauf les Sélaciens, la vertèbre résulterait de la combinaison de pièces arcuales fondamentales constantes (basidorsal, basiventral, interdorsal, interventral) et serait un arcocentre, il semble que l'on doive aussi renoncer à trouver des pleurocentres et des intercentres homologues chez tous les Vertébrés tétrapodes. En effet, ces éléments peuvent être d'origine arcuale ou périchor-

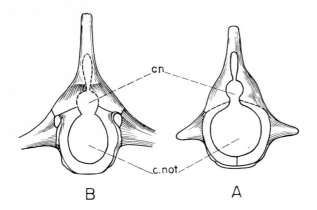

Fig. 1. Vertèbres d'*Eusthenopteron* (*A*) et d'*Ichthyostega* (*B*); d'après Jarvik. *cn*, canal neural; *c.not*, canal notochordal.

dale. Ils sont d'origine arcuale (au moins essentiellement) chez *Eusthenopteron* et de même chez *Ichthyostega*. Dans ces deux cas (Fig. 1), le canal notochordal est volumineux et il n'y a pas d'anneau osseux complet autour de la notochorde puisque le canal notochordal communique vers le haut avec le canal neural. Naturellement, puisque l'embryologie de ces formes fossiles n'est pas connue, il est impossible d'affirmer que l'ossification périchordale n'ait joué aucun rôle dans la formation de la pièce considérée par Jarvik comme un basiventro-interventral fusionné, mais le fait qu'un basidorsal et un interdorsal distincts existent montre nettement que ces vertèbres sont, au moins en ce qui concerne ces éléments, d'origine arcuale. Chez les Rachitomes, les pièces vertébrales centrales sont plus ou moins épaisses et ont même parfois l'aspect d'arceaux lamelleux minces (*Trimerorachis*, *Lydekerrina*, pleurocentres d'*Archegosaurus*); mais le plus souvent ces pièces sont épaissies (*Eryops*, intercentres d'*Archegosaurus*). Chez les Embolomères, pleurocentres et intercentres sont représentés en général par des disques massifs à part une perforation notochordale au milieu du disque, mais, chez *Pholidogaster*, Romer (1964) a montré que l'intercentre était un anneau creux ouvert dorsalement et que les pleurocentres étaient deux demi anneaux symétriques; cette disposition peut être qualifiée en français de schizomérique (d'après le terme « Schizomeri » proposé par Romer), tandis que celle de *Diplovertebron*, comparable mais à pleurocentre en anneau complet, peut être désignée comme diplomérique (Fig. 2). Il est peu probable que le mode de formation embryologique des intercentres et pleurocentres lamelleux soit le même que celui des intercentres et pleurocentres massifs; les premiers peuvent être d'origine essentiellement arcuale ou arcualo-périchordale, tandis que les seconds sont vraisemblablement d'origine exclusivement périchordale (ou autocentrique) ou même d'origine périchordalo-chordale. Si intercentre et pleurocentre n'ont pas dans ces deux cas même nature du point de vue embryologique, ces formations ne seraient pas dans ces deux cas homologues mais analogues. Naturellement ces considérations sur la nature embryologique des

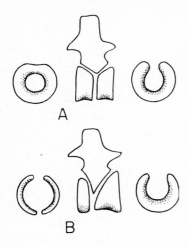

Fig. 2. Schéma structural des vertèbres diplomériques (*A*) et schizomériques (*B*); d'après Romer. À gauche pleurocentre vu de l'avant; à droite, intercentre vu de l'arrière.

vertèbres des Amphibiens fossiles sont hypothétiques; elles montrent cependant qu'il existe un problème et que la recherche d'intercentres et de pleurocentres homologues des Poissons aux Tétrapodes puis aux Amniotes peut très bien n'avoir pas grand sens. Quand Williams admet que le corps vertébral des Anoures est un pleurocentre c'est par comparaison avec les Amniotes; mais ce pleurocentre des Amniotes est-il bien homologue à celui d'*Ichthyostega* défini comme correspondant à l'interdorsal des Poissons, cela est loin d'être évident.

Dans ces conditions il semble bien téméraire de fonder un *tableau phylétique* des premiers Tétrapodes sur des considérations portant sur les vertèbres (et à plus forte raison sur ces seules considérations comme dans le travail de Williams, 1959). Dans ce tableau (Fig. 3), les Anoures, sans que leur origine soit précisée, sont rapprochés des Embolomères en raison de l'interprétation de leur centre vertébral considéré comme un pleurocentre. Or, par ailleurs, rien dans l'anatomie des Anoures ne rappelle les Embolomères; c'est d'ailleurs ce que soulignent ultérieurement Parsons & Williams (1963: 33) : « If the modern amphibians retain only pleurocentra, they might reasonably be thought related to these forms » [les Anthracosaures]. « Unfortunately this is in conflict with the evidence from the skull, since the labyrinthodonts which most resemble modern forms in palatal features are the temnospondyls ». Rappelons à ce propos que les Anthracosaures ont un palais fermé et de très grands ptérygoïdes et que notamment Säve-Söderbergh (1936: 610) et Piveteau (1937; 1955) ont mis en évidence de nombreux caractères communs aux Anoures et aux Temnospondyles; de même Jarvik (1967) souligne les différences fondamentales de disposition anatomique de l'encéphale chez les Reptiliomorphes et les Batrachomorphes.

Dans un autre travail Parsons & Williams (1962) admettent que les dents

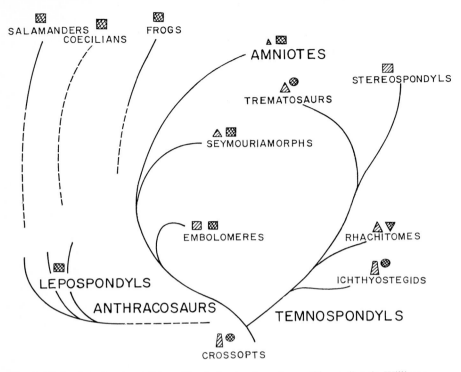

Fig. 3. Phylogénie des Amphibiens d'après la structure des vertèbres; d'après Williams.

des Amphibiens actuels sont morphologiquement assez homogènes et que les Amphibiens par suite ne peuvent être considérés comme diphylétiques; selon ces auteurs en effet, les dents des Amphibiens actuels seraient caractérisées par une couronne présentant un apex articulé sur un pédicelle. Il résulterait de cette constatation que les Amphibiens actuels seraient bien une unité phylétique puisque Parsons & Williams pensent qu'un tel caractère n'a pu apparaître par évolution parallèle. Cette théorie soulève diverses objections :

(1) D'après Parsons & Williams (1962) eux-mêmes, la présence de cette articulation entre apex de la couronne et pédicelle n'est pas générale.

(2) On sait qu'il existe chez les Amphibiens deux types de dents, des monocuspides et des bicuspides. Les dents monocuspides des Anoures, chez *Ceratophrys* tout au moins, ne semblent pas posséder cette articulation (observation personnelle). Les monocuspides apparaissent chez *Ceratophrys* après les bicuspides, sauf chez *C. dorsata* chez laquelle on n'observe que des monocuspides même chez le jeune (Noble, 1931). Le fait que le caractère invoqué n'existe pas dans les monocuspides en réduit la portée.

(3) Morphologiquement les dents des Amphibiens peuvent être assez différentes et il ne semble pas que la structure dentaire représente un critère bien

Fig. 4. *Rana esculenta*, dent; d'après Meyer. L'os est hachuré.

important chez les Amphibiens actuels; rappelons que, selon le sexe, les dents peuvent parfois être mono- ou bicuspides (Noble, 1931).

(4) Les bicuspides n'ont pas même aspect chez les Anoures et les Urodèles, semble-t-il, d'après les travaux publiés. Il est regrettable que l'on ne possède qu'une littérature scientifique insuffisante à ce sujet; mais, chez les Anoures l'apex de la couronne présente des cuspides en lèvres superposées et recourbées en crochet (*Rana esculenta* : Fig. 4; voir aussi Meyer, 1944; *Racophorus leucomystax, Alytes obstetricans* (à 3 cuspides), *Hylodes binotatus, Borlorocoetes taeniatus, Discoglossus pictus* : Oltmanns, 1952, figs. 2, 16, 20, 23), tandis que chez les Urodèles les cuspides correspondent plutôt à des pointes ou à des angles saillants des bords de la couronne (*Salamandra maculata* : Hertwig, 1874, pl. 3, fig. 1; Kerr, 1960; *Salamandra maculata, Amphiuma means, Desmognathus fuscus, Salamandrina perpicillata, Proteus anguineus* : Oltmanns, 1951, figs. 4, 15, 17, 19, 22).

(5) Si l'articulation de la couronne paraît bien exister toutefois chez les bicuspides dans la majorité des cas, cela ne veut pas dire que toutes les dents présentant cette disposition soient pour autant comparables; car l'histologie est différente au moins semble-t-il dans les cas trop peu nombreux où elle a été récemment étudiée : Oltmanns (1952) distingue trois types de structure histologique dans les dents d'Amphibiens actuels : (a) type I, la couronne est formée d'un capuchon de dentine surmonté d' « émail » entourant une cavité pulpaire; (b) type II (*Proteus, Discoglossus*), la structure est la même mais le vide de la cavité pulpaire est en partie rempli par des trabécules de tissu osseux; (c) type III, il n'y a plus de cavité pulpaire celle-ci est entièrement remplie par des trabécules osseux et la dentine serait de la plicidentine (*Ceratophrys*). Bien que je ne pense pas d'après mes observations que l'on puisse parler dans ce cas de plicidentine, il n'en demeure pas moins qu'Oltmanns a démontré que les dents des Amphibiens actuels présentaient des structures histologiques diverses.

Si l'on compare d'ailleurs les figures de Gilette (1955) concernant *Rana*

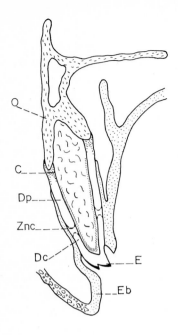

Fig. 5. *Rana pipiens*, coupe labiolinguale d'une dent; d'après Gilette.
C, cément; *Dc*, dentine de la couronne; *Dp*, dentine pédestale; *E*, émail; *Eb*, épithelium buccal; *O*, os; *Znc*, zone non calcifiée.

pipiens (Fig. 5) et de Kerr (1960) représentant une coupe de dent de *Salamandra* (Fig. 6), on remarque que chez l'Anoure en question sous la dentine pédestale existe du cément, tandis que chez l'Urodèle la dentine pédestale n'est présente que du côté lingual, l'apex de la couronne étant soutenu directement par de l'os du côté labial. Les deux structures sont donc différentes.

(6) Les Anoures se distinguent des Urodèles par l'absence de dents calcifiées avant la métamorphose (les Têtards n'ont que des dents cornées), tandis que les Urodèles ne possèdent pas de dents cornées juvéniles.

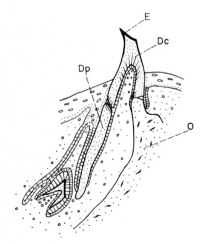

Fig. 6. Coupe d'une dent de Salamandre; d'après Kerr.
Dc, dentine de la couronne; *Dp*, dentine pédestale; *E*, émail; *O*, os.

(7) Dans l'ensemble, les Anoures ont des dents palatales beaucoup moins développées que les Urodèles et certains Anoures (Discoglossidae, Pelobatidae, Hylinae, Cystignatinae, Discophynae, Raninae) ne possèdent pas de dents mandibulaires et même sont parfois dépourvus aussi de dents sur les mâchoires supérieures (Bufonidae, Dendrophryniscinae, Engystomatinae, Dendrobatinae). L'absence de dents mandibulaires est probablement corrélative du mode d'insertion de la langue sur l'angle antérieur des mandibules chez les Anoures; autrement dit, cette absence correspond à la différence fondamentale de nature de la langue des Anoures et de celle des Urodèles (voir Jarvik, 1963).

(8) Il est peu probable que la denture représente un caractère phylétique important sauf à partir du niveau de l'ordre; il s'agit en effet d'un caractère directement adaptatif lié au régime alimentaire.

Il ne nous apparaît donc pas, dans ces conditions, que l'existence de dents à charnières chez les Amphibiens actuels puisse être interprétée comme démontrant l'unité phylétique de ces animaux. De toute façon d'ailleurs, comme les Ostéolépiformes et les Porolépiformes présentent des dispositions dentaires assez voisines (sauf dans la région symphysaire), cet argument, à supposer que contrairement à notre discussion il soit valable, ne saurait être décisif.

Bibliographie

Buistrov, A. P. (1944). *Kotlassia prima* Amalitsky. *Bull. geol. Soc. Amer.*, **55**: 379–416.
Buistrov, A. P. & Efremov, J. A. (1940). *Benthosuchus sushkini* Efr., a labyrinthodont from the Eotriassic of Sharzhenga River. *Trud. Inst. palaeont. Acad. Sci. U.S.S.R.*, **10**: 1–152.
François, Y. (1967). Structure vertébrale des Actinoptérygiens. *Colloques int. Cent. natn. Rech. Scient.*, **163**: 155–172.
Gilette, R. (1955). The dynamics of continuous succession of teeth in the frog (*Rana pipiens*). *Am. J. Anat.*, **96**: 1–36.
Goodrich, E. S. (1930). *Studies on the structure and development of vertebrates.* London: Macmillan.
Hertwig, O. (1874). Ueber das Zahnsystem der Amphibien, und seine Bedeutung für die Genese des Skelets der Mundhöhle. Eine vergleichend anatomische, entwicklungsgeschichtliche Untersuchung. *Arch. mikr. Anat.*, **11** (*Supplh.*): 1–208.
Jarvik, E. (1952). On the fish-like tail in the ichthyostegid stegocephalians, with descriptions of a new stegocephalian and a new crossopterygian from the Upper Devonian of East Greenland. *Medd. Grønland*, **114**: 1–90.
Jarvik, E. (1955). Ichthyostegalia. Dans *Traité de Paléontologie*, éd. Piveteau, J. **5**: 53–66. Paris: Masson.
Jarvik, E. (1963). The composition of the intermandibular division of the head in fish and tetrapods and the diphyletic origin of the tetrapod tongue. *K. svenska VetenskAkad. Handl.*, (4) **9**: 1–74.
Jarvik, E. (1967). The homologies of frontal and parietal bones in fishes and tetrapods. *Colloques int. Cent. natn. Rech. Scient.*, **163**: 181–213.
Kerr, T. (1960). Development and structure of some actinopterygian and urodele teeth. *Proc. zool. Soc. Lond.*, **133**: 401–422.

Lehman, J.-P. (1952). Étude complémentaire des poissons de l'Eotrias de Madagascar. *K. svenska VetenskAkad. Handl.*, (4) **2**: 1–201.
Lehman, J.-P. (1961). Les Stégocéphales du Trias de Madagascar. *Annls Paléont.*, **47**: 1–46.
Meyer, P. (1944). Beiträge zur Kenntniss des Gebisses von *Rana esculenta* L. mit besonderer Berücksichtigung des Zahnwechsels. *Vjschr. naturf. Ges. Zürich*, **89**: 1–37.
Nielsen, E. (1942). Studies on Triassic fishes from East Greenland. 1. *Glaucolepis* and *Boreosomus*. *Palaeozool. Groenland.*, **1**: 1–403.
Nielsen, E. (1949). Studies on Triassic fishes from East Greenland 2. *Australosomus* and *Birgeria*. *Palaeozool. Groenland.*, **3**: 1–309.
Noble, G. K. (1931). *The biology of Amphibia.* New York: MacGraw-Hill.
Oltmanns, E. (1952). Zur Morphologie der Zähne rezenter Amphibien. *Anat. Anz.*, **98**: 369–389.
Panchen, A. L. (1967). The homologies of the labyrinthodont centrum. *Evolution*, **21**: 24–33.
Parsons, T. S. & Williams, E. E. (1962). The teeth of Amphibia and their relation to amphibian phylogeny. *J. Morph.*, **110**: 375–383.
Parsons, T. S. & Williams, E. E. (1963). The relationships of modern Amphibia: a re-examination. *Q. Rev. Biol.*, **38**: 26–53.
Piveteau, J. (1937). Un amphibien du Trias inférieur. Essai sur l'origine et l'évolution des amphibiens anoures. *Annls Paléont.*, **26**: 135–177.
Piveteau, J. (1955). Stereospondyli. Dans *Traité de Paléontologie*, éd. Piveteau, J., **5**: 136–172. Paris: Masson.
Romer, A. S. (1964). The skeleton of the Lower Carboniferous labyrinthodont *Pholidogaster pisciformis*. *Bull. Mus. comp. Zool. Harv.*, **131**: 129–159.
Säve-Söderbergh, G. (1936). On the morphology of Triassic stegocephalians from Spitzbergen and the interpretation of the endocranium in Labyrinthodontia. *K. svenska VetenskAkad. Handl.*, (3) **16**: 1–181.
Williams, E. E. (1959). Gadow's arcualia and the development of tetrapod vertebrae. *Q. Rev. Biol.*, **34**: 1–32.

The development of bones in the skull roof of Amphibia

By Natalie S. Lebedkina

Embryological Laboratory, Zoological Institute of the Academy of Science of the USSR, Moscow, USSR

One of the basic changes in the skull during the transition from crossopterygians to tetrapods was the reduction and stabilization in number of dermal bones in the anterior part of the cranial roof. In this area in crossopterygians there is considerable variation in the number of dermal bones, as shown by their fusion into the fronto-ethmoidal shield of some forms, and by their disintegration into series of postrostrals and nasals between the premaxillaries and frontals of others. When many bones are present, variability is so great that different bone numbers are observed not only among species of the same genus (Vorobyeva, 1962), but also among specimens of the same species (Jarvik, 1948).

Beginning with ichthyostegids among tetrapods, the number of bones in this area of the skull became reduced and stabilized. The only bones preserved are the premaxillary, nasal and frontal. The single postrostral is infrequently encountered in stegocephalians (Säve-Söderbergh, 1932; 1935; Huene, 1956) and among the recent amphibians in *Siren* (Holmgren, 1949), *Xenopus* and *Calyptocephalus* (Paterson, 1939; Reinbach, 1939; Jarvik, 1948). There are indications suggesting fusion between the postrostral and nasal of stegocephalians (Säve-Söderbergh, 1932) and primitive urodeles (Lebedkina, 1964b), and that the premaxillary in tetrapods, like that in rhipidistians, is a complex bone (naso-rostro-premaxillary, Jarvik, 1942).

What is the significance of this transformation observed in the skull roof bones of tetrapods? I have attempted to answer this question by tracing the development of skull roof bones of amphibians. The investigation was based on the study of microscopic serial sections with the use of graphic and waxplate reconstructions.

The shape of bone anlage was examined with the aid of a vital stain (alizarin red "S") and the dynamics of bone growth in urodeles by repeated injections of alizarin "S" into the body of growing animals (Lebedkina, 1964b). The structure of bone and cartilage was investigated by the "Spaltlinien" method (Benninghoff, 1925). For this purpose were used four urodele species (from hatching to the end of metamorphosis): *Ranodon sibiricus*, 89 larvae; *Ambystoma tigrinum*, 102 larvae; *Pleurodeles waltlii*, 112 larvae; and *Triton cristatus karelini*, 70 larvae; and four anuran species (from fully formed larvae to the end of metamorphosis): *Rana ridibunda*, 68 larvae; *R. temporaria*, 37 larvae; *Hyla japonica*, 16 larvae; and *Bombina orientalis*, 13 larvae. Larval stages were determined for *R. sibiricus* according to Lebedkina (1964a), for *A. tigrinum* according to Harrison (from Vorontsova, Liozner *et al.*, 1952), for *P. waltlii* according to Gallien & Durocher (1957), and for *T. cristatus karelini* according to Glücksohn (from Vorontsova, Liozner *et al.*, 1952). In addition material illustrating skull development of some fishes was used for comparative purposes.

Before reviewing skull development in urodeles, it would be appropriate to trace the development of those fish skull bones pertinent to the problem at hand. Skull development in larvae of holosteans and brachiopterygians shows that growth of the premaxillary is related to the gradual development of its support by the chondrocranium. The mode of support varies in connection with the structural diversity of the ethmoidal region in the skull of these fishes. In holosteans, with their reduced nasal cavity, the premaxillary forms an outgrowth on the dorsal side of the ethmoidal plate. It is well developed in 11 mm *Amia calva* larvae in which teeth begin to attach to the premaxillary. This outgrowth is still greater in 12 mm larvae which already are actively feeding, and continues to increase in subsequent stages (13, 16 and 22 mm), during which more teeth are attached to the bone. Thus, for this bone the area of support by cartilage increases progressively. A similar condition was observed in *Lepisosteus* larvae of 14, 20 and 24 mm.

A comparative study of 24 mm *Polypterus senegalus* larvae (Pehrson, 1947, fig. 28) and of 45 mm. *Polypterus* sp. larvae showed that for the premaxillary the area of support by the chondrocranium increases as the bone grows, spreading along the ventral side of the ethmoidal plate. Moreover, the canal bones (rostral and tectal, Jarvik, 1942; rostral and antorbital, Pehrson, 1947) fuse with the premaxillary. These bones, in connection with the complete development of the nasal cavity, become supported by the cartilage of the anterolateral wall, and consequently allow for an increase in the total supporting area of the premaxillary.

In fishes, the direction of initial growth of the frontal is determined by the course of the supraorbital sensory canal. Soon after, the frontal anlage becomes supported by the supraorbital cartilage. For example, in 12 mm *Amia* larvae the frontal is connected to the cartilage by a plexus of collagen fibres.

In fishes, one observes a definite sequence in the development of dermal bones and of those parts of the chondrocranium supporting them. For example, in *Amia* the premaxillary teeth appear after chondification of the anterior trabecles (9 mm larvae). The premaxillary develops only after the trabecles, having been connected by a cross-piece, are united into a monolithic

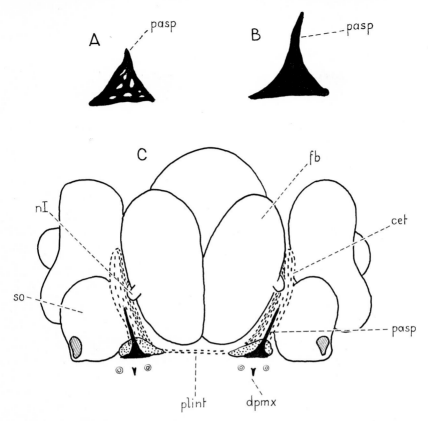

Fig. 1. *Ranodon sibiricus*, development of premaxillary in IIIrd stage larvae. *A*, 19 mm; *B*, 21 mm; *C*, anterior view of the head in 17.5 mm larva. Cartilage indicated by dots, mesenchyme by interrupted lines, tooth rudiments by double circles.
cet, columna ethmoidalis; *dpmx*, tooth of premaxillary; *fb*, forebrain; *nI*, n.olfactorius; *pasp*, proc.ascendens of premaxillary; *plint*, planum internasale; *so*, olfactory sac.

entity, the ethmoidal plate. In this case the premaxillary is connected to the ethmoidal plate also by means of collagen fibres (10 mm larvae). By the time the frontal develops the supraorbital cartilage already is formed, but the frontal attaches to it only after the cartilage has formed an arch fixed anteriorly in the ethmoidal region and posteriorly in the auditory capsule.

Thus, in fishes the development of the cartilages precedes the development of the dermal bones which attach to them. This refers not only to the differentiation of cartilaginous tissue, but also to the development of the cartilaginous structures forming parts of the skull.

During the study of the development of the skull in urodeles, no separate ossification centres were detected in the premaxillary, which grows very rapidly. For example, among 22 larvae of *R. sibiricus* (IIIrd stage) some had for the premaxillary an early anlage consisting of a plexus of several bone trabeculae,

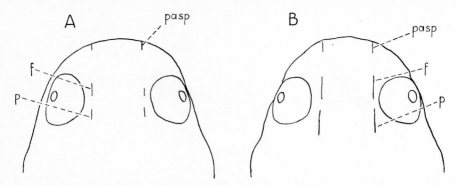

Fig. 2. *Ranodon sibiricus*, development of bones of the skull roof in IIIrd stage larvae, *A*, 20.5 mm; *B*, 19 mm.

f, frontal; *p*, parietal; *pasp*, proc. ascendens of premaxillary.

and exhibiting a primordium of the proc. ascendens (*pasp*, Fig. 1 A); others had an anlage whose process attained a level amounting to one-half of the height of the nasal sac (*pasp, so*, Fig. 1 C). A similar rapid growth of the proc. ascendens was traced in 19 larvae of *A. tigrinum* (38th stage), 21 larvae of *P. waltlii* (36th stage) and 7 larvae of *T. cristatus karelini* (46th stage).

The anlage of the frontal appears at the same stage as that of the premaxillary, or somewhat later. In all the examined species the shape of the frontal anlage and the direction of its growth are identical: the bone primordium has the shape of a thin rod (*f*, Fig. 2 A) and grows forward towards the end of the proc. ascendens of the premaxillary. In a short while both bones grow towards each other and their ends become connected by collagen fibers of the bone suture (in *R. sibiricus* at the VIth stage, in *A. tigrinum* at the 39th stage, in *P. waltlii* at the 41st stage). The posterior end of the frontal grows backwards in the direction of the parietal anlage (*p*, Fig. 2 A), grows up to this bone, and connects with it by a suture, as observed in *R. sibiricus* (VIth stage), *A. tigrinum* (43rd stage), *P. waltlii* (42nd stage) and *T. cristatus karelini* (50th stage).

What are the relations between the developing dermal bones and the chondrocranium of urodeles? It has been noticed that only *R. sibiricus* in the development of the pars dentalis of the premaxillary has a sequence of processes of bone and cartilage development like that observed in fishes. But *R. sibiricus* exhibits a certain shift towards accelerated development of bone in relation to cartilage: the fully developed teeth of the premaxillary appear when the anterior end of the trabecles are still formed by prochondral tissue, and the rudiment of the premaxillary develops before chondrification of the transverse bridge fixing the anterior ends of the trabecles (*plint*, Fig. 1 C). In more highly organized urodeles the development of the pars dentalis of the premaxillary takes place still earlier relative to the development of the chondrocranium: in

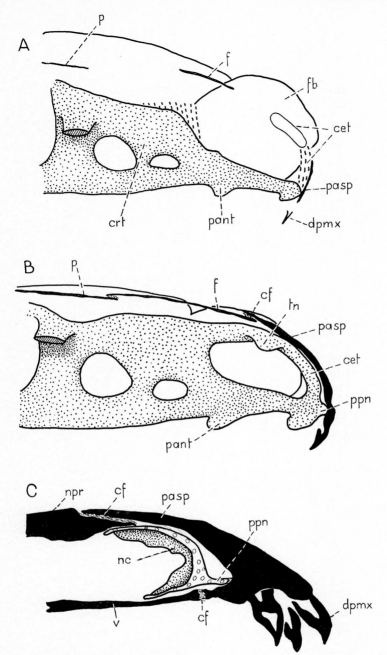

Fig. 3. *Ranodon sibiricus*. *A, B*, lateral view of the anterior part of the skull (palatoquadrate removed) in *A*, IIIrd stage larva, 19.8 mm, and *B*, VIth stage larva, 23 mm. *C*, position in lateral view of dermal bones in relation to the ethmoidal region of the chondrocranium after metamorphosis (91 mm). Cartilage indicated by dots and mesenchyme by interrupted lines; prochondral tissue white.

cet, columna ethmoidalis; *cf*, collagen fibres of suture; *crt*, crista trabeculae; *dpmx*, tooth of premaxillary; *f*, frontal; *fb*, forebrain; *npr*, naso-postrostral; *p*, parietal; *pant*, proc. antorbitalis; *pasp*, proc. ascendens of premaxillary; *ppn*, proc. praenasalis inferior lateralis; *tn*, tectum nasi; *v*, vomer.

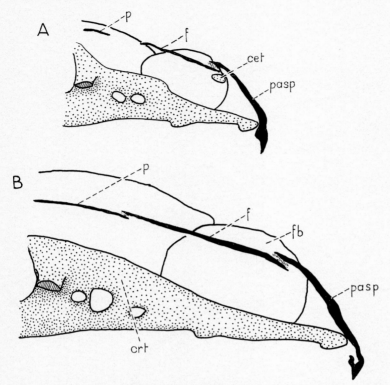

Fig. 4. Lateral view of the anterior part of the skull (palatoquadrate removed) in *A*, *Amblystoma tigrinum* 42th stage larva, 18 mm; and *B*, *Pleurodeles waltlii* 42th stage larva, 15 mm.
  cet, columna ethmoidalis; *crt*, crista trabeculae; *f*, frontal; *fb*, forebrain; *p*, parietal; *pasp*, proc.ascendens of premaxillary.

A. tigrinum and *P. waltlii*, not only does the bone appear but teeth begin to attach to it before the development of the planum internasale; in *T. cristatus karelini* the bone appears when the anterior end of the trabecles are still formed of prochondral tissue.

With respect to the proc. ascendens of the premaxillary, the homologue of the pars facialis of the "premaxillary" of crossopterygians (Jarvik, 1942), it may be noted that only in a primitive form such as *R. sibiricus* the process develops with support by the primordium of the prenasal wall, the columna ethmoidalis. But even in this case, when the proc. ascendens begins to grow upwards, the columna ethmoidalis still is formed of prochondral or even mesenchymal tissue (*pasp, cet*, Figs. 1 C, 3 A). However, by the beginning of active feeding (VIth stage) the columna ethmoidalis is fused with the crista trabeculae, and forms a support for the proc. ascendens (Fig. 3 B). In more highly organized urodeles the proc. ascendens grows without any support by this cartilage (Fig. 4 B), for the appearance of the columna ethmoidalis shifts to later developmental stages; in *A. tigrinum* it develops during the 41–42nd

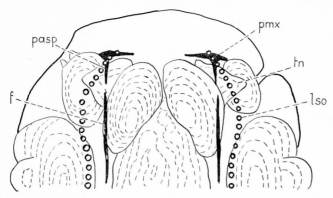

Fig. 5. *Ranodon sibiricus*, dorsal view of the anterior part of the head in Vth stage larva, 23 mm. Supraorbital sensory line indicated by circles.

f, frontal; *lso*, supraorbital sensory line; *pasp*, proc.ascendens of premaxillary, *pmx*; *tn*, tectum nasi.

stages, when the proc. ascendens has already reached a considerable development (*cet, pasp*, Fig. 4 A) and in *P. waltlii* and in *T. cristatus karelini* after about $1\frac{1}{2}$–$2\frac{1}{2}$ months, respectively, following the appearance of the premaxillary. In urodele larvae the frontal arises on the brain membrane in total isolation from the chondrocranium (*f*, Figs. 3 A, B, 4 A, B). This is true also for the parietal, but sometime later the parietal becomes supported by cartilage.

Earlier we stressed that the enlargement of the premaxillary in fish larvae is related to an increase in the area of the cartilage supporting it. In urodele larvae the growth of the proc. ascendens appears to be unexplainable from this point of view, for it starts its growth upwards without support by cartilage. The same condition is observed for the frontal.

In tetrapods the relation between bones and sensory lines is lost. Indeed, the direction of initial growth of the frontal and of the proc. ascendens of the

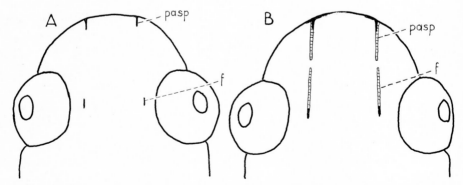

Fig. 6. *Pleurodeles waltlii*, development of bones of the skull roof. *A*, 38th stage larva, 13.5 mm; *B*, 41th stage larva, 14 mm. Bone stained by alizarin injection at 38th stage black; bone growth between 38th and 41 stages indicated by oblique hatching.

f, frontal; *pasp*, proc.ascendens of premaxillary.

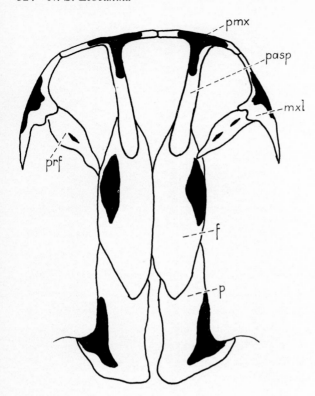

Fig. 7. Skull roof of axolotl killed 9 months after bones were stained with alizarin red "S". Stained parts of bones black; bone growth after injection white.

f, frontal; *mxl*, maxillary; *p*, parietal; *pasp*, proc.ascendens of premaxillary, *pmx*; *prf*, prefrontal.

premaxillary (with originates from canal bones) fails to correspond with the course of the supraorbital line of urodeles (*f, lso*, Fig. 5); these bones grow towards each other.

Since the earliest stages of their development, however, these bones continually "draw apart" owing to the growth of the chondrocranium. A connection between these bones is possible only under the condition of positive allometry of their growth in relation to the growth of the skull. As shown in Fig. 6 B the growth increments of the proc. ascendens and of the frontal of *P. waltlii* larvae are greater than the distance separating these bones at stage 38 (Fig. 6 A), but the bones have not yet united (in *P. waltlii* and especially in *T. cristatus karelini* the growth of the anterior region of the neurocranium is so rapid that the connection between premaxillary and frontal occurs at a relatively late stage as compared with the condition in the axolotl and in *R. sibiricus*). Throughout the subsequent period of growth of the cranium, these bones continue their growth in opposite directions. Such compensatory growth maintains contact between bones and also increases the suture area connecting overlapping bones. Fig. 7 shows in an axolotl how far the premaxillary and frontal, which were connected at the stage alizarin staining was performed, have drawn apart from each other. In hynobiids the growth of the proc. ascendens throughout meta-

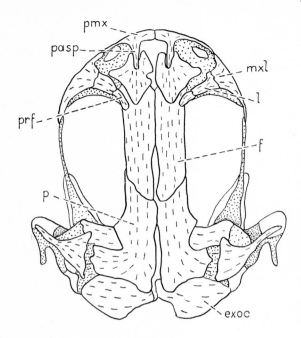

Fig. 8. Metamorphosed *Ranodon sibiricus* (92 mm), distribution of "Spaltlinien" in bones of the skull roof. *exoc*, exoccipital; *f*, frontal; *l*, lacrymal; *mxl*, maxillary; *p*, parietal; *pasp*, proc. ascendens of premaxillary, *pmx*; *prf*, prefrontal.

morphosis is sharply reduced, but between the proc. ascendens and the frontal is located the newly developed naso-postrostral (Lebedkina, 1964a).

Hence contact between the premaxillary and the bones of the skull roof is a result of opposite growth, started during the earliest stages of their development and continuing for the entire period of skull growth.

What is the significance of this trend of growth of the proc. ascendens of the premaxillary and the frontal of urodele larvae? The early growth of these bones results in a rapid establishment of connection between them, a connection formed during the stages when the larvae begin active feeding (in *R. sibiricus* at VIth stage, *A tigrinum* 39th stage, *P. waltlii* 38th stage, and *T. cristatus karelini* 48th stage). Thus a bone chain develops without any support by cartilage, and serves the purpose of supporting the proc. ascendens of the premaxillary.

Bone fibres are oriented in the direction of pressure imposed on the bone. The "Spaltlinien" method of Benninghoff (1925) reveals the orientation of fibres within the bone, and allows the bone to be regarded as a functional unit.

The application of the "Spaltlinien" method to the study of the urodele skull shows a structural relation between the premaxillary and the bones of the skull roof: the "Spaltlinien" of the premaxillary pass into those of the naso-postrostral, then into the frontal (in hynobiids, Fig. 8), or proceed from the premaxillary directly into the frontal (in the more highly organized urodeles). From the frontal the "Spaltlinien" pass into the parietal and spread into the replacement ossifications of the auditory region (and in larvae, into the cartilaginous auditory capsule).

Hence, urodele larvae develop an integrated system of bones, unified by the function of conducting pressure from the premaxillary onto the bones of the skull roof. The maxillary, appearing during metamorphosis of urodeles, is connected with the bones of the skull roof in a similar manner, that is through the prefrontal and lacrymal bones (*mxl*, *prf*, *l*, Fig. 8; see also Lebedkina, 1964*a*). A similar system of support of the upper jaw, for pressure conducted from the premaxillary and maxillary onto the dermal bones of the skull roof, is characteristic also of amniotes including mammals.

In urodeles, the formation of the support system for the premaxillary is due to a disruption in early larval stages in the relations between the bones and the cartilage of the ethmoidal region. This leads also in adult animals to reduced contact between the cartilage of the ethmoidal region and the premaxillary bone, such that forces of pressure pass from the premaxillary onto the bones of the skull roof, to a large extent above the chondrocranium (Fig. 3 C).

The study of skull development within urodeles, ranging from hynobiids to more highly organized forms, reveals a process of gradual reduction in the number of dermal bones in the anterior part of the skull roof. For the support system of the premaxillary the naso-postrostral was reduced, and for the support system of the maxillary the prefrontal was reduced, and at times also the lacrymal. The analysis of the significance of these reductions points to the disappearance of intermediate links in the lines of pressure transmission from the upper jaw onto the skull roof (Lebedkina, 1964*a*).

Could the tendency for a more simplified bone system of pressure conductivity, revealed in the course of evolution of the urodeles, be regarded as the endpoint of an old process which began in ancient tetrapods? In crossopterygians this part of the skull was covered by canal and "anamestic" bones. One could suppose that an incorporation of these elements into a bone system unified by functions of pressure conductivity might have led to a drastic reduction in number, but more importantly to a stabilization of their position and interrelations in ancient tetrapods.

The reduction of endochondral ossification in the olfactory capsules, preserved only in the ancient ichthyostegids (Jarvik, 1955) and possibly also in plesiopods (Eaton & Stewart, 1960; Tatarinov, 1964) may be regarded as good evidence in favour of the view that a similar system had already formed in the ancestors of recent amphibians. The formation of a dermal bone system, conducting pressure from the upper jaw onto the skull roof above the endocranium, permitted the reduction of these ossifications, which were strongly developed in rhipidistians.

The study of this trend in the development of the skull of urodeles, which

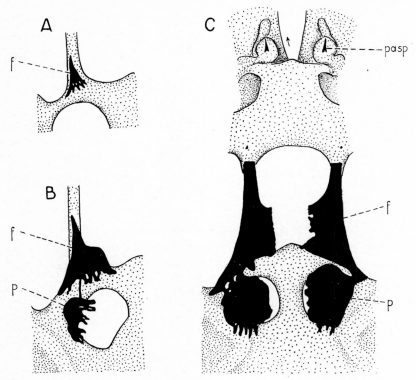

Fig. 9. *Rana ridibunda*, development of fronto-parietal. *A*, 56 mm larva; *B*, 57 mm larva; *C*, dorsal view of skull roof in 53 mm larva. Arrow indicates direction of growth of the premaxillary.

f, frontal; *p*, parietal; *pasp*, proc.ascendens of premaxillary.

are tetrapods having larvae with teeth on the upper jaw bones (similar in this respect to stegocephalian larvae) is important. It suggests that the development of a new type of support for the premaxillary was significant for the purpose of larval adaptation: the disruption of the relations between dermal bones and cartilage provided ample possibilities for the development of adhesion between the premaxillary and the dermal bones of the skull roof, hence for the change to active feeding by the larvae at a relatively early stage of chondrocranial development.

Referring back to skull development in anurans one may pay attention to the great differences compared with the condition observed in urodeles. In connection with the development of the rostral cartilages, the appearance of the anlage of the premaxillary is shifted to a later stage of development, toward the end of the premetamorphosis period. The direction of growth of this bone is directly opposite to that observed for urodeles: the bone does not grow posteriorly toward the skull roof, but rostrally (Fig. 9 C). The proc. ascendens is the first to appear; it grows dorsoventrally along the surface of the cartilago praenasalis superior and then it passes over onto the cartilago praenasalis inferior, having formed an extension of the pars dentalis, whereas teeth appear only somewhat later (if at all). The differentiation process is completed in the cartilago praenasalis by the time the premaxillary develops. The frontal bones (or their combined anlage with the parietals) appear before the premaxillaries, as has been shown for *Rana ridibunda*, *R. temporaria*, *Hyla japonica* and *Bombina orientalis*; an even greater disjunction in the development of these bones is in *Calyptocephalus* (Reinbach, 1939) and *Xenopus* (Sedra & Michael, 1957). The initial growth of these bones, not influenced as in urodeles by the premaxillary, proceeds in a different direction, for they show extensive growth along the cartilaginous surface providing support for them. The directions of initial growth vary: the frontal sometimes grows along the edge of the side wall of the neurocranium (*B. orientalis*, *H. japonica*) or in the medial direction, towards the tectum transversum (*R. ridibunda*, *R. temporaria*; f, Fig. 9 A).

Neither larval nor adult anurans show any contact between premaxillary and frontal, and consequently lack the means of transmission of pressure from the premaxillary onto the bones of the skull roof. This leads to an additional change related to the fact that the cartilage of the ethmoidal region acts as a support for the frontal and nasal bones. The cartilage is not so feeble as that of urodeles, and sometimes develops even replacement ossifications. These sometimes lie in the septum internasale, in the roof and in the floor of the nasal cavity (Parker, 1881; Ramaswami, 1935; Badenhorst, 1945; Baldauf, 1955).

A similar and profound disturbance in the development of the skull roof bones of anurans is caused by the development of the rostral apparatus. However, in the assumed ancestors (eoanurans) of the anurans the skull had structural features common to all stegocephalians: contact between premaxillary and bones of the skull roof, and no replacement ossifications in the nasal capsules. This allows the conclusion that the larvae of these forms had a toothed premaxillary.

It appears, therefore, that a larval stage with a rostral apparatus developed later. But only future finds of intermediate forms in the ancestry of the anurans will establish the evolutionary level at which arose a tadpole with such an apparatus.

References

Badenhorst, Cornelia E. (1945). Die Skedelmorfologie van die neotropiese anure *Atelopus moreirae* de Mirando-Ribeiro. *Ann. Univ. Stellenbosch*, 23A: 1–19.

Baldauf, R. J. (1955). Contributions to the cranial morphology of *Bufo w. woodhousei* Girard. *Texas J. Sci.*, 7: 275–311.

Benninghoff, A. (1925). Spaltlinien am Knochen, eine Methode zur Ermittelung der Architektur der platten Knochen. *Anat. Anz.* (*Ergänzh.*), **60**: 189–206.

Eaton, T. H. Jr. & Stewart, P. L. (1960). A new order of fishlike Amphibia from the Pennsylvanian of Kansas. *Univ. Kansas Publ. Mus. nat. Hist.*, **12**: 217–240.

Gallien, L. & Durocher, M. (1957). Table chronologique du développement chez *Pleurodeles waltlii* Michah. *Bull. biol. France Belg.*, **91**: 97–114.

Holmgren, N. (1949). Contribution to the question of the origin of tetrapods. *Acta zool. Stockh.*, **30**: 459–484.

Huene, F. v. (1965). *Paläontologie und Phylogenie der niederen Tetrapoden.* Jena: Fischer.

Jarvik, E. (1942). On the structure of the snout of crossopterygians and lower gnathostomes in general. *Zool. Bidr. Uppsala*, **21**: 235–675.

Jarvik, E. (1948). On the morphology and taxonomy of the Middle Devonian osteolepid fishes of Scotland. *K. svenska VetenskAkad. Handl.*, (3) **25**: 1–301.

Jarvik, E. (1955). Ichthyostegalia. In *Traité de Paléontologie*, éd. Piveteau, J., **5**: 53–66. Paris: Masson.

Lebedkina, Natalie S. (1964a). The development of the dermal bones of the basement of the skull in Urodela (Hynobiidae). *Trud. Inst. zool. Acad. Sci. U.S.S.R.*, **33**: 75–172. In Russian.

Lebedkina, Natalie S. (1964b). Razvitie nosovuikh kostei u khvostatuikh amfibii (The development of the nasal bones in Urodela). *Dokl. Akad. Sci. U.S.S.R.*, **159**: 219–222. In Russian.

Parker, W. K. (1881). On the structure and development of the skull in Batrachia. 3. *Phil. Trans. R. Soc.* **172**: 601–669.

Paterson, N. F. (1939). The head of *Xenopus laevis. Q. J. micr. Sci.*, **81**: 161–234.

Pehrson, T. (1947). Some new interpretations of the skull in *Polypterus. Acta zool. Stockh.*, **28**: 399–455.

Ramaswami, L. S. (1935). The cranial morphology of some examples of Pelobatidae (Anura). *Anat. Anz.*, **81**: 65–96.

Reinbach, W. (1939). Untersuchungen über die Entwicklung des Kopfskeletts von *Calyptocephalus gayi* (mit einem Anhang über das Os supratemporale der anuren Amphibien). *Jena. Z. Naturw.*, **72**: 211–362.

Säve-Söderbergh, G. (1932). Preliminary note on Devonian stegocephalians from East Greenland. *Medd. Grønland.*, **94**: 1–105.

Säve-Söderbergh, G. (1935). On the dermal bones of the head in labyrinthodont stegocephalians and primitive Reptilia, with special reference to Eotriassic stegocephalians from East Greenland. *Medd. Grønland.*, **98**: 1–211.

Sedra, S. N. & Michael, M. I. (1957). The development of the skull, visceral arches, larynx and visceral muscles of the South African clawed toad, *Xenopus laevis* (Daudin) during the process of metamorphosis (from stage 55 to stage 66). *Verh. Akad. Wet. Amst.*, (2) **51**: 1–80.

Tatarinov, L. P. (1964). Otriad Plesiopoda. In *Osnovui paleontologii* (*Fundamentals of Palaeontology*) ed. Orlov, I. A., **12**: 123–124. Moscow: Acad. Sci. U.S.S.R.

Vorobyeva, Emilia (1962). Rizodontnuie kisteperuie ruibui glavnoge devonskogo polia SSSR (Rhizodont crossopterygians from the Devonian Main Field of the U.S.S.R.). *Trud. Inst. palaeont. Acad. Sci. U.S.S.R.*, **104**: 1–108. In Russian.

Vorontsova, M. A., Liozner, L. D., et al. (1952). *Triton and axolotl.* Moscow: Acad. Sci. U.S.S.R. In Russian.

Die Homologie des Jacobsonschen Organs bei Anura und Urodela

Von Irene M. Medvedeva

Embryologisches Laboratorium, Zoologisches Institut der Akademie der Wissenschaften der U.d.S.S.R., Moskau, U.d.S.S.R.

Die Homologie des Gebildes, das nur bei Tetrapoda vorhanden ist und als Jacobsonsches Organ bezeichnet wird, bleibt nach wie vor nicht nur hinsichtlich der Amphibien selbst, sondern auch hinsichtlich der Amphibien einerseits und der höheren Wirbeltiere andererseits eine Streitfrage. Die Schwierigkeit der Feststellung der Homologie dieses Gebildes besteht darin, dass der Bau, die Lage, manchmal auch die Entwicklung des Jacobsonschen Organs bei verschiedenen Gruppen ausserordentlich mannigfaltig und die Funktion dieses Gebildes in mehreren Fällen unklar ist. Inzwischen ist die Lösung dieser Frage hinsichtlich der Amphibien für das Problem einer mono- oder polyphyletischen Herkunft der Landwirbeltiere von Bedeutung.

In diesem kurzen Bericht habe ich keine Möglichkeit auf die Geschichte der Frage über die Homologie des Jacobsonschen Organs bei Amphibien einzugehen. Es ist aber darauf hinzuweisen, dass dieses Gebilde bei Amphibien hauptsächlich nach seiner Lage zu unterscheiden ist: medial bei Anura, lateral bei Urodela und Apoda. Die Lösung dieser Frage ist also in den Besonderheiten der Entwicklung des Jacobsonschen Organs, besonders auf frühen Entwicklungsstadien, zu suchen.

Zu diesem Zweck ist die Untersuchung bei den Vertretern von zwei Urodela-Familien: Salamandridae (*Triton taeniatus, T. cristatus karelini*) und Hynobiidae (*Hynobius keyserlingii, Ranodon sibiricus*), sowie von zwei Anura-Familien: Ranidae (*Rana temporaria, R. esculenta*) und Discoglossidae (*Bombina bombina, B. orientalis*) durchgeführt. Dabei stellte sich heraus, dass wenn das Jacobsonsche Organ bei höheren Anura und Urodela eine völlig entgegengesetzte Lage einnimmt, so bei *Bombina*, und Urodela (Abb. 1, 3 B, 4 B, 5 B, 6 B), es nach lateral gerichtet ist, was mit den entsprechenden Angaben von Helling (1938) übereinstimmt. Wenn man in Betracht zieht, dass bei der erwachsenen *Pipa americana* das Jacobsonsche Organ nach Bancroft (1895) und Thrams (1936) auch lateral liegt, kann man behaupten, dass es wenigstens bei manchen niederen Anura im Vergleich mit Urodela keinen Unterschied in der Lage des Organs gibt.

Was der frühen Anlage des Jacobsonschen Organs anbetrifft, so werden die Angaben mehrerer Autoren (Burckhardt, 1891; Seydel, 1895; Hinsberg, 1901;

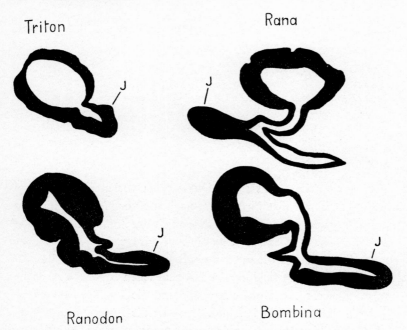

Abb. 1. Querschnitte durch das Geruchsorgan an der Ableitung des Jacobsonschen Organs bei älteren Larven.
J, das Jacobsonsche Organ.

u.a.) über ihre Verschiebung im Prozess der Entwicklung der Urodelen von der Medial- zur Lateralseite bestätigt (Abb. 4, 5). Das Gleiche geschieht auch bei *Bombina* (Abb. 3). Praktisch lässt der Vergleich der frühen Anlage des Jacobsonschen Organs (Abb. 2, 3 A, 5 A) mit seiner weiteren Entwicklung bei

Abb. 2. Querschnitte durch das Geruchsorgan an der Anlage des Jacobsonschen Organ.
J, das Jacobsonsche Organ.

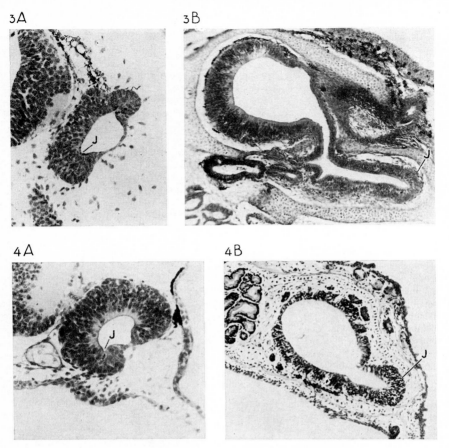

Abb. 3. Querschnitte des rechten Geruchsorgans bei *Bombina bombina*. *A*, Larve 9 mm; *B*, Larve 50 mm.
J, das Jacobsonsche Organ.

Abb. 4. Querschnitte des rechten Geruchsorgans bei *Triton taeniatus*. *A*, Larve 12 mm; *B*, Molch 30 mm.
J, das Jacobsonsche Organ.

Bombina und Hynobiidae keinen Zweifel über die Homologie dieses Gebildes: so überraschend ähnlich verläuft dieser Prozess. Von allen untersuchten Arten wurde nur bei *Rana* eine ursprüngliche Lage des Jacobsonschen Organs erhalten (Abb. 6).

Jedoch war die Verschiebung der Anlage des Jacobsonschen Organs von der medialen zur lateralen Seite bei Urodela und seine laterale Lage bei manchen niederen Anura schon längst bekannt. Dies war jedoch vorläufig nicht überzeugend genug, um die Homologie dieses Gebildes bei Anura und Urodela allgemein anerkannt werden zu lassen. Aus diesem Grunde muss man die Untersuchungen zur Lösung dieser Frage auf einer anderen Ebene durchführen, und zwar:

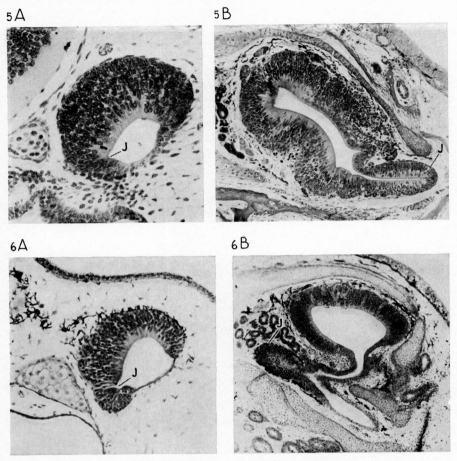

Abb. 5. Querschnitte des rechten Geruchsorgans bei *Ranodon sibiricus*. *A*, 21–22 mm; *B*, Larve 74 mm.
J, das Jacobsonsche Organ.

Abb. 6. Querschnitte des rechten Geruchsorgans bei *Rana*. *A*, Larve von *R. esculenta*, 10,5 mm; *B*, Larve von *R. temporaria*, 36 mm.
J, das Jacobsonsche Organ.

(1) Die Lokalisation des presumptiven Materials des Jacobsonschen Organs bei Anura und Urodela zu bestimmen.

(2) Die Ursachen der Verschiebung der Anlage des Jacobsonschen Organs von der medialen zur lateralen Seite feststellen; das heisst klären, ob in der geschichtlichen Entwicklung der Vorfahren der Amphibien jene Drehung des Nasensackes stattfand, auf die die Verschiebung der Anlage des Jacobsonschen Organs zeigt, und was solch eine Drehung hervorrufen könnte.

(3) Eine reale Erklärung der Ursachen einer so grossen Varietät des Jacobsonschen Organs (auch völliges Fehlen) bei verschiedenen Landwirbeltieren,

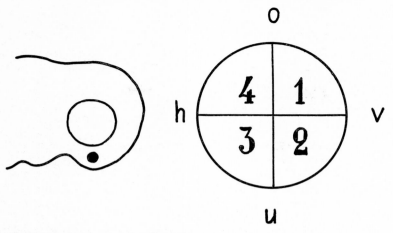

Abb. 7. Erläuterungen zu den Operationen.
o, v, u, h, obere, vordere, untere und hintere Hälften der Riechplakode; *1–4* Quadranten.

sowie des gleichzeitigen Vorhandenseins der Anlage dieses Gebildes bei allen ihren untersuchten Vertretern zu geben.

Erst wenn alle diese Fragen beantwortet sind, dürfen wir die festgestellte Homologie des Jacobsonschen Organs bei Urodela und Anura beurteilen.

Die Bestimmung der Lokalisation des presumptiven Materials des Jacobsonschen Organs

Zu diesem Zweck wurden die Vertreter der Familien Salamandridae, Hynobiidae und Discoglossidae experimentell studiert. Die Versuche bestanden in der stückweisen Entfernung der Riechplakode und bildeten vier Serien: die Entfernung ihrer Vorder-, Hinter-, Ober- und Unterhälften. Diese Methodik ermöglichte die Lokalisation des presumptiven Materials in der Riechplakode bis auf einen Quadranten genau zu bestimmen (Abb. 7). Das Fehlen dieses oder jenes Teils des Geruchsorgans bei der Entfernung sowohl der Ober-, als auch der Vorderhälfte der Plakode zeugt davon, dass das Material dieses Teils im Quadranten 1 sich befindet. Dementsprechend befindet sich das Material im Quadranten 2, wenn dieser Teil bei der Entfernung sowohl der Vorder-, als auch der Unterhälfte fehlt. Für die Unter- und Hinterhälfte der Plakode ist der Quadrant 3 gemeinsam, die Hinter- und Oberhälfte haben als gemeinsam Quadrant 4. Diese Ergebnisse sind an 330 Larven gewonnen; das heisst 141 Exemplare von *Hynobius keyserlingii*, 96 von *Triton taeniatus* und *T. cristatus*, und 93 von *Bombina bombina*.

Die Versuche ergaben unterschiedliche Resultate, abhängig davon, welche Plakodenhälfte entfernt worden war, und zwar 175 Fälle mit einem normal

Entfernte Hälfte	Defektes Geruchsorgan			Das Jacobsonsche Organ fehlt		
	Triton	Hynobius	Bombina	Triton	Hynobius	Bombina
Vordere	1	2	5	0	0	0
Untere	1	36	11	0	27	10
Hintere	20	25	15	11	19	12
Obere	12	1	1	11	0	0

Abb. 8. Quantitative Tabelle der Bildungszahl eines defekten Geruchsorgans sowie des Fehlen eines Jacobsonschen Organs.

entwickelten Geruchsorgan, und 130 Fälle eines defekten Geruchsorgans auf Kosten der Abwesenheit oder des Mangels einiger Teile des Geruchsorgans: des Nasenloches, des Einführungskanals, des Choanenkanals und des Jacobsonschen Organs. Ausserdem gab es noch weitere 25 Fälle mit einer teilweisen oder völligen Degeneration des Geruchsorgans, was im vorliegenden Bericht nicht erörtert werden wird.

Die Tabelle (Abb. 8) zeigt die quantitativen Verhältnisse der Fälle mit der Bildung eines defekten Geruchsorgans nach der Entfernung verschiedener Hälften der Riechplakode, sowie dieselben für das Jacobsonsche Organ. Aus der Tabelle ist ersichtlich, dass bei höheren Urodela die Mehrzahl aller Fälle (32 von 34) auf die Serien mit der Entfernung der Hinter- und Oberhälfte der Riechplakode entfällt. Dasselbe trifft auch für das Jacobsonsche Organ zu. Für die Hinter- und Oberhälften ist der Quadrant 4 gemeinsam, also gerade im Quadranten 4 befindet sich das presumptive Material der Hauptteile des Geruchsorgans (exklusive die Haupthöhle) und des Jacobsonschen Organs auch (Abb. 9 C).

Wenn wir zu primitiven Urodela übergehen, so können wir sehen, dass wiederum die Mehrzahl der Fälle (61 von 64), auf die Serien mit der Entfernung zweier Plakodenhälften, aber diesmal die andere Plakodenhälfte und zwar die Unter- und die Hinterhälften, entfällt. Gemeinsam in diesem Fall ist der Quadrant 3. Dasselbe Verhältnis gilt auch für das Jacobsonsche Organ. Bei niederen Urodela also befindet sich das presumptive Material des Nasenloches, des Einführungskanals, des Choanenkanals und des Jacobsonschen Organs, wie auch bei höheren in einem bestimmten Quadrant der Riechplakode, dieser Quadrant aber liegt an anderer Stelle der Plakode.

Betrachten wir die niederen Anura (Abb. 8), so stellen wir dieselben Verhältnisse wie auch bei niederen Urodela fest; das Material des Nasenloches, des Einführungskanals, des Choanenkanals und des Jacobsonschen Organs ist im Quadranten 3 zu finden (Abb. 9 B). Dass das presumptive Material des Jacobsonschen Organs bei niederen Anura und Urodela gleich lokalisiert ist,

Die Homologie des Jacobsonschen Organs 337

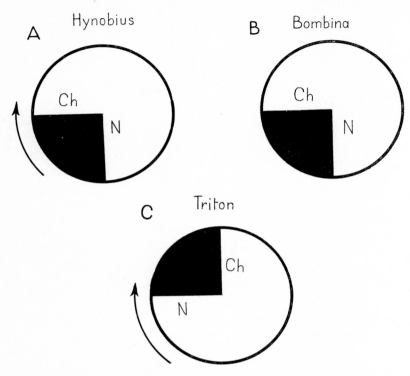

Abb. 9. Schema der Lokalisation des presumptiven Materials (Nasenloch, Einführungskanal, Choanenkanal, Jacobsonsches Organ) bei niederen Urodela und Anura und bei höheren Urodela.

Ch, das presumptive Material des Choanenkanals; *N*, das presumptive Material des Nasenloches und des Einführungskanals.

darf als ein weiterer Beweis für die Homologie dieses Gebildes in beiden Ordnungen eingeschätzt werden.

Einzelfälle der Enstehung eines defekten Geruchsorgans bei der Entfernung jener Plakodenhälften, in denen es kein entsprechendes Material gibt, können durch eine zufällige Ergreifung des Materials einer benachbarten Hälfte erklärt werden. Diese Fälle erlauben uns, die Lokalisation des Materials des Nasenloches, des Einführungskanals, einerseits, und des Choanenkanals, andererseits, genauer zu bestimmen. Bei *Triton* gibt es zum Beispiel ausser den erwähnten Fällen noch zwei weitere Fälle der Enstehung eines defekten Geruchsorgans und zwar: einer bei der Entfernung der vorderen und der andere bei der Entfernung der unteren Hälfte. Beide Fälle unterscheiden sich nach den Resultaten: bei der Entfernung der unteren Hälfte fehlt das Nasenloch und der Einführungskanal ist beschädigt; bei der Entfernung der vorderen Hälfte sind das Nasenloch und der Einführungskanal normal entwickelt, aber dafür fehlt der Choanenkanal in diesem Fall. Das Material des Nasenloches liegt also im Quadranten 4 an der Grenze mit der Unterhälfte und das Material des Choanenkanals — an der Grenze mit der Vorderhälfte (Abb. 9 C).

Bei niederen Urodela ist das Nasenloch und der Einführungskanal in zwei Fällen bei der Entfernung der Vorderhälfte defekt entwickelt; der Choanenkanal fehlt in einem Fall der Entfernung der Oberhälfte der Plakode. *Bombina* weist dieselben Verhältnisse wie bei Hynobiidae auf: die Fälle mit der Entfernung der Vorderhälfte beziehen sich auf das Nasenloch und den Einführungskanal, und der einzige Fall mit der Entfernung der oberen Hälfte ist mit dem fehlenden Choanenkanal verbunden. Bei niederen Urodela und Anura ist also das Material des Nasenloches und des Einführungskanals im Quadranten 3 an der Grenze mit der Vorderhälfte, das Material des Choanenkanals an der Grenze mit der Oberhälfte zu finden (Abb. 9. A, B).

Auf diesem Schema ist gezeigt worden (Abb. 9), wie das Material bei allen drei Gruppen lokalisiert ist. Bei niederen Anura (B), wie auch bei niederen Urodela (A) ist das Material völlig gleich lokalisiert, sowohl nach seiner Lage in der Plakode, als auch nach dem Verhältnis des Materials verschiedener Teile zueinander. Wenn wir das mit höheren Urodela vergleichen (C), so kann man sehen, dass das Verhältnis zwischen dem Material verschiedener Teile dasselbe bleibt, der ganze Komplex aber wird aus dem Quadranten 3 in den Quadranten 4 verschoben. Mit anderen Worten ist das Verhältnis des Materials verschiedener Teile des Geruchsorgans bei allen Amphibien gleich, die Plakode selbst aber ist bei höheren Urodela im Vergleich mit niederen Urodela und Anura ungefähr um 90 Grad gedreht worden.

Eine entwicklungsgeschichtliche Drehung des Nasensackes der Vorfahren der Amphibien ist wirklich geschehen, es ist nicht nur in der Verschiebung der Anlage des Jacobsonschen Organs sondern auch in der Drehung der Plakode selbst fixiert worden. In der Entwicklung der höheren Urodela ist das Stadium einer primären Lage des Materials in der Plakode schon ausgefallen, und nur die Untersuchung von mehr primitiven Amphibien hilft uns, solch eine Drehung nachzuweisen.

Die Frage über die Ursache der phyletischen Drehung des Nasensackes beantwortet die Hypothese von Schmalhausen (1957; 1958) über die Herkunft der Choane von Tetrapoda aus dem hinteren Nasenloch von Osteichthyes, das von der Seitenfläche des Kopfes in die Mundhöhle sich verschob. Bei dieser Versetzung musste das hintere Nasenloch unvermeidlich das hintere Ende des Nasensackes hinter sich herziehen und dieses um die Längsachse drehen. Zugunsten dieser Hypothese spricht auch eine gemeinsame Lage des Materials des Nasenloches und des Choanenkanals in der Plakode der Amphibien, die mit der einheitlichen Anlage beider Nasenlöcher bei Fischen ähnlich ist. Die Lokalisation des presumptiven Materials des Choanenkanals in jenem Teil der Riechplakode, an den bei niederen Urodela (Hynobiidae) die Flexura lacrymalis der Infraorbitallinie grenzt, bestätigt die Hypothese, dass das Hineinwachsen der Infraorbitalplakode zwischen der Riechplakode und der Mundgrube eine Ursache der Abtrennung des Choanenkanals bei den Vorfahren der Urodela und Anura ist (Medvedeva, 1961; 1964). Diese Abtrennung hat die Veränderung der Entwicklung des Geruchsorgans bei Amphibia verursacht.

Mögliche Ursachen der Divergenz und der Varietät des Jacobsonschen Organs

Man muss noch mögliche Ursachen der Enstehung einer so grossen Varietät des Jacobsonschen Organs von Tetrapoda betrachten. Eine der Ursachen stellt die entwicklungsgeschichtliche Drehung des Nasensackes dar, die mit der Versetzung des hinteren Nasenloches in die Mundhöhle verbunden ist, und die durch den vorliegenden Bericht bestätigt wird. Die deutlichen Spuren dieser Drehung werden durch die laterale Lage des Jacobsonschen Organs bei Urodela und bei einigen primitiven Anura erhalten.

Jedoch erklärt die vollzogene Drehung weder die beobachtete Varietät des Jacobsonschen Organs der gegenwärtigen Tetrapoden, noch den Umstand, dass dieses Organ wenigstens in Form einer rudimentären Anlage bei allen ihren Vertretern vorhanden ist. Die Lösung dieses Gegensatzes ist mit der Frage über die Funktion des Jacobsonschen Organs, genauer über den Wechsel der Funktion im Prozess der phyletischen Entwicklung verbunden.

Das Fehlen des Jacobsonschen Organs bei den Fischen, seine Entstehung nur bei den Landwirbeltieren, seine Reduktion bei den Amphibien, die zum Wasserleben wiederum zurückkehren, wie bei Proteidae (Seydel, 1895), und bei *Cryptobranchus* (Fleissig, 1906), zeugen davon, dass dieses Organ für das Wasserleben nicht nötig ist. Andererseits, eine ganze Reihe von Landwirbeltieren (Vögel, einige Säugetiere) büssten es sekundär ein. Daraus kann man folgern, dass das Jacobsonsche Organ nur in der Übergangsperiode, vom Wasser- zum Landleben lebenswichtig war. Wie könnte damals seine Funktion gewesen sein? Ursprünglich war das Tier, dass sein Wasserleben beendete und erstmalig ein Landleben begann, völlig hilflos, weil seine Sinnesorgane zum Wahrnehmen der Empfindungen im Luftmilieu nicht angepasst waren. Das trifft auch fürs Geruchsorgan zu. Eine trockene Schleimhaut der Nase konnte keinen Geruch wahrnehmen, und die Drüsen, die die Schleimhaut benetzen sollen, waren noch nicht da. Wenn jedoch das Tier in seiner Nasenhöhle einen akzessorischen Nasensack hatte, der sich von dem Boden dieser Höhle abzweigte, so konnte das Wasser gerade darin bleiben und eine normale Funktion des Riechepitels dieses Abschnittes gewährleisten. Zugunsten dieser Behauptung spricht auch die Tatsache, dass akzessorische Nasensäcke des Geruchsorgans bei mehreren Fischen mannigfaltig vertreten sind, und einer von diesen Nasensäcken, der von dem Boden der Nasenhöhle abgeleitet ist, als Grundlage für das Jacobsonsche Organ dienen könnte. Davon zeugt auch die unbedingte Ableitung des Jacobsonschen Organs der Tetrapoden von der Ventralwand der Nasenhöhle, obwohl seine Struktur und seine Lage sehr verschieden sein können. Solch ein Nasensack bot dem Tiere die Möglichkeit, beim Landleben noch das ans Wasserleben angepasste Geruchsorgan zu benutzen. Als die Drüsen sich entwickelten, büsste das Jacobsonsche Organ diese seine lebenswichtige Bedeutung ein. Die ersten Landwirbeltiere, als sie das Land besiedelten, gingen den Weg einer breiten adaptiven Radiation und Divergenz ein. Dementsprechend wurde das Jacobsonsche Organ im Zusammenhang mit der Änderung seiner Funktion entweder umgebildet, oder behielt zum Teil seine anfängliche Funktion, oder erwies sich überhaupt unnötig. Ein eigenartiger Weg der Entwicklungsgeschichte dieses Gebildes bedingte seine grosse Varietät bei gegenwärtigen Tetrapoden hinsichtlich seiner Funktion, und erschwerte dadurch die Lösung der Frage seiner Homologie.

Literaturverzeichnis

Bancroft, J. R. (1895). The nasal organs of *Pipa americana*. *Bull. Essex Inst.*, **27**: 101–107.
Burckhardt, R. (1891). Untersuchungen am Hirn und Geruchsorgan von *Triton* und *Ichthyophis*. *Z. wiss. Zool.*, **52**: 369–403.
Fleissig, J. (1906). Zur Anatomie der Nasenhöhle von *Cryptobranchus japonicus*. *Anat. Anz.*, **35**: 48–54.
Helling, H. (1938). Das Geruchsorgan der Anuren, vergleichend-morphologisch betrachtet. *Z. Anat. EntwGesch.*, **108**: 587–643.
Hinsberg, V. (1901). Die Entwicklung der Nasenhöhle bei Amphibien. *Arch. mikr. Anat.*, **58**: 411–482.
Medvedeva, Irene M. (1961). K voprosu o proiskhozhdenii khoan amfibii (On the question of the origin of the choanae in Amphibia). *Dokl. Akad. Sci. U.S.S.R.*, **137**: 468–471. Russisch.
Medvedeva, Irene M. (1964). The development, origin and homology of the choanae and choanal canal in Amphibia. *Trud. Inst. zool. Akad. Sci. U.S.S.R.*, **33**: 173–211. Russisch.
Schmalhausen, I. I. (1957). On the seismosensory system of urodels in connection with the problem of the origin of the tetrapods. *Zool. Zh.*, **36**: 100–112. Russisch mit englischer Zusammenfassung.
Schmalhausen, I. I. (1958). Nostrils of fishes and their fate in terrestrial vertebrates. *Zool. Zh.*, **37**: 1710–1718. Russisch mit englischer Zusammenfassung.
Seydel, O. (1895). Über die Nasenhöhle und das Jacobson'sche Organ der Amphibien. *Morph. Jb.*, **23**: 453–543.
Thrams, O. K. (1936). Das Geruchsorgan von *Pipa americana*. *Z. Anat. EntwGesch.*, **105**: 678–694.

The second somite with special reference to the evolution of its myotomic derivatives

By Hans C. Bjerring

Section of Palaeozoology, Swedish Museum of Natural History, Stockholm, Sweden

The present paper is one of a series dealing with the anterior four somites of the vertebrate body. It has been excerpted from an investigation in progress on the basic morphology of the vertebrate head with particular reference to the posterior myodome (see also Bjerring, 1967).

Description

As is well known, the process of gastrulation in *Squalus acanthias* results in a flat embryo consisting of three layers of cells (Fig. 1): an external layer, the ectoderm, and two internal layers, one of which, the mesoderm, is next to the ectoderm, whereas the other one, the entoderm, faces the yolk material. Of these layers, the mesoderm exhibits an initiated differentiation into median and lateral portions. In the sequel these portions are referred to respectively as notohylic and pleurohylic mesoderm (definitions in Bjerring, 1967).

Following gastrulation the embryo rises above the extraembryonic tissues and achieves a cylindroid body form. Concurrently, the pleurohylic mesoderm becomes hollow and throughout its upper parts undergoes a partition into a paired series of segments.

Trunk somites. Once pleurohylic segments are established in the developing trunk, they all differentiate according to the same basic pattern (Fig. 2). First the parachordal parts of their mesial walls produce migratory mesenchyme. This spreads out as a paired sheet applied to the sides of both notochord and neural tube opposite the segments. Each of these mesenchymal sheet later becomes converted into a dorsal and a ventral longitudinal series of arcual elements ultimately entering into the formation of the vertebral column (cf. van Wijhe, 1922). As this proliferation of mesenchyme ceases, the lowermost parts of the pleurohylic trunk segments separate themselves from the overlying mesoderm. Later these separated parts contribute to the development of the excretory system (cf. Nelsen, 1953). Meanwhile, as these separations occur, the dorsal parts of the mesial walls of the pleurohylic trunk segments rapidly enlarge and

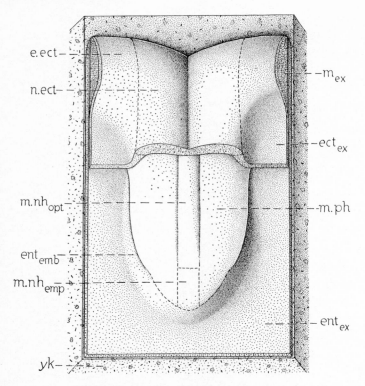

Fig. 1. *Squalus acanthias*. Anterodorsal view of late gastrula, with anterior half of ectoderm removed.

e.ect, epidermal ectoderm; ect_{ex}, extraembryonic ectoderm; ent_{emb}, embryonic entoderm; ent_{ex}, extraembryonic entoderm; m_{ex}, extraembryonic mesoderm; $m.nh_{emp}$, emprosthial portion of notohylic mesoderm; $m.nh_{opt}$, opisthial portion of notohylic mesoderm; *m.ph*, pleurohylic mesoderm; *n.ect*, neural ectoderm; *yk*, yolk material.

develop into segmentally arranged masses of striated muscle tissue (cf. Scammon 1911), the myomeres, which eventually form the trunk division of the somatic musculature. Finally, after the differentiation of the mesial walls of the pleurohylic trunk segments, the dorsal parts or their lateral walls, by splitting up into disconnected cells, proliferate mesenchyme ultimately contributing to the formation of the deeper portions of the integument.

From the above it appears that each pleurohylic segment in the trunk of *Squalus acanthias* is composed of four presumptive organ-forming regions: (1) A ventral region, the nephrotome, from which arises a nephric unit or nephron; (2) a parachordal region, the sclerotome, which develop into dorsal and ventral arcual elements; (3) a dorsomesial region, the myotome, from which originates a somatic muscle segment or myomere; and (4) a dorsolateral region, the dermatome, which enters into the formation of the hypodermal part of the skin. Of these four regions the sclerotome, myotome, and dermatome are collectively spoken of as a somite. Consequently, in the trunk the pleurohylic mesoderm is divisible into an upper somitic portion and a lower parasomitic portion. The parasomitic portion dorsally contains the segmentally arranged nephrotomes.

Second somite. The second pleurohylic segment of *Squalus acanthias* is fully established at about the end of neuralization (cf. Bjerring, 1967). It is larger than any other pleurohylic segment, a circumstance probably attributable to its position near the well pronounced cephalic flexure in the region of the developing midbrain. Ventrally it merges into the mesoderm immediately anterior to the second (spiracular) branchial (pharyngeal or visceral) pouch.

The first observable differentiation of the segment is a thickening of the parachordal part of its mesial wall (Fig. 3 A 1, A 2). Shortly this thickening undergoes extensive proliferation and gives rise to migratory mesenchyme ingressing into the spaces surrounding the anterior end of the notochord and the forming midbrain (Fig. 3 B 1, B 2). This mesenchyme later differentiates into the antotic pila and the ipsilateral half of the acrochordal which eventually enter into the formation of the posterior lower part of the orbitotemporal region of the endocranium (cf. Holmgren, 1940). Immediately after this cell proliferation ceases, an accelerated growth starts in the anterodorsal part of the mesial wall of the segment. Soon this activety results in a tubular evagination extending forwards dorsomesially to the developing eye. Later, the cells of the anterior portion of this evagination differentiate into myoblasts (cf. Holmgren, op. cit. fig. 32) ultimately forming the superior oblique eye muscle. During the above evagination, the original posterodorsal part of the mesial wall of the segment produces a small group of mesenchymal cells (Fig. 3 C). These cells, located just above the upper part of the efferent pseudobranchial artery (cf. Holmgren, 1943 fig. 24), gradually aggregate into a distinct blastema representing the primordium of the polar cartilage. (Incidentally it may be noted that the embryonic structure of *Squalus acanthias* referred to as the "e muscle" by Platt (1891), Lamb (1902), and Neal (1918) most likely represents the polar cartilage primordium). With time this blastema is transformed into a cartilage exhibiting similar relations to adjacent structures as those depicted for the polar cartilage of *Amia calva* in Fig. 5. Later, by fusing anteriorly with the rear of the trabecle (i.e., the probably infrapharyngobranchial portion of the second or mandibular visceral arch), mesially with the acrochordal, and posteriorly with the front end of the basiotic lamina, the polar cartilage is incorporated into the lower posterolateral part of the orbitotemporal region of the endocranium. Meanwhile, as the polar cartilage blastema develops, the original lowermost part of the second pleurohylic segment becomes compressed transversally (cf. Platt, 1891; Lamb, 1902). This compressed part lies immediately above the forming masticatory musculature and is soon converted into mesenchyme. Finally, concurrent with the events

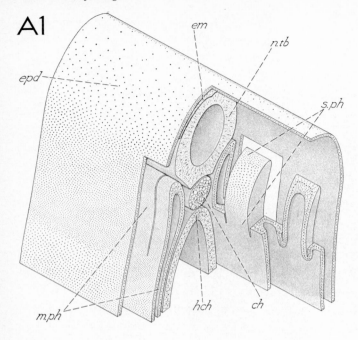
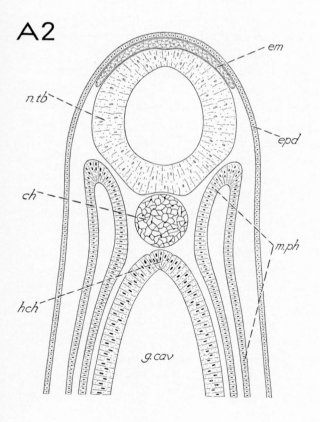

Fig. 2 A.

The second somite 345

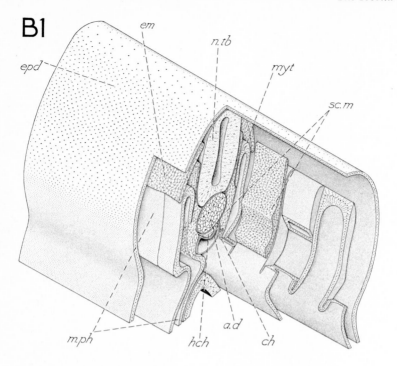

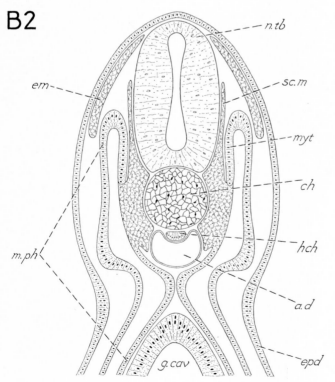

Fig. 2 B.

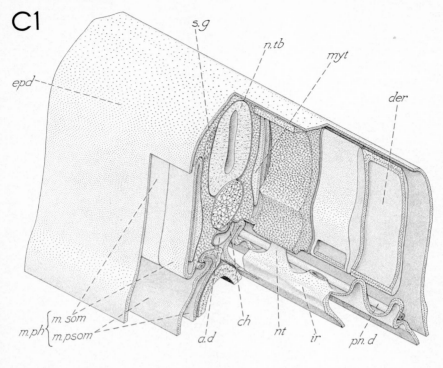
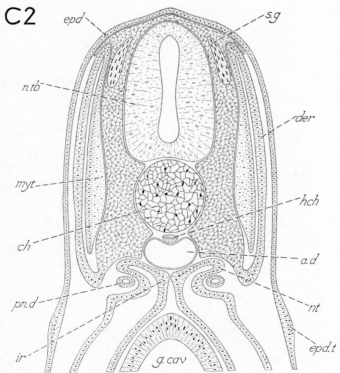

Fig. 2C.

just described, the dorsolateral wall of the segment breaks up into mesenchyme later differentiating into sclerotic and other orbital connective tissues.

On the basis of these observations it appears that the second pleurohylic segment of *Squalus acanthias* possesses the following five presumptive organ-forming regions: (1) A lower region which develops into connective tissue; (2) a mesial parachordal region from which arise the antotic pila and the adjacent half of the acrochordal; (3) an anterior dorsomesial region which produces the superior oblique eye muscle; (4) a posterior dorsomesial region which develops the polar cartilage; and (5) a dorsolateral region from which originate sclerotic and other orbital connective tissues.

Thus, from the above descriptions, the antotic pila and the ipsilateral half of the acrochordal form from a region of the second pleurohylic segment corresponding to the sclerotome of a trunk somite. The superior oblique eye muscle and the polar cartilage form from a region corresponding to the myotome of a trunk somite. The sclerotic and other orbital connective tissues forming from the dorsolateral wall of the second pleurohylic segment arise from a region corresponding to the dermatome of a trunk somite. Finally, the lowermost part of the second pleurohylic segment would appear to correspond to the nephrotome of a pleurohylic segment in the trunk.

It follows, therefore, that the second somite would be that portion of the second pleurohylic segment from which arise the antotic pila, the adjacent half of the acrochordal, the superior oblique eye muscle, the polar cartilage, and sclerotic and other orbital connective tissues. Also, the lowermost portion of the second pleurohylic segment, which develops into connective tissue, together with the underlying cell material, from which originates masticatory musculature, would constitute the foremost part of the parasomitic mesoderm.

Conclusions

(1) From the foregoing account it appears that the antotic pilae together with the acrochordal derive from the sclerotomes of the second pair of somites and are serially homologous to the metamerically arranged arcualia of the vertebral

---

Fig. 2. *Squalus acanthias*. Stereograms in oblique anterodorsal views and transverse sections of developing trunk, showing mesoderm differentiation. $A1$, $A2$, embryo of 3.75 mm in length; $B1$, $B2$, embryo of 5.5 mm in length; $C1$, $C2$, embryo of 9.0 mm in length.

a.d, dorsal aorta; *ch*, notochord; *der*, dermatome; *em*, ectomesenchyme; *epd*, epidermis; *epd.t*, epidermal thickening representing potential fin-fold; *g.cav*, gut cavity; *hch*, hypochord; *ir*, thickened part of parasomitic mesoderm representing primordium of interrenal body; *m.ph*, pleurohylic mesoderm (epimeric, mesomeric, and hypomeric mesoderm in current terminology); *m.psom*, parasomitic mesoderm (mesomeric and hypomeric mesoderm in current terminology); *m.som*, somitic mesoderm; *myt*, myotome; *n.tb*, neural tube; *nt*, nephrotome; *pn.d*, pronephric duct; *s.g*, differentiating spinal ganglion; *s.ph*, pleurohylic segment; *sc.m*, sclerotomic mesenchyme.

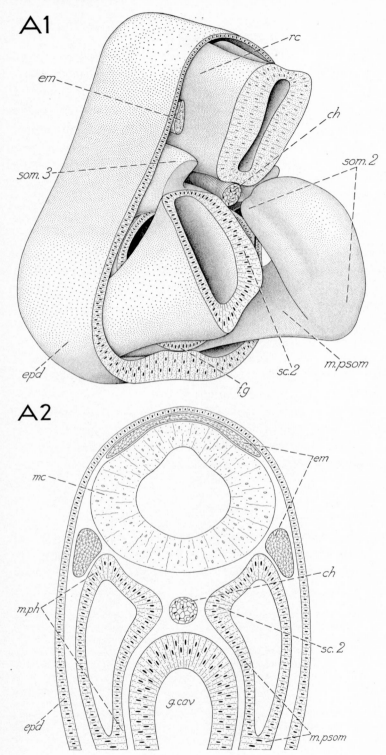

Fig. 3 A.

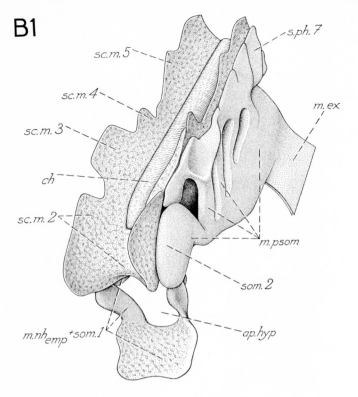
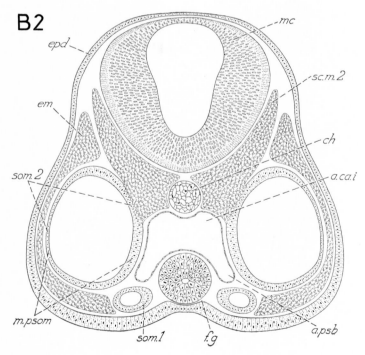

Fig. 3 B.

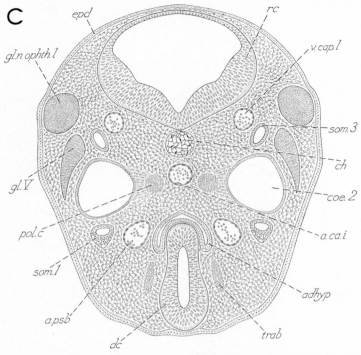

Fig. 3. *Squalus acanthias*. Stereograms in oblique anterodorsal views and transverse sections of developing head, showing differentiation of second somite. *A1, A2*, embryo of about 4.5 mm in length; *B1, B2*, embryo of about 5.5 mm in length; *C*, embryo of about 16.5 mm in length.

a.ca.i, internal carotid artery; *a.psb*, efferent pseudobranchial artery; *adhyp*, adenohypophysis; *ap.hyp*, fenestration in foremost part of mesoderm through which stomodaeal and neural ectoderm meet (cf. Hoffmann, 1896); *ch*, notochord; *coe.2*, coelom of second pleurohylic segment; *dc*, diencephalic division of developing brain; *em*, ectomesenchyme; *epd*, epidermis; *f.g*, foregut; *g.cav*, gut cavity; *gl.n.ophth.l*, ganglion of ophthalmic nerve of lateralis system; *gl.V*, ganglion of trigeminal nerves; *m.ex*, extraembryonic mesoderm; $m.nh_{emp} + som.1$, emprosthial portion of notohylic mesoderm and first pair of somites (Platt's vesicles or anterior head-cavities and premandibular somites); *m.ph*, pleurohylic mesoderm; *m.psom*, parasomitic mesoderm; *mc*, mesencephalic division of developing brain; *pol.c*, polar cartilage blastema; *rc*, rhombencephalic division of developing brain; *s.ph.7*, seventh pleurohylic segment; *sc.m.2–sc.m.5*, sclerotomic mesenchyme derived from second to fifth somite; *sc.2*, sclerotome of second somite; *som.1, som.2, som.3*, first, second, and third somite, respectively; *trab*, trabecular primordium; *v.cap.l*, lateral capitis vein.

column; also, that the superior oblique eye muscles and the polar cartilages originate from the myotomes of the second pair of somites and are serially homologous to the trunk myomeres; finally, that the connective tissues produced by the lateral walls of the second pair of somites are serially homologous to the dermatomic contributions of the trunk somites to the deeper part of the integument.

(2) As exemplified by the embryology of *Squalus acanthias*, the antotic pilae and the acrochordal enter into the formation of the posterior part of the orbitotemporal region of the endocranium. It is known also that the corresponding

part of the endocranium in coelacanthiforms, osteolepiforms, and porolepiforms exhibits a complicated juncture apparatus (cf. Bjerring, 1967). Thus, this intracranial articulatory complex, met with only in these three groups (the situation in struniiforms is not yet known), occupies an intrametameric position, similar to that of the articulations between adjacent vertebrae. It follows, therefore, that the intracranial juncture apparatus is of the same nature as, and serially homologous to, the intrametamerically arranged articulations of the vertebral column. This means that the intracranial juncture apparatus is a relatively primitive vertebrate feature, which among recent forms persists only in one species, the coelacanthiform *Latimeria chalumnae*.

(3) Whereas the superior/posterior oblique eye muscles probably occur throughout the entire craniate vertebrate group, the polar cartilages are known only in those forms in which the intracranial juncture apparatus is secondarily lacking. Accordingly, the absence of an intracranial juncture apparatus seems to be correlated with the presence of polar cartilages. Besides, since the intracranial juncture apparatus is a primitive feature, the presence of polar cartilages may be an advanced one.

(4) A basicranial muscle spanning the intracranial juncture apparatus has been demonstrated recently in various coelacanthiforms, osteolepiforms, and porolepiforms (Bjerring, 1967). It has been shown also that this muscle in coelacanthiform phylogeny probably underwent a progressive development correlated with a number of changes in both the head skeleton and the brain. Furthermore, it has been stated that the basicranial muscle in osteolepiforms and porolepiforms was comparatively short, and in the main situated along the lower posterior part of the orbitotemporal region of the endocranium. That being so, and since all tetrapods lack a basicranial muscle as well as an intracranial juncture apparatus, it follows that the condition of the basicranial muscle encountered in osteolepiforms and porolepiforms would have to be regarded as primitive relative to that of the corresponding muscle in coelacanthiforms.

(5) A comparison of the basicranial muscle in the osteolepiform *Eusthenopteron foordi* (Fig. 4) with the polar cartilage (Fig. 5) shows that these two structures apparently have the same relative position in the head. Thus, they both extend from the anteroventral part of the auditory capsule to the posterolateral part of the hypophysial division of the cranium. Moreover, dorsolaterally they are in relation with the palatine ramus of the facial nerve and the lateral capitis vein, dorsomesially with the abducens nerve, and ventrolaterally with the internal carotid artery and the horizontal infrapharyngeal dental plate (or its equivalent in the parasphenoid) of the third or hyoid visceral arch. Finally, anterodorsally they are crossed by the musculature innervated by the abducens nerve and in a certain degree also by the pituitary vein. On this basis, and inasmuch as the basicranial muscle acts on an articulatory complex within the second

metamere and the polar cartilage takes its origin from the myotome of the second somite, the basicranial muscle and the polar cartilage appear to be homologous structures. Besides, no doubt the basicranial muscle is phylogenetically older than the polar cartilage.

(6) Owing to what has been set forth above, it seems likely that during vertebrate evolution the prospective significance or fate of the myotome of the second somite has changed more than once from the superior/posterior oblique eye muscle and the basicranial muscle to the superior/posterior oblique eye muscle and the polar cartilage. Apparently this change was correlated with the ankylosis of the articulatory complex of the second metamere. Consequently, the prospective significance of the myotome of the second somite of coelacanthiforms, osteolepiforms, and porolepiforms would have to be regarded as more primitive than that of all other known vertebrates.

(7) Unlike that of gnathostomes, the posterior/superior oblique eye muscle of the lampreys occupies a postocular position and is innervated by neurons whose cytonal portions are located dorsally in the ipsilateral as well as the contralateral half of the anterior part of the rhombencephalic brain division. Ac-

---

Fig. 4. *Eusthenopteron foordi*. Attempted restoration of certain head structures in lateral view, to show the relative position of the basicranial muscle; approx. ×2. (Cf. Bjerring, 1967).

Eb.4, epibranchial portion of fourth visceral arch; *Eb.5*, epibranchial portion of fifth visceral arch; *Hy*, hyomandibula (= epibranchial portion of third or hyoid visceral arch); *Podp*, paraotic dental plate (= horizontal infrapharyngobranchial dental plate complex of third or hyoid visceral arch); *Psph*, parasphenoid; *Sb.4*, suprapharyngobranchial portion of fourth visceral arch; *Sb.5*, suprapharyngobranchial portion of fifth visceral arch; *a.ca.i*, internal carotid artery; *a.ophth.mg*, ophthalmic magna artery: *art.ptm*, area articulating with paratemporal process of palatoquadrate; *ast.pch*, branch of branchial trunk of trigeminus I anastomosing with palatine ramus of facial nerve (*r.pal*) posterior to choanal tube; *ast.V_1–gl.cil*, anastomosis between dorsal ramus of trigeminus I (V_1) and ciliary ganglion; *c.r.ot.l*, canals for posterior ramus of otic nerve of lateralis system; *ch*, notochord; *fos.br*, fossa bridgei; *gl.cil*, ciliary ganglion; *i.j.a*, intracranial juncture apparatus (= intra-arcual articulation of second metamere); *m.bc*, basicranial muscle; *m.bur*, bursal muscle; *m.o.inf*, inferior oblique eye muscle; *m.o.sup*, superior oblique eye muscle; *m.pro.hy*, protractor muscle of hyomandibula; *m.r.ext*, external rectus eye muscle; *m.r.inf*, inferior rectus eye muscle; *m.r.int*, internal rectus eye muscle; *m.r.sup*, superior rectus eye muscle; *m.ret.bul*, retractor bulbi muscle; *n.buc.l*, buccal nerve of lateralis system; *n.hy.VII*, hyoid branch of facial nerve; *n.mand.VII*, mandibular branch of facial nerve; *n.ophth.l*, ophthalmic nerve of lateralis system; *n.ophth.V*, dorsal ramus of trigeminus II; *o.b.s*, sclerotic parts of ocular bulb; *pr.bp*, basipterygoid process of ethmosphenoid; *pr.op*, opercular process of hyomandibula; *pr.sp*, suprapterygoid process of ethmosphenoid; *r.io.V_2*, infraorbital branch of branchial trunk of trigeminus I; *r.pal*, palatine ramus of facial nerve; *r.pn.V_2*, palatonasal branch of branchial trunk of trigeminus I; *spic*, spiracular canal; *sy.ot–oc.d*, dorsal otico-occipital synchondrosis (= upper part of embryonic metotic fissure—superior occipitocapsular fissure in human embryo—and probably representing vestigial intra-arcual articulation of fifth metamere); *v.cap.l*, lateral capitis vein; *v.cer.m*, middle cerebral vein; *v.io*, infraorbital vein; *v.pit*, pituitary vein; *v.so*, supraorbital vein; *v.subhy*, subhyomandibular vein; *II*, optic nerve; *III*, oculomotor nerve; *IV*, trochlear nerve; V_1, dorsal ramus of trigeminus I (profundus division of trigeminal nerve in current terminology); V_2, branchial trunk of trigeminus I (maxillary division of trigeminal nerve in current terminology); V_3, branchial trunk of trigeminus II (mandibular division of trigeminal nerve in current terminology); *VI*, abducens nerve; *X*, vagus nerve.

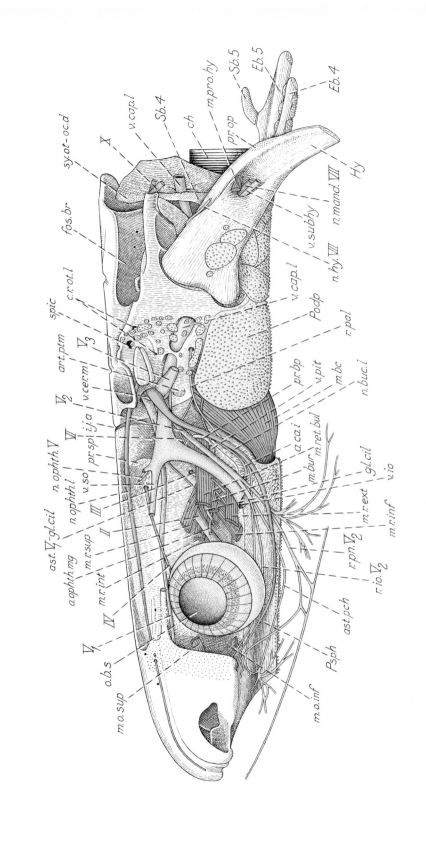

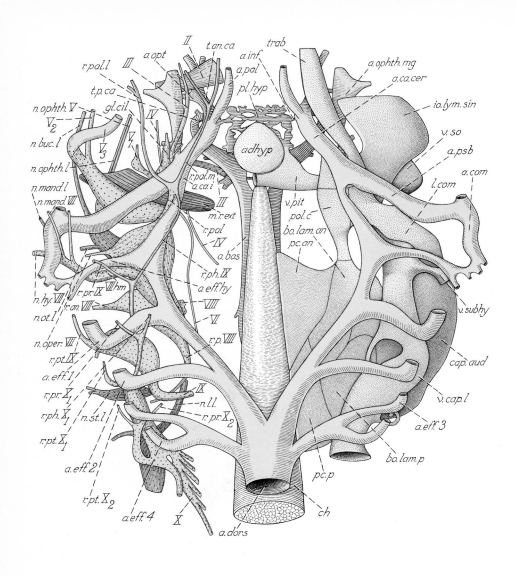

cording to Damas (1944) this muscle apparently takes its origin from the myotome of the second somite. As may be gathered from the description by Johnels (1948) of the lamprey, the embryonic elements entering into the formation of that part of the endocranium adjacent to the posterior/superior oblique eye muscle exhibit positional relations similar to the skeletal derivatives of the second somite of *Squalus acanthias* (the trabecular primordium of Johnels = the polar cartilage; the connective tissue differentiating into the cartilage surrounding the anterior end of the notochord = the acrochordal; and the cartilaginous pila situated immediately in front of the foramen for the trigeminal nerves = the antotic pila). It seems likely, therefore, that the position and innervation of the posterior/superior oblique eye muscle in the lampreys are more primitive than those of the corresponding muscle in gnathostomes.

(8) Inasmuch as the eye is a vertebrate specialization, the posterior/superior oblique eye muscle—as well as the other ocular muscles—are themselves specia-

---

Fig. 5. *Amia calva*. Drawing based on graphic and wax-plate reconstructions, showing various structures in middle and posterior parts of head in embryo of 8 mm in length. Ventral view; approx. × 100. (Cf. Bjerring, 1967; Bertmar, 1959.)

a.bas, basilar artery; *a.ca.cer*, cerebral carotid artery; *a.ca.i*, internal carotid artery; *a.com*, commissure between efferent artery of hyoid arch and efferent pseudobranchial artery; *a.dors*, dorsal aorta; *a.eff.hy*, efferent artery of hyoid arch; *a.eff.1*, efferent artery of first gill arch; *a.eff.2*, efferent artery of second gill arch; *a.eff.3*, efferent artery of third gill arch; *a.eff.4*, efferent artery of fourth gill arch; *a.inf*, infundibular artery; *a.ophth.mg*, ophthalmic magna artery; *a.opt*, optic artery; *a.pal*, palatine artery; *a.psb*, efferent pseudobranchial artery; *adhyp*, adenohypophysis; *bo.lam.an*, anterior element of basiotic lamina (derived from myotome of third somite); *bo.lam.p*, posterior element of basiotic lamina (produced by myotome of fourth somite); *cap.aud*, auditory capsule; *ch*, notochord; *gl.cil*, ciliary ganglion; *io.lym.sin*, infraorbital lymph sinus; *l.com*, lateral commissure of otical region of endocranium (= suprapharyngobranchial portion of third or hyoid visceral arch); *m.r.ext*, external rectus eye muscle; *n.buc.l*, buccal nerve of lateralis system; *n.hy.VII*, hyoid branch of facial nerve; *n.l.l*, lineal nerve of lateralis system; *n.mand.l*, mandibular nerve of lateralis system; *n.mand.VII*, mandibular branch of facial nerve; *n.oper.VII*, opercular nerve; *n.ophth.l*, ophthalmic nerve of lateralis system; *n.ophth.V*, dorsal ramus of trigeminus II; *n.ot.l*, otic nerve of lateralis system; *n.st.l*, supratemporal nerve of lateralis system; *pc.an*, anterior parachordal element (= ventral arcual of third metamere); *pc.p*, posterior parachordal element (= ventral arcual of fourth metamere); *pl.hyp*, vascular plexus between adenohypophysis and brain; *pol.c*, polar cartilage; *r.an.VIII*, anterior branch of acoustic nerve; *r.p.VIII*, posterior branch of acoustic nerve; *r.pal*, palatine ramus of facial nerve; *r.pal.l*. lateral branch of palatine ramus of facial nerve; *r.pal.m*, mesial branch of palatine ramus of facial nerve; *r.ph.IX*, pharyngeal branch of glossopharyngeal nerve; *r.ph.X*, pharyngeal branch of vagus nerve; *r.pr.IX*, pretrematic branch of glossopharyngeal nerve; *r.pr.X$_1$*, first pretrematic branch of vagus nerve; *r.pr.X$_2$*, second pretrematic branch of vagus nerve; *r.pt.IX*, posttrematic branch of glossopharyngeal nerve; *r.pt.X$_1$*, first posttrematic branch of vagus nerve; *r.pt.X$_2$*, second posttrematic branch of vagus nerve; *t.an.ca*, anterior trunk of cerebral carotid artery; *t.p.ca*, posterior trunk of cerebral carotid artery; *trab*, trabecle (= infrapharyngobranchial portion of second or mandibular visceral arch); *v.cap.l*, lateral capitis vein; *v.pit*, pituitary vein; *v.so*, supraorbital vein; *v.subhy*, subhyomandibular vein; *II*, optic nerve; *III*, oculomotor nerve; *IV*, trochlear nerve; V_1, dorsal ramus of tringeminus I (profundus division of trigeminal nerve in current terminology); V_2, branchial trunk of trigeminus I (maxillary division of trigeminal nerve in current terminology); V_3, branchial trunk of trigeminus II (mandibular division of trigeminal nerve in current terminology); *VI*, abducent nerve; *VII hm*, branchial trunk of facial nerve; *VIII*, acoustic nerve; *IX*, glossopharyngeal nerve; *X*, vagus nerve.

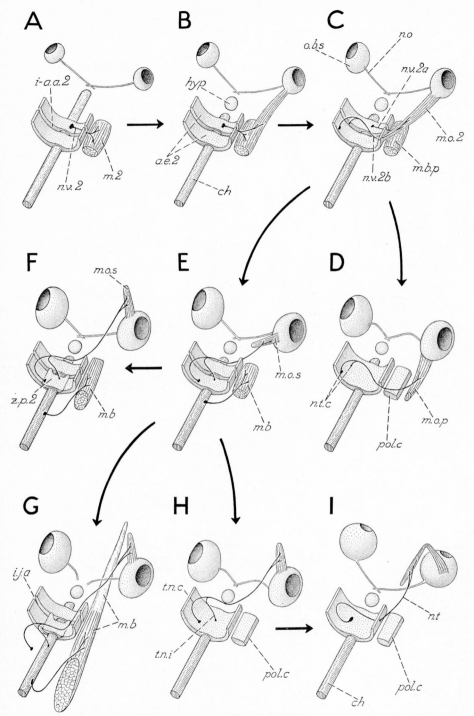

Fig. 6. Diagram showing probable evolution of various structures of second metamere of vertebrate body. Arrows indicate lines of succession.

A, hypothetic initial stage (to some extent suggestive of amphioxus) displaying intra-arcual articulation 2 (i.e. precursory intracranial juncture apparatus), unmodified myomere 2 (which is functionally a part of the locomotor mechanism), and unparted ventral nerve 2 (the cytonal portions of which are located ventrally in the ipsilateral half of the evolving brain opposite the middle of myomere 2).

B, hypothetical transitional stage in which myomere 2 as well as ventral nerve 2 are undergoing bipartition.

C, hypothetical stage showing complete bipartition of both myomere 2 (into a primitive basicranial muscle acting on an unconsolidated portion within the middle of the evolving cranium and a precursory superior/posterior oblique eye muscle functioning as an ocular retractor) and ventral nerve 2. Note that the cytonal portions of the nerve division for the primitive basicranial muscle occupy their original location in the forming brain, whereas those of the nerve division for the precursory superior/posterior oblique eye muscle (i.e. the cytones of the trochlear nerve) have migrated to the dorsal part of the evolving brain where they are in part ipsilateral and in part contralateral in position; as judged from the connexions of the trochlear nerve in a lamprey (cf. Johnston, 1902; Clark, 1906; Tretjakoff, 1909; Heier, 1948) it appears probable that this migration was caused by a neurobiotactic influence of both ipsilateral and contralateral impulses from the acoustico-lateralis system (cf. Addens, 1933; Larsell, 1947).

D, stage characteristic of a lamprey exhibiting an ankylosed intra-arcual articulation 2, a polar cartilage evolved from the primitive basicranial muscle, a posterior oblique eye muscle in a postocular position, and a trochlear nerve having its cytonal portions located dorsally in the ipsilateral and contralateral halves of the brain.

E, hypothetical protognathostome stage in which the following modifications are taking place: (1) The superior oblique eye muscle is moving forwards dorsomesially to the eye (possibly this positional change was correlated with the process of transformation of the two foremost visceral arches into the gnathostome type of biting mechanism); (2) the ipsilateral and contralateral parts of the cytonal portions of the trochlear nerve are shifting from a dorsal to a ventral location in the brain (since the trochlear nerve of a lamprey receives its majority of impulses from the acoustico-lateralis system and that of a gnathostome is mainly supplied by the medial longitudinal fasciculus, it appears probable that an increase of impulses from the medial longitudinal fasciculus in early gnathostome phylogeny has effected a neurobiotactic influence causing a downward shifting of the cytonal portions of the trochlear nerve); (3) the cytonal portions of the nerve division for the basicranial muscle are migrating backwards in the ipsilateral half of the brain (this migration was very likely due to a neurobiotactic influence of impulses from evolving gustatory brain centers; thus stimulations from the mouth region found expression on the motor side in intracranial movements executed by the basicranial muscles).

F, stage representative of osteolepiforms and porolepiforms.

G, stage seen in present-day coelacanthiforms; observe the well developed basicranial muscle which is innervated by neurons whose axons are passing with the vagus nerve (cf. Millot and Anthony, 1965).

H, stage characteristic of modern noncoelacanthiform gnathostomes in which the intracranial juncture apparatus is ankylosed and the basicranial muscle converted into a polar cartilage.

I, stage showing the condition in man; note the backward growth of the superior oblique eye muscle.

a.e.2, sclerotomic derivatives of second somites; *ch*, notochord produced by notohylic mesoderm; *hyp*, adenohypophysis; *i–a.a.2*, intra-arcual articulation of second metamere (precursory intracranial juncture apparatus); *i.j.a*, intracranial juncture apparatus; *m.2*, myomere of second metamere formed from myotome of second somite; *m.b*, basicranial muscle produced by myotome of second somite; *m.b.p*, precursory basicranial muscle; *m.o.2*, ocular muscle of second metamere (precursory posterior/superior oblique eye muscle) formed from myotome of second somite; *m.o.p*, posterior oblique eye muscle; *m.o.s*, superior oblique eye muscle; *n.o*, optic nerve; *n.t.* trochlear nerve; *n.t.c*, cytonal portions of trochlear nerve; *n.v.2*, original ventral nerve of second metamere; *n.v.2a*, nerve division innervating basicranial muscle; *n.v.2b*, nerve division supplying ocular muscle of second metamere (precursory trochlear nerve); *o.b.s*, sclerotic part of ocular bulb; *pol.c*, polar cartilage produced by myotome of second somite; *t.n.c*, contralateral part of trochlear nerve; *t.n.i*, ipsilateral part of trochlear nerve (met with in lampreys, coelacanthiforms, and teleosts); *z.p.2*, suprachordal zygal plate of second metamere (cf. Bjerring, 1967).

lizations. This implies that the posterior/superior oblique eye muscle and probably also the basicranial muscle were secondary, not primitive myotomic derivatives of the second somite. In the nonoculate and incraniate vertebrate amphioxus the myotome of the second somite produces one muscle solely (cf. Hatschek, 1881), the foremost myomere of the body, which is functionally a part of the locomotor effector system and which is supplied by the most anterior ventral nerve. In this light, it seems reasonable to assume that a simple myomere was the original contribution by the myotome of the second somite of the evolving vertebrate body.

(9) From the above it is likely that in the course of cephalization the myotomic derivatives of the second somite have undergone an evolutionary progression as follows (cf. Fig. 6):

    (a) A simple myomere, which functioned as a component part of the locomotor effector system, became converted into

    (b) two muscles, viz. a primitive eye muscle executing ocular retractions and a primitive basicranial muscle acting on the middle of the evolving cranium; then,

    (c) in cyclostome phylogeny the primitive ocular retractor gave rise to the posterior oblique eye muscle and the primitive basicranial muscle became transformed into a polar cartilage; on the other hand, in gnathostome phylogeny the primitive ocular retractor produced the superior oblique eye muscle and the primitive basicranial muscle either underwent a progressive development into the basicranial muscle proper of coelacanthiforms, osteolepiforms, and porolepiforms or became modified into a polar cartilage.

Thus, in that the polar cartilage apparently has evolved independently within various major groups of vertebrates, this structure would have to be regarded as polyphyletic.

(10) Synchronized with this developmental progression of the elements derived from the myotome of the second somite, also the original ventral nerve of the second metamere underwent a number of changes. These modifications, which may be gathered from Fig. 6, will be further discussed in a forthcoming paper.

(11) Finally, what it all adds up to is simply this: *There appears to be no fundamental difference between the second somite and the trunk somites.*

References

Addens, J. L. (1933). The motor nuclei and roots of the cranial and first spinal nerves of vertebrates. 1. Introduction. Cyclostomes. *Z. Anat. EntwGesch.*, **101**: 307–410.
Ariëns Kappers, C. U., Huber, G. C. & Crosby, E. C. (1960). *The comparative anatomy of the nervous system of vertebrates, including man.* New York: Hafner.

Bertmar, G. (1959). On the ontogeny of the chondral skull in Characidae, with a discussion on the chondrocranial base and the visceral chondrocranium in fishes. *Acta zool. Stockh.*, **40**: 203–364.

Bjerring, H. C. (1967). Does a homology exist between the basicranial muscle and the polar cartilage? *Colloques int. Cent. natn. Rech. Scient.*, **163**: 223–267.

Clark, W. B. (1960). The cerebellum of *Petromyzon fluviatilis*. *J. Anat. Physiol. Lond.*, **40**: 318–325.

Damas, H., 1944. Recherches sur le développement de *Lampetra fluviatilis* L. Contribution à l'étude de la céphalogenèse des vertébrés. *Arch. Biol. Liége*, **55**: 1–284.

Hatschek, B. (1881). Studien über Entwicklung des *Amphioxus*. *Arb. zool. Inst. Wien*, **4**: 1–88.

Heier, P. (1948). Fundamental principles in the structure of the brain. A study of the brain of *Petromyzon fluviatilis*. *Acta anat. Suppl.*, **8**: 1–213.

Hoffman, C. K. (1896). Beiträge zur Entwicklungsgeschichte der Selachii. *Morph. Jb.*, **24**: 209–286.

Holmgren, N. (1940). Studies on the head in fishes. Embryological, morphological, and phylogenetical researches. 1. Development of the skull in sharks and rays. *Acta zool. Stockh.*, **21**: 51–267.

Holmgren, N. (1943). Studies on the head of fishes. An embryological, morphological, and phylogenetical study. 4. General morphology of the head in fish. *Acta zool. Stockh.*, **24**: 1–188.

Johnels, A. G. (1948). On the development and morphology of the skeleton of the head of *Petromyzon*. *Acta zool. Stockh.*, **29**: 139–279.

Johnston, J. B. (1902). The brain of *Petromyzon*. *J. comp. Neurol.*, **12**: 1–86.

Lamb, A. B. (1902). The development of the eye muscles in *Acanthias*. *Am. J. Anat.*, **1**: 185–202.

Larsell, O. (1947). The nucleus of the IVth nerve in petromyzonts. *J. comp. Neurol.*, **86**: 447–466.

Millot, J. & Anthony, J. (1965). *Anatomie de Latimeria chalumnae*. 2. *Système nerveux et organes des sens*. Paris: Cent. natn. Rech. Scient.

Neal, H. V. (1918). The history of the eye muscles. *J. Morph.*, **30**: 433–453.

Nelsen, O. E. (1953). *Comparative embryology of the vertebrates*. New York: Blakiston.

Platt, Julia B. (1891). A contribution to the morphology of the vertebrate head, based on a study of *Acanthias vulgaris*. *J. Morph.*, **5**: 79–112.

Scammon, R. E. (1911). Normal plates of the development of *Squalus acanthias*. *NormTaf. EntwGesch. Wirbeltiere*, **12**: 1–140.

Tretjakoff, D. (1909). Das Nervensystem von *Ammocoetes*. 2. Gehirn. *Arch. mikr. Anat.*, **74**: 636–779.

Wijhe, J. W. van (1922). Frühe Entwicklungsstadien des Kopf- und Rumpfskeletts von *Acanthias vulgaris*. *Bijdr. Dierk.*, **22**: 271–298.

The origin of vertebrate calcified tissues

By Melvin L. Moss

Dept. of Anatomy, College of Physicians and Surgeons, Columbia University, New York, N.Y. USA

The primary source of data in vertebrate paleontology consists almost exclusively of calcified tissues. To a great extent the corpus of the literature in this field consists of painstakingly accurate descriptions of both the gross and the microscopic structure of the dermal, cranial and post-cranial skeletal tissues, teeth and calcified cartilages. To a surprising degree vertebrate paleontology, as a discipline, seems rather unaware of the data currently available on the formation, growth, structures and composition of similar tissues in recent vertebrates. To an even greater extent, the relevant data of modern descriptive and experimental embryology dealing with the ontogenesis of calcified tissues is generally excluded from discussions of vertebrate phylogeny.

Admittedly each branch of science increasingly requires an almost monastic dedication to increasingly restricted areas of knowledge, if even nominal mastery of a given field is to be gained. Despite this, it remains true that vertebrate paleontologists, however unknowingly, share a common field of interest with large numbers of academic and clinical research workers in many fields. At a time when traditional departmental boundaries are becoming increasingly blurred, and interdisciplinary activities are experiencing a current vogue, it seems appropriate that a conference on lower vertebrate phylogeny should include a consideration of the pertinent data and concepts currently operative in the active areas of calcified tissue research.

This paper is a review of a number of the basic theoretical biological and biochemical requirements necessary for calcified tissues to occur in vertebrate phylogeny. It is intended to be neither exhaustive in detail nor monographic in extent.

The general theory of biological mineralization

The formation of calcified tissues in vertebrates is not a unique biological event, but rather a specific example of a more generalized phenomenon, biological mineralization. This is the deposition of any crystallite or amorphous solid material in or on a pre-existent biologically produced organic matrix. The basic postulates of a General Theory of Biological Calcification and, similarly, of a General Theory of Biological Mineralization, have been published by (Moss, 1964a; 1968a) and Isenberg, Moss & Lavine (1968). While the former dealt with the general requirements for the formation of any calcium salt (calcium carbonate, calcium phosphate, calcium magnesium carbonate, etc.), the later theory concerned itself with all types of biological mineralization in plants as well as in animals. Here we deal not only with the various inorganic ions (calcium, strontium, magnesium, iron, silica), but also with the several varieties of organic deposits (oxalates, bilirubinates, cholesterol, uric acids). There is a logical hierarchy of terms involved here, passing from the more general to the more specific: *mineralization* (any deposit); *calcification* (calcium carbonate or phosphate); and *ossification* (the formation of hydroxyapatite on a collagenous matrix in the histologically defined bony tissues of vertebrates). In a very real sense we may consider that this same verbal sequence represents a similar phylogenetic sequence; with vertebrate ossification arising as a specific example of a generalized series of pre-existent calcification processes, which in turn arose as a specific example of a still more basic series of biological mineralizations.

The formal statement of the General Theory of Biological Mineralization has six postulates:

(1) All biological mineralizations are cellular processes. They require the prescence of a cell type, generically termed a *scleroblast*. This cell type, which may have many different specific forms of cytodifferentiation, is concerned with the elaboration of the organic matrix of the mineralized structure (tissue, organ). In vertebrates these scleroblasts are variously termed osteoblasts, chondroblasts (some), odontoblasts, cementoblasts, and ameloblasts. It is critical to note that, in the vertebrates, cells derived from the ectoderm as well as the mesoderm, and most importantly, the ectomesoderm (neural crest) are included in the generic term scleroblast.

(2) All biological mineralizations consist of an organic and an inorganic phase. In all cases the organic matrix is formed first, by the vital activity of the scleroblast. The inorganic phase is formed after some finite interval in or on the pre-existent organic phase. The role of the scleroblast in the second, inorganic phase, is either indirect or non-existent. While there are considerable data that the scleroblasts may, in some cases, play a role in transportation of the several ions or radicals of the inorganic phase, the operative formation of the inorganic phase seems able to occur without the direct mediation of vital cell activity (intracellular mineralizations may not be exceptions to this statement in all cases either).

In vertebrates, the organic matrices of bone, calcified cartilage, enamel, dentin, cementum, etc., are examples of this postulate.

(3) The organic phase, in turn, has two components; a fibrous, organized phase and an amorphous phase. This unorganized "ground substance" is usually a viscous, highly hydrated, polyanionic colloid.

(4) The form of the mineralized tissue is determined by the organic phase. The correctness of this postulate is self-evident when we consider the retention of osseous form following total decalcification of bone which has a relatively large amount of fibrous organic material (roughly 50 per cent). It is less evident, but nevertheless still true, in such matrix-poor tissues as dental enamel.

(5) All mineralizable fibers are composed of complex polymers of proteins and carbohydrates. Vertebrate collagen does not exist as such *in vivo*, but rather as a complex with the slight but critical amounts of carbohydrate.

(6) Mineralization is not a simple chemical precipitation. It is the current consensus that the organic phase plays an important role in biological mineralizations. In vertebrates the specific nature of this role (or roles) is being actively investigated and, as yet, no single hypothesis has been accepted generally. Various laboratories favor differing portions of the organic matrix; some the fiber, others the ground substance, while still others emphasize the combination of the two as being the operative component. One point of general agreement is that calcification may be initiated by epitactic mechanisms. This theory stresses the role of the matrix in providing unique spatial and electrochemical properties which, of themselves, are capable of inducing the nucleation of the crystalline apatitic salt. These matrices presumably are capable of determining the primary crystallographic axes of the salt as well as the type and species of the crystal. The recent work of Posner's laboratory showing that vertebrate ossification may involve the preliminary formation of an amorphous calcium phosphate prior to its transformation into crystalline hydroxyapatite, differs only in suggesting that short-order repeat units (amorphous, albeit still crystalline in a sense) occurs prior to the long-order repeat (crystalline) stage discussed above. The reader is referred to Moss (1964a; 1968a) and Isenberg, Moss & Lavine (1968) for a more complete statement of this general theory and a complete bibliography.

The general theory of biological calcification presented here has recently been utilized successfully in the construction of a model for mineralization in the calcareous Foraminifera (Towe & Cifelli, 1967).

Invertebrate calcification processes

All vertebrate calcified tissues (in bones, teeth, scales, etc.) fulfil the six postulates of biological mineralization presented above. The origin of vertebrates required neither the addition of any new postulate (or process) nor the loss or substitution of any postulate. All that did happen was that a *specific* type of salt (hydroxyapatite) was deposited upon either of two specific types of organic matrices (collagen or the keratin-like matrix of enamel). This particular selection of salt and matrix did not occur *de novo*, but rather was a selection from a pre-existent spectrum of invertebrate modes of calcification. This being so, we may profitably review such invertebrate processes for some indication of how the vertebrate selection came to be.

(1) *Site*. Calcified tissues are found in invertebrates in either one of two possible sites, or matrices: these of *mesodermal* or of *ectodermal* origin. No known invertebrate calcified (or mineralized) tissue has a dual origin; i.e., is a composite of ectodermal and mesodermal sites. Such composite calcified tissues are uniquely vertebrate, represented by teeth (dermal and oral) and by tooth-like derivatives (scales, fin-rays, etc., see Moss, 1968b, for a comprehensive review). Invertebrate mesodermal calcified structures are typified by the test of echinoderms, while the shells of molluscs are typical invertebrate ectodermal calcified tissues.

It is evident that vertebrates did not have to evolve either the formation of calcified mesodermal or of calcified ectodermal tissues. What they did specifically evolve was the ability to combine these two in a composite tissue. This ability, as will be shown later, was determined by the appearance of several specifically vertebrate tissues and tissue interactions.

(2) *Cell types*. Invertebrate calcified tissues are formed as the result of the vital activity of specific cell types. The invertebrate ectodermal calcified tissues are derived from the activity of epithelial cells whose cytomorphological differentiations during this process are strikingly similar to thse noted in vertebrate amelogenesis. Flat or cuboidal epithelial cells typically become elongated and columnar, with a polarization of intracellular organelles when actively secreting a calcifiable matrix. A typical example is seen in the cytodifferentiation of the mantle epithelium accompanying shell formation in molluscs: these changes are at least analogous, if not homologous, with the changes in functioning vertebrate ameloblasts. Invertebrate mesodermal scleroblasts generally are less well differentiated than their mammalian counterparts. A typical example is observed in echinoderms, where the process of test formation involves the prior function of a highly organized organic matrix and its subsequent mineralization by calcium carbonate (Moss & Murchison, 1966; Moss & Meehan, 1967). There is some indication that in the invertebrates we observe a general phylogenetic sequence in which the scleroblasts of the lower forms may actually have a portion of their cytoplasm utilized as the fibrous portion of their organic matrices, while in the higher forms these matrices become totally extracellular in location.

(3) *Matrix fibers*. There are two principal types of organized fibers in invertebrates; the first is predominantly carbohydrate with a small, but important,

protein prosthetic group (S-p). Chitin is an excellent example, consisting of n-acetylglucosamine together with such protein components as sclerotin or arthropodin. The second group is predominantly a protein (collagen) bound to a carbohydrate moiety (P-s). Characteristic invertebrate collagens are tectin, gorgonin, spongin and conchiolin. We note that these invertebrate collagens differ in certain characteristic structural ways from vertebrate collagens, and additionally may be non-mineralizable. An instructive example of the applicability of the general theory in the invertebrates is taken from the work of Watabe & Wilbur, using a series of *in vivo* cross-implants of decalcified molluscan shells, in which one normally was calcified with aragonite and the other with calcite. In each case, when the decalcified organic matrix was placed in the shell defect site of the host, it recalcified with the species of calcium carbonate characteristic of the donor and not of the host. In other words, it seems that the matrix itself determined the species of calcification and not the ambient environment of the host; nor did the surrounding areas of calcification exert any localized effect (Watabe & Wilbur, 1960).

All such invertebrate fibers normally are bound up with the surrounding amorphous ground substance into a complex of relatively high molecular weight. While the precise chemical composition of the specific ground substances are not known in many invertebrates, it may be said with some confidence that they are all generically related sulfated mucopolysaccharides (see: Moss & Meehan, 1967).

(4) *Inorganic salts*. Invertebrate calcified tissues have a broad spectrum of composition. They may be calcium carbonate (of a variety of species); calcium phosphate (as an apatite in some Brachiopoda); and a mixed carbonate-phosphate, as in the lobster, *Homarus americanus*. We are not able, at this time, to describe the possible biochemical or biophysical bases for these different sorts of calcium salt depositions. For our present purposes, it is sufficient only to note that neither phosphate salts in general nor apatite specifically is a unique property of vertebrate calcified tissues.

Vertebrate processes

This brief review of the pertinent features of invertebrate calcification process places the phylogenetic origin of vertebrate calcified tissues in proper perspective. Wherever the first vertebrates arose, and whatever their environment, the first appearance of vertebrate calcified tissues of any sort required that two interrelated conditions must have existed: (a) that there be present an organic matrix capable of calcifying; and (b) that there be an internal environment that permits calcification.

(A) *Calcifiable organic matrices*. It was not necessary for the first vertebrate with calcified tissues to acquire a significantly new type of organic matrix. It was sufficient, in a sense, to make a selection from the pre-existing invertebrate repertoire. If indeed the first vertebrates were uncalcified, then this selection process occurred somewhat later. In this case one would suppose that vertebrate organic matrices were selected first, and that the acquisition of calcifiability was a secondary event. On the other hand, if the first vertebrate had calcified tissues, then a one stage selective process occurred. This may have required only a slight, but highly significant, shift in molecular structure.

The data presented here are consistent with recent statements on the molecular evolution of cartilage (Mathews, 1966), and on the macromolecular evolution of connective tissue (Mathews, 1967).

Fibers. Here again there are two theoretical possibilities. A shift from a predominantly carbohydrate fiber (S-p) to a predominantly protein fiber (P-s) is one possible method. However, it is easier to believe that the shift was from a non-calcifiable protein fiber, (P-s), to a calcifiable one, (P-s). This change in collagen may have occurred in several ways; the collagen molecule itself may be rearranged or reactive sites may now become formed or exposed, or alternatively, both may have occurred. We know, for example, that calcifiable mammalian collagens differ from non-calcifiable collagens by the possession of free ε-lysine reactive end groups. The recent work of Glimcher, in taking normally uncalcified rat tendon collagen, dissolving and reprecipitating it after which the collagen is capable of initiating nucleation of hydroxyapatite, is analogous (Glimcher, 1960). This experiment did not necessarily imply that any change occurred in the collagen, per se: it might be explained on the basis of a freeing, or exposure, of reactive sites which normally had been covered by the complexed ground substance.

Amorphous ground substance. We may now postulate that a similar molecular shift occurred in the composition of the ground substances, again utilizing the pre-existent invertebrate materials. We have relatively less data available on the amorphous than on the organized phase of the organic matrix. Accordingly we are left with a somewhat greater area of speculation. While it is possible that the acquisition by vertebrates of the property of calcifiability was sequential in the case of the organized and amorphous phases of the organic matrix, it seems more probable, to me, that the matrix complex evolved as a unit, since it is a fibrous-amorphous complex that it functions. If this was the case, we still do not know if this change occurred simultaneously with the origin of vertebrates, or whether it was subsequent to this event. If the sequential theory is correct, then we must postulate uncalcified vertebrate ancestors.

(B) *The internal environment.* No organic matrix of whatever composition or structure will calcify if the internal fluid environment bathing that matrix will not permit mineralization to occur. It is theorized currently that the operative parameter of the internal fluid is not related to the concentration of the relevant mineral ions, but rather to its ionic strength, a more meaningful parameter of ionic dissociation and consequent reactivity.

The critical point is that if the ionic strength of internal fluids exceeds a certain critical limit, calcification will not occur. All recent teleost and elasmobranch fishes have similar serum concentrations of calcium and phosphate ions. Obviously the serum concentration of these ions does not, of itself, determine whether bone or cartilage will be formed in the skeletal system. Recent agnathic fishes have significantly higher ionic strengths than do either recent teleosts or elasmobranchs. Accordingly it seems reasonable to speculate that the fossil agnatha, with calcified tissues present, may well have had a lower ionic strength than the recent agnatha; and that this lower ionic strength was another biochemical prerequisite for the first appearance of calcified tissues in vertebrates.

(C) *Conditions in the earliest vertebrates.* Combining our knowledge of requirements for a calcifiable matrix with the seemingly critical factor of serum ionic strength we may postulate the following statement: If there were early vertebrates *without* calcified tissues, one of three possible conditions may have existed: (a) there may have been a potentially calcifiable matrix together with a high ionic strength; (b) a non-calcifiable matrix may have existed together with an equally high ionic strength; or (c), a lower ionic strength may have been found together with a non-calcifiable matrix. In all three situations, calcified skeletal tissues would have been impossible. The only combination of conditions which could have permitted vertebrate calcified skeletal tissues to form is a lower serum ionic strength, together with a potentially calcifiable matrix.

Cartilaginous precursors

One current area of controversy concerns the possibility that cartilaginous skeletal tissues preceded the phylogenetic appearance of osseous tissues. Although the ontogenetic sequence of the endoskeleton generally follows this sequential order, it is by no means certain that this is a valid concept in vertebrate evolution. Romer (1942; 1963) has been strong in his opposition to this idea, and the data he has marshalled in support of his views of cartilage as a neotenic tissue are impressive. Denison (1963), on the other hand, is one of a group who feel that cartilage may well have a longer phylogenetic history than bone, having appeared first.

The nature of available paleontological material being as it is, no one is able to decisively end this discussion to the satisfaction of all concerned. My only contribution to the problem is to suggest an eclectic position. The argument runs as follows: Cartilage usually (but not always) appears only in the endoskeleton of vertebrates, while only bone usually is found in the vertebrate dermal skeleton. No one has ever suggested, nor do I, that a cartilaginous dermal (exo-) skeleton ever existed in any form. Similarly there is general agreement that the processes of osseous replacement (endochondral ossification) of the endoskeletal cartilages arose somewhat later in vertebrate history. In essence, I suggest that by differentiating between dermal and endoskeletal sites the problem is capable of resolution. Romer, in this sense, is right. Cartilage did not precede bone in the dermal skeleton. Denison is right also. Cartilage did precede bone in the endoskeleton.

The vertebrate processes of ossification

The appearance of osseous tissues in vertebrates required the phylogenetic acquisition of four unique processes: (1) the formation of composite calcified tissues; (2) inductive interaction between neural crest derivatives and a competent epidermal cell layer; (3) delamination; and (4) epidermal co-participation. These morphogenetic events, when added to the biochemical parameters discussed above, produce the full spectrum of vertebrate calcified skeletal tissues.

(A) *Composite tissues.* Invertebrate calcified tissues may be either ectodermal (mollusc shells) or mesodermal in origin (echinoderm test and spicules). No invertebrate, fossil or recent, possesses a calcified tissue which is a composite, formed by the combination of tissues of these two origins. Such composite skeletal tissues are uniquely vertebrate. Dental tissues are a prime example of such composite structures. It is characteristic of these skeletal composites that they form at restricted sites topographically, superficially at the junction of the epithelium (ectodermal or endodermal) with the subjacent mesoderm (ectomesenchyme). At some stage there is always a basement membrane to be found between these two tissue layers. The dermal plates of the early agnathic vertebrates clearly belong to this category, as do palaeoniscoid and lepidosteoid scales. Many other, if not all, dermal skeletal strutures share this composite structure—a point discussed below.

(B) *Inductive interaction.* The neural crest is a unique vertebrate structure. Our interest in this tissue is restricted to those derivative ectomesenchymal cells which are involved in the production of cartilage and bone (cf. Hörstadius, 1950). Correlated with this skeletogenic activity of the ectomesenchyme is the competency of the basal layer of epidermal cells, at specific sites, to interact with these neural crest derivatives in a well regulated sequence of inductive interactions. The site of this interaction is always at the point of junction between these two tissues; i.e., at the epidermal basement membrane (or lamella). These inductive processes require two pre-existent conditions; topographic relationship and tissue competency. The first factor, discussed above, states only that the two interacting tissues have an intimate spatial relationship, while the second factor implies that both tissues must be able (competent) to respond to an inductive stimulus.

The topographic relationship may be a function of the ectomesenchyme. In this case it is the selective migration, and latter site specific concentration, of these cells subjacent to the epidermal basal cell layer that determines which areas of overlying epithelium will become competent to interact with it. On the other hand, it is possible also that it is a site specific activity of the epithelium which then determines where the migrating ectomesenchymal cells will accumulate.

We do not possess data to answer this problem, nor are we on firm ground when dealing with subsequent stages of their inductive interactions. Some feel that the underlying dermal mesodermal tissues determine not only where a given type of dermal structure will form, but more importantly, *what* type of structure will form (Rawles, 1965). However, Sengel & Abbott (1965) present data (in the chick) which suggest that the epidermis plays a significant role in the determination of dermal structural specificity (see Dodson, 1967, for a recent statement).

The phylogenetic aspect of these processes is interesting. It is possible that the definitive vertebrate topographic relationship between the epithelium and the migrating ectomesenchyme could have been achieved before the acquisition of inductive competency between these two tissues. This would have led inevitably to a non-osseous dermal skeleton. It is more difficult to conceive of the phylogenetic appearance of inductive competency prior to the attainment of the topographic relationship between these two tissues. In essence, without definitive proof, it seems preferable to opt for a simultaneous evolutionary appearance of these two prerequisites for dermal ossification, rather than to postulate a more difficult stepwise sequence.

The viewpoint has been expressed previously that the earliest ossified vertebrates possessed the intrinsic (genetic) capacity to produce the entire histological spectrum of vertebrate skeletal tissues; bone (cellular as well as acellular), dentins of various types, enamel and the several types of cartilage (Moss, 1964*a*; 1964*b*). While it is undoubtedly true that some part of this tissue differentiation (or indeed a great part of it) may reflect adaptation to varying extrinsic (environmental) functional demands, there are other possible causes for both the range of tissues seen in early vertebrates, as well as for the tissue specificity. In terms of inductive interactions between the epidermis and ectomesenchyme we note four factors that may have been as significant in early vertebrates as they are in recent forms (see Moss, 1960; 1968*a*). These are: (1) the *strength* of the inductive stimulus; (2) the *duration* of the inductive stimulus; (3) the *specificity* of the inductive stimulus; and (4) the *degree of competency* of one tissue to respond to a stimulus from the other.

Concerning the histological identification of fossil vertebrate skeletal tissues, one precautionary remark here will not be amiss. Ground sections are the usual materials dealt with (see Moss, 1961 *a*, for a technique of examining decalcified fossil bones). As such the interpretation of all spaces in a calcified tissue, either found empty or filled with secondarily intrusive matrix materials, become of importance. The greatest precision in interpretation of fossil skeletal material is made possible by an intimate knowledge of the structure of recent skeletal tissues of similar classes. One recent example is the suggestion of Tarlo (1963) that the acellular bone (aspidin) of some fossil Agnatha contained enclosed cells, termed by him aspidinocytes. In support of this view, he published illustrations of many irregularly oval oblate spaces which were considered to represent osteocytic lacunae. I cannot accept this identification. My previous work on the osseous tissues of recent teleost fishes (Moss, 1961 *b*; 1961 *c*; 1962 *a*; 1962 *b*; 1963 *a*; 1963 *b*; 1964 *b*; 1965) has demonstrated beyond doubt that many such bones contain large numbers of large diametered *uncalcified* collagen bundles. Any oblique sectioning of such material will produce precisely the apparent image observed by Tarlo.

(C) *Delamination.* The morphology of the dermal skeleton of early vertebrates is well known. The basic structural plan is a basal layer of vascular bone (variously arranged to be sure) surmounted by a dental mass consisting of dentin, surrounding some sort of pulp (neurovascular) chamber and covered externally with an enamel layer. In its next more complicated state of organization, this dento-osseous complex is found to be repetitively formed in a vertical manner, so that several successive generations of this complex are visualized in a single area of the dermal skeleton. A unitary developmental process underlies the formation of all of the varieties of the dermal skeleton. This process is delamination, whose acquisition was another of the events necessary for the origin of vertebrate hard tissues. Postulated first by Holmgren (1940) and later expounded by Jarvik (1959) delamination discussed the morphogenetic movements of the tissue products formed at the basement membrane as the result of the inductive interaction between the ectomesenchyme and the basal epidermal cell layers. In essence, this theory proposes that each generation of tissues so produced gradually "sinks deeper" into the dermis with age, while the next succeeding generation of tissue begins to form above it at the basement membrane, only to sink down in its turn. It is noted that the types of tissues in succeeding generations need not be identical, thus allowing bone to be formed first, followed by more superficial dental tissues, which in turn may be again covered by bone in a repetitive cycle. Composite vertebrate dermal skeletal tissues arose in this manner.

Of the greatest importance is the fact that delamination is involved in the formation of both endoskeletal as well as exoskeletal structures. As Jarvik (1959) stated, "Hence it follows that there cannot be any fundamental difference between ... the endoskeleton and the exoskeleton or between the various types of exoskeletal and endoskeletal formations."

Delamination is a unique process which evolved very early in the history of the vertebrates, probably simultaneously with the ability of the ectomesenchyme and the basal epidermal layers to interact inductively. This process is ubiquitous in vertebrates, being involved in the formation of all dermal derivatives. The recent work of Edds (1964) and of Hay (1964) in providing a demonstration of the ultrastructural basis for delamination leads us to the consideration of the fourth, and final, unique vertebrate process required for the appearance of skeletal tissues.

(D) *Epidermal co-participation.* In essence this hypothesis suggests that the basal epidermal cells play a significant role in the formation of all vertebrate integumental skeletal structures. The inclusive definition of the term skeletal encompasses all tissues which play a protective-supporting role. This permits us to include both calcified as well as non-calcified derivatives in this hypothesis. Extensively reviewed recently (Moss, 1968b), the co-participation hypothesis is derived when the operational consequences of the other three vertebrate calcification processes are integrated. This co-participation of the epidermal basal cell layer with the subjacent ectomesenchymal cells is capable of a varied morphological expression, in which all vertebrate dermal skeleton tissues may be grouped in a tripartite classification: structured epidermal derivatives (hair, horn, keratinized teeth, etc.); structured epidermal-ectomesenchymal derivatives (teeth, scales, scutes, fin-rays, etc.); structured mesodermal derivatives (dermal bone, antlers, osteoderms, dermal collagen fibers). It is not necessary to go into the details of the process at the present time (see: Moss, 1968b). It is sufficient to state here that neither the formation of composite calcified tissues, nor inductive interaction, nor delamination, either singly or in any combination, could account for the first appearance of vertebrate skeletal tissues. The addition of epidermal co-participation provides the final, integrative, process which makes them all operative.

Throughout this review I have consistently emphasized the role of the epidermal basal cell in many critical skeletogenic events. In closing, it is this point I wish to stress. Most previous thought on the origin of vertebrate skeletal tissues has centered primarily on the mesodermal tissues. I wish to redress the balance by pointing to the importance of genetic (evolutionary) alterations in the epidermis, which is seen now as more than a passive, enclosing tissue layer, but rather as a dynamic participant in those biological processes which first produced the spectrum of vertebrate skeletal tissues, and whose basic parameters have remained basically unchanged.

This study was aided, in part, by grant No. DE-01715, and a training grant No. 5-TO1-DE-00132, from the National Institute of Dental Research.

References

Denison, R. H. (1963). The early history of the vertebrate calcified skeleton. *Clin. Orthop.*, **31**: 141–152.
Dodson, J. W. (1967). The differentiation of epidermis. 1. The interrelationship of epidermis and dermis in embryonic chicken skin. *J. Embryol. exp. Morph.*, **17**: 83–105.
Edds, M. V., Jr. (1964). The basement lamella of developing amphibian skin. In *Proceed. Conf. Small Blood Vessel Involvement in Diabetes Mellitus*, ed. Siperstein, M. D., Colwell, A. R. & Meyer, K.: 245–252. Am. Inst. Biol. Sci.
Glimcher, M. J. (1960). Specificity of the molecular structure of organic matrices in mineralization. In *Calcification in Biological Systems*, ed. Sognnaes, R. F., *Am. Ass. Adv. Sci. Publ.*, **64**: 421–488.
Hay, Elizabeth D. (1964). Secretion of a connective tissue protein by developing epidermis. In *The Epidermis*, ed. Montagna, W. & Labitz, W. Jr.,: 97–116. New York: Academic Press.
Holmgren, N. (1940). Studies on the head in fishes. Embryological, morphological and phylogenetical researches. 1. Development of the skull in sharks and rays. *Acta zool. Stockh.*, **21**: 51–267.
Hörstadius, S. (1950). *The neural crest.* Oxford: Oxford University Press.
Isenberg, H., Moss, M. L. & Lavine, L. (1968). An introduction to the comparative morphology of mineralized tissues. *Adv. Clin. Orthop.* In press.
Jarvik, E. (1959). Dermal fin-rays and Holmgren's principle of delamination. *K. svenska VetenskAkad. Handl.*, (4) **6**: 1–51.
Mathews, M. B. (1966). The molecular evolution of cartilage. *Clin. Orthop.*, **48**: 267–283.
Mathews, M. B. (1967). Macromolecular evolution of connective tissue. *Biol. Rev.*, **42**: 499-551.
Moss, M. L. (1960). Experimental induction of osteogenesis. In *Calcification in Biological Systems*. ed. Sognnaes, R. F., *Am. Ass. Adv. Sci. Publ.*, **64**: 323–348.
Moss, M. L. (1961a). The initial phylogenetic appearance of bone: an experimental hypothesis. *Trans. N. Y. Acad. Sci.*, (2) **23**: 495–500.
Moss, M. L. (1961b). Osteogenesis of acellular teleost bone. *Am. J. Anat.*, **108**: 99–110.
Moss, M. L. (1961c). Studies of the acellular bone of teleost fish. 1. Morphological and systematic variation. *Acta anat.*, **46**: 343–362.
Moss, M. L. (1962a). Studies of the acellular bone of teleost fish. 2. Response to fracture under normal and acalcemic conditions. *Acta anat.*, **48**: 46–63.
Moss, M. L. (1962b). Studies of the acellular bone of teleost fish. 3. Intraskeletal heterografts in the rat. *Acta anat.*, **49**: 266–280.
Moss, M. L. (1963a). The biology of acellular teleost bone. *Annls N.Y. Acad. Sci.*, **109**: 337–350.
Moss, M. L. (1963b). Studies of the acellular bone of teleost fish. 4. Inorganic content. *Acta anat.*, **53**: 1–8.
Moss, M. L. (1964a). Phylogeny of mineralized tissues. *Int. Rev. gen. exp. Zool.*, **1**: 297–331.

Moss, M. L. (1964b). Development of cellular dentin and lepidosteal tubules in the bowfin (*Amia calva*). *Acta anat.*, **58**: 333–354.

Moss, M. L. (1965). Studies of the acellular bone of teleost fish. 5. Histology and mineral homeostasis of fresh water species. *Acta anat.*, **60**: 262–275.

Moss, M. L. (1968a). Bone, dentin and enamel and the evolution of vertebrates. In *Biology of Oral Tissues*, ed. Person. P., *Am. Ass. Adv. Sci. Publ.* In press.

Moss, M. L. (1968b). Comparative anatomy of vertebrate dermal bone and teeth. 1. The epidermal co-participation hypothesis. *Acta anat.* In press.

Moss, M. L. & Meehan, M. (1967). Sutural connective tissues in the test of an echinoid (*Arbacia punctulata*). *Acta anat.*, **66**: 279–304.

Moss, M. L. & Murchison, E. (1966). Calcified anal teeth and pharyngeal ring in a holothurian (*Actinopyga mauritania*). *Acta anat.*, **64**: 446–461.

Rawles, M. E. (1965). Tissue interactions in the morphogenesis of the feather. In *Biology of the Skin and Hair Growth*, ed. Lyne, A. G. & Short, B. F.,: 105–128. New York: Elsevier.

Romer, A. S. (1942). Cartilage an embryonic adaptation. *Am. Nat.*, **76**: 394–404.

Romer, A. S. (1963). The ancient history of bone. *Annls N.Y. Acad. Sci.*, **109**: 168–176.

Sengel, P. & Abbot, V. K. (1965). *In vitro* studies with the scaleless mutant: interactions during feather and scale differentiation. *J. Hered.*, **54**: 254–262.

Tarlo, L. B. H. (1963). Aspidin: the precursor of bone. *Nature, Lond.*, **199**: 46–48.

Towe, K. M. & Cifelli, R. (1967). Wall ultrastructure in the calcareous foraminifera: crystallographic aspects and a model for calcification. *J. Paleont.*, **41**: 742–762.

Watabe, N. & Wilbur, K. M. (1960). Influence of the organic matrix on crystal type in molluscs. *Nature, Lond.*, **188**: 334.

The dermal skeleton; general considerations

By Tor Ørvig

Section of Palaeozoology, Swedish Museum of Natural History, Stockholm, Sweden

The study of the dermal skeleton in lower vertebrates embraces a rich variety of organisms at different levels of organization and involves much debated problems in such fields as morphology, histology, ecology and physiology. Without covering all of these, the following is an account of the initial formation of this skeleton and some phyletic trends in its subsequent history. Naturally, some of the suggestions made here are tentative; they may serve, if nothing else, to stimulate interest and discussion from other points of view.

The rise of the dermal skeleton

There has been much discussion concerning the organizational level at which, the circumstances under which, and the reasons for which hard tissue originally emerged in the chordate stock; these matters, however, cannot yet be said to be anywhere near their solution. At any rate it seems to be a reasonable assumption (on which the following considerations are based) that in early chordates the initial site of mineralization was the corium, i.e. that a dermal skeleton of some sort was the first to form (see e.g. Denison, 1963: 150; the same is probably indicated by the delamination theory, see Jarvik, 1959). Mineralizations of endoskeletal tissues, not dealt with in any detail here, in all probability belonged to subsequent, and more advanced, phyletic stages.

(1) At present, there is no telling exactly *when*, in early chordates or their predecessors (Moss, 1961; 1964), the ability to form integumental hard tissues was acquired and mineralized parts of some sort arose. Comparisons between the dermal skeleton in various ostracoderms and an arthropod exoskeleton like that of *Limulus* (Patten, 1912), the lime-plates of echinoderms such as carpoids and cystoids (Haeman in Spencer, 1938; Gregory, 1946; Caster & Eaton, 1956; see also Jefferies, 1967: 201), and the ectodermal shell in nautiloids (Sillman, 1960), yield hardly any information in this respect. The enigmatic conodonts, because they consist of calcium phosphate and sometimes exhibit a structure not unlike that of acellular bone tissue as e.g. in anaspids, have been held to represent some otherwise unknown, early group of chordates (Gross, 1954; see Rhodes & Wingard, 1957), but this is entirely conjectural (see e.g.

Lindström, 1964). Finally, none of the various pre-Carboniferous fossils which have been interpreted with a varying degree of credibility as some sort of prochordates, throw any light on early, dermal hard tissue formation (Graham-Smith, 1935; Scourfield, 1937; White, 1946; Lehman, 1957; 1964; Tarlo 1960; etc.).

(2) In the present state of knowledge, the question *why* did hard tissues, and thus mineralized parts, first emerge in the chordates is a purely theoretical one. According to one school of thought, the vertebrates evolved, during or after migration of their unossified, marine ancestors into fresh water, where they adjusted themselves to new environmental conditions; one result of this, it is claimed, was the acquisition of the power to develop a dermal armour as a means to meet specific demands or insure survival, either as a means of protection against eurypterid predators (Romer, 1933; 1946), as an impervious external covering which, in the absence of an effective internal osmoregulating mechanism, could counteract edema, and ultimately death as a result of osmotic inflow of the surrounding water with a salt concentration lower than that of the body fluids (Smith, 1939), or as a receptacle for excreted calcium and phosphorus of which, after transition to lake or river habitats, there would in some cases be excessive amounts in the blood, and with which, at that organizational level, the renal system was unable to cope (Berrill, 1955). These theories have been criticized—and justly so—from various points of view; the only objection to them which needs mention here concerns the basic conception on which they were all founded, that the vertebrates first acquired their armour while living in fresh water surroundings, a conception which is hardly tenable any longer because all evidence indicates that the earliest vertebrates were *marine* animals (see Denison, 1956; White, 1958; Spjeldnæs, 1967; etc.) and included forms such as the Ordovician Heterostraci with a thick and resistant dermal carapace. Whatever new uses the early vertebrates could put their dermal skeleton to after migration into fresh water, the power to develop such a skeleton surely existed well before that migration (as, incidentally, also a glomerular kidney probably was successfully produced as a preadaptation before entering fresh water; see Robertson, 1959). A further theory is that of Westoll according to whom (1942; 1945) the formation of dermal armour in the earliest ostracoderms could have been instigated on reaching, or closely approaching, maturity as a result of glandular readjustments during some sort of metamorphosis; once skeleton production had commenced, it could not be checked for the lack of an effective mechanism regulating calcium metabolism, with the result that the ensuing armour reached disproportionate dimensions. Armour reduction during subsequent phyletic stages (regressive phase, see below) might then have been the result of a gradually developed control of calcium metabolism. Even this theory needs no further comment because, like those of Romer, Smith and

Berrill already referred to, it is based on the earlier and wide-spread notion that the acquisition of a heavily developed dermal armour represents a truly primitive stage in early vertebrate history, and there is reason to believe that this was, in fact, not so (assimilative phase, see below). Finally, there is also the suggestion made by Pautard (1961) and sustained by Tarlo (1964), that the initial hard tissue formation in vertebrates may have taken place in response to the need within the organism of developing a store of salts such as the essential phosphates which could be freely recirculated whenever the supply from external sources diminished or ceased. Comments have been made (Denison, 1963; Urist, 1964; Moss, 1964; Spjeldnæs, 1967) on the extent to which, in various ostracoderms, mineralized parts were physiologically active, and the extent to which such activity could have been of importance for homeostatic control. But however this may be, the condition in such respects in the ostracoderms known from the fossil record, surely in their own way highly specialized animals (see Jarvik, 1964), does not provide information about the very *origin* of hard tissues in earlier and more primitive ancestors. Once they had been evolved, hard tissues could clearly take over, or get adjusted to, physiological functions of various kinds (such as acting as a readily available mineral reservoir), but this does not necessarily mean that hard tissues originally formed *because* of such functions and their potential fulfillment.

The point just made is one which, in the opinion of the writer, should be emphasized even in a broader sense. In early chordate history, hard tissues did not necessarily spring forth once the circumstances, external or internal, arose which to our mind make such substances useful. Those who so believe tend to underrate the specificity and complexity of the interacting factors which are—and presumably always were—prerequisites for inotropic calcification, that is the formation of hard tissues as bone tissue, dentinous tissue and dentine (see Denison, 1963) and which, in the opinion of Moss (1961; 1964), make there hard tissues unique in the animal kingdom. For those unfamiliar with the subject, these prerequisites may be briefly stated as (a) differentiation of special scleroblasts producing organic matrix containing collagen fibrils and mucopolysaccharides, (b) specific properties of the collagen, possibly also of the mucopolysaccharides, necessary for crystallite nucleation and growth, (c) functioning of a control mechanism by which the body fluids could furnish the matrix with the necessary salts, including phosphate, in proper concentrations, and (d) probably supply also of agents such as specific hormones, enzymes or vitamins. This rather special set of circumstances would hardly be liable to arise automatically once hard tissue were advantageous in one way or another. It is more conceivable that when hard tissue actually did form at some early stage in chordate history, it was at first non-adaptive, perhaps so remaining for a considerable span of time, and only *subsequently* assumed adaptive significance by reaching a more comprehensive level of development, spreading to new parts of the organism and serving, to an increasing extent, a variety of purposes having to do with the external or internal environment. The emergence in this way of mineralized parts, like those of the dermal skeleton, with definite functions to fulfill, or the potentiality to take over entirely new functions in new habitats, may well have been of importance in stages of vertebrate history prior to the appearance of the typical ostracoderms (see Denison, 1963).

(3) Only a few comments have been made so far on what *kind* of hard tissue that could have been the first to form in the chordate stock; in this connection one may note that although bone tissue, as maintained by Stensiö (1927) and others, may have been a common heritage in the various lineages leading from the eocraniates to the ostracoderms on the one hand, and the Preichthyes, and eventually the different groups of fishes on the other, the possibility certainly exists also that it may have evolved independently in two or more lineages (Denison, 1963; Spjeldnæs, 1967), and that the bone tissue emerging on one of these occasions was not exactly the same as that which did on another, except perhaps in basic design. However this may be, the point of interest here is mainly by what sort of calcification hard tissues may initially have formed in early chordates.

The mechanism of inotropic calcification referred to above is clearly very old in vertebrate history, existing as far back as we have evidence of the animals themselves. However, another matter is what warrants the belief that this mechanism dates right back to the very beginnings of vertebrate hard tissue formation (Moss, 1961); it may have done so, naturally, but for all one knows it could be also the product of early specialization, and have had a forerunner in another, previously established and in some respects different calcification mechanism which it gradually supplanted, or replaced, prior to the emergence of bone tissue, dentinous tissues and dentine as these substances are developed in the dermal skeleton of ostracoderms. It is well known that in dentine ontogeny, calcification may begin by being spheritic (globular dentine) and then turn into inotropic; reference may also be made to the circumstance that in the cartilage of elasmobranchs, spheritic (globular) calcification is phyletically, and in various instances and to a varying degree also ontogenetically, the precursor of inotropic (prismatic) calcification (Ørvig, 1951; despite criticism by Moss, 1964; 313, the writer still maintains that these conclusions are warranted by a careful study of fossil elasmobranch material, taking into consideration, of course, the effects on hard tissues of "post-mortem geological mineralizations"). Spheritic calcification, furthermore, is manifest also in mineralized parts belonging to certain of the Ordovician vertebrates, and is thus at least as old as inotropic calcification in vertebrate history (Ørvig, 1951: 381, 415, 433, $glob_d$, fig. 22 B; that figure shows globular structures in aspidin, not in dentine as then maintained by the writer, see Ørvig, 1967a: 82). The full significance of this, and other, evidence may not yet be apparent, but a possible implication could be that spheritic calcification does not represent a mere "preliminary and transient phase" in hard tissue ontogeny (Moss, 1964: 313), and is not only the result of "überstürzter Kristallisation" either (Schmidt in Schmidt & Keil, 1958: 67), but has a specific significance from the point of view of phylogeny. More precisely, of the two kinds of calcification, the spheritic may represent the more primitive which in early phyletic stages was responsible for hard tissue formation, and the inotropic the more advanced, that is a later acquisition which, as it evolved, eventually came to predominate in most hard tissues, at any rate those of the dermal skeleton (Ørvig, 1967a: 50–51). In so assuming, even such an important, and in early vertebrates widely distributed, hard tissue as bone tissue, where inotropic calcification normally is the condition throughout ontogeny, and spheritic calcification a comparatively rare occurrence (Fig. 1; see also Ørvig, 1967a: 50), could have been derived phyletically from some sort of related, more primitive mineralized substance with the structure of spheritic calcification (a "globular bone" as one might say). Although this cannot yet be proved by any evidence, it is nevertheless a theoretical possibility meriting consideration. If one postulates spheritic calcification in the corium of early chordates as a possible forerunner of typical dermal bone tissue, this calcification presumably needed no active participation of the collagen in the organic matrix where it took place; it probably neither required a matrix especially produced by scleroblasts for the sole purpose of receiving the lime-salts. Consequently hard tissue could, in early chordates, initially have formed *without* two important prerequisites necessary for inotropic calcification (see above), viz. (a) the differentiation of matrix-producing scleroblasts, and (b) the ability of such scleroblasts to produce that particular collagen which could interact with calcium and phosphorus ions for the nucleation of apatite crystallites (see e.g. Moss, 1961: 499). In these circumstances, it is *possible* to envisage an initial mechanism of hard tissue formation of a comparatively simple kind, not too far removed from the condition of some, not nearer definable, ancestors (perhaps already at invertebrate level), in which features like apatite as a calcification material (Moss, 1961; 1964), the spheritic mode of calcification, the mesoderm as the site of calcification, or all this simultaneously, could have been foreshadowed to some extent. Furthermore, it is also *possible* to imagine that after an initial period of spheritic hard tissue formation in the corium of early chordates, the era of typical vertebrate dermal hard tissue eventually dawned by the introduction of an entirely new element in histogenesis, namely the *active scleroblastic cell* which could exert a definite influence on matrix composition and matrix properties, and through this indirectly also on the succeding mineralization, yielding as an end-product the inotropic pattern of crystallites orien-

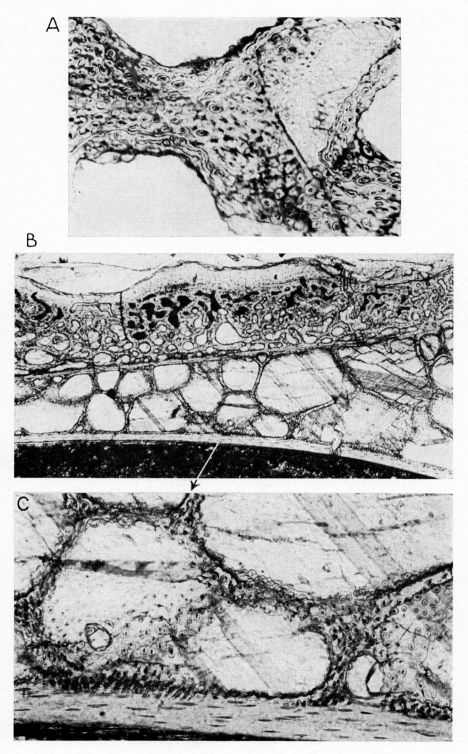

tated in alignment with the collagen fibrils. When viewed in this purely speculative way, typical vertebrate hard tissue may not have developed suddenly, once the prerequisites for inotropic calcification were there (although this alternative also exists), but could instead represent the ultimate stage in a phyletic process of increasing complexity of the calcification mechanism in early forms so far unknown. Such a chain of events, incidentally, may also be interpreted in terms of what has been pointed out above concerning the *origin* of hard tissue on the one hand as opposed to the *functions* of such substance on the other. For a variety of reasons which cannot be mentioned here, a sprinkling of spheritic hard tissue—if that was what first existed in the corium of early chordates—could hardly have been physiologically active to any noticeable extent or of any particular significance either for mechanical support or protection; hence such hard tissue, at this stage, presumably had little, if any, adaptive value. Inotropic dermal hard tissue, whenever that arose, may first also have been essentially non-adaptive, but once it had formed, the *potentiality* existed for it to be utilized, in later phyletic stages, for functions related to external or internal factors of various kinds.

It remains to be discussed, finally, what kind of hard tissue the mechanism of inotropic calcification, whenever and in what way it may have arisen, initially was able to produce. This has to be based on an evaluation (to the extent possible) of the degree of primitiveness, or specialization, of structural features in the hard tissues known to exist in the dermal skeleton of early vertebrates. Without going into detail here, there are two main points that deserve consideration: (a) the phyletic relations between dentine and dentinous tissues on the one hand, and bone tissue on the other, and (b) the presence or absence of osteocytes as a primary character in bone tissue.

Because in ostracoderms, as is well known, the dermal skeleton frequently is made up superficially of dentine or dentinous tissue and elsewhere of bone tissue, one might assume (as has been done) that this was indeed so as far back as a skeleton of this kind existed. However, as pointed out on another occasion (Ørvig, 1967a), there is reason to believe, at any rate in the opinion of the writer, that dentine as a hard tissue is *not* truly primitive, but one which achieved its acellularity by specialization from precursors among the dentinous tissues, i.e. substances characterized by the enclosure during ontogeny of a larger or smaller number of scleroblasts. By tracing phyletic changes in dentinous tissues of groups such as cephalaspids, arthrodires and some acanthodians (Ørvig,

---

Fig. 1. Occurrence of globular structures in elsewhere normal bone tissue of the dermal skeleton; *A*, in a cephalaspid, *Alaspis macrotuberculata* Ø. from the Lower Upper Devonian of Escuminac Bay, Canada; Section No. S 1363; ×450; *B, C*, in an antiarch, *Bothriolepis canadensis* (Whiteaves) from the same locality; *B*, section of a dermal bone, Section No. S 448; ×37.5, and *C*, more highly magnified detail of *B* (at arrow); ×150.

1958; 1967a; 1967b), furthermore, there is reason to believe also that the dentinous tissues, in their turn, had a common origin with, and are derivatives from, some sort of osteocyte-bearing bone tissue which, according to this chain of reasoning, would be more primitive that either the dentinous tissues or dentine itself (Ørvig, 1967a: 78–79). Another opinion has recently been expressed by Tarlo (1963: 46), according to whom dentine has been derived from acellular aspidin; for certain reasons, however, this is unlikely (Ørvig, 1967a: 78).

As stated elsewhere (Ørvig, 1951; 1957b; 1965; 1967a; 1967b), whenever it is possible to trace phyletic changes in the cytoplasmatic content of bone tissue, viz. in the cephalaspids, climatiid acanthodians and various actinopterygians, the tendency is towards the *loss* of osteocytes and not the opposite, i.e. from acellularity to a subsequent acquisition of enclosed cells (for comments, see e.g. Moss, 1964: 309–310). The interpretation of aspidin still poses a problem because this hard tissue was completely acellular already in the earliest, Ordovician, representatives of the Heterostraci, a circumstance which has on occasion been taken to be primitive (Denison, 1963: 145–146; for comments see e.g. Ørvig, 1965). In view of all other evidence, however, there is reason to conclude that the possession of osteocytes is indeed a primitive condition in bone tissue of lower vertebrates in general.

From the above, it is possible to reach the conclusion that the inotropic hard tissue first to form in the corium of early chordates in all probability was a cell-bearing bone tissue. More precisely, there is reason to believe that once cells became active as scleroblasts in histogenesis, possibly after an initial period of spheritic calcification (see above), these cells, upon termination of their scleroblastic activities, became enclosed in the hard tissue whose matrix they had produced, and persisted there as osteocytes. Exactly what took place during the hypothetical transition from an early spheritic hard tissue (a "globular bone") to a cell-bearing bone tissue is, of course, difficult to say; regardless of this uncertainty, however, all evidence seems to indicate that it was by differentiation from cell-bearing bone tissue that other characteristic hard tissues of the dermal skeleton arose, that is on the one hand dentinous tissue and eventually dentine, and on the other acellular bone tissue.

Stages of phyletic assimilation and regression

The opinions expressed from time to time during more than a hundred years about the phylogeny of dermal elements have been reviewed on various occasions and need no repetition here. One contention (shared by the writer) which has long been in existence but has gained new ground in recent decades, is that (expressing matters at this point with a certain simplification) large dermal elements arose phyletically, in part or completely, by the coalescence of smaller plates and scales (for comments on the different "morphogenetic" processes which are involved here, see Westoll, 1967). As concerns ostracoderms, this is now maintained by several writers (Obruchev, 1945; Buistrov, 1955; Stensiö, 1961; 1964; Ørvig, 1961; Tarlo, 1962; 1967, and other papers; etc.; see also comments by Denison, 1964); much the same may well be true of lower vertebrates in general. A step-wise building-up of large dermal elements from smaller ones may be referred to as the *assimilative phase* in the history of the dermal skeleton. Well known is also the phyletic process working in the opposite direction, the *regressive phase*, during which dermal elements are subject to reduction in a variety of ways. Previously, emphasis has frequently been put on one or the other of these phases, and the many modifications of the dermal skeleton accordingly interpreted mainly as expressions either of assimilation or of regression. There can be little doubt, however, that both phases have existed, and that the phyletic development of the dermal skeleton has taken place as an interaction between them.

Skeletal assimilation, and perhaps above all regression, are processes which with regard to the degree they reached and the way they acted were subject to considerable variation in different groups. Returning to some of these matters in the sequel, it may be of interest, at this point, to reconstruct a *complete cycle* of these processes, both carried to their extreme. Such a cycle possibly consisted of the following stages (adopting here a simple terminology which refers to the *relative size* of the dermal elements, not to their growth or other properties). After an initial *naked* stage, presumably having existed at some early time in vertebrates or vertebrate ancestors, the assimilative phase in all probability commenced with a *primary micromeric stage* in which the dermal skeleton consisted throughout of minute, evenly distributed scales. Then followed a *primary mesomeric stage* in which plates and scales of larger size emerged, and finally a concluding *macromeric stage* in which the earlier existing mesomeric elements on the head and on a larger or smaller portion of the trunk provided building material for a system of dermal bones or in some cases, as e.g. in the cephalaspids, a continuous dermal armour. In the regressive phase, following sooner or later upon the assimilative one, the dermal skeleton, by increasing disintegration of its component parts, may have passed once more through somewhat similar stages as before, only this time in the reverse order of their appearance. Thus, from the macromeric stage, it may first have entered into a *secondary mesomeric stage*, then into a *secondary micromeric stage* and may eventually be reduced to such an extent that vestiges of dermal elements were lost altogether, and a *secondary naked stage* was reached.

Examples of some of these hypothetical stages of assimilation and regression may be chosen among the ostracoderms. In Ordovician Heterostraci such as e.g. *Astraspis*, the dermal carapace throughout its extent is made up of a mosaic of fairly small, independent plates. As is generally agreed, this condition in so ancient forms, can hardly be taken as anything else than a primary mesomeric stage in skeletal assimilation, a phyletic precursor to the macromeric stage of so many of the geologically younger members of the group in which large plates or discs are formed. A similar, primary mesomeric stage probably occurs also in the dermal skeleton of the Downtonian *Tesseraspis* (see e.g. Ørvig, 1961: 517) as a forerunner of the macromeric condition of discs and other plates of later psammosteids. These circumstances have been analyzed on more occasions than one, and need no further discussion here. More controversial, on the other hand, is the evidence in the Heterostraci of the succeeding phase of skeletal development, that of regression. It is possible that some of these forms, by complete reduction of their dermal skeleton, gave rise to secondarily naked forms more or less like the myxinoids of the present day, but this for various reasons is uncertain; it is possible also that the myxinoids, for instance, are primarily naked forms which at no time during their past history possessed any dermal elements (see comments in Spjeldnæs, 1967). For illustrations of skeletal regression in ostracoderms, one may instead turn to the Osteostraci.

In the early representatives of the Osteostraci, the dermal skeleton (Fig. 2 A) is well known from earlier descriptions (Stensiö, 1927; 1932; Denison, 1951; etc.); it consists of (a) a single, large cephalic shield which, although subdivided into polygonal tesserae, is nevertheless a continuous, macromeric formation, and (b) a squamation of comparatively large, mesomeric scutes on the trunk. In one of the youngest cephalaspids, the Lower Upper Devonian *Alaspis macrotuberculata* Ø. (Fig. 2 B), the cephalic shield in contrast is modified in such a way that it consists throughout of small, independently growing, suturally interconnected, polygonal plates (Fig. 2 C), which correspond to the tesserae in the shield of earlier forms, although they are not in this case continuous with each other. Evidently, this condition in *Alaspis* corresponds to a *secondary mesomeric stage* of skeletal regression into which, late in the history of the group, the cephalic shield had entered. The squamation of *Alaspis* (Fig. 2 D), like that of contemporary forms, consists of a great many, minute scales, and has thus, as compared with the condition of the earlier cephalaspids, been reduced to the extent of entering a *secondary micromeric stage* of skeletal regression. Apart from demonstrating that skeletal regression did, in fact, take place in the cephalaspids, a point on which some doubt has been expressed (Denison, 1951: 172; Wängsjö, 1952: 52), the condition of *Alaspis* indicates also that this process acts differently on the dermal skeleton in different parts of the organisms, here (as in the contemporary *Escuminaspis*) having had a more pronounced effect on the squamation than on the cephalic shield itself. The cephalaspids are unknown before the Upper Silurian (or almost so, see Martinsson, 1966), but it seems unlikely that the very first members of the group appeared out of the

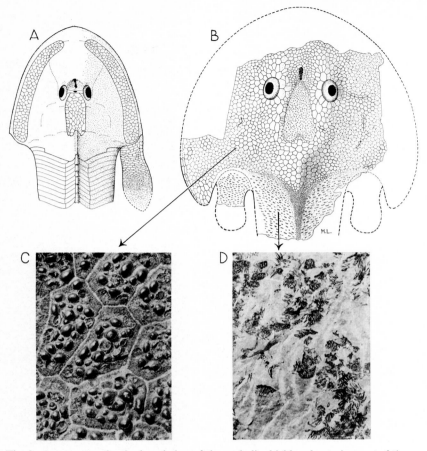

Fig. 2. *A*, reconstruction in dorsal view of the cephalic shield and anterior part of the squamation in an early (Downtonian) cephalaspid, *Hemicyclaspis murchisoni* (Egerton); from Stensiö, 1932. *B–D*, the dermal skeleton in one of the latest representatives of the group, *Alaspis macrotuberculata*; figures based on the holotype No. 8402 belonging to the Museum of Comparative Zoölogy, Harvard University; *B*, similar reconstruction as in *A*, from Ørvig, 1957a; *C*, polygonal plates of the cephalic shield; approx. ×4.5; *D*, trunk scales; approx. ×5.

blue with a fully formed macromeric cephalic shield an a mesomeric squamation. Although direct evidence is lacking in this respect, it seems more conceivable that the early history of this group embraced successive stages of skeletal assimilation. If so, these probably took place over lengthy periods of time and were similar, by and large, to those we have reason to believe existed during the history of other ostracoderms. Thus, possibly early in this process, there were cephalaspids, or cephalaspid ancestors, with a primary micromeric dermal skeleton of about the same kind as that secondarily reappearing by regression in the squamation of *Alaspis*, and possibly at a subsequent stage, there were forms with a primary mesomeric cephalic shield consisting of independent plates corresponding to those again met with after regression in the shield of *Alaspis*.

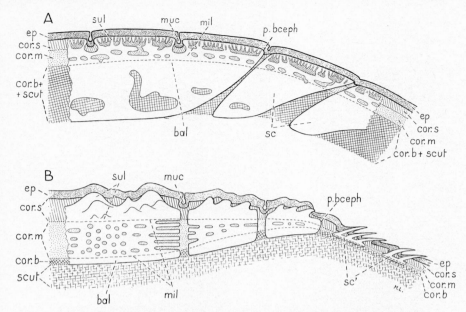

Fig. 3. Diagrammatical vertical section (with adjoining soft tissues added) of the most posterior part of the cephalic shield and the anterior part of the squamation in *A*, an early representative of the cephalaspids (e.g. *Hemicyclaspis*; cf. figures in Stensiö, 1932), and *B*, *Alaspis*.

bal, basal layer of dermal skeleton; *cor.b*, *cor.m*, *cor.s*, basal, middle and superficial parts of the corium, respectively; *ep*, epidermis; *mil*, middle layer of dermal skeleton; *muc*, mucuous canals (circumareal canals in *A*, lying entirely in soft tissue in *B*); *p.bceph*, posterior border of cephalic shield; *sc*, scales, and *sc'*, scales in a regressive state of development; *scut*, subcutaneous tissue; *sul*, superficial layer of dermal skeleton.

It is unknown what happened to the cephalaspids beyond Lower Upper Devonian times, but because *Alaspis* (and the contemporary *Escuminaspis*) seem definitively to have entered on the road of skeletal regression, further advance along these lines could eventually have lead to the loss of the dermal skeleton with secondarily naked cyclostomes of some sort as descendants. If so, the cephalaspids during their history would have passed through a complete cycle of skeletal assimilation and regression, which according to what is now known presumably is more than can be said of any other single group of lower vertebrates known from the fossil record. A direct passage by loss of the dermal skeleton (and, of course, endoskeletal hard tissue) from the late cephalaspids to more or less *Petromyzon*-like cyclostomes is not *necessarily* what actually took place, but the following circumstance is worth noting in this connection.

In *Alaspis*, the polygonal plates of the cephalic shield consist of the same 3 layers, superficial, middle, and basal, as those of the entire dermal skeleton, cephalic shield as well as squamation, in earlier forms (*sul, mil, bal*, Fig. 3), In *Alaspis*, however, passing from the posterior part of the shield into the spuamation, the middle, vascular layer thins out; in the small scales there is little, if anything left of it, even though it is so distinctly developed even in the scutes of the trunk of earlier forms (*mil*, Fig. 3 A, right side). This circumstance is suggestive of a phyletic reduction in the cephalaspids of the middle, vascular layer of the corium on the trunk behind the cephalic shield (*cor.m*, Fig. 2), a reduction possibly somehow correlated with the process of regression in the dermal skeleton itself. In *Petromyzon* (Johnels, 1950), the corium of the head is split up during metamorphosis into 3 layers, including a middle, vascular one, whereas behind the head it retains its embryonic structure of a single, undivided aponeurosis and accordingly is devoid of a middle, vascular layer. As far as concerns the structure of the corium, therefore, the condition of *Alaspis* could easily give rise to that of e.g. *Petromyzon*.

Apart from the various cases which with reasonable certainty seem to involve either skeletal assimilation or regression, there are also others whose significance in this respect is obscure and might represent either stages of smaller elements coalescing into larger ones or, reversely, stages of larger elements disintegrating into smaller. One may, for instance, refer to the condition of the dermal skeleton in forms such as psammosteids and radotinid arthrodires for which both interpretations have been suggested (for comments and references, see Westoll, 1967). Without here going into the details of these particular forms, it may be of interest to discuss the interpretation of the dermal skeleton of a group such as the thelodonts (including here *Turinia*). Theoretically, with their dermal skeleton consisting throughout of very small scales, the thelodonts could be situated either at the beginning or almost at the end of a full cycle of skeletal assimilation and regression. Thus, on the one hand, they could be primitive in this respect, as both Obruchev (1945: 268–269) and Buistrov (1955: 477, 480) seem to believe (see also Traquair, 1899; Goodrich, 1907: 747–755 1909: 195–196), surviving successfully in a primary micromeric stage of skeletal development long after allied forms within the ostracoderms passed into a macromeric stage by acquiring a heavy dermal armour. On the other hand, they could be also descendants of earlier armour-bearing ostracoderms, as maintained by Stensiö (1927: 332–334; 1932: 196; 1964: 176, 271, 371; see also Hoppe, 1931: 80; Gross, 1947: 99), and consequently be specialized to the extent of reaching a secondary micromeric stage of skeletal regression. In weighing these two alternatives one against the other, the reason why the former on the whole seems to be the more probable lies in the circumstance that, as already indicated with regard to the cephalaspids, the dermal skeleton as a general rule is not affected by regression in precisely the same way and to exactly the same extent in all parts of the organism. Thus, as shown in various lineages of lower vertebrates, the process of reduction and disintegration acted earlier and more thoroughly on the squamation than on the dermal skeleton of the head and shoulder girdle. If a form such as the late cephalaspid *Alaspis* had descendants, in these the remnants of the squamation by continued regression would likely have disappeared completely long before the polygonal plates of the cephalic shield. One may mention also a variety of examples of selective action of skeletal regression in fishes, for instance among actinopterygians in such forms as e.g. *Tarrasius* (Moy-Thomas, 1934) or *Birgeria* (Stensiö, 1921; Nielsen, 1949) in which the squamation is incompletely developed, or almost lost, whereas the dermal bones of the head and shoulder girdle still exist. Further examples from teleostomians show how dermal bones were reduced and subject to disintegration in certain regions much more than in others (cf. e.g. the condition of the chondrostean *Errolichthys:* Lehman, 1952, figs. 80, 82; Nielsen, 1955). The point of this line of argument is that, as all experience goes to prove, regression once begun seems to work differently on different parts of the dermal skeleton. If so, this process, even when far advanced, probably could not produce a dermal skeleton like that of the thelodonts, consisting, from the anterior part of the snout to the tip of the tail, of a nearly uniform system of minute scales and containing no vestiges, however small, of dermal plates on the head or anywhere else. From these considerations there is reason to believe that the condition of the dermal skeleton in such forms as the thelodonts is, in fact, primitive, representing a primary micromeric stage of skeletal development, and that dermal elements of a larger size never existed in their ancestral forms. When Stensiö (1964: 176; see also 1927: 334), in support of this view that the dermal skeleton of thelodonts is regressively developed, states with particular reference to *Turinia* that "la réalité de cette régression est corroborée entre autres par la continuité de l'endosquelette de l'armure qui avait essentiellement même étendue par example chez les *Pteraspida* et les *Cyathaspida*", this argument in the opinion of the present writer is not entirely convincing as long as we have no means of telling whether phyletically speaking a bulky endoskeleton of the head like that of the Heterostraci, is responsible for the formation of a strong encasing dermal armour or vice versa (if any correlation exists here at all); a large continuous head endoskeleton, made up of some sort of cartilage proper or cartilaginous connective tissue, for all we know possibly existed already in early stages of dermal skeleton assimilation, as for instance in thelodonts. The contention, it may be added finally that the thelodonts are larval ostracoderms whose scales represent only the superficial part of a strong dermal armour forming later in ontogeny (Westoll, 1945: 346–348, 353; cf. also Parrington, 1958: 124–127; etc), is admirable in theory but unsupported by the facts now known. The circumstance cited by Westoll, for instance, that in *Ateleaspis robusta* (Kiær) there are a great many small scales on the ventral side of the cephalic shield and trunk (Heintz, 1939), proves nothing in the present respect because it is not known whether these scales really are of the thelodont-kind (to be so, they have to fulfill qualifications other than just being small;

cf. e.g. Gross, 1967). Space does not allow further discussion of this and other matters relating to the thelodonts (on the subject of which much more could be said); it does seem, however, that throughout their span of existence, these forms were in the possession of a dermal skeleton of a primitive kind.

As far as concerns elasmobranchs, much the same conclusion as in the case of the thelodonts, and for similar reasons, probably applies to the dermal skeleton of many of the sharks (not discussing in this connection the dermal skeleton of the rays and holocephalians where other interpretations may be indicated). Although sharks possibly are related in some general way to arthrodires, their direct ancestors in all probability were not arthrodires equipped with powerful dermal bones and scales, because such dermal bones and scales by regression would hardly yield as an end-product a typical shark squamation of small scales, evenly distributed all over the body. In these circumstances, the existence in the head of selachian embryos of specific mesenchymatic fields might not have the significance maintained by Holmgren (1940), viz. that of indicating the site of dermal bones with an arthrodiran pattern in ancestral forms. To avoid misunderstanding here, it should perhaps be added that although, as recently pointed out by Stensiö (1961), the scales of post-Permian selachians with regard to ontogeny and structure are probably of a specialized kind relative to those of Devonian and at least certain Carboniferous forms (Ørvig, 1966; Zangerl, in this Volume), the sharks (or at any rate the great majority of shark lineages) seem never to have passed beyond a primary micromeric stage of skeletal development as far as concerns the relative size and distribution of the constituent parts of their squamation (the only indication of further skeletal assimilation being, in this case, the development of fin spines, whenever such occur). A third group likely to have the dermal skeleton in a primitive stage of development includes the acanthodians, although in certain of these forms there is a tendency toward development of dermal plates in certain parts of the head and shoulder girdle (Watson, 1937; Miles, 1966, figs. 15, 18; Ørvig, 1967*b*, fig. 2). In any case, it seems evident from the above that the degree of skeletal assimilation attained within the various groups of lower vertebrates is by no means dependent on or correlated with the level of organization of the animals concerned. This process, thus, reached a very advanced stage in most ostracoderms but a far inferior one in gnathostomes such as e.g. sharks and acanthodians.

Although processes of assimilation and regression are traceable to varying degrees and in various ways in the dermal skeleton of lower vertebrates, it is an entirely different matter to explain *why* these processes have taken place, and despite much discussion, their possible adaptive or physiological implications still remain obscure. Sometimes one gets the impression that once set in motion these processes tend to proceed on their own momentum further than might seem reasonable in light of any functional significance (whenever, that is, functional aspects enter into the picture). Les us return for a moment to the examples among ostracoderms considered earlier. It is tempting to interpret the transition from a carapace, made up of many mesomeric plates like that of the Ordovician *Astraspis*, to large plates and discs, like those of the later Heterostraci, as a phyletic trend towards the formation of a dermal skeleton with an enhanced ability to withstand mechanical stress, or with a greater protective effectiveness. This is doubtful, however, for the reason that the carapace of *Astraspis* is in itself a very firm and resistent structure because the mesomeric plates of which it is composed are firmly tied to each other by fibre systems running from the marginal faces of the individual plates to the corresponding faces of the adjoining ones. There can in this case have been as little moveability of the plates in relation to each other as there is in the cuirass of the ostraciontids among the recent teleosts. And once a carapace like that of *Astraspis*—or if one prefers that of *Tesseraspis*—had come into existence, there would mechanically speaking have been little reason for a further assimilative process leading to a macromeric dermal armour like that of pteraspids, cyathaspids etc. The cephalic shield of the late cephalaspid *Alaspis* may be viewed in a similar manner. In this case, one may think that the process of regression acting on the dermal skeleton served the purpose of increased moveability. This possibly is true of the trunk which, with its squamation of very small scales, surely could move more easily both in the horizontal and vertical planes than that of the earlier forms with their squamation of thick, inbricating scutes. But it is hardly true of the cephalic shield in which the subdivision into small plates means little from a mechanical point of view, these plates like those of *Astraspis* being firmly interconnected by fibres and in addition also firmly attached to the substratum by very strong fibre systems. As far as one can tell, *Alaspis* had much the same mode of life and existed in surroundings exerting similar physiological demands, as the earlier members of the group, and regression of the dermal skeleton can in this case hardly be seen in any perspective relevant to those conditions; that in the late cephalispads the process was in action, seems to be all one can say. By this is not suggested that regression in the dermal skeleton of other forms, for instance various teleostomians, cannot be correlated with particular external conditions but only that in the cephalaspids at any rate, as presumably also in the anaspids, it is difficult to see any such correlation and hence to explain the phyletic trend with which one is here concerned.

Odontodes as constituent parts of dermal elements

Structures of the dermal skeleton in many lower vertebrates below tetrapod level, previously called "dermal teeth" and referred to by the writer as *odontodes*

(Ørvig, 1967a: 47; more fully discussed and defined in Ørvig, 1968), show a great variety of shapes and sizes. They agree with jaw-teeth (in animals where such occur) in that they consist of dentine or dentinous tissue and in most cases of superficial enameloid substance, and that they arise ontogenetically, each of them, in a single, undivided dental papilla adjoined peripherally by an epithelial dental organ of epidermis cells; on the other hand, they are also distinguishable from teeth in the proper sense by characters such as their relative order of size, their positions in the dermal skeleton, and their site of formation in the superficial part of the corium instead of in an ingrowing dental lamina (Ørvig, 1967a, fig. 14).

The odontodes of special interest here are those developed as tubercles or ridges ornamenting the external face of dermal elements in a great number of fossil lower vertebrates and a few recent ones (*Latimeria:* Roux, 1942; Millot & Anthony, 1958; dermal bones of *Polypterus:* Ørvig, 1967a, fig. 12). When analysed with regard to their distribution pattern in relation to the growth of the dermal elements to which they belong, odontodes of this kind might yield results of much phyletic interest; so far, however, these matters have been taken into consideration in some of the ostracoderms (Obruchev, 1945; Stensiö, 1964; Denison, 1964; Tarlo, 1967; etc.) but have been largely ignored in gnathostomes. The conclusions reached by the writer in this respect will be fully presented and illustrated elsewhere (Ørvig, 1968); in the present connection a few points may be briefly mentioned.

In lower vertebrates, as is well known, stages of growth of dermal elements in a horizontal direction (areal growth) is frequently indicated by the distribution of odontodes in consecutive, concentrically arranged, *areal zones of growth* ("zones aréales proprement dites", Stensiö, 1964: 178–179). Increase in thickness of dermal elements, furthermore, in many cases involves *superpositional* growth, i.e. the apposition in a superficial direction of successive *generations* of odontodes upon each other. One may also encounter a combined *areosuperpositional* growth (as in, for instance, the scales of some holoptychiid crossopterygians: Ørvig, 1957a: 402, and in those of *Latimeria*: Ørvig, 1968) in which the odontodes of superpositional growth within each successive generation in their distribution tend to conform to the pattern of areal growth, i.e. spreading out from, and lying concentrically arranged around, the central (ontogenetically oldest) portion of the dermal elements.

With regard to the distribution of odontodes in the dermal bones of various Mesozoic ganoid fishes (Ørvig, 1968), two circumstances are worth noting: (a) the tendency for odontodes of consecutive generations to develop above each other, thus forming vertical or oblique columns (t_1–t_7, Fig. 4; Ørvig, 1967a, fig. 13; also displayed in dermal elements of various other lower vertebrates as illustrated in Gross 1930; 1961; Buistrov, 1939; Ørvig, 1951; 1957a; etc.), and

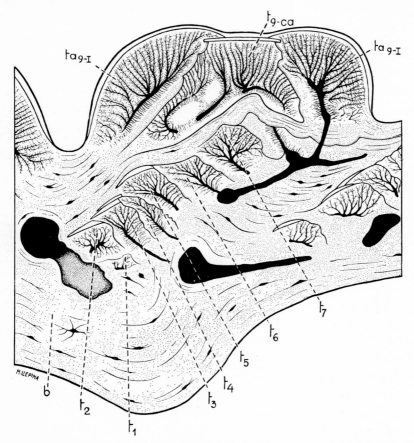

Fig. 4. Vertical section of a palatal dermal bone in a Triassic palaeoniscid, *Boreosomus piveteaui* Nielsen; Section No. S 1663; ×100; from Ørvig, 1968.

b, bone tissue; t_1–t_7, column of odontodes of the first to seventh generation (the odontode of the eight generation topping this column has here disappeared as a result of resorption); t_{9-ca}, odontode of the ninth generation acting as a centre of local areo-superpositional growth, and ta_{9-I}, another odontode of the same generation representing the first areal zone of this growth.

(b) the condition of *local* areo-superpositional growth (an occasional occurrence in certain forms, a more usual one in others) meaning that the dermal bones not only exhibit *one* centre of growth (corresponding to the centre of radiation), but also a number of other, separate, strictly localized growth centres represented by single odontodes (in any stage of superpositional growth and in any position) around which, or in specific topographical relation to which, other odontodes of the *same* generation, and on the same horizontal level, form one or a few areal zones of growth (t_{9-ca}, ta_{9-I}, Fig. 4). All this seems to indicate that although in ganoids, odontodes clearly are integral parts of dermal bones, they nevertheless in some cases tend to lead a sort of separate existence of their own and to reflect in some ways the behaviour of single micromeric dermal elements like those that once (in the remote past of chordate history) were the sole constituents of the dermal skeleton. Whenever displayed, the columnar arrangement of odontodes belonging to consecutive generations (Ørvig, 1967a, fig. 13) implies that the distribution of these units of one generation is not random but somehow governed by that of the corresponding units of the preceeding one or, to express matters differently, that each individual odontode of one generation possess the ability to induce the formation of another of the next generation superficial to it: the columns of odontodes now under consideration are reminiscent of successive generations of small, micromeric dermal elements with similar topographical relations to each other. Odontode complexes formed by local areo-superpositional growth are also significant; these all in probability represent micromeric dermal elements which have retained that much of their original individuality that they show areal growth of their own, entirely independent of the growth of the dermal bones as a whole.

In view of the evidence of ganoid fishes now briefly referred to, a dermal bone can hardly be taken as that sort of fully integrated structure it is generally assumed to be. It is thus a *complex* formation, consisting of (a) odontodes in its upper part which may conform to the pattern of areal and superpositional growth and, in cases, also tend to behave as separate, minor dermal elements, and (b) the whole mass of bone tissue underneath. This composition is surely not coincidental, but likely reflects certain aspects of the history of the dermal skeleton. One may thus envisage an early stage in this history where only micromeric elements corresponding in a broad sense to the superficial odontodes of dermal bones (more precisely to such odontodes provided with thin bony basal plates) were developed, and subsequent stages where another component was added which originally had no real relation to these micromeric elements, namely the thick layer of bone tissue underneath them. This layer may largely be interpreted as a secondary aquisition which arose by ossification of the whole middle and basal parts of the corium long after the first superficial micromeric elements emerged, and which in binding together groups of these micromeric elements by providing a thick, common base for them, gave rise to dermal elements of the mesomeric, and eventually macromeric, kind. This conclusion is in agreement with what e.g. Goodrich assumed (1907: 755) with special reference to ostracoderms; the new evidence touched upon here (see Ørvig, 1968) gives further support to, and since teleostomians also are involved, a wider perspective of, this contention (see also comments in Westoll, 1967).

In addition to what has now been said concerning ganoid fishes, brief reference may finally be made to another example of local areo-superpositional growth with possible phyletic implications, viz. the formation of blisters—isles of second (sometimes third) generation odontodes—on the external face of dermal elements in certain psammosteids (Fig. 5 A–C; see also Agassiz, 1844,

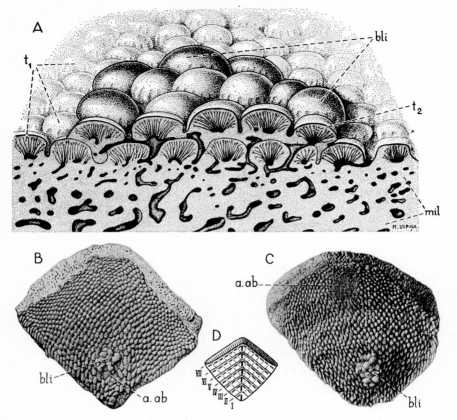

Fig. 5. *A–C*, *Psammolepis paradoxa* Ag., a representative of the Heterostraci from the Baltic Devonian. *A*, three-dimensional diagram of a blister of second generation odontodes on a dermal bone; *B*, a scale on the posterior part of which such a blister has formed on an abraded area, and *C*, another scale where the blister occupies a similar position as in *B*, and the abraded area lies anteriorly; all figures from Ørvig, 1968. *D*, *Drepanaspis gemuendenensis* Schlüter, another representative of the same group of the Heterostraci, diagram of a scale showing position of centre of growth; from Gross, 1963.

a.ab, abraded area; *bli*, blister; *mil*, middle layer of dermal skeleton; t_1, t_2, odontodes of first and second generation; *I–VII*, areal zones of growth.

pl. 27, fig. 3; Gross, 1935, pl. 3, fig. 3; Heintz, 1957, fig. 3, pl. 9, figs. 2–6; Ørvig, 1961, fig. 9; Obruchev & Mark-Kurik, 1965, pl. 54, fig. 1 b; and others). It has been maintained (Gross, 1935: 16; see also comments by Buistrov, 1955: 505; Heintz, 1957: 160; Tarlo & Tarlo, 1961) that these blisters arose as a means of compensating injuries inflicted upon the dermal skeleton, forming in places where the first generation odontodes had been damaged or lost by abrasion. Various evidence show, however, that although the blisters frequently fulfilled such a regenerative function, they did not *primarily* arise for that specific purpose (see further Ørvig, 1968). They existed already in the Lower Devonian *Psephaspis* (Ørvig, 1961, fig. 9) where the dermal skeleton in its entirety seems to have been unaffected by abrasion, and cannot there be explained as anything else than an expression of superpositional odontode growth, that growth which, as one has every reason to believe, originally was the common property of dermal elements of lower vertebrates in general (as indicated also by the delamination theory, see Jarvik, 1959). In later forms as for instance *Psammolepis*, they frequently coincide with areas of abrasion but by no means always; when these matters are analysed in detail one finds that the blisters tend to occupy that position on scales, or scales incorporated as scale-areas in large plates, which corresponds to the *centre of growth* of these elements, *regardless* of whether that part of the surface on which they formed had previously been subject to abrasion or not (*bli, a.ab*, Fig. 5 B, C). In view of this, blister formation in e.g. *Psammolepis* may signify a trend towards the development of new generations of mesomeric dermal elements superficially to those existing either as independent scales or as scale-areas in large plates; it is even conceivable that blisters forming on dermal plates where no subdivision into scale-areas is perceivable (as e.g. in *Psephaspis*) have a similar phyletic significance, indicating the position of mesomeric components of those plates which have completely lost their original individuality except for the ability they still possess of inducing the formation of isles of new odontodes superficially to them.

The evidence of ganoid fishes and psammosteids now touched upon demonstrate the complexity of macromeric dermal elements. Such elements do not always behave as single units with regard to their growth, but may contain vestiges of micromeric or mesomeric constituent parts which have retained in some measure an existence of their own, and a behaviour of their own. In this way a dermal bone may reflect the phyletic history of the dermal skeleton from a micromeric, over a mesomeric, to a final macromeric stage.

References

Agassiz, J. L. R. (1844). *Monographie des poissons fossiles du Vieux Grès Rouge ou système Dévonien (Old Red Sandstone) des îles Britanniques et de Russie.* Neuchâtel.

Berrill, N. J. (1955). *The origin of vertebrates.* Oxford: Oxford University Press.

Buistrov, A. P. (1939). Zahnstruktur der Crossopterygier. *Acta zool. Stockh.*, **20**: 283–338.

Buistrov, A. P. (1955). Mikrostruktura pantsiuria bescheliustnuikh pozvonochnuikh silura i devona (The microstructure of the shield of Agnatha, the jawless vertebrates from the Silurian and Devonian). In *L. S. Berg Mem. Vol.*, ed. Pavlovskii, E. N.,: 427–523. Moscow: Acad. Sci. U.S.S.R. In Rusisan.

Caster, K. E. & Eaton, T. H. Jr. (1956). Microstructure of the plates in the carpoid echinoderm *Paranacystis*. *J. Paleont.*, **30**: 611–614.

Denison, R. H. (1951). The exoskeleton of early Osteostraci. *Fieldiana:Geol.*, **11**: 199–218.

Denison, R. H. (1956). A review of the habitat of the earliest vertebrates. *Fieldiana: Geol.*, **11**: 359–457.

Denison, R. H. (1963). The early history of the vertebrate calcified skeleton. *Clin. Orthop.*, **31**: 141–152.

Denison, R. H. (1964). The Cyathaspididae. A family of Silurian and Devonian jawless vertebrates. *Fieldiana:Geol.*, **13**: 307–473.

Goodrich, E. S. (1907). On the scales of fishes, living and extinct, and their importance in classification. *Proc. zool. Soc. Lond.*, **1907**: 751–774.

Goodrich, E. S. (1909). Cyclostomes and fishes. In *A Treatise on Zoology*, ed. Lankester, E. R., vol. **9**: Vertebrata Craniata, fasc. 1. London: A. & C. Black.

Graham-Smith, W. (1935). *Scaumenella mesacanthi*, gen. et sp.n., a peculiar organism from the Upper Devonian of Scaumenac Bay, P. Q., Canada. *Annls Mag. nat. Hist.*, (10) **16**: 473–476.

Gregory, W. K. (1946). The roles of motile larvae and fixed adults in the origin of vertebrates. *Q. Rev. Biol.*, **21**: 348–364.

Gross, W. (1930). Die Fische des mittleren Old Red Süd-Livlands. *Geol. paläont. Abh. Berl.*, (N.F.) **18**: 1–34.

Gross, W. (1935). Histologische Studien am Aussenskelett fossiler Agnathen und Fische. *Palaeontographica*, (A) **83**: 1–60.

Gross, W. (1947). Die Agnathen und Acanthodier des obersilurischen Beyrichienkalks. *Palaeontographica*, (A) **96**: 91–158.

Gross, W. (1954). Zur Conodonten-Frage. *Senckenb.leth.*, **35**: 73–85.

Gross, W. (1961). Aufbau des Panzers obersilurischer Heterostraci und Osteostraci Norddeutschlands (Geschiebe) und Oesels. *Acta zool. Stockh.*, **42**: 73–150.

Gross, W. (1963). *Drepanaspis gemuendenensis* Schlüter. Neuuntersuchung. *Palaeontographica* (A) **121**: 133–155.

Gross, W. (1967). Über Thelodontier-Schuppen. *Palaeontographica*, (A) **127**: 1–67.

Heintz, A. (1939). Cephalaspida from Downtonian of Norway. *Skr. norske Vidensk-Akad. Oslo, Mat.-naturv. Kl.*, **1939**: 1–119.

Heintz, A. (1957). The dorsal shield of *Psammolepis paradoxa* Agassiz. *J. palaeont. Soc. India*, **2**: 153–162.

Holmgren, N. (1940). Studies on the head in fishes. Embryological, morphological and phylogenetical researches. 1. Development of the skull in sharks and rays. *Acta zool. Stockh.*, **21**: 51–267.

Hoppe, K.-H. (1931). Die Coelolepiden und Acanthodier des Obersilurs der Insel Ösel. Ihre Paläobiologie und Paläontologie. *Palaeontographica*, **76**: 35–94.

Jarvik, E. (1959). Dermal fin-rays and Holmgren's principle of delamination. *K. svenska VetenskAkad. Handl.*, (4) **6**: 1–51.

Jarvik, E. (1964). Specialization in early vertebrates. *Annls Soc. R. zool. Belg.*, **94**: 11–95.

Jefferies, R. P. S. (1967). Some fossil chordates with echinoderm affinities. In *Echinoderm Biology*, ed. Millott, N., *Symp. zool. Soc. Lond.*, **20**: 163–208.

Johnels, A. G. (1950). On the dermal connective tissue of the head of *Petromyzon*. *Acta zool. Stockh.*, **31**: 177–185.

Lehman, J.-P. (1952). Étude complémentaire des poissons de l'Eotrias de Madagascar. *K. svenska VetenskAkad. Handl.*, (4) **2**: 1–201.

Lehman, J.-P. (1957). Un problème non resolu: l'origine des vertébrés. *Nature, Paris*, **1957**: 174–177.

Lehman, J.-P. (1964). L'origine des vertébrés; le milieu des premiers vertébrés. In *Traité de Paléontologie*, ed. Piveteau, J., **4:1**: 78–91. Paris: Masson.

Lindström, M. (1964). *Conodonts*. Amsterdam: Elsevier.

Martinsson, A. (1966). Beyrichiacean ostracodes associated with the earliest Silurian vertebrates from Gotland. *Geol. Fören. Stockh. Förh.*, **88**: 327–339.

Miles, R. S. (1966). The acanthodian fishes of the Devonian Plattenkalk of the Paffrath trough in the Rhineland, with an appendix containing a classification of the Acanthodii and a revision of the genus *Homalacanthus*. *Ark. Zool.* (2) **18**: 147–194.

Millot, J. & Anthony, J. (1958). *Anatomie de Latimeria chalumnae*. 1. *Squelette, muscles et formation de soutien*. Paris: Cent. natn. Rech. Scient.

Moss, M. L. (1961). The initial phylogenetic appearance of bone: an experimental hypothesis. *Trans. N.Y. Acad. Sci.*, (2) **23**: 495–500.

Moss, M. L. (1964). The phylogeny of mineralized tissues. *Int. Rev. gen. exp. Zool.*, **1**: 297–331.

Moy-Thomas, J. A. (1934). The structure and affinities of *Tarrasius problematicus* Traquair. *Proc. zool. Soc. Lond.*, **1934:2**: 367–375.

Nielsen, E. (1949). Studies on Triassic fishes from East Greenland. 2. *Australosomus* and *Birgeria*. *Palaeozool. Groenland.*, **3**: 1–309.

Nielsen, E. (1955). Notes on Triassic fishes from Madagascar. 1. *Errolichthys mirabilis* Lehman. *Medd. dansk geol. Foren.*, **12**: 563–578.

Obruchev, D. V. (1945). Evoliutsii agnatha (On the evolution of Agnatha). *Zool. Zh.*, **24**: 257–272. In Russian.

Obruchev, D. V. & Mark-Kurik, Elga (1965). *Devonian psammosteids (Agnatha, Psammosteidae) of the U.S.S.R.* Tallinn: Geol. Inst. Acad. Sci. Est. SSR. In Russian with English summary.

Ørvig, T. (1951). Histologic studies of placoderms and fossil elasmobranchs. 1. The endoskeleton, with remarks on the hard tissues of lower vertebrates in general. *Ark. Zool.*, (2) **2**: 321–454.

Ørvig, T. (1957a). Remarks on the vertebrate fauna of the Lower Upper Devonian of Escuminac Bay, P. Q., Canada, with special reference to the porolepiform crossopterygians. *Ark. Zool.*, (2) **10**: 367–426.

Ørvig, T. (1957b). Paleohistologic notes. 1. On the structure of the bone tissue in the scales of certain Palaeonisciformes. *Ark. Zool.*, (2) **10**: 481–490.

Ørvig, T. (1958). Tänderna och tandvävnaderna genom tiderna (The teeth and their hard tissues through the ages). *Zool. Rev. Stockh.*, **1958**: 30–39, 46–63. In Swedish with English summary.

Ørvig, T. (1961). Notes on some early representatives of the Drepanaspida (Pteraspidomorphi, Heterostraci). *Ark. Zool.* (2) **12**: 515–535.

Ørvig, T. (1965). Palaeohistological notes. 2. Certain comments on the phyletic significance of acellular bone tissue in early vertebrates. *Ark. Zool.*, (2) **16**: 551–556.

Ørvig, T. (1966). Histologic studies of ostracoderms, placoderms and fossil elasmobranchs. 2. On the dermal skeleton of two late Palaeozoic elasmobranchs. *Ark. Zool.*, (2) **19**: 1–39.

Ørvig, T. (1967a). Phylogeny of tooth tissues: evolution of some calcified tissues in early vertebrates. In *Structural and Chemical Organization of Teeth*, ed. Miles, A. E. W., vol. **1**: 45–110. New York & London: Academic Press.

Ørvig, T. (1967b). Some new acanthodian material from the Lower Devonian of Europe. In *Fossil Vertebrates*, ed. Patterson, C. & Greenwood, P. H., *J. Linn. Soc. (Zool.)*, **47**: 131–153.

Ørvig, T. (1968). Palaeohistological notes. 3. On the microstructure and growth of the dermal skeleton, particularly the dermal bones, in some ganoid fishes from the early Mesozoic of East Greenland, Sweden and Central Europe, with comparative remarks on processes of growth in dermal elements in certain other lower vertebrates. Manuscr.

Parrington, F. R. (1958). On the nature of Anaspida. In *Studies on Fossil Vertebrates*, ed. Westoll, T. S.,: 108–128. London: Athlone Press.

Patten, W. (1912). *The evolution of the vertebrates and their kin*. Philadelphia: Blakiston.

Pautard, F. (1961). Calcium, phosphorus and the origin of backbones. *New Scientist*, **12**: 364–366.
Rhodes, F. H. T. & Wingard, P. S. (1957). Chemical composition, microstructure, and affinities of the Neurodontiformes. *J. Paleont.*, **31**: 448–454.
Robertson, J. D. (1959). The origin of vertebrates—marine or freshwater? *Adv. Sci.*, **61**: 516–520.
Romer, A. S. (1933). Eurypterid influence on vertebrate history. *Science*, **78**: 114–117.
Romer, A. S. (1946). The early history of fishes. *Q. Rev. Biol.*, **21**: 33–69.
Roux, G. H. (1942). The microscopic anatomy of the *Latimeria* scale. *S. Afr. J. Med. Sci. (Biol. Suppl.)*, **7**: 1–18.
Schmidt, W. J. & Keil, A. (1958). *Die gesunden und die erkrankten Zahngewebe des Menschen und der Wirbeltiere im Polarisationsmikroskop. Theorie, Methodik, Ergebnisse der optischen Strukturanalyse der Zahnhartsubstanzen samt ihrer Umgebung*. München: Hanser Verl.
Scourfield, D. J. (1937). An anomalous fossil organism, possibly a new type of chordate, from the Upper Silurian of Lesmahagow, Lanarkshire,—*Ainiktozoon loganense*, gen. et sp. nov. *Proc. R. Soc. Lond.*, (B) **121**: 533–547.
Sillman, L. S. (1960). The origin of vertebrates. *J. Paleont.*, **34**: 540–544.
Smith, H. W. (1939). *Studies in the physiology of the kidney*. Lawrence: Univ. Kansas.
Spencer, W. K. (1938). Some aspects of evolution in Echinodermata. In *Evolution, Essays on Aspects of Evolutionary Biology*, ed. de Beer, G. R.,: 287–303. Oxford: Oxford University Press.
Spjeldnæs, N. (1967). The palaeoecology of the Ordovician vertbrates of the Harding Formation (Colorado, U.S.A.). *Colloques int. Cent. natn. Rech. Scient.*, **168**: 11–20.
Stensiö, E. A. (1921). *Triassic fishes from Spitsbergen*. 1. Vienna.
Stensiö, E. A. (1927). The Downtonian and Devonian vertebrates of Spitsbergen 1. Family Cephalaspidae. *Skr. Svalbard Nordishavet*, **12**: i-xii, 1–391.
Stensiö, E. A. (1932). *The cephalaspids of Great Britain*. London: Br. Mus. (nat. Hist.).
Stensiö, E. A. (1961). Permian vertebrates. In *Geology of the Arctic*, ed. Raasch, G. O., vol. **1**: 231–247. Toronto: University of Toronto Press.
Stensiö, E. A. (1964). Les cyclostomes fossiles ou ostracodermes. In *Traité de Paléontologie*, ed. Piveteau, J., **4:1**: 96–382. Paris: Masson.
Tarlo, L. B. H. (1962). Lignées évolutive chez les ostracodermes hétérostracées. *Colloques int. Cent. natn. Rech. Scient.*, **104**: 31–37.
Tarlo, L. B. H. (1963). Aspidin, the precursor of bone. *Nature, Lond.*, **199**: 46–48.
Tarlo, L. B. H. (1964). The origin of bone. In *Bone and Tooth, Proc. First Europ. Symp.*, ed. Blackwood, H. J. J.,: 3–15. Oxford: Pergamon Press.
Tarlo, L. B. H. (1967). The tessellated pattern of the dermal armour in the Heterostraci. In *Fossil Vertebrates*, ed. Patterson, C. & Greenwood, P. H., *J. Linn. Soc. (Zool.)*, **47**: 45–54.
Tarlo, L. B. H. & Tarlo, Beryl J. (1961). (Exhibition of) histological sections of the dermal armour of psammosteid ostracoderms. *Proc. geol. Soc. Lond.*, **1593**: 3–4.
Traquair, R. H. (1899). Report on fossil fishes collected by the Geological Survey of Scotland in the Silurian rocks in the South of Scotland. *Trans. R. Soc. Edinb.*, **39**: 827–864.
Urist, M. R. (1964). Further observations bearing on the bone-body fluid continuum: composition of the skeleton and serums of cyclostomes, elasmobranchs, and bony vertebrates. In *Bone Biodynamics, Henry Ford Hosp. int. Symp.*, ed. Frost, H. M.,: 151–179. Boston: Little, Brown.
Wängsjö, G. (1952). The Downtonian and Devonian vertebrates of Spitsbergen. 9. Morphologic and systematic studies of the Spitsbergen cephalaspids. Results of

Th. Vogt's Expedition 1928 and the English–Norwegian–Swedish Expedition 1939. *Skr. norsk Polarinst.*, **97**: 1–615.

Watson, D. M. S. (1937). The acanthodian fishes. *Phil. Trans. R. Soc.*, (B) **228**: 49–146.

Westoll, T. S. (1942). The earliest *Panzergruppen*. *Aberdeen Univ. Rev.*, **29**: 114–122.

Westoll, T. S. (1954). A new cephalaspid fish from the Downtonian of Scotland, with notes on the structure and classification of ostracoderms. *Trans. R. Soc. Edinb.*, **61**: 341–357.

Westoll, T. S. (1967). *Radotina* and other tesserate fishes. In *Fossil Vertebrates*, ed. Patterson, C. & Greenwood, P. H., *J. Linn. Soc. (Zool.)*, **47**: 83–98.

White, E. I. (1946). *Jamoytius kerwoodi*, a new chordate from the Silurian of Lanarkshire. *Geol. Mag.*, **83**: 89–97.

White, E. I. (1958). Original environment of the craniates. In *Studies on Fossil Vertebrates*, ed. Westoll, T. S.,: 212–234. London: Athlone Press.

The morphology and the developmental history of the scales of the Paleozoic sharks *Holmesella*? sp. and *Orodus*

By Rainer Zangerl

Dept. of Geology, Field Museum of Natural History, Chicago, Illinois. USA

The Pennsylvanian sharks *Holmsella?* sp. and *Orodus* are still very inadequately known. *Holmesella* is based on dermal denticles; still meager, but associated remains presumably belonging to this genus were recently described by Ørvig (1966). An additional specimen from the Field Museum's Logan Quarry (Zangerl & Richardson, 1963) in West-central Indiana is also far from adequate, showing many thousands of scales and a short piece of vertebral column consisting of simple, calcified, cartilaginous neural (probably) or haemal (less probably) elements, as is the case in other sharks of Paleozoic age.

The genus *Orodus* was proposed by Agassiz (1843) for isolated teeth of fairly characteristic shape. Many teeth of this type have since been collected both in Europe and North America, but no skeletons have come to light until 1958 when a large specimen of extraordinary preservation was excavated in the Logan quarry of Indiana. In this specimen the shagreen of the skin is preserved intact over much of the body, so that large dorsal, smaller lateral and minute ventral scales may be studied, as well as developmental stages, and irregular sclerifications of the skin that are tentatively interpretated as having formed in areas where the skin had been injured.

The squamation of *Orodus* and *Holmesella?* sp. is similar in a number of respects. In both forms variable numbers of denticles combine to form complex scales on the dorsal side of the hide, more simple ones on the ventral side.

In referring to specimens in the sequel, Field Museum of Natural History has been abbreviated FMNH.

Orodus

Ventral scales. The ventral scales are tiny, measuring from 60 to 800 microns in antero-posterior direction (Fig. 1 a–f). They consist of a number of finger-shaped, hollow denticles that are fused at the bases in much the same way as the compound denticles of *Agassizodus?* sp., figured by Zangerl (1966, fig. 26). The number of finger-shaped denticles and their lengths vary considerably within a single scale and there are probably no two scales that are exactly

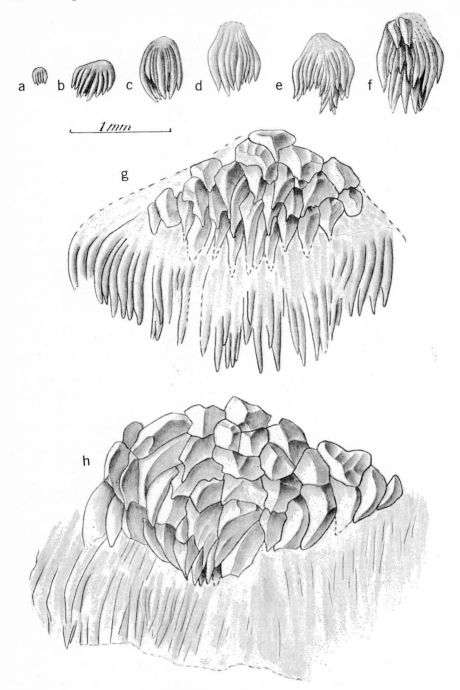

Fig. 1. Compund scales of the skin (behind the pectoral fins) of a large specimen of *Orodus* (FMNH PF 2201). *a–f*, scales from the ventral side of the body; *g*, from the flanks; *h*, dorsal scale. The upper margin of illustration points anterior.

Fig. 2. Ventral scale of *Orodus* (FMNH PF 2201, section no. 4720) in vertical section. On the right side, finger-shaped denticles point posteriorly. Their bases are fused and form a thorn-like projection that extended into dermis in life. The shape of this scale probably resembled that shown in Fig. 1 f.

alike (Fig. 1). In cross section (Fig. 2), the fused bases of the finger-shaped denticles form a thorn-like projection that extended into the dermis at a right angle to the surface of the skin and an anterior, bony projection parallel to the surface of the skin.

Lateral scales. These scales differ from the ventral scales in that a structural element appears that is absent in the ventral scales: at the anterior scale margin ᴧlarger, massive and differently-shaped denticles develop (Fig. 1 g). These denticles have small pulp cavities and thick walls in contrast to the finger-shaped denticles whose very thin walls surround relatively large pulp cavities. There is also a difference in the basal areas of these scales between the basal portion belonging to the finger-shaped denticles and that belonging to the massive denticles. The former consists of "bone" (see discussion) containing relatively few, large "bone" cells with irregular, long canaliculi, but no connective tissue fibers; the latter consists of an apparently acellular tissue (perhaps bone) that contains vast numbers of connective tissue fibers, as described by Ørvig,(1966) for *Holmesella?* sp. This substance occupies a morphologically comparable position to the substance called bone by Stensiö (1962, pl. 1, fig. X) in complex scales of a Permian edestid from Greenland, but in this case the substance was said to contain bone cell lacunae. The two described basal components, the one containing bone cell spaces and lacking fibers, and the acellular, fibrous one are continuous but are easily distinguished on sections by their very different appearance and they have, moreover, a different ontogenetic history as will be shown below.

The scales of this area vary in size from about 1000 to 2500 microns in antero-posterior direction. The larger the scales, the more fingershaped denticles are present both in internal-external direction and on the sides of the scales (dorsal and ventral as oriented in the skin of the animal) and massive denticles on the anterior scale margin. The shape of the massive denticles which tend to form sharp ridges in antero-posterior direction, is so highly variable that no two scales appear to be alike (Fig. 1).

Dorsal scales. The dorsal scales are the largest and most complex ones, and the major portion of each consists of massive denticles and a large, cushion-shaped, concentrically growing base. The finger-shaped denticles along the posterior scale margin are still present and prominent, but because they are very delicate, they are usually broken off, or at best, preserved in broken condition (Fig. 1 h). There are no important histological differences between the larger lateral scales and the dorsal ones which may reach 3000 microns in length.

Antero-posterior cross sections of lateral and dorsal scales show the posterior finger-shaped denticles with their common, bony base that consists of cellular "bone", lacking connective tissue fibers, and the massive anterior denticles with their fibrous concentrically growing cushion base that appears to lack bone cell lacunae (Fig. 3). Early developmental stages of the massive denticles were never observed (though the number of sectioned scales is large); a late developmental stage is represented by the most anterior denticle of the scale illustrated in Fig. 3, enlarged in Fig. 4, where the pulp cavity is still notable and where the branching dentinal tubules show relatively large lumina. In the fully formed massive denticles the dentinal tubules are extremely thin and barely visible. The connective tissue fiber arrangement in the large cushion-shaped base appears to be exactly as described by Ørvig (1966).

Immature scales. Thin sections across patches of fully differentiated dorsal and lateral scales revealed the presence of immature scales in various stages of growth apparently in places where old scales had dropped off. The smallest ones consist of a few finger-shaped denticles and their fused bases that together form a long process which must have extended into the dermis approximately at a right angle as in the ventral scales. As far as I can determine these developing scales are identical with the smallest ventral scales. In slightly more advanced stages, additional finger-shaped denticles seem to have been added to the posterior margin of the scale and beneath the already present denticles hence lengthening the nail-shaped dermal prong. Furthermore, one or two massive denticles have developed at the anterior scale margin (Fig. 5); these already have thick dentine walls in all cases observed. Beneath them the acellular tissue

---

Fig. 3 (above). Dorsal scale of *Orodus* (FMNH PF 2201, section no. 5103) in antero-posterior, vertical section. Finger-shaped posterior denticles broken off; their fused bases are sharply set off from the large, cushion-shaped scale base by the lack of fibers. In front of finger-shaped denticles a series of nine massive denticles the anterior-most of which is relatively young (see Fig. 4). The pulp cavities in the massive denticles are reduced to small, basal canals.

Fig. 4 (middle). Higher magnification of the anterior-most massive denticle of the scale illustrated in Fig. 3. Note the dentinal tubules with their relatively large lumina, and the branching pattern.

Fig. 5 (below). A young growth stage of a dorsal scale of *Orodus* (FMNH PF 2201, section no. 5103) amidst fully developed scales. Only a single massive denticle is present.

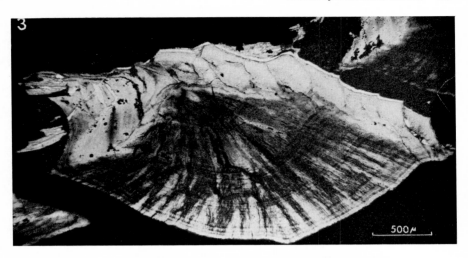

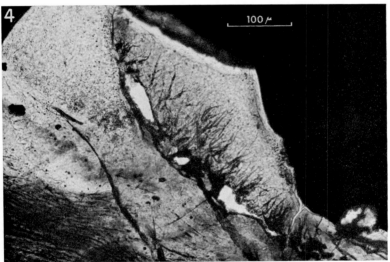

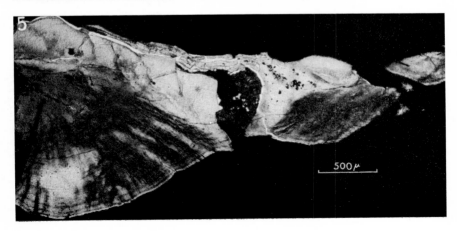

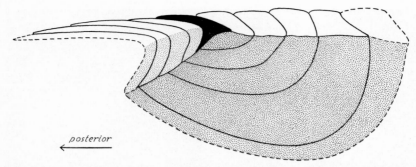

Fig. 6. Diagram to illustrate the mode of growth of the scales in *Orodus* (and *Holmesella*?). Black, the first appearing scale element, a finger-shaped denticle of lepidormorial simplicity. Additional finger-shaped elements develop posteriorly and massive denticles anteriorly. Dotted, the latest additions to the scale. Stippled area, the fused base of the finger-shaped elements; dark tone, the fibrous cushion base of the massive denticles.

cushion begins to be formed, and this contains connective tissue fibers already in the first layer of hard tissue deposition.

The evidence strongly suggests that the scales grow both along the posterior and anterior margins and that the growth starts from a single lepidomorial denticle whose base later forms the boundary zone between the two basal components described earlier, and whose tiny crown lies theoretically outermost and anterior-most of all finger-shaped denticles of the posterior scale fringe (Fig. 6). It appears that the crowns of the older finger-shaped denticles have been worn off in the fully developed scales (Fig. 3).

Hard tissues in injured patches of the hide. In some areas of the otherwise intactly preserved shagreen the typical scale arrangement is absent; instead there occurs an irregular spread of globular bodies of a hard tissue that is not seen in regular scales. These hard tissue bodies are of irregular shape and vary greatly in size. Their surfaces, on section, often have a scalloped outline, and the tissue contains vascular canals of varying diameter but no connective tissue fibers. The tissue adjacent to the canals contains cell processes which might belong either to odontoblasts (in which case the tissue should be called a trabecular dentine)

Fig. 7 (above). Irregular hard substances located in a patch of skin of *Orodus* (FMNH PF 2201, section no. 5120) that may have been injured and subsequently healed. Section is parallel to body surface. The identification of the substance as bone, or trabecular dentine, remains uncertain.

Fig. 8 (middle). Similar structures as seen in Fig. 7, but in a section at right angle to the skin surface (FMNH PF 2201, section no. 5113). At upper side of picture beginning of development of atypical denticles.

Fig. 9 (below). Atypical scale of *Orodus* (FMNH PF 2201, section no. 5099) that may have developed from structures similar to those shown in Figs. 7, 8. The denticles along the outer margin of the scale differ markedly from those of the typical scales of this fish.

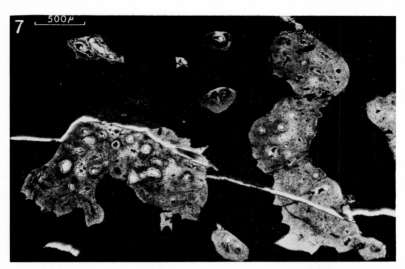
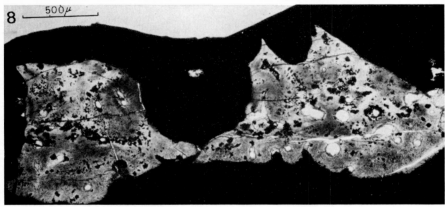

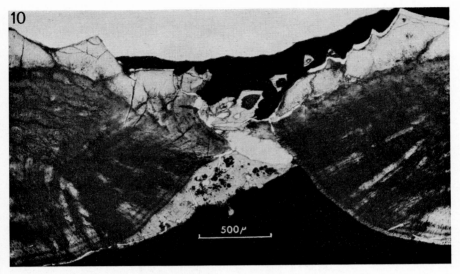

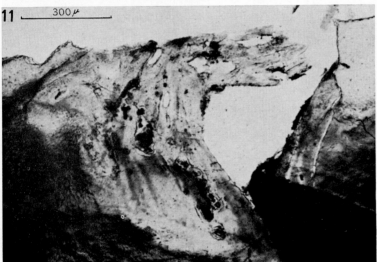

Fig. 10 (above). Two adjacent dorsal scales of *Orodus* (FMNH PF 2201, section no. 5115) in transversal, vertical section. Along the base of the left scale, wedge of bright, vascularized hard tissue that lacks fibers. Black spots within this tissue are sulfides. Significance of this accessory tissue is not understood.

Fig. 11 (below). Posterior, finger-shaped denticles and their fused base of *Orodus* (FMNH PF 2201, section no. 5117) enlarged to show irregular areas containing opaque sulfide crystals. These are interpreted as resorption spaces.

or to retreating osteoblasts in which case it should be regarded as a kind of bone (Fig. 7), typically developed in *Ornithoprion* (Zangerl, 1966). In a few cases vertical sections through globular bodies of this sort show what looks like the development of massive denticles on the outer (comparable to the crown surface of the scales) surface of the bodies (Fig. 8). These differ, however, from

Fig. 12. Dorsal scale of *Holmesella*? sp. (FMNH PF 2631, section no. 5129) in an obliquely horizontal section. In center of upper part of picture the spoon-shaped, hollow posterior denticles; to either side of them massive denticles; beneath the denticles the fibrous cushion-shaped base of the massive denticles, which contains some cell spaces; see area in outlined rectangle enlarged in Figs. 16, 17.

the massive denticles on ordinary scales by irregular shape, differing size and lesser mineralization. It is probable that the scale illustrated in Fig. 9 originated in this unusual fashion.

The highly irregular shape and distribution of these structures suggests that they developed in places of the skin that had been injured and where scar tissue had formed.

Accessory hard substances on regular scales. In two thin sections ground vertically and transversally through normal, fully developed scales, small accessory hard substances were noted along the lower edge of the dorsal or ventral scale side (Fig. 10). These are adjacent to the cushion-shaped base that contains large numbers of connective tissue fibers, but they themselves contain none and are thus sharply differentiated from the cushion by bright yellow color (under the microscope), by the fact that they contain vascular canals and by the presence in them of tiny canals that extend inward from the surface and might represent the canaliculi of retreating bone cells, or, possibly dentinal tubules. This tissue is thus similar to that of the irregular bodies described above. The curious aspect of these accessory hard tissues lies in the fact that they are located where the collagenous fibers of the cushions should have extended into the surrounding connective tissue in the living animal, in the manner suggested by Ørvig (1966). Yet they contain no fibers. Since Ørvig's interpretation is certainly correct in view of similar anchorage of dermal denticles and teeth of modern sharks within the dermis, it would appear that in this case the connective tissue fibers were destroyed as the accessory hard tissue was formed. In view of the fact that such accessory calcifications are rare, it seems certain that they are not part of the normal scale structure and may be related to histological changes in the dermis that may have taken place during the shedding process of the scales, or in response to traumatic or parasitic lesions.

Resorption phenomena. Ørvig (1966: 9) could not find any evidence of resorption in his material of *Holmesella?* sp. In the scales of *Orodus* here described there appear—here and there, though not commonly—peculiar cavities that show a narrow, bright halo around the irregular outlines of the cavities. Inside of these spaces there are usually fairly coarse sulfide crystals. Since these cavities have no structural relationships to the surrounding tissues, they are best interpreted as resorption spaces (Fig. 11). Most commonly, these are seen in the cellular bases of the finger-shaped denticles and in the bases of abnormal scales as described above (Fig. 9).

Holmesella? sp.

In principle the scales of *Holmesella?* sp. have the same basic construction as those of *Orodus*. The present material (FMNH PF 2631) confirms Ørvig's (1966) description based on the Kansas material, but permits some additional observations.

The posterior margin of these scales consists, as in *Orodus*, of lepidomorial denticles which are so delicate that they tend to be broken off except in optimally preserved material. In the Kansas specimen they were mostly broken away and hence could not be studied by Ørvig. The shape of these lepidomoria differs from that in *Orodus* in that the elements are rather flat, relatively wide, spoon-shaped structures with thin crown walls and large pulp cavities (Fig. 12). Together they form a thin shelf around the posterior half of the main body of the scale and the posterior end of this is more or less pointed. Hence the overall appearance of these scales is quite different from those of *Orodus*. The micro-

Scales of Paleozoic sharks 409

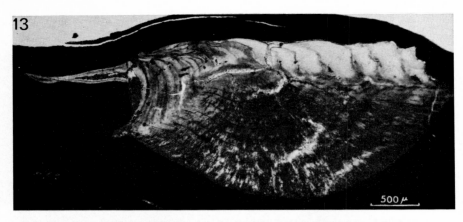

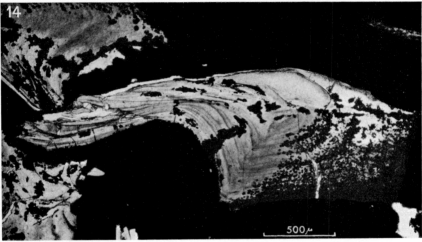

Fig. 13 (above). Antero-posterior, vertical section through dorsal scale of *Holmesella*? sp. (FMNH PF 2631, section no. 5138), spoon-shaped denticles at left, massive denticles at right. Compare with Fig. 3.

Fig. 14 (below). Young developmental stage of a dorsal scale (or possibly a fully developed, ventro-lateral scale) of *Holmesella*? sp. (FMNH PF 2631, section no. 5127). Compare with Fig. 5.

scopic structure and the basal region are indistinguishable, however (Fig. 13), and the mode of growth is the same in both scale types (Fig. 14). As in *Orodus*, the fused basal region of the lepidomorial denticles contains bone cells (or rather their bitumen-filled cavities) and lacks connective tissue fibers (Figs. 13, 15). The massive denticles in front of the lepidomorial fringe, as in *Orodus*, tend to vary in shape considerably, both individually and in different parts of the hide. As in *Orodus* they overlap one another in antero-posterior direction and tend to be broader than in the compared genus. The histology is as described by Ørvig.

The cushion-shaped base beneath the massive denticles was described by

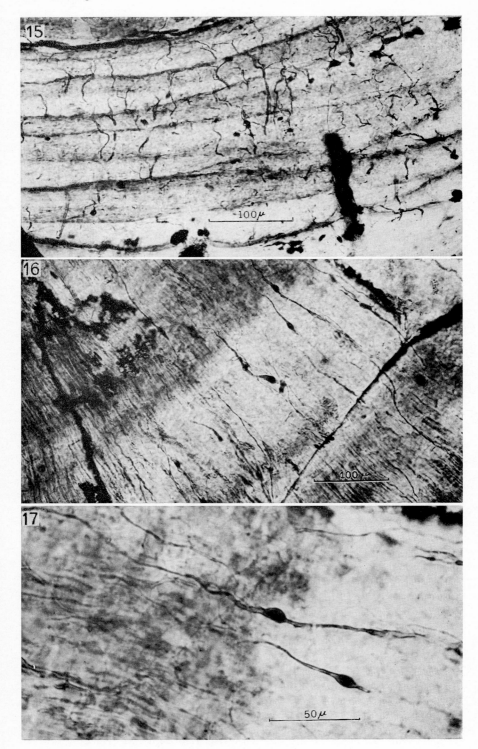

Ørvig as consisting of acellular bone. In one of our sections there is a scale (Fig. 12) that shows cells extending transversal to the growth zones of this region with long canaliculi approximately parallel to the course of the connective tissue fibers, and shorter ones distributed in the tissue in a more irregular fashion (Figs. 16, 17). In most of the cushion area the connective tissue fibers are, however, so tightly crowded, that bone cells, even if they were there, could hardly be identified beyond doubt.

In summary, the differences between *Orodus* and *Holmesella?* sp. scales are shape differences in the posterior lepidomorial fringe and in the massive anterior denticles of the composite scales. The histological structure and the mode of growth appear to be identical in the two genera. I agree with Ørvig that this scale type resembles very closely that described by Gross (1938) in his redescription of *Protacrodus vetustus* Jaekel from the late Devonian of Wildungen. Unfortunately, the histological structure of the Wildungen material is not known.

In connection with the description of the Pennsylvanian edestid shark *Ornithoprion*, I suggested, on the basis of much less evidence than I have at this time, that composite scales of the type here described and which occur (with variations in shape) also in *Ornithoprion, Agassizodus?* sp. and the edestids from East Greenland, that such scales might be characteristic of the Edestidae. Since then I have had the opportunity to study scales in as yet undescribed Paleozoic sharks that are clearly not edestids. These also consist of a group of lepidomorial denticles that are mainly fused at the bases, though the detailed histology remains to be further studied. At any rate my suggestion that this scale type might be characteristic of the Edestidae clearly seems to have been premature. Paleozoic sharks seem to display an amazing variety of scale types and it remains to be seen to what extent the scales have any taxonomic significance.

More important and puzzling is the histogenetic complexity of the scales in *Orodus* and *Holmesella?* sp. Why do two rather different kinds of denticles combine to form one scale: one forming on the anterior side of the scale, the other on the posterior side? Why are the bone cells (presently their cell spaces) oriented with their axes transversal to growth increment lines (as is true of odon-

---

Fig. 15 (above). Base of a dorsal scale of *Holmesella*? sp. (FMNH PF 2631, section no. 5129) beneath the posterior, spoon-shaped denticles; in horizontal section. Numerous cell spaces and cell processes are visible; most of them are oriented at right angle to the growth increment lines which are very pronounced.

Fig. 16 (middle). Enlargement of area marked off on Fig. 12, showing the shape and distribution of mostly bipolar cell spaces and their large and very long processes. arranged at right angle to the growth increment lines. Black material inside of cell spaces and sharp outlines of processes have probably been produced by coagulated bitumina.

Fig. 17 (below). Further enlargement of some of the cell spaces and their processes (Fig. 16) to show the nature of the preservation of the material.

toblasts and their processes), rather than parallel with them as for example in osteonal bone? It is at least possible that these cells were not truly osteocytes, but cells of intermediate character between odontoblasts and osteoblasts and that the tissue they formed is not really bone, but a hard substance intermediate between bone and dentine.

References

Agassiz, J. L. R. (1833–43). *Recherches sur les poissons fossiles.* 5 vols. Neuchâtel.

Gross, W. (1938). Das Kopfskelett von *Cladodus wildungensis* Jaekel. 2. Der Kieferbogen. Anhang: *Protacrodus vetustus* Jaekel. *Senckenbergiana*, **20**: 123–145.

Ørvig, T. (1966). Histologic studies of ostracoderms, placoderms and fossil elasmobranchs. 2. On the dermal skeleton of two Palaeozoic elasmobranchs. *Ark. Zool.* (2) **19**: 1–39.

Stensiö, E. A. (1962). Origine et nature des écailles placoïdes et des dents. *Colloques int. Cent. natn. Rech. Scient.*, **104**: 75–85.

Zangerl. R. (1966). A new shark of the family Edestidae, *Ornithoprion hertwigi* from the Pennsylvanian Mecca and Logan Quarry Shales of Indiana. *Fieldiana: Geol.*, **16**: 1–43.

Zangerl, R. & Richardson, E. S. (1963). The paleoecological history of two Pennsylvanian black shales. *Fieldiana:Geol. Mem.*, **4**: 1–352.

Les os dermiques crâniens des Poissons et des Amphibiens; points de vue embryologiques sur les "territoires osseux" et les "fusions"

Par Ch. Devillers et J. Corsin

Laboratoire d'Anatomie et d'Histologie Comparées, Université de Paris, Paris, France

Le terme de « territoire d'un os », très utilisé dans les comparaisons entre crânes de diverses lignées de Poissons ou d'Amphibiens surtout, recouvre une notion mal définie et discutée.

Certains auteurs attribuent à chaque élément dermique crânien une aire d'occupation, un « territoire » strictement déterminé, caractéristique. Pour expliquer alors les variations, importantes, d'extension de tel ou tel élément, observables entre différentes espèces, il leur faut admettre qu'elles sont essentiellement dues à des fusions entre éléments adjacents.

Pour d'autres auteurs la détermination de la forme, de l'étendue des os, serait plus labile, épigénétique. Les fusions existent bien mais les os ont la capacité de s'étendre, par poussée autonome, sur le territoire d'un élément adjacent. Le territoire ne relève plus alors d'une propriété intrinsèque de l'ébauche osseuse mais, plus souvent, d'une compétition au sein d'une population d'ébauches dont les vitesses de croissance sont différentes intra- et inter-spécifiquement.

Un résumé de ces discussions, sur lesquelles nous ne nous étendrons pas, a été donné, par exemple par Parrington (1949).

Les discussions, jusqu'à maintenant, portent presque exclusivement sur le crâne achevé dans lequel les limites inter-osseuses sont établies. Il n'apparaît pas qu'il ait été porté aux processus de l'organogenèse osseuse toute l'attention qu'ils mériteraient puisqu'en définitive ce sont les facteurs décisifs du modelage de la forme. C'est de ces facteurs, mal connus encore, que nous voudrions traiter ici, en limitant notre exposé aux Téléostéens et aux Amphibiens Urodèles actuels.

Les facteurs de l'organogenèse osseuse

On peut, de façon arbitraire puisqu'ils sont étroitement liés, les séparer en quatre catégories :

(1) facteurs déterminant l'apparition de l'initiale mésenchymateuse de l'os dermique;

(2) facteurs de multiplication des ostéoblastes, qui conditionnent la vitesse de croissance et l'éventuelle compétition entre ébauches;

(3) conditions spatiales dans lesquelles s'effectue la croissance de l'ébauche;

(4) facteurs d'affinités cellulaires, positive ou négative, qui permettent aux ébauches de se fusionner ou de garder leur individualité.

La détermination des ébauches. La première indication d'un os dermique est une agglomération d'ostéoblastes, un blastème.

Cette initiale est unique, cas fréquent, ou multiple comme pour certains os liés au système latéral céphalique (os à canaux, canal-bones; Pehrson 1922; 1940; Devillers, 1947).

Pour les os à canaux d'*Amia* (Pehrson), la seule observation du développement fait soupçonner que le neuromaste, d'abord superficiel, puis invaginé, joue le rôle d'un inducteur du blastème qui s'organise à son niveau, rôle qui a été expérimentalement démontré sur *Salmo* par la technique des ablations et des greffes (Devillers; Guinnebault, 1953). Les premières expériences sur *Salmo* reviennent à Moy-Thomas (1941) mais le rôle inducteur n'avait pu être démontré.

Pourtant, la relation d'induction neuromaste/blastème n'est pas un processus général : chez *Salmo* par exemple le corps du dentaire ou du préoperculaire se développe à partir d'une seule ébauche qui s'accroît de façon autonome sans aucune intervention des neuromastes qui se bornent à induire, plus tard, la formation du tunnel osseux soudé au corps et qui abrite le canal sensoriel. Ce mode de formation autonome est celui de tous les os à canaux des Cyprinidés (Devillers; Lekander, 1949).

Le rôle ostéogène du système latéral, au moins chez les formes vivantes, paraît donc assez limité puisque, selon les cas, certains de ses segments, ou la totalité du système, est privée de capacité organisatrice dans la genèse du corps de l'os, seul élément qui nous intéresse (le tunnel osseux n'est qu'un élément secondaire ajouté, sans importance architecturale).

Les organes des fossettes sensorielles (pit-organs) qui restent superficiels sont normalement dépourvus d'activité, comme le sont aussi les neuromastes superficiels de l'Amphibien Urodèle *Pleurodeles waltlii* puisque l'excision des placodes latérales n'altère en rien la formation du crâne dermique (expériences de Corsin).

Il est par contre curieux de déceler l'intervention des neuromastes otiquespostotiques au cours de la formation du pariétal pourtant situé hors du réseau des canaux céphaliques. Cette intervention, soupçonnée par Pehrson (1940) a été démontrée par les expériences d'ablation de Guinnebault: si l'excision des neuromastes est effectuée avant toute apparition des blastèmes du ptérotique et du pariétal, plus tardif, les deux os ne se forment pas; si l'excision a lieu lorsque les blastèmes ptérotiques sont déjà en place, un pariétal réduit se constitue. La nature de cette liaison morphogénétique est encore inexpliquée.

Nous ne connaissons rien encore du déterminisme d'apparition des autres os dermiques. Il n'est pas exclu que certaines des grandes structures céphaliques puissent exercer un rôle inducteur au cours de la formation de certains d'entre eux si l'on tient compte des résultats obtenus dans d'autres groupes (J. Corsin, 1968 b, sur l'embryon de *Pleurodeles*; Schowing, 1961, sur l'embryon de Poulet).

Puisque les os à canaux et le système latéral céphalique ne sont pas nécessairement liés, au cours du développement, par une relation d'induction, la raison d'être de la règle d'Allis (1904) sur la constance des rapports entre canaux céphaliques et os dermiques nous échappe. Bien qu'elle soit très utilisée, la valeur de cette règle a pu être mise en doute par exemple par Parrington (1949). Reconnaissons pourtant que son application conduit à des résultats cohérents, ce qui constitue un critère de sa validité. Il faut alors chercher une explication.

Une solution, proposée par Parrington (1949), renverse le sens de la relation os/canaux, les os déterminant l'emplacement des canaux; l'action, de nature inductrice, serait exercée, non par les ébauches osseuses, qui apparaissent après les neuromastes, mais par leurs précurseurs, invisibles, les champs morphogénétiques correspondants. Cette conception, toute théorique, contredit ce que nous savons des effets inducteurs, toujours exercés par des structures mises en place.

Les rapports constants os/canaux, n'étant pas toujours explicables par une liaison d'induction chez les formes vivantes (et, vraisemblablement chez les fossiles), pourraient-ils être le fait de la seule coïncidence. Répétée un si grand nombre de fois, une telle coïncidence paraît invraisemblable. N'oublions pas que l'architecture générale de la tête, la course des canaux sensoriels, manifestent, dans l'ensemble des Poissons, une impressionnante constance.

Dans l'organogenèse de la tête les constituants essentiels, cerveau et organes des sens, pharynx, nerfs, vaisseaux, muscles, chondrocrâne sont mis en place bien avant les os. Ceux-ci se développent alors dans un espace structuré qui restreint leur degré de liberté et doit leur imposer certaines voies de croissance immuables ce qui pourrait expliquer la constance des rapports entre les os et les plus importantes des structures céphaliques.

Modes et conditions de croissance des ébauches. Sous un aspect purement géométrique le problème de croissance des os dermiques a été analysé par Parrington (1956). Compte tenu de la disposition d'une population d'ébauches, selon un réseau rectangulaire ou oblique, compte tenu de la croissance, uniforme dans toutes les directions (isotrophe, représentée alors par un cercle) ou anisotrope (représentée par une ellipse), compte tenu enfin des vitesses de croissance propres à chacun des blastèmes, il serait possible d'expliquer, par ces considérations simples, le tracé des lignes de sutures.

La coïncidence entre dessins des lignes de suture, théoriquement analysés et réellement observés sur le crâne de *Seymouria* par exemple, est frappante. On peut toutefois se demander si le modelé du contour des os relève d'un processus exclusivement épigénétique ou si, pour une part plus ou moins importante, il n'est pas intrinsèquement déterminé.

Nous avons tenté d'examiner cette question au cours du développement des frontaux et des pariétaux chez *Pleurodeles* (Fig. 1). Le matériel n'est guère favorable : les os ne butent pas franchement lorsqu'ils viennent en contact, comme le faisaient les os épais des formes fossiles mais glissent l'un sur l'autre de sorte qu'il est difficile d'affirmer qu'ils se limitent mutuellement dans leur croissance.

Toutefois l'examen de quatre stades permet quelques observations :
— les taux de croissance ne sont pas réguliers mais s'accélèrent vers les stades finaux puisque entre A et B il s'est écoulé 30 jours, 15 jours seulement entre B et D.
— le pariétal s'accroît plus vite que le frontal; en effet le second est apparu 30 jours avant le stade figuré en A et le premier 20 jours seulement, or, il est plus grand que le frontal;
— au cours de l'extension, le dessin, irrégulier des contours n'annonce guère celui des sutures dont le modelé final, assez simple, ne se réaliserait qu'avec l'articulation des os. Le frontal émet, en direction orbitaire, un processus qui, chez les animaux âgés, se lie par ligament au squamosal. La formation de ce processus extériorise l'existence d'une zone limitée où la croissance est plus importante que dans le reste de l'os. Le tracé de son contour relève ici d'une propriété intrinsèque de l'ébauche osseuse, non d'une interaction avec les éléments voisins.

De cette observation limitée il semble qu'on puisse conclure que la détermination des contours (et sutures) fait intervenir à la fois un processus de détermination génétique traduit par la distribution des vitesses de croissance dans l'ébauche (comme l'admet obligatoirement la conception de Parrington) et, à un degré difficile à apprécier, une détermination épigénétique en fin de croissance. L'étude des ablations d'os mettra mieux en évidence la réalité de ce second aspect.

La question importante est alors de déceler les facteurs qui conditionnent les vitesses et les directions de croissance des ébauches. En s'appuyant sur certains résultats de Weiss (1929), quelques suggestions pourraient être avancées. Une culture de fibroblastes sur coagulum de plasma se développe également dans toutes les directions; si l'on exerce une tension, constante, sur le coagulum, une croissance plus importante se produit alors dans la direction de la tension. Weiss attribue ce fait à ce que les fibrilles du milieu, orientées par la traction, guident les migrations cellulaires.

Le blastème osseux se développe dans le conjonctif dont les fibres collagènes, très probablement, ne sont pas disposées au hasard; les contraintes exercées par les structures céphaliques en croissance devraient orienter les fibres en faisceaux (structuration de l'espace) qui constitueraient autant de voies privilégiées pour la migration et le groupement des ostéoblastes. Cette idée n'a pas encore été soumise directement à l'épreuve de l'expérience mais il se pourrait que certaines des altérations observées sur la croissance des os après ablation d'ébauches sensorielles résultent de remaniements non décelables dans l'organisation du conjonctif céphalique. D'une façon analogue Massler & Schour (1951; cités d'après Hoyte, 1966) attribuent la formation d'indentations aux sutures par l'existence de lignes de tension dans le tissu intercalé entre les os.

Expérimentalement, le problème de la détermination intrinsèque ou extrinsèque du territoire peut être abordé de deux façons :
— mise en culture, *in vitro*, d'ébauches osseuses;
— ablations précoces d'ébauches permettant d'étudier le comportement des os adjacents ainsi débarrassés de leurs concurrents pour l'occupation d'un espace donné.

L'ébauche du frontal de *Pleurodeles* (Corsin en collaboration avec Mme Hubert), cultivée sur un coagulum de plasma de Poule, où elle dispose de tout l'espace, va-t-elle maintenir sa forme caractéristique (détermination intrinsèque) ou va-t-elle croître au-delà des limites normales (détermination épigénétique).

Les premiers résultats (Fig. 1; culture à partir du stade 50 pendant trois semaines) ne sont pas encore concluants : l'ébauche s'accroît dans les premiers temps en conservant un contour proche de la normale (Fig. 2) mais ensuite la forme s'altère.

Il est vrai que les conditions de culture sont ici fort éloignées de celles du développement : le réseau conjonctif n'existe plus; le milieu de culture, probablement trop riche, favoriserait une prolifération anormale des ostéoblastes (Wolff, 1952). L'expérience sera à reprendre avec des milieux variés, moins riches. Soulignons cependant que, dans les mêmes conditions, une ébauche cartilagineuse comme celle d'un fémur de Poulet (Fell & Robison, 1929) s'accroît en conservant sa forme normale qui est donc, dans ce cas, propriété intrinsèque de l'ébauche.

La méthode d'excision donne des résultats plus intéressants. Elle a été pratiquée sur le frontal et le pariétal droits de la larve de *Pleudodeles* aux stades 48 à 52; les animaux opérés ayant été fixés environ 3 mois après (Corsin 1968 b). Les résultats, concordants, montrent que ces deux os ont la possibilité d'acqué-

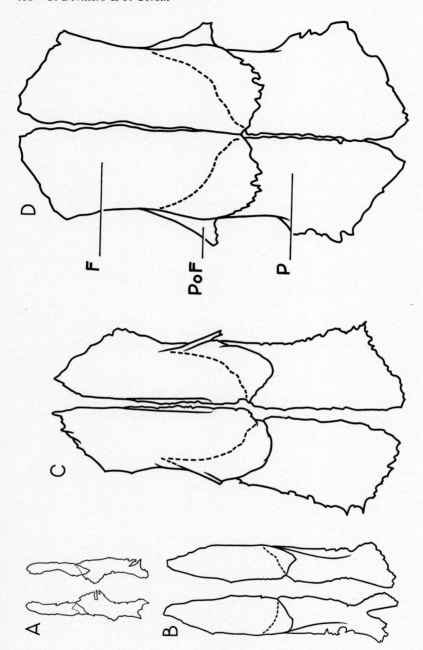

Fig. 1. Quatre stades de développement des frontaux et pariétaux de *Pleurodeles waltlii* (coloration à l'alizarine). *A* — stade 50 (longueur totale du corps 2,5 cm); *B* — stade 55 A (longueur totale du corps 4,5 cm); *C* — stade 55 B (longueur totale du corps 6,0 cm); *D* — stade 55 C (longueur totale du corps 7,0 cm).

F, frontal; *PoF*, processus orbitaire du frontal; *P*, pariétal.

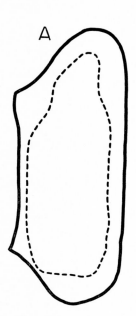 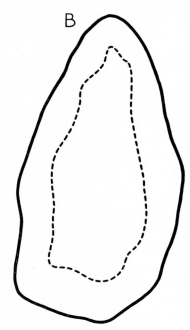

Fig. 2. Culture de blastèmes osseux. *A*, schéma montrant l'accroissement normal d'un frontal gauche resté en place entre les stades 51 et 54. *B*, schéma montrant l'accroissement d'un frontal mis en culture pendant 3 semaines : la croissance a lieu également dans toutes les directions donnant à l'os une forme anormale.

rir une extension plus grande (croissance totale) que celle normalement observée (croissance réelle), ainsi :
— en l'absence du frontal droit, son territoire est en grande partie envahi par ses voisins les plus immédiats, le frontal gauche, le pariétal et le préfrontal droits (une très faible régénération de l'os excisé n'est pas exclue). De même, en l'absence du pariétal droit, son symétrique et le frontal droit colonisent largement l'espace devenu disponible. Le squamosal droit qui aurait lui-aussi la possibilité de s'étendre ne le fait pourtant pas de façon appréciable; vers la région pariétale au moins, ses croissances, réelle et totale, sont les mêmes.

La capacité d'expansion ne serait donc pas une propriété de tous les os mais d'un certain nombre d'entre eux et probablement même, dans certaines directions seulement, comme le montre le comportement du squamosal dans deux situations différentes.

Des phénomènes comparables d'envahissement ont été observés après suppression unilatérale précoce des ébauches des vésicules olfactive et optique. Dans ce cas l'organisation spatiale du milieu dans lequel se développent les os environnants est encore plus profondément altérée : par la disparition de la structure sensorielle dont l'existence imposait aux os certaines voies de croissance; par un possible remaniement du réseau conjonctif; par des altérations du chondrocrâne; par la disparition de certains os dermiques dont le développement doit être induit par la structure excisée.

Ainsi, en l'absence du sac nasal, le nasal et le processus dorsal du prémaxillaire manquent; le préfrontal est très réduit tandis que le maxillaire poussant de façon incontrôlée, envahit la zone de la narine externe et, vers l'arrière, émet un processus qui peut se souder au frontal.

En l'absence de l'œil, le maxillaire s'allonge vers l'arrière, jusqu'à rejoindre le squamosal, établissant un arc sous-oculaire qui n'existe plus chez les Urodèles actuels mais était présent chez les Labyrinthodontes, avec une structure différente.

Le squamosal extériorise, en direction du maxillaire, une possibilité de croissance qu'il ne possédait pas en direction du pariétal.

Tatarko (1934) a montré que l'operculaire de *Cyprinus carpio* possédait une capacité qu'il qualifie d' « illimitée » (des expériences comparables à celles exposées ci-dessus devraient être entreprises sur le crâne des Téléostéens). Pour les Mammifères, Troitsky (1932), au contraire, ne trouvait qu'une extra-croissance (croissance totale) très faible des os dermiques de la voûte crânienne du Rat. En réalité, comme l'a montré Moss (1954), les sutures ne sont pas des limites infranchissables; à la condition que l'excision de l'ébauche pariétale soit réalisée à un stade assez précoce le frontal peut alors envahir le domaine libre.

De ces investigations découle la conclusion que l'occupation, par un os A (chez une espèce X) du territoire tenu par un os B chez une autre espèce Y ne relève pas nécessairement d'une fusion (A + B). Rien ne s'oppose à ce que A (et d'autres os) puisse coloniser le territoire, vide, de B.

La disparition de B pourrait revêtir trois modalités différentes :

(1) son ébauche ne se forme plus. Cette situation est comparable à celle où, dans nos expériences, maxillaire et squamosal reconstruisent un arc sous-orbitaire malgré l'absence du jugal qui existait chez les Labyrinthodontes;

(2) l'ébauche de B pousse dans une espèce X à une vitesse plus réduite que chez Y; il n'y a donc pas ici réellement de disparition de B mais réduction extrême de son territoire. Ce serait approximativement ce qui se produit dans le cas du pariétal de *Polypterus* (Pehrson, 1947) dont le territoire se fait coloniser par le ptérotique.

(3) l'ébauche de B pourrait encore se constituer mais elle débute trop tard et, à ce moment, son « territoire » est déjà complètement envahi.

Cette dernière possibilité nous conduit à poser le problème des compétitions embryonnaires entre ébauches pour l'utilisation d'un même substrat ou de matériel cellulaire (Spiegelman, 1945), dont certaines conséquences ont déjà été analysées dans une autre publication (Devillers, 1965). Il n'est pas évident que le matériel squelettogène nécessaire au développement d'ébauches soit disponible en quantité illimitée. Si cet approvisionnement est au contraire limité (artificiellement dans les expériences de Bretscher, Bretscher, Tschumi & Tschumi sur le développement des doigts chez les Amphibiens et dans celles de Hampe & Schue sur le Poulet; voir bibliographie dans Devillers, 1965), la compétition entre ébauches s'extériorise avec une certaine acuité. Certaines initiales, ou trop « faibles » (sans que nous puissions définir ce terme de façon plus précise), ou d'apparition trop tardive, voient tout le matériel squelettogène accaparé par les initiales, plus « fortes » ou plus précoces qui, seules, se développent tandis que les autres disparaissent totalement ou brusquement.

Sans connaître le déterminisme des causes initiales de compétition, cette dernière est un fait d'observation qui devrait permettre de comprendre certains processus de disparition d'ébauches et d'annexion de leurs territoires par les os adjacents.

L'examen du phénomène d'affinité cellulaire, positive ou négative, conduit à analyser les problèmes des fusions (ou des non-fusions) à des stades très précoces.

Rôle de l'adhésivité cellulaire. L'affinité cellulaire, bien analysée dans ses manifestations, par exemple lors de la ségrégation des feuillets au cours de la gastrulation (Holtfreter, 1939) reste mystérieuse en ce qui concerne sa nature (Devillers, 1955; Curtis, 1962).

Dans la série des blastèmes supra-orbitaires, tous semblables, d'*Amia* ou de *Salmo*, les plus antérieurs ont une affinité cellulaire positive révélée par le fait qu'ils se soudent en initiale du corps du nasal; les autres blastèmes s'unissent en corps du frontal mais les blastèmes 3 et 4 ont entre eux une affinité négative grâce à laquelle les ébauches des deux os restent séparées.

L'affinité, négative, conduisant à la formation d'une suture entre deux ébauches A et B n'est pas immuable d'une espèce à l'autre ou toute la vie dans une même espèce : ainsi les cellules des blastèmes frontal et pariétal ont entre elles une affinité négative chez les Anoures, une affinité positive chez les Urodèles conduisant à une fusion en un fronto-pariétal; deux os séparés chez un jeune animal peuvent se souder chez les individus âgés comme cela est fréquent dans le crâne des Mammifères ou dans celui des Vertébrés volants. Ces indiscutables fusions interviennent à un stade tardif où les os ont leur localisation précise, leur forme caractéristique. Passant à des stades plus précoces, pouvons-nous conserver au terme de fusion, la même signification et l'appliquer à l'union des blastèmes d'un os à canal? Dans ce cas cela conduirait à admettre l'existence d'un stade ancestral où chacun de ces blastèmes devenait une ossification distincte. Cela revient au fond aux idées exprimées, d'une façon moins radicale peut-être, par Jarvik (1948) par exemple. Les Dipneustes primaires, comme *Dipterus* par exemple, donneraient une représentation de cette mosaïque primitive bien qu'il paraisse étonnant que ce groupe, si tôt spécialisé, ait pu, à un tel degré, conserver le plan ancestral.

Assigner à cette union de blastèmes la valeur de fusions (avec signification phylogénétique) soulève des difficultés d'interprétation puisque, par exemple, le nombre des blastèmes formateurs d'un os déterminé n'est pas le même dans des formes différentes comme *Salmo*, *Amia*, les Cyprinidés et Amphibiens (un seul blastème); les frontaux de ces formes ne seraient plus alors strictement homologues. La notion d'homologie perd ainsi toute signification.

Chez *Salmo* ou *Amia* la position d'un neuromaste dans la série supraorbitaire, assigne au blastème qu'il va induire une destinée déterminée, nasale ou frontale mais cette détermination peut devenir plus fluctuante dans la section infra-orbitaire de *Salmo* (Devillers, 1947). La chaîne osseuse y comporte, à l'état adulte : un antorbitaire, un lacrymal, un infra- et trois post-orbitaires. L'antorbitaire excepté, ces os se constituent par association de blastèmes nés sous chacun des neuromastes mais la destinée de certains d'entre ces blastèmes peut varier selon les individus et parfois, sur un même individu, d'un côté sur l'autre de la tête (Fig. 3) : celui de rang n allant, selon les cas, s'adjoindre à l'infra-orbitaire ou au post-orbitaire inférieur.

Du fait de l'existence de ces blastèmes « vagabonds », la constitution des os, la position des sutures ne sont pas rigoureusement fixées : selon les hasards des « soudures » aurons-nous un infra-orbito-post-orbitaire et un postorbitaire ou un infraorbitaire et post-orbito-infra-orbitaire? La question est dépourvue de sens puisqu'elle implique, au stade des ébauches, une stricte détermination qui n'existe pas.

L'extrême variabilité de la chaîne infra-orbitaire dans l'ensemble des poissons (voir aussi Moy-Thomas, 1938) fait soupçonner que chez les formes vivantes et fossiles, les mêmes éléments d'incertitude existent dans la détermination des ébauches.

L'organogenèse osseuse le long de la section otique-post-otique soulève des difficultés comparables.

De façon caractéristique les Crossoptérygiens possèdent dans cette région un intertemporal et un supratemporal et les Actinoptérygiens, un ptérotique appelé inter-temporo-supratemporal par certains auteurs (Holmgren & Stensiö, 1936). Cette dernière dénomination revient à considérer l'état Crossoptérygien comme représentatif d'une disposition primitive d'où dériverait, par fusion, celle des Actinoptérygiens. Une disposition à deux os séparés, comparables à l'inter- et au supratemporal a été décrite par Le Danois (1959) chez les Téléostéens Orbiculates.

Lors de la formation du ptérotique de *Salmo* (Fig. 4; Devillers, 1947) les blastèmes formés sous les trois premiers neuromastes se soudent (ébauche intertemporale selon la nomenclature de Pehrson, (1940); le 4ème blastème

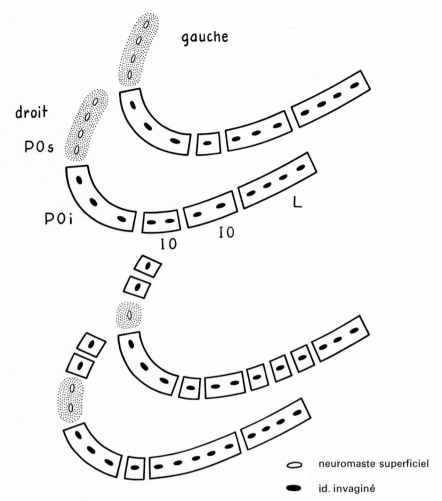

Fig. 3. Schémas du développement de la chaine circum-orbitaire chez deux alevins de *Salmo irideus*. L'antorbitaire n'a pas été représenté. En pointillé : blastème du corps de l'os; contours épais : ossifications (d'après Devillers, 1947, modifié).

IO, infra-orbitaire; *L*, lacrymal; *POi*, *POs*, post-orbitaires inférieur et supérieur.

garde un temps son individualité (blastème supratemporal) puis s'unit à l'ébauche antérieure et l'ossification débute; les mêmes aspects se retrouvent chez *Amia* et *Polypterus* (Pehrson, 1940; 1944*a*; 1947). Ce même stade mésenchymateux à deux composants devait exister dans l'embryon de Crossoptérygien mais il se maintenait avec l'ossification.

L'homologie serait complète entre ptérotique et intertemporal + supratemporal et le nom composé se justifierait si nous ne considérions que ces cas (dont *Acipenser* (Pehrson, 1944*a*) est encore assez proche bien que les deux ébauches ne résultent pas de la fusion d'autant de blastèmes qu'il y a de neuromastes à

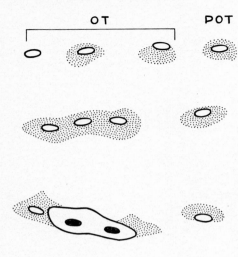

Fig. 4. Trois stades de developpement du corps du ptérotique de *Salmo irideus*. En pointillé : blastèmes du corps du ptérotique; contour épais : ossification (d'après Devillers, 1947, modifié).
OT, neuromastes otiques; *POT*, neuromaste postotique.

leur niveau (respectivement 3 et 4)); mais avec *Esox* (Pehrson, 1944*b*), seule se constitue l'ébauche « intertemporale » qui, poussant vers l'arrière, envahit la zone du neuromaste « supratemporal », inactif; enfin (Devillers, 1947; Lekander, 1947), l'unique ébauche des Cyprinidés se développe par croissance autonome sans liaison avec les neuromastes. Il existe donc parmi les quelques Actinoptérygiens jusqu'ici étudiés, une variété de cas que nous retrouvons aussi pour le nasal, formé par plusieurs blastèmes qui s'unissent chez *Salmo* et *Amia* tandis qu'ils conservent leur individualité après ossification chez *Polypterus* (Pehrson, 1947) et plusieurs formes fossiles.

Conclusions

Pour expliquer l'évolution du plan des os dermiques du crâne, des auteurs comme Säve-Söderbergh, Stensiö, Jarvik, etc. sont conduits à utiliser une hypothèse de travail que l'on peut résumer de la façon suivante.

Des formes les plus primitives aux plus récentes des lignées on assiste à une diminution (qui n'est pas nécessairement régulière) du nombre des unités osseuses. Cette diminution n'est, pour sa majeure partie, qu'apparente et résulte en réalité d'une perte d'individualité d'éléments adjacents qui se fusionnent.

De ce fait, des os, en apparence simples, peuvent, en réalité, être des complexes et doivent porter des noms composés qui rappellent leurs homologies réelles.

Statistiquement, on constate en effet une tendance à la diminution du nombre des os crâniens au cours de l'évolution, constatation que Gregory (1935) a voulu ériger en « loi » de Williston, loi qui comporte d'ailleurs de nombreuses exceptions.

Ceci étant admis, le problème que nous avons cherché à examiner ici avec l'aide de l'embryologie, est celui de la valeur des homologies établies en suivant le principe ci-dessus énoncé.

Le bilan de cet examen est assez décevant. Aux questions posées par le morphologiste : qu'est-ce que le territoire d'un os? qu'est-ce qu'une fusion? l'embryologie n'apporte guère de réponse immédiate mais conduirait plutôt à utiliser avec prudence, et dans certaines limites, la notion même d'homologie.

Prenons la seule catégorie des os à canaux : chaque tentative pour étayer sur des bases embryologiques les homologies des morphologistes, pour atteindre un degré plus grand de précision dans la détermination des homologies, conduit à l'évanouissement de la notion même d'homologie. Ceci résulte de ce que les neuromastes n'ont pas partout un rôle inducteur; que, même s'ils sont inducteurs, leur nombre varie dans le territoire d'un os donné, d'une espèce à l'autre; de ce que les propriétés d'affinités, positives ou négatives, des ostéblastes constitutifs des blastèmes, manifestent une variabilité imprévisible et que de ce fait les limites interblastèmes (inter-osseuses) peuvent être indéterminées.

Ainsi, des os morphologiquement homologues dans le crâne adulte, se sont en réalité constitués selon des voies dissemblables : ce sera par exemple le cas des trois nasaux de *Polypterus* correspondant chacun à un blastème de neuromaste (Pehrson, 1947), du nasal unique d'*Amia* ou de *Salmo*, né de l'union de plusieurs blastèmes et du nasal des Cyprinidés, formé à partir d'un seul blastème apparu sans intervention des neuromastes.

Dans la région otique-post-otique le nombre des os qui s'y forment, moins variable, devient caractéristique des lignées : intertemporal et supratemporal des Crossoptérygiens, ptérotique des Actinoptérygiens.

Nous ne connaissons les aspects embryonnaires que chez quelques Téléostéens mais il est possible d'admettre que *Salmo* ou *Amia* nous donnent une représentation vraisemblable de ce qui existait chez les embryons de Crossoptérygiens. On pourrait être tenté d'attribuer la différence entre les deux lignées à la seule affinité entre les cellules des blastèmes « intertemporal » et « supratemporal », négative pour les Crossoptérygiens (deux os séparés), positive pour les Actinoptérygiens dont l'unique ptérotique pourrait alors être homologué à un intertemporal+supratemporal. Mais dans plusieurs cas, nous l'avons vu, toute indication d'une origine complexe a disparu dans le ptérotique.

Enfin, l'examen de la chaîne infra-orbitaire nous a placés devant un cas où la détermination de certains blastèmes, devenue fluctuante, interdit pratiquement l'établissement d'homologies strictes.

S'il n'est donc pas possible, embryologiquement, de dépasser une limite dans le degré de finesse de nos recherches d'homologies, faute de quoi la notion perd tout sens, pouvons-nous alors conserver les comparaisons « point par point » telles que les auteurs ont cherché à les établir entre les os crâniens à l'intérieur d'une même lignée et surtout entre lignées différentes?

A notre avis il serait plus conforme aux faits de nous satisfaire d'homologies plus larges, plus globales et de faire porter les comparaisons sur l'ensemble des os appartenant à un territoire donné, à une section déterminée du système latéral céphalique, etc. Aux os, considérés individuellement, nous substituons la notion de série de recouvrement, formée dans ces zones et les noms des os seront alors seulement fondés sur leurs emplacements respectifs dans la série.

Ainsi, dans la série circum-orbitaire, le premier élément sera l'antorbitaire, le second, le lacrymal, les suivants des infra-orbitaires, puis des post-orbitaires; la série otique-post-otique comporte, selon les cas, un ou deux éléments qui, pris en bloc, sont homologues.

Il ne peut être question de nier l'existence des fusions osseuses, les preuves de leur existence sont trop nombreuses mais on peut discuter de leur valeur comme facteur explicatif, prépondérant, des variations du plan crânien.

Là encore, lorsque nous remontons vers des stades de plus en plus précoces nous trouvons, comme pour l'homologie, que, passée une certaine limite, la notion perd son sens habituel, celui donné par les anatomistes. Considérer les unions de blastèmes comme des « fusions » conduit à des difficultés inextricables, à des pseudo-problèmes d'homologies. Force est donc de considérer ces unions comme des processus purement embryologiques, sans signification structurale ou phylogénétique.

La véritable fusion entre ébauches ne peut intervenir que lorsque chacune d'elles est constituée par une population d'ostéoblastes, en principe unique, aux contours assez bien définis pour qu'elle puisse, sans équivoque possible, être reconnue comme l'initiale d'un os bien déterminé.

Il semblerait, mais ceci n'est qu'une hypothèse, qu'au cours de l'évolution, au moins dans le cas des os liés au système latéral céphalique on assiste à une simplification progressive du mode de formation des ébauches. Au type, supposé primitif, qui fait intervenir l'union de plusieurs blastèmes constitués chacun sous un neuromaste, se serait progressivement substituée une ébauche unique, à croissance autonome, telle qu'on la rencontre chez des Téléostéens évolués comme les Cyprinidés et chez les Amphibiens.

Quel que soit son mode de formation, l'initiale osseuse individualisée, va s'accroître. Possède-t-elle, dès le départ, une stricte détermination, génétique, de son territoire?

Les facteurs, multiples, de cette croissance, sont encore trop peu connus pour autoriser une réponse générale. Quelques indications partielles montrent au moins que les croissances d'une même initiale, dans les différentes directions, n'obéissent pas nécessairement toutes à un seul et même déterminisme, génétique, épigénétique. Le frontal, le pariétal, etc. de *Pleurodeles* poussent d'abord, lorsqu'ils sont encore bien séparés de leurs voisins, selon un « plan de croissance » déterminé qui leur confère une forme caractéristique. Dans le modelé final de leurs lignes de sutures interviennent certainement des compétitions avec leurs voisins, des limitations d'expansion, de caractère épigénétique, dont la réalité est bien montrée du fait que les croissances réelle et totale ne sont pas superposables.

Ce modelé final n'est pourtant pas, dans sa totalité, épigénétique : le processus orbitaire du frontal se dessine sans intervention de compétition, le squamosal ne vient pas envahir le territoire d'un pariétal absent, etc.

Quelle que soit la nature des facteurs déterminants de la croissance nous savons, pour les quelques cas étudiés, que le « territoire » d'un os n'est pas une propriété immuable. Une unité osseuse a la possibilité de s'étendre au-delà des limites effectivement réalisées; ses variations de superficie d'une espèce à l'autre ne relèvent donc pas, de façon prépondérante, de fusions avec des éléments adjacents. Dans la réalisation des multiples modèles de plans crâniens du monde des Poissons et des Amphibiens, soudures et colonisation ont joué conjointement. Quelles sont, selon les cas, leurs importances respectives? Nous ne pouvons en décider au seul examen des crânes achevés.

Bibliographie sommaire

Allis, E. P. Jr. (1904). The latero-sensory canals and related bones in fishes. *Int. Mschr. Anat. Physiol.*, **21**: 401–502.
Corsin, J. (1968*a*). Influence des placodes olfactives et des ébauches optiques sur la morphogenèse du squelette crânien. *Annls Embryol. exp. Morphog.*, **1**. Sous presse.
Corsin, J. (1968*b*). Rôle de la compétition osseuse dans la forme des os du toit crânien des urodèles. *J. Embryol. exp. Morph*, **19**: 103–108.
Curtis, A. S. G. (1962). Cell contact and adhesion. *Biol. Rev.*, **37**: 82–129.
Devillers, C. (1947). Recherches sur la crâne dermique des téléostéens. *Annls Paléont.*, **33**: 1–94.
Devillers, C. (1955). Adhésivité cellulaire et morphogènese. Dans *Problèmes de structures, d'ultrastructures et de fonctions cellulaires*, ed. André-Thomas, J.,: 139–166. Paris: Masson.
Devillers, C. (1965). The role of morphogenesis in the origin of higher levels of organization. *Syst. Zool.*, **24**: 259–271.
Fell, Honor B. & Robison, R. (1929). The growth, development and phosphatase activity of embryonic avian femora and limb-buds cultivated *in vitro*. *Biochem. J.*, **23**: 767–784.
Guinnebault, M. (1953). Recherches expérimentales sur la genèse de quelques os dermiques chez *Salmo fario* L. Diplôme d'Etudes Supérieures, Paris. Non publié.
Holmgren, N. & Stensiö, E. A. (1936). Kranium und Visceralskelett der Akranier, Cyclostomen und Fische. Dans *Handbuch der vergleichenden Anatomie der Wirbeltiere*, hrsg. Bolk, L., Göppert, E., Kallius, E. & Lubosch, W., **4**: 233–500. Berlin & Wien: Urban & Schwarzenberg.
Holtfreter, J. (1939). Gewebeaffinität, ein Mittel der embryonalen Formbildung. *Arch. exp. Zellf.*, **23**: 169–209.

Hoyte, D. A. N. (1966). Experimental investigation of skull morphology and growth. *Int. Rev. gen. exp. Zool.*, **2**: 345–407.

Jarvik, E. (1948). On the morphology and taxonomy of the Middle Devonian osteolepid fishes of Scotland. *K. svenska VetenskAkad. Handl.*, (3) **25**: 1–301.

Le Danois, Yseult (1959). Etude ostéologique, myologique et systématique des poissons du sous-ordre des Orbiculates. *Annls Inst. Océonogr.*, **36**: 104–120.

Lekander, B. (1949). The sensory-line system and the canal bones in the head of some Ostariophysi. *Acta zool. Stockh.*, **30**: 1–131.

Moss, M. L. (1954). Growth of the calvaria in the rat. *Am. J. Anat.*, **94**: 333–358.

Moy-Thomas, J. A. (1938). The problem of the evolution of the dermal bones in fishes. Dans *Evolution, Essays on Aspects of Evolutionary Biology*, ed. de Beer, G. R.,: 305–319. Oxford: Oxford University Press.

Moy-Thomas, J. A. (1941). Development of the frontal bones of the rainbow trout. *Nature, Lond.*, **147**: 681–682.

Parrington, F. R. (1949). A theory of the relations of lateral lines to dermal bones. *Proc. zool. Soc. Lond.*, **119**: 65–78.

Parrington, F. R. (1956). The patterns of dermal bones in primitive vertebrates. *Proc. zool. Soc. Lond.*, **127**: 389–411.

Pehrson, T. (1922). Some points in the cranial development of teleostomian fishes. *Acta zool. Stockh.*, **3**: 1–63.

Pehrson, T. (1940). The development of dermal bones in the skull of *Amia calva*. *Acta zool. Stockh.*, **21**: 1–50.

Pehrson, T. (1944a). Some observations on the development and morphology of the dermal bones in the skull of *Acipenser* and *Polyodon*. *Acta zool. Stockh.*, **25**: 27–48.

Pehrson, T. (1944b). The development of latero-sensory canal bones in the skull of *Esox lucius*. *Acta zool. Stockh.*, **25**: 135–157.

Pehrson, T. (1947). Some new interpretations of the skull in *Polypterus*. *Acta zool. Stockh.*, **28**: 400–455.

Schowing, J. (1961). Influence inductrice de l'encéphale et de la chorde sur la morphogenèse du squelette crânien chez l'embryon de poulet. *J. Embryol. exp. Morph.*, **9**: 326–334.

Spiegelman, S. (1945). Physiological competition as a regulatory mechanism in morphogenesis. *Q. Rev. Biol.*, **20**: 121–146.

Tatarko, K. (1934). Restitution des Kiemendeckels des Karpfens. Ein Versuch des Studiums der Formbildung durch Analyse der Wechselbeziehungen zwischen Form und Funktion bei der Restitution. *Zool. Jb. (Allg. Zool.)*, **53**: 461–500.

Troitsky, W. (1932). Zur Frage der Formbildung des Schädeldaches. *Z. Morph. Anthrop.*, **30**: 504–532.

Weiss, P. (1928). Erzwingung elementarer Strukturverschiedenheiten am *in vitro* wachsender Gewebe. *Arch. EntwMech. Org.*, **116**: 438–554.

Wolff, E. (1952). La culture d'organes embryonnaires *in vitro*. *Rev. sci. Paris*, **3317**: 189–198.

The gular plates and branchiostegal rays in *Amia*, *Elops* and *Polypterus*

By Hans Jessen

Section of Palaeozoology, Swedish Museum of Natural History, Stockholm, Sweden

Supporting dermal bones of the gill-cover of acanthodians, crossopterygians, actinopterygians and dipnoans have been investigated on various occasions. According to current opinion these dermal bones constitute an operculo-gular series—or the remains of such a series—primitively associated with the hyoidean gill-cover.

By definition the dermal bones of the hyoidean gill-cover should be supported by the hyoid arch either through articulation, contact or connective tissue attachment. As far as known, this is normally the case for the opercular bones and the branchiostegal rays. In contrast, the antero-ventrally placed gular plates frequently seem to have no direct relation with the hyoid arch. In order to gain some information on these matters, an investigation of the gular plate in *Amia* and *Elops* was undertaken. For comparison the condition of *Polypterus* was also considered. All observations were made on alizarin-stained material.

Description

The gular plates are of different shapes in *Amia*, *Elops* and *Polypterus* (Figs. 1, 2 A, C; Goodrich, 1930, figs. 297, 301, 304). A feature they have in common is their location between the lower jaw rami. Furthermore the gular plates are situated ventral to the intermandibular and anterior interhyoidean musculature, but have no firm attachment to these muscles.

The roughly quadrangular gular plate of *Amia* (Figs. 1 A, 2 A) extends forwards to the symphysis of the lower jaw. Strong connective tissue (*ct*) ties the gular plate at its anterior and antero-lateral margins to the ventro-medial boundary of the anterior end of the lower jaw. The intermandibular musculature (*m.im.a*, *m.im.p*; see Allis, 1897, pl. 31) is situated above the anterior half of the gular plate. The m. interhyoideus anterior runs from the ceratohyal in an antero-medial direction and meets its antimere above the gular plate between the hinder part of the m. intermandibularis posterior and the m. geniobranchialis (*m. ih.a*, *m.gbr*, Figs. 1 A, 3 A). Posteriorly, the gular plate covers an antero-lateral area of the anterior branchiostegal rays and the antero-medial part of the hyoi-

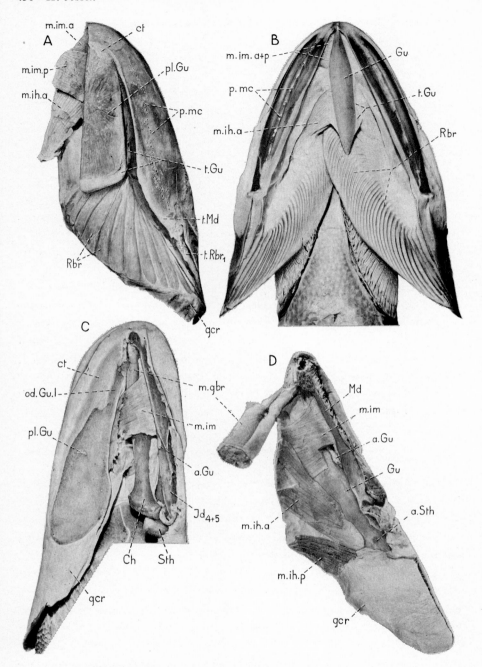

Fig. 1. The intermandibular division in various teleostomians. *A*, *Amia calva*, head in ventral view with skin, hyoidean gill-cover, half gular plate and ventral parts of visceral arches of right side removed; nat. size. *B*, *Elops saurus*, ventral view of head with dilated gill-covers; skin underneath intermandibular musculature and m. interhyoideus anterior removed; nat. size. *C*, *Polypterus lapraedi*, head in ventral and slightly lateral view; gular plate with surrounding skin and m. interhyoideus anterior of left side removed; ×3/2. *D*, *P. bichir*, right

dean gill-cover (*od.Gu*, *Rbr*, *gcr.m*). Lateral to the postero-lateral corner of the gular plate the associated skin (*t.Gu*) fuses with the hyoidean gill-cover (*gcr*) at the ventro-lateral border of the m.interhyoideus anterior.

At its anterior end the comparatively long and narrow gular plate in *Elops* (Fig. 1 B; cf. *Megalops* and *Albula*: Nybelin, 1960; Liem, 1967) extends forwards to the lower jaw symphysis where a narrow band of strong connective tissue connects the gular plate to the ventro-medial boundary of the anterior end of the lower jaw (Fig. 3 B). The intermandibular musculature is situated above the anterior third of the gular plate. The m. interhyoideus anterior takes a similar course as in *Amia* and its anterior portion meets its fellow of the opposite side in the median line immediately behind the m. intermandibularis posterior and above the posterior two-third of the gular plate. As in *Amia*, the posterior portion of the gular plate covers parts of the anterior branchiostegal rays and the hyoidean gill-cover. The skin associated with the gular plate at the m. interhyoideus anterior passes over into the antero-lateral part of the hyoidean gill-cover.

The paired gular plate of *Polypterus* is roughly triangular in shape (Figs. 1 C–D, 2 C, 3 C; Allis, 1922) with its anterior end situated a short distance behind the lower jaw symphysis. The gular plate is firmly attached to the ventro-medial boundary of the lower jaw by strong connective tissue at its antero-lateral margin and an adjoining area of its outer side. The m. intermandibularis (*m.im*) is situated above the anterior half of the gular plate. The m. interhyoideus anterior behaves similarly as in *Amia* and *Elops* and its anterior portion, located above the posterior half of the gular plate, extends forwards between the m. intermandibularis and m. geniobranchialis. Posteriorly, the skin associated with the gular plate fuses with the hyoidean gill-cover. However, an antero-medial part of the hyoidean gill-cover (*gcr.m*) does not pass over into this skin but runs in an antero-dorsal direction above the postero-medial corner of the gular plate.

In *Amia* and *Elops* the connection between the skin associated with the gular plate and the hyoidean gill-cover is found underneath the m. interhyoideus

---

lower jaw and gular plate with parts of the musculature of the intermandibular division in dorsal view; anterior portion of m. geniobranchialis turned to the left; ×3/4.

Ch, ceratohyal; *Gu*, gular plate; Id_{4+5}, posterior infradentaries; *Md*, lower jaw; *Rbr*, branchiostegal rays; *Sth*, stylohyal.

a.Gu, articulation of gular plate with lower jaw; *a.Sth*, shallow depression on inner side of gular plate for the supporting ventral head of stylohyal; *ct*, connective tissue; *gcr*, hyoidean gill-cover; *m.gbr*, musculus geniobranchialis; *m.ih.a*, *m.ih.p*, musculi interhyoideus anterior and posterior; *m.im*, musculus intermandibularis; *m.im.a*, *m.im.a+p*, *m.im.p*, musculi intermandibularis anterior and posterior; *od.Gu.l*, area overlapped by left gular plate; *pl.Gu*, pit-line of gular plate; *p.mc*, pores of mandibular sensory canal; *t.Gu*, *t.Md*, $t.Rbr_1$, skin associated with gular plate, lower jaw and dorsal branchiostegal ray.

anterior, lateral to the free, posterior part of this plate. In *Polypterus*, the corresponding skin is fused with the hyoidean gill-cover in a similar manner, even at the posterior margin of the gular plate. In all three forms, an antero-medial part of the hyoidean gill-cover remains independent of the gular plate and the skin associated with that plate. This part is rather broad in *Amia* and *Elops* and contains the anterior branchiostegal rays (Figs. 1 A, B, 3 A, B).

Among recent actinopterygians with a gular plate only *Amia* displays plate-like branchiostegal rays. The 10 lower branchiostegal rays of the right side of *Amia* are of different shapes and arranged in a special way within the row (Fig. 2 A). The posterior ray of the row (Rbr_2, second ray counted from the opercular bones) has the largest dimensions. In its dorsal and posterior parts it is overlapped by the first, dorsalmost ray ($od.Rbr_1$). The anterior margin of the second ray runs in an antero-dorsal direction and this ray has an antero-dorsal protruding corner, which is partly concealed by the skin associated with the first ray ($od.t.Rbr_1$; Jarvik, 1963, fig. 1 B). This skin flap extends in an antero-ventral direction attaining contact with the skin fold at the ventro-medial boundary of the lower jaw ($t.Rbr_1$, $t.Md$, Fig. 1 A). Together they form a skin fold, which covers the antero-lateral, somewhat dorsally bent, ends of the branchiostegal rays 3 to 8 (Rbr_{3-8}, $od.t.$ $Rbr_1 + Md$). The third ray of the row is similar to the fourth. In its posterior part it is overlapped by ray 2. The same kind of overlapping occurs throughout the row from rays 2 to 10 ($od.Rbr_{1-9}$). In the same direction, the overlapped areas diminish in length and width. Between the rather narrow, anterior part of the third ray and the corresponding portion of the second, there is a distinct notch without a covering dermal bone situated directly behind the lower jaw. Anteriorly at its lower margin the third ray is overlapped by the fourth ray. Throughout the row rays 3 to 10 show a similar way of overlapping (od_1Rbr_{4-11}; *Birgeria*: Nielsen, 1949: 249, fig. 75), with a gradual increase in the length of the overlapped areas. The ornamentation is indistinct on the three anterior rays (Rbr_{9-11}), a circumstance which is explained by the increasing degree of antero-lateral and postero-medial overlap of the three rays by the gular plate with associated skin and the opposed gill-cover, respectively ($od.Gu$, $od.gcr.l$).

Similar observations are difficult to make on the lower branchiostegal rays of *Elops* because these rays show no or only slight mutual overlapping (Figs. 1 B, 3 B). In fossil actinopterygians information in this respect is practically nil. In *Palaeoniscus?* sp. from the Permian of Germany certain conditions of overlap could be recognized throughout the row of branchiostegal rays, but the exact way of overlapping is difficult to make out with certainty and seems to be rather complicated particularly in the area behind the lower jaw. As a whole the manner of overlap seems in this case to be somewhat different from that in *Amia*. In contrast, the shape and way of overlapping of the branchio-

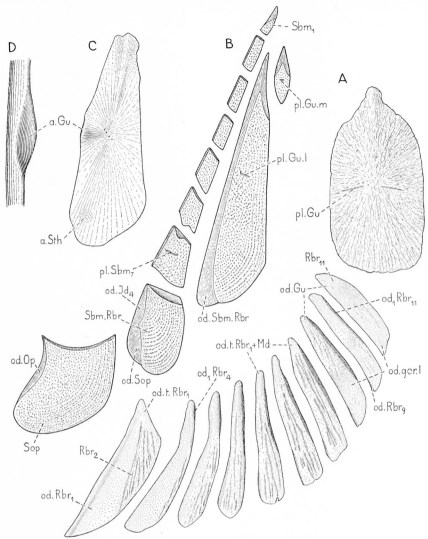

Fig. 2. *A–B*, Dermal bones of the hyoidean and mandibular gill-covers, spread out on the same level. *A*, *Amia calva*, ventral view; ×2/3. *B*, *Eusthenopteron foordi*, compilation based on Jarvik, 1944, Fig. 9 C–D; ×8/9. *C–D*, *Polypterus lapraedi*, gular plate in internal aspect (*C*); ×3/2 and area of gular plate articulating with lower jaw, lateral view (*D*); ×4,5.

Rbr_2, Rbr_{11}, branchiostegal rays 2 and 11; Sbm_1, anterior submandibular; *Sbm.Rbr*, submandibulo-branchiostegal plate; *Sop*, subopercular.

a.Gu, articulation of gular plate with lower jaw; *a. Sth*, shallow depression on inner side of gular plate for the supporting ventral head of stylohyal; *od.gcr.l*, area overlapped by left hyoidean gill-cover; *od.Gu*, area overlapped by gular plate; $od.Id_4$, area overlapped by posterior infradentary; *od.Op*, area overlapped by opercular; $od.Rbr_1$, $od.Rbr_9$, areas overlapped by dorsal branchiostegal ray and branchiostegal ray 9; od_1Rbr_4, od_1Rbr_{11}, areas overlapped by branchiostegal ray 4 and 11; *od.Sbm.Rbr*, area overlapped by submandibulo-branchiostegal plate; od.Sop, area overlapped by subopercular; $od.t.Rbr_1$, $od.t.Rbr_1 + Md$, areas overlapped by skin associated with dorsal branchiostegal ray and lower jaw; *pl.Gu*, pit-line of gular plate; *pl.Gu.l*, *pl.Gu.m*, pit-lines of lateral and median gular plate; $pl.Sbm_7$, pit-line of submandibular 7.

stegal rays of *Caturus, Pachycormus* and *Hypsocormus* (Lehman, 1949) agree closely with those of *Amia*.

Among early, fossil teleostomians the shape and position of the dermal bones of the hyoidean gill-cover and intermandibular region are known in detail in *Eusthenopteron* (Fig. 2 B; Jarvik, 1944; 1963). A comparison with *Amia* reveals some interesting similarities. The supporting elements of the mandibular gill-cover (Sbm_{1-7}, $Sbm.Rbr$, Fig. 2 B; "lateral gulars" of Regan, 1904) overlap in a posterior direction within the row. This way of overlapping approaches the condition at the lower or antero-medial margin of the branchiostegal rays in *Amia* (od_1 Rbr_{4-11}). Between the subopercular and the presumed posterior submandibular element of the submandibulo-branchiostegal plate of *Eusthenopteron*, a corner is formed ($od.Id_4$) which, like the anterior notch between branchiostegal ray 2 and 3 in *Amia*, is situated directly behind the ventro-caudal end of the lower jaw. The supporting dermal elements of the hyoidean gill-cover overlap within the series in an antero-ventral direction ($od.Op$, $od.Sop$, $od.Sbm.Rbr$) comparable to the overlapping at the upper or postero-lateral margin of the branchiostegal rays in *Amia* ($od.Rbr_{1-9}$). These similarities might mean that the nine anterior branchiostegal rays of *Amia* are of a compound nature. One could imagine that these rays in the direction of the gular plate contain an increasing amount of elements belonging to the mandibular gill-cover and a decreasing amount of elements of the hyoidean gill-cover. However, this hypothesis cannot be proved at present; hence the branchiostegal rays of *Amia* may still be regarded as supporting elements of the hyoidean gill-cover because of their connection with the hyoid arch (for *Australosomus* see Nielsen, 1949, fig. 1, sections 50, 80).

On removing the gular plate in *Amia* and *Elops*, it is seen that the antero-medial part of the hyoidean gill-cover extends in an antero-dorsal direction independent of this plate with associated skin (*gcr.m*, Fig. 3 A, B). This part contains the antero-ventral portion of the m. interhyoideus posterior, which anteriorly is continued by a medial and a lateral fibre bundle (*cf*, fig. 3 A, B; *Amia*: McMurrich, 1885; Allis, 1897, figs. 43–45; *Esox*: Vetter, 1878, pl. 13, fig. 10; *Trigla*: Dietz, 1914, fig. 27). The lateral of those two fibre bundles extends to the hypohyal of the same side; the medial one cross its antimere, approaches the lateral bundle of the opposite side and follows the latter to the hypohyal of that side (Holmqvist, 1911; Edgeworth, 1935). Thus, the gular plate of *Amia* and *Elops* seems to be completely independent of the hyoid arch. The gular plate is situated antero-ventral to the lower portion of the hyoid arch, separated from it by the intermandibular musculature and the anterior part of the m. interhyoideus anterior. Moreover, the fusion between the skin associated with the gular plate and the hyoidean gill-cover underneath the m. interhyoideus anterior is the only connection between these two structures. Under these cir-

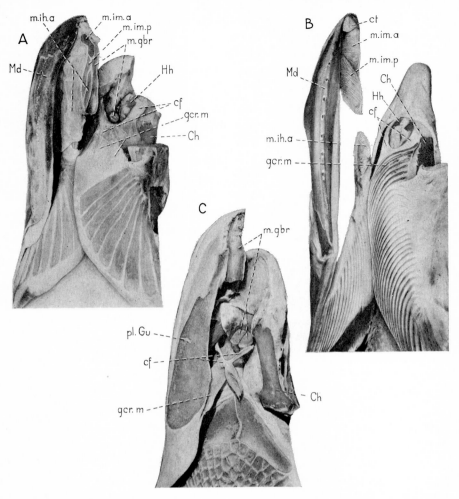

Fig. 3. The intermandibular division in ventral and slightly lateral view with skin, left lower jaw and parts of the musculature of the intermandibular division removed. *A*, *Amia calva*; nat. size. *B*, *Elops saurus*; ×5/4. *C*, *Polypterus lapraedi*; ×3/2.

Ch, ceratohyal; *Hh*, hypohyal; *Md*, lower jaw.
cf, crossing fibre bundles; *ct*, connective tissue; *gcr.m*, medial part of hyoidean gill-cover; *m.gbr*, musculus geniobranchialis; *m.ih.a*, musculus interhyoideus anterior; *m.im.a*, *m.im.p*, musculi intermandibularis anterior and posterior; *pl.Gu*, pit-line of gular plate.

cumstances the gular plate only arbitrarily can be considered a supporting element of the hyoidean gill-cover. To do so would imply that the gular plate is formed by anterior, secondarily separated dermal elements of the hyoidean gill-cover. This would involve a complete detachment of such elements from the hyoidean gill-cover and their subsequent forward displacement underneath the intermandibular musculature. *An alternative explanation is that the gular plate includes supporting dermal elements or vestiges of such elements of the man-*

dibular gill-cover. Supporting this hypothesis may be the existence of strong connective tissue between the anterior part of the gular plate and the lower jaw. If this is correct, the gular plate would be developed from anterior submandibular elements which are secondary enlarged in a posterior direction (cf. the descriptions of *Amia* by Allis, 1889; and Pehrson, 1940). Because the gular plate of *Caturus*, *Pachycormus* and *Hypsocormus* is similar to that in *Amia* in shape and location it most likely also would include elements of the mandibular gill-cover. To which gular plate of fossil palaeoniscoids (e.g. *Moythomasia*: Jessen, 1968) the plate of *Amia* and *Elops* should be compared is difficult to decide. Both the median and the lateral gular plates in palaeoniscoids possess a pit-line which could correspond to that in *Amia* (*pl.Gu*).

In this connection it is of interest that the hyoidean gill-cover of *Polypterus*, as mentioned above, also possesses an antero-medial part, which is independent of the gular plate and the tissue associated with this plate (*gcr.m*, Fig. 3 C). Unlike that of *Amia* and *Elops*, this part is narrow, carries no branchiostegal rays and contains a weak m. interhyoideus posterior (Pollard, 1892; Allis, 1922, fig. 50). Nevertheless this medial part meets its antimere above the midpoint of the medial margin of the gular plate and here both send off bundles of fibres to the hypohyals as in *Amia* and *Elops*. Consequently, the hyoidean gill-cover of *Polypterus* shows some similarity with that in *Amia* and *Elops*, but this does not immediately warrant an interpretation of the crossopterygian-like gular plate in *Polypterus* as a submandibular element.

In *Polypterus*, contrary to the condition of *Amia* and *Elops*, the tissue associated with the gular plate passes over into the hyoidean gill-cover along its whole posterior breadth. Moreover, the gular plate is supported on the inner side by the stylohyal in a shallow, postero-laterally situated depression (*a.Sth*, Figs. 1 D, 2 C) and has connection also with the ceratohyal by fibrous tissue. This may indicate that at least the posterior portion of the gular plate belongs to the hyoidean gill-cover. Yet the possibility cannot be entirely rejected that submandibular elements might have aquired a secondary relationship to the hyoid arch. Thus the attachment of the gular plate to the hyoid arch may be of the same nature as the articulation in *Eusthenopteron* of the submandibular 7 and the submandibulo-branchiostegal plate with the anterior and posterior ceratohyal respectively (Jarvik, 1963, figs. 8 C, 12 A). Attention should also be paid to the cartilage-covered prominent process at the lower border of the ceratohyal in *Latimeria* (Millot & Anthony, 1958, figs. 17, 20, pl. 41), which possibly, as in *Polypterus*, constitutes a support for the gular plate.

Finally, it is a remarkable fact that the gular plate of *Polypterus* articulates with the ventro-medial boundary of the posterior infradentaries (*a.Gu*, Id_{4+5}, figs. 1 C, D, 2 C, D). This articulation is situated behind the strong connective tissue between the gular plate and the lower jaw and directly ventro-caudal to the lateral part of the m. intermandibularis. *These circumstances tend to show that at least the anterior portion of the gular plate of Polypterus belongs to the mandibular gill-cover.*

Conclusions

It is not certainly known that a mandibular gill-cover occurred in the most primitive actinopterygians and brachiopterygians. In the recent *Amia, Elops* and *Polypterus* a mandibular gill-cover with its own series of submandibular elements like that of the porolepiforms (*Holoptychius*: Jarvik, 1963, fig. 11 A, B) certainly is not present. Furthermore, in these recent forms it is impossible to demonstrate an independent submandibular series of the kind found in the intermandibular region of osteolepiforms and dipnoans (Jarvik, 1963; 1967, fig. 5 B). However, the overlap of the intermandibular branchiostegal rays in *Amia*, as well as their development behind the lower jaw are suggestive of those found in the submandibular and operculo-gular series of *Eusthenopteron*. That these branchiostegal rays, as presumably the submandibulo-branchiostegal plate of *Eusthenopteron*, contain submandibular components is a plausible hypothesis, but one without decisive support. It would seem for the moment that these branchiostegal rays, because of their clear relations with the hyoid arch, should be considered as elements of the hyoidean gill-cover. In contrast, the gular plate of *Amia* and *Elops* has little or no relation with the hyoid arch. Seemingly independent of the hyoidean gill-cover, it is here suggested to belong to a mandibular gill-cover.

The paired gular plate of *Polypterus* is connected both with the hyoid arch and the lower jaw, and accordingly may be a composite derivate, formed of dermal elements from both a mandibular and a hyoidean gill-cover. In any event, this plate suggestive of the gular plates of crossopterygians, may conceivably contain submandibular components at least in its anterior portion.

In summary, it would seem that the dermal elements of the intermandibular region of recent actinopterygians and brachiopterygians collectively are of a nature too complex to be interpretable entirely as derivates of the hyoidean gill-cover. Present evidence suggests that the gular plate is the most likely element to represent dermal components of the mandibular gill-cover or at least to comprise vestiges of such components.

This study was supported by grant No. 213–34 from the Swedish Natural Science Research Council (Statens naturvetenskapliga forskningsråd), Stockholm.

References

Allis, E. P. Jr. (1889). The anatomy and development of the lateral line system in *Amia calva. J. Morph.*, **2**: 463–564.
Allis, E. P. Jr. (1897). The cranial muscles and cranial and first spinal nerves in *Amia calva. J. Morph.*, **12**: 487–808.
Allis, E. P. Jr. (1922). The cranial anatomy of *Polypterus*, with special reference to *Polypterus bichir. J. Anat.*, **56**: 189–294.
Dietz, P. A. (1914). Beiträge zur Kenntnis der Kiefer- und Kiemenbogenmuskeln der Acanthopterygier. *Mitt. zool. Stn. Neapel*, **22**: 99–162.
Edgeworth, F. H. (1935). *The cranial muscles of vertebrates*. Cambridge: Cambridge University Press.
Goodrich, E. S. (1930). *Studies on the structure and development of vertebrates*. London: Macmillan.
Holmqvist, O. (1911). Studien in der von den NN. Trigeminus und Facialis innervierten Muskulatur der Knochenfische. *K. fysiogr. Sällsk. Lund Handl.*, **22**: 1–79.
Jarvik, E. (1944). On the dermal bones, sensory canals and pit-lines of the skull in *Eusthenopteron foordi* Whiteaves, with some remarks on *E. säve-söderberghi* Jarvik. *K. svenska VetenskAkad. Handl.*, (3) **21**: 1–48.
Jarvik, E. (1963). The composition of the intermandibular division of the head in fish and tetrapods and the diphyletic origin of the tetrapod tongue. *K. svenska VetenskAkad. Handl.*, (4) **9**: 1–74.
Jarvik, E. (1967). On the structure of the lower jaw in dipnoans: with a description of an early Devonian dipnoan from Canada, *Melanognathus canadensis* gen. et sp. nov. In *Fossil Vertebrates*, ed. Patterson, C. & Greenwood, P. H., *J. Linn. Soc. (Zool.)*, **47**: 155–185.
Jessen, H. (1968). *Moythomasia nitida* Gross und *M.* cf. *striata* Gross, devonische Palaeonisciden aus dem Oberen Plattenkalk der Bergisch-Gladbach-Paffrather Mulde (Rheinisches Schiefergebirge). *Palaeontographica*, (A) **128**: 87–114.
Lehman, J.-P. (1949). Étude d'un *Pachycormus* du Lias de Normandie. *K. svenska VetenskAkad. Handl.*, (4) **1**: 1–44.
Liem, K. F. (1967). A morphological study of *Luciocephalus pulcher*, with notes on gular elements in other recent teleosts. *J. Morph.*, **121**: 103–133.
McMurrich, J.P. (1885). The cranial muscles of *Amia calva* (L.), with a consideration of the post-occipital and hypoglossal nerves in the various vertebrate groups. *Stud. biol. Lab. Baltimore*, **3**: 121–153.
Millot, J. & Anthony, J. (1958). *Anatomie de Latimeria chalumnae. 1. Squelette, muscles et formations de soutien*. Paris: Cent. natn. Rech. Scient.
Nielsen, E. (1949). Studies on Triassic fishes from East Greenland 2. *Australosomus* and *Birgeria. Palaeozool. Groenland.*, **3**: 11–309.
Nybelin, O. (1960). A gular plate in *Albula vulpes* (L.). *Nature, Lond.*, **188**: 78.
Pehrson, T. (1940). The development of dermal bones in the skull of *Amia calva. Acta zool. Stockh.*, **21**: 1–50.
Pollard, H. B. (1892). On the anatomy and phylogenetic position of *Polypterus. Zool. Jb. (Anat.)*, **5**: 387–428.
Regan, C. T. (1904). The phylogeny of the teleostomi. *Nat. Hist. Mag.*, **13**: 329–349.
Vetter, B. (1878). Untersuchungen zur vergleichenden Anatomie der Kiemen- und Kiefermusculatur der Fische. *Jena. Z. Naturw.*, **12**: 431–550.

The dentition in the mouth cavity of *Elops*

By Orvar Nybelin

Natural History Museum, Gothenburg, Sweden

Many primitive characters exhibited by species of the recent genus *Elops*, such as the presence of a large gular plate, a well developed ethmoidal commissure, a bucco-hypophysial canal, a processus coronoideus on the Meckelian cartilage etc. are already mentioned in the literature. Here I should like to call attention to a further character which likewise seems to me to be primitive, namely the abundance of dental plates in the mucosa lining the mouth and pharynx.

The dentition in the mouth cavity of *Elops saurus* L. is mentioned by Ridewood (1904), but obviously his material consisted of dried skeletons only and the description is incomplete. With the use of alizarin staining many more details can be observed.

The presence of a well-developed dentition in the mouth cavity, also including dental plates, has been described for some thoroughly investigated holostean species but as far as I know, no illustrations of the dentition in its entirety have been published. The only complete analysis of these structures seems to be that of Jarvik (1954) for *Eusthenopteron*. In spite of the quite different systematic position of this crossopterygian, it is the only form with which *Elops* can be compared. This comparison must consequently be very unequal with regard to many details, but I hope that it will give some idea of the primitiveness of *Elops* in this respect also.

A comparison between the dentition in the roof of the mouth in *Elops* (Fig. 1) and in *Eusthenopteron* (according to Jarvik's reconstruction, 1954, figs. 25, 27 C) shows that there are some striking similarities regarding the tooth-bearing bones. The exposed ventral part of the parasphenoid, pierced by the bucco-hypophysial canal, and the ventral surface of the large entopterygoid carry a fine dentition and the ectopterygoid and the dermopalatine have somewhat larger teeth, but of course not as large as in *Eusthenopteron*. The vomer is an unpaired bone in *Elops*, not paired as in *Eusthenopteron* but its teeth are arranged in two distinct patches, possibly indicating a paired origin of the bone.

In *Elops* as well as in *Eusthenopteron* there are a lot of small dental plates on the gill-arches; their arrangement is to a large extent almost the same.

The infrapharyngobranchial part of the first branchial arch in *Eusthenopteron* is covered by some rather large dental plates, named by Jarvik ascending and

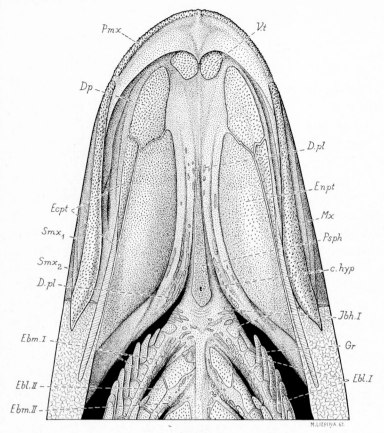

Fig. 1. *Elops saurus* L. The roof of the mouth and anterior part of pharynx.

Dp, dermopalatine; *D.pl.* dental plates medial to the entopterygoid; *Ebl.I, Ebl.II*, lateral epibranchial dental plates on first and second branchial arches; *Emb.I, Emb.II*, medial epibranchial dental plates on first and second branchial arches; *Ecpt*, ectopterygoid (partly cut through); *Enpt*, entopterygoid; *Gr*, gill-rakers; *Ibh.I*, horizontal infrapharyngeal dental plates on first branchial arch; *Mx*, maxillary; *Pmx*, premaxillary; *Psph*, exposed ventral part of parasphenoid; Smx_1, Smx_2, supramaxillaries 1 and 2; *V.t*, patch of vomerine teeth; *c.hyp*, hypophysial canal.

horizontal infrapharyngeal dental plates (Jarvik, 1954, figs. 8 A, 22, 23 A). Small dental plates on the infrapharyngobranchial part of the first branchial arch in *Elops* may be equivalent to these plates (*Ibh.I*, Fig. 1; *Iba.I, Ibh.I*, Fig. 2). As in *Eusthenopteron* the suprapharyngobranchial has no direct relation to any tooth-bearing element.

The dental plates on the epibranchial part of the first branchial arch in *Elops* are not easy to identify. On the medial side of the arch there are numerous dental plates along the base of the gill-lamellae (*Ebm.I*, Fig. 1); they may be equivalent to the medial epibranchial dental plates in *Eusthenopteron*. Along the ventral or pharyngeal side of the epibranchial there is a row of dental plates (*Ebl.I*, Figs. 1, 2) medial to, and alternating with, the gill-rakers which

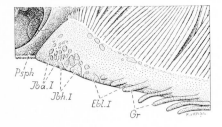

Fig. 2. *Elops saurus* L. Antero-dorsal part of first branchial arch in lateral view.
Ebl.I, lateral epibranchial dental plates; *Gr*, gill-rakers; *Iba.I*, ascending infrapharyngeal dental plates; *Ibh.I*, horizontal infrapharyngeal dental plates; *Psph*, parasphenoid.

may also be modified dental plates (no gill-rakers are found in *Eusthenopteron*). But as these plates are continuous with the horizontal infrapharyngeal dental plates they may be equivalent to the lateral epibranchial dental plates in *Eusthenopteron* in spite of their placement on the ventral side of the epibranchial and not on its lateral side. On epibranchials 2–4 the ventrally situated dental plates are considerably larger than those on the first epibranchial (*Ebl.II*, Fig. 1). On the medial side of these arches there are also rows of medial epibranchial dental plates (*Ebm.II*, Fig. 1). Dental plates are, moreover, found on the dorsal (pharyngeal) side of the ceratobranchials; on the anterior ceratobranchial they become larger anteriorly, and on this part of the first and second gill-arches the medial dental plates are comparatively large.

In addition to the dental plates on the branchial arches, numerous plates of this kind also occur on the medial side of the gill-cover, including the hyomandibular, and on the ceratohyals, i.e. on the lateral wall of the first branchial cleft (Fig. 3). The *Elops* species possess a well developed pseudobranch (*Ps*, Fig. 3) and the dental plates are mainly situated along the base of the pseudobranch and may perhaps be a serial homologue to the medial epibranchial dental plates on the gill-arches. Delicate dental plates are scattered over the medial surface of the gill-cover ventral to the posterior part of the pseudobranch in *Elops lacerta* Cuv. & Val. In *Elops saurus* the medial side of the gill-cover shows, however, many more small dental plates (Fig. 3). Small plates are situated dorsal to the pseudobranch; the area ventral to the posterior part of the pseudobranch is much richer in dental plates and similar plates also occur ventrally to the anterior part of this branch. Small elongate plates are, moreover, present more ventrally along the posterior branchiostegal rays. All these dental plates or at least some of them may perhaps correspond to the medial hyomandibular dental plates in *Eusthenopteron*. On the medial side of the ceratohyals there are also small rounded dental plates scattered over the surface, more numerous and partly larger in *E.saurus* than in *E.lacerta*, and on the anterior ceratohyal there is, dorsally, an almost continuous row of fused dental plates (*Chm*, Fig. 3); these obviously are equivalent to the medial ceratohyal dental plates in *Eusthenopteron* (*Chm*, Jarvik, 1954, fig. 8 B). On the lateral side of the ceratohyals, however, I have not seen any dental plates.

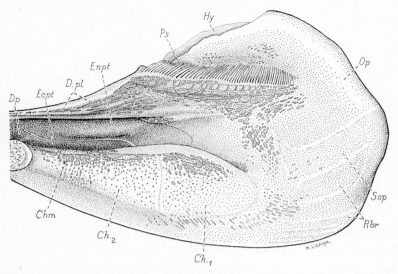

Fig. 3. *Elops saurus* L. Right gill-cover in medial view. Ch_1, Ch_2, posterior and anterio ceratohyals; *Chm*, medial ceratohyal dental plates; *Dp*, dermopalatine; *D.pl*, dental plates medial to the entopterygoid; *Ecpt*, ectopterygoid; *Enpt*, entopterygoid; *Hy*, hyomandibular; *Op*, operculum; *Ps*, pseudobranch; *R.br.* branchiostegal rays; *Sop*, suboperculum.

On horizontally sectioned heads of *Elops saurus* some small dental plates can be observed between the entopterygoid and the exposed part of the parasphenoid (that is lateral to the first branchial cleft), posteriorly forming a more distinct group lateral to the processus ascendens of the parasphenoid (*D.pl*, Figs. 1, 3). The position of this group is consequently about the same as that of the spiracular dental plates in *Eusthenopteron*, and because of this it was tempting to interpret these dental plates as remnants of spiracular dental plates originally situated at the internal opening of the spiracular tube which is, of course, totally obliterated in *Elops*. But, as they seem to form a direct continuation of the dental plates ventral to the anterior end of the pseudobranch (Fig. 3), this interpretation is rather uncertain. Perhaps they may be some kind of horizontal hyomandibular dental plates, possibly corresponding to the outer dentigerous layer of the paraotic dental plates, the inner, bony parts of which being incorporated in the parasphenoid. According to Jarvik (1954: 42) the long axis of the paraotic dental plates forms a proximal continuation of the long axis of the hyomandibula. In *Elops* the dental plates in question are situated more anteriorly in relation to the parasphenoid than are the paraotic dental plates in *Eusthenopteron* but this seems to me to be of minor importance. The small scattered dental plates medial to the anterior part of the entopterygoid might perhaps be isolated remnants of dentigerous plates belonging to the infrapharyngomandibular. In any case all these small dental plates in the roof of the mouth medial to the entopterygoid must be remnants of a former general presence of dental plates in the mucuous lining the mouth, perhaps without direct homologies in *Eusthenopteron*. For the moment I prefer to leave the question regarding their identification unanswered.

Be my interpretation of some of the dental plates in *Elops* as it may, the main purpose of this small communication has been to show that, as far as I know, dental plates are more abundant in the mouth cavity of *Elops* than in teleosts in general. On the medial surface of the gill-cover in the likewise primitive megalopids I have found only the row of fused dental plates along the dorsal margin of the anterior ceratohyal in a small specimen of *Tarpon* and in a rather large specimen of *Megalops*. These two species lack a pseudobranch which might signify that the presence of dental plates stands in certain relation to the gills. In no other teleostean hitherto examined by me can the medial ceratohyal dental plates be observed.

The similarities regarding the dental plates of *Elops* and *Eusthenopteron* mentioned here do not, of course, reflect a closer relationship between these genera but may depend upon the development from a common, much more primitive stage. In his analysis of the visceral skeleton in *Eusthenopteron* Jarvik starts from "a hypothetic primitive stage, showing numerous denticles in the mucuous lining of the dorsal parts of the visceral arches and the walls of the intervening gill-slits". Such a primitive stage was certainly characteristic of the primitive teleostomians, and was followed by a stage with a rich array of small dental plates. These plates then have partly fused into larger plates or bones, and partly become reduced in various ways in the different phyletic lines of development of the teleostomians. Their presence to such an extent as shown here for *Elops* seems to me to underline the primitiveness of this genus.

References

Jarvik, E. (1954). On the visceral skeleton in *Eusthenopteron* with a discussion of the parasphenoid and palatoquadrate in fishes. *K.svenska VetenskAkad.Handl.*, (4) **5**: 1–104.

Ridewood, W. G. (1904). On the cranial osteology of the fishes Elopidae and Albulidae, with remarks on the morphology of the skull in the lower teleostean fishes generally. *Proc.zool.Soc.Lond.*, **1904**: **2**: 35–81.

Evolution of cell size in lower vertebrates

By Henryk Szarski

Hoyer Department of Comparative Anatomy, Jagiellonian University, Kraków, Poland

Vertebrate cell size has been the subject of numerous publications (see summaries by e.g. Levi, 1925; Teissier, 1939). In more recent times some ensuing problems were discussed by Rensch (1954). There is nevertheless a lack of generalizations in this field. The reason is to be found in the fact that only the size of erythrocytes is known in a large amount of species, and the mammals are the only class in which the cell sizes of many different organs are known. The tacit assumption that the erythrocyte size is correlated with that of tissue cells is not always correct. Mammalian erythrocytes, deprived of nuclei, are distinctly smaller than tissue cells. In nonmammalian vertebrates this difference is generally less pronounced. In *Tragulus*, the erythrocytes are the smallest known in any vertebrate (1,5 microns in diameter; Duke, 1963), whereas the liver cells have an average size for a mammal, and the kidney cells are even above that average (Levi, 1925). It is also known that the visual cells of birds are distinctly smaller than these of other vertebrates. Such facts have sometimes been disregarded in order to connect cell size with some other factors. In the majority of cases the correlation of the cell size of various tissues seems indeed to be rather close, permitting the following generalizations.

It is usually assumed that dipnoans and caudate amphibians have the largest cells among vertebrates. Elasmobranchii and Cyclostomata have somewhat smaller cells, and they are still smaller in Anura and Reptilia. Finally, the smallest cells characterize teleosts, birds, and mammals.

It is sometimes stated that there is no correlation between the cell size and the total size of the animal. This is rarely true, although Goss (1964) rightly points out that small elements in the organism tend to vary in number rather than size. In a few cases the comparison of two related species of different size demonstrates that the total number of cells is similar and the entire difference of size could be accounted for by the difference in cell size. This was claimed by Kaufman (1930) for the pigeon and the chick. In the majority of cases the cells of a larger species within a genus, or a larger specimen within a species, are somewhat larger and simultaneously much more numerous than in smaller ones (e.g. Szarski & J. Czopek, 1965; Szarski & G. Czopek, 1966). It is not uncommon, however, to find similar cell size in animals widely differing in weight. Thus, for instance, the liver cells of diminutive *Acris crepitans* have a size similar to those of the huge *Rana catesbeiana* (Szarski & J. Czopek, 1965), and there is no difference in liver cell size between a cow and a rat (Rensch, 1954). But in two cow breeds, the larger has larger muscle cells (Górski, 1965), the same being found in two breeds of mice (Berezowski, 1910). An interesting example of correlation between cell and animal size was described recently by Gauthier & Padykula (1966) in the muscles of the mammalian diaphragm. In this case the correlation probably results from the fact that the diaphragm is a respiratory muscle and that large mammals, having a comparatively low tissue metabolism, respire less frequently.

Very varied are the fates of cell sizes in ontogeny. During segmentation the cell size obviously goes down (e.g. Marrable, 1965). This process is usually prolonged during larval life (e.g. Szarski & Cybulska, 1967). Amphibian metamorphosis is accompanied by a considerable decrease in cell size (see e.g. Church, 1961; Szarski & J. Czopek, 1965; Fox, 1966). The final stages of development on the other hand, are often characterized by an increase in cell size. The tissues may be classified in three groups (Messier & Leblond, 1960; Bertallanfy, 1964) : (1) *static tissues* (e.g. nervous and contractile tissues), in which cells are unable to divide and growth takes place only by cellular hypertrophy or by accumulation of intercellular material; (2) *expanding tissues* (illustrated by many glands), which enlarge by cell divisions; and (3) *renewing tissues* (various epithelia, blood cells), which must proliferate to replace cells being lost. Even in expanding and renewing tissues an increase in cell size was often observed during growth of the organism (e.g. Illing, 1905; Church, 1961; Szarski & J. Czopek, 1965; Szarski & G. Czopek, 1966).

How is it possible to account for these facts? It has often been stated that large cell size must limit the cell metabolism through the surface to mass ratio, but few attempts have been made to demonstrate this relation experimentally. One can quote H. M. Smith (1925) who showed that amphibians possessing large erythrocytes produce less CO_2 per gram than those with smaller ones. Vernberg (1955*a*; 1955*b*) observed that *Plethodon cinereus* is more active, shows a higher glucose blood level, and has smaller cells than the related *P. glutinosus*. We have tried to correlate the liver cell size of several amphibian species with the amount of respiratory capillaries per gram of body weight (Szarski & J. Czopek, 1965). It was found that all species possessing large cells have a small amount of respiratory capillaries per gram, but among animals with small cells the amount of respiratory capillaries varies considerably. Thus, for instance, *Siren intermedia* has small cells and few respiratory capillaries per gram. Liver cells have, however, a storage function and are accordingly easily influenced by the nutritional state of the individual (Morgulis, 1911; Bush, 1963). The correlation of the amount of respiratory capillaries per gram with the kidney cell size appears closer (Sianko, in preparation).

It has been demonstrated (Henneman & Olson, 1965; Henneman, Somjen & Carpenter, 1965a; 1965b) that the largest neurons and the largest muscle fibers in mammals have the lowest metabolism and are rarely active, whereas the smallest neurons and the smallest muscle fibres are constantly in an active state producing the muscle tone. The smallest muscle fibres have a high level of ATP-ase, contain a large amount of mitochondria, and are always close to blood capillaries, whereas the thick fibres are poor in ATP-ase, contain few mitochondria, and have an irregular contact with capillaries.

These examples suggest that cell size is often inversely correlated with cell metabolism.

The cell size may be also influenced by the fluctuations in osmotic pressure of body fluids. When a marine animal enters fresh or brackish waters it usually swells and its body fluids become diluted. In euryhaline species the swelling diminishes after a time when the animal becomes adapted to the lower salinity and its nephridia excrete excess water (Krishnamoorthi & Krishnaswamy, 1967). The animal may then return to the initial concentration of body fluids or retain their lowered concentration. Any change in the composition of body fluids influences the cells which may either reduce their internal osmotic pressure or actively expel water. The ability to withstand the variations of the internal medium is probably a function of the surface to volume ratio of the cell, therefore an increase in cell size may be advantageous for sea animals invading brackish waters.

It is sometimes maintained that the ancestors of vertebrates were marine chordates and that vertebrate origin was connected with the invasion of brackish and fresh waters. If so, a simultaneous increase in cell size can be regarded as probable: the tunicates and *Amphioxus* have very small cells, the cells of lower vertebrates are distinctly larger. Even Myxinoidea, which are isoosmotic with sea-water, have rather large cells. These animals, however, may be secondarily marine, and have had freshwater ancestors.

Another factor which possibly is influential on cell size is the difference between the ionic composition of the cell fluid and of the extracellular fluids. In the cell membrane there is a continously acting "sodium pump" which works against the forces tending to equalize the ionic concentrations. The difference in ions between the two faces of the cell membrane is the cause of membrane potential which forms the basis of cell excitability. The work of the sodium pump requires energy, the amount of energy depending on the surface to mass ratio of the cell. It follows that an increase in cell size may be advantageous in certain circumstances as it reduces the expenditure of energy.

As the ratio of the nucleus to plasma fluctuates within narrow limits, although it is always shifted in favour of plasma in large cells, very large cells contain a considerable amount of DNA. There are few examples of polyploid species among vertebrates (e.g. Uzell, 1964), but polyploid cells are common in many tissues (e.g. Teissier, 1939; Zubina, 1963; Zhorno & Ovchinnikova, 1964). The nuclei of polyploid cells presumably contain many sets of genetic information. It is, however, unknown whether the enormous amount of DNA present in nuclei of e.g. urodels (Goin, Goin & Bachmann, 1967), contain redundant genetic information (see discussion by Commoner, 1964; Beermann, 1966). If so, then the animals possessing large cells may be less susceptible to injury by mutagenic agents. The possibility of DNA repair has so far been demonstrated, however, only in Bacteria (Hanawalt & Haynes, 1967).

Among various invertebrate groups it has been demonstrated several times that tissue cells contain less DNA than the cell of the germ-line. The somatic cells are in these cases often very small (e.g. in some insects). The cells of vertebrates are always larger and a reduction of chromatin in somatic cells has never been observed. If in vertebrates the tissue cells differ in their DNA content from the cells of the germ-line it is always a result of polyploidy. Polyploidism of tissue cells is probably the simplest way to increase the cell volume. It occurs when the maintenance of volume is the principal cell function (e.g. in the chorda, Wallace, 1963), or when the cells are employed in storage, as in vertebrate liver, or in the fat body of insects (*Lucilia coesar;* Teissier, 1939). On the other hand, it probably decreases metabolism by diminishing the relative cell surface.

Study by means of the electron microscope opens the way to connect cell size with the size and amount of cellular organelles. Kruszyński & Boothroyd (1964) found that in the axolotl, the mitochondria and desmosomes of gigantic epithelial cells are not larger than usual, whereas very large mitochondria occur in the small cells of insect flight muscles (Chapman, 1954). Horridge (1964), however, demonstrated that the largest mitochondria known are present in extremely large cells of Ctenophora. Further studies of this relation could give very interesting results.

It can be inferred from the above that cell size is probably related in several ways to cell function. So far, these complicated relations are incompletely understood. Some tentative hypotheses can, however, be formulated as a frame for further research. I suggest that as the ancestors of early vertebrates invaded first brackish and later fresh waters, their dimensions enlarged, their metabolic level decreased, and their cell size increased. The large cell size persisted in groups having low metabolism, as in cyclostomes and elasmobranchs, and probably even increased in animals which retained a low metabolism and were often subject to variations of osmotic concentration of body fluids, such as Dipnoi (H. W. Smith, 1953) and Amphibia (Bentley, 1966). In teleosts and amniotes evolution increased the level of metabolism and thus the cell size decreased.

So much for the comparison of the cell size of larger systematic units. When one compares the cell size of closely related species it is possible to draw some conclusions about the evolution of size within the group. The working hypothesis starts from the fact than any evolutionary increase in the size of the species is usually accompanied by a decrease in metabolic level, and any decrease in size by an increase in metabolism (Hemmingsen, 1960; Szarski, 1964; Gould, 1966). One of the consequences of a change in size of the animal will therefore be a tendency to change also in cell size. The forces that are working for the adjustment of cell size to the metabolic level probably have a very low efficiency. Therefore, within smaller systematic units the cell size will probably be correlated with the size of the ancestor.

Thus, for instance, as mentioned above, the liver cells of the huge *Rana catesbeiana* are not larger than those of the diminutive *Acris crepitans* but larger than those of the medium-sized *Scaphiopus couchi*, *Hyla arborea*, and small *H. crucifer*. In *Rana catesbeiana* the ratio of respiratory capillaries to body size is rather low. We conclude that, first, *R. catesbeiana* ancestors were much smaller, second, that the increase in the size of the species is fairly recent, third, that the ancestors of *Acris* were larger, and fourth, that the reduction of size in this line is more recent than in that leading to *Hyla crucifer*.

Let us now turn to the diversification of cell sizes in a vertebrate organism. Small cell dimensions characterize those tissues in which the cells are continuously replaced during the life of the organism. The largest cells are among the striated muscle fibres and neurons which multiply only during the embryonal development. This is probably a consequence of developmental mechanisms.

The neuroblasts multiply as long as they have no processes. They first assume their relative positions in the nervous system and then develop the processes which interconnect them into a definite pattern. In this stage the organism is still small, as are the neuroblasts. The glial cells are few, and in consequence the developing cell processes have only a short way to pass in order to reach their targets. Later the embryo enlarges, the glial cells multiply, and the formation of new interconnections between nerve cells becomes an impossibility. The creation of new nerve cells is then useless. They only enlarge their dimensions.

A similar process may be responsible for the cessation of multiplication of striated muscle fibres in ontogeny. Every striated muscle cell must be individually innervated for its activity and existence. The nerve processes reach the fibres and assume their characteristic distribution when the fibres are small. The formation of the new connections between nerve cells and muscle fibres is later impossible, so that muscle cells cannot multiply but only enlarge.

The multiplication of connective tissue cells, of gland cells, etc. remains possible through life because these cells are not innervated individually. The nerve endings reach the cell groups and the chemical mediators stimulate their targets by diffusion. This is possible, as the timing of reactions may be less precise. The necessity of an immediate reaction is the original cause of the early cessation of mitoses in nerve cells and in the striated muscles.

The diversification of nerve cell size probably developed during vertebrate evolution. This leads us to another problem. The cell size limits metabolism by the change in the surface to mass ratio. The large nerve cells solve the difficulty by an extraordinary development of the cell membrane into a complicated pattern of processes and invaginations. Even endocellular capillaries have been described (Nakajima, Pappas & Bennet, 1965). The T-system of membrane infoldings in the muscle cells allows some diffusion of substances but does not compensate for the decreased surface to mass ratio. This is proved by a calculation performed by Shoshenko (1966) on data collected by G. Czopek (1963). It was demonstrated that during the growth of the frog muscle the ratio of the surface of muscle cells to the surface of blood capillaries remains constant, although the diameters of muscle cells are greatly enlarged.

The muscles constitute up to 50 % of the mass of a vertebrate animal and form the principal site of metabolism. It could be asked : why do not all the cells enlarge their size simultaneously with the muscle cells? If it is true that muscle metabolism is limited by the muscle cell size, the metabolism of other cells whose principal role consists in providing the muscle cells with metabolic materials may perhaps be decreased in a similar degree. We know, however, that the cells of vegetative tissues usually remain quite small, even in very large animals. The answer may be the following:

First, muscle metabolism is fully active only in short periods of high activity. Such periods are neither common nor prolonged and are separated by periods of rest. During bursts of activity the muscle cells work in an anaerobic way, forming an oxygen debt which is paid off during rest. Hence, the limitation of muscle metabolism by the size of cells may perhaps not be as acute as is suggested by their surface to mass ratio.

Secondly, the metabolic level of muscle tissues is of great importance for the individual, but even more crucial for the species is the metabolism of the female reproductive system, responsible for the elaboration of yolk or for supplying the embryos with food and oxygen. The efficiency of the reproductive system depends on the work of all vegetative organs, therefore the small size of the cells may be favoured.

Summary

Cell size is probably influenced by many diverse circumstances, such as the size of the animal, its usual level of metabolism, the fluctuations in the osmotic pressure of its body fluids, the level of membrane potential of its cells, etc.

It is suggested that during vertebrate emergence from marine chordates the cells gained in size, reaching the dimensions seen in recent Cyclostomata and Elasmobranchii. A further increase in cell size took place during the evolution of Dipnoi and Urodela. Teleostei and Amniota are generally characterized by a progressive increase in metabolism accompanied by a decrease in cell dimensions.

The comparison of cell size in closely related animals may give information about the total size of the common ancestor.

The early cessation in ontogeny of neuroblast multiplication and the inability of muscle fibres to divide in late ontogenetic stages are consequences of developmental mechanisms.

References

Beermann, W. (1966). Differentiation at the level of the chromosomes. In *Cell Differentiation and Morphogenesis*: 24–54. Amsterdam: North Holland Publ. Co.

Bentley, P. J. (1966). Adaptations of amphibia to arid environments. *Science*, **152**: 619–623.

Berezowski, A. (1910). Studien über die Zellgrösse. 1. Über das Verhältnis zwischen der Zellgrösse und der Gesamtgrösse des wachsenden Organismus. *Arch. Zellf.*, **5**: 375–384.

Bertallanfy, L. v. (1964). Basic concepts in quantitative biology of metabolism. *Helgoland wiss. Meeresunt.*, **9**: 5–38.

Bush, F. M. (1963). Effects of light and temperature on the gross composition of the toad, *Bufo fowleri*. *J. exp. Zool.*, **153**: 1–13.

Chapman, G. B. (1954). Electron microscopy of ultra-thin sections of insects flight muscle. *J. Morph.*, **95**: 237–262.

Church, G. (1961). Auxetic growth of the Jawanese toad, *Bufo melanostictus*. *Science*, **133**: 2012–2014.

Commoner, B. (1964). DNA and the chemistry of inheritance. *Am. Scientist*, **52**: 365–388.

Czopek, G. (1963). The distribution of capillaries in muscles of some Amphibia. *Studia Soc. Sci. Torun.*, (E) **7**: 60–98.

Duke, K. L. (1963). Erythrocyte diameter in *Tragulus javanicus*, the chevrotain or mouse deer. *Anat. Rec.*, **147**: 239–241.

Fox, H. (1960). Early pronephric growth in *Neoceratodus* larvae. *Proc. zool. Soc. Lond.*, **134**: 659–663.

Gauthier, G. F. & Padykula, H. A. (1966). Cytological studies of fiber types in skeletal muscle. A comparative study of the mammalian diaphragm. *J. Cell Biol.*, **28**: 333–354.

Goin, O. B., Goin, C. J. & Bachmann, K. (1967). The nuclear DNA of a caecilian. *Copeia*, **1967**: 233.

Górski, A. (1965). Układ ważniejszych elementów w komórce mięśniowej u bydła rasy nizinnej czarno-białej i rasy czerwonej polskiej odmiany górskiej. *Zootechnika Wrocław*, **13**: 37–45. In Polish with English Summary.
Goss, R. J. (1964). *Adaptive growth.* New York & London: Academic Press.
Gould, S. J. (1966). Allometry and size in ontogeny and phylogeny. *Biol. Rev.*, **41**: 587–640.
Hanawalt, P. C. & Haynes, R. H. (1967). The repair of DNA. *Sci. Amer.* **216**: 36–43.
Hemmingsen, A. M. (1960). Energy metabolism as related to body size and respiratory surfaces, and its evolution. *Rep. Steno Mem. Hosp.*, **9**: 1–110.
Henneman, E. & Olson, C. B. (1965). Relations between structure and function in the design of skeletal muscles. *J. Neurophysiol.*, **28**: 581–598.
Henneman, E., Somjen, G. & Carpenter, D. O. (1965a). Functional significance of cell size in spinal motoneurons. *J. Neurophysiol.*, **28**: 560–580.
Henneman, E., Somjen, G. & Carpenter, D. O. (1965b). Excitability and inhibitibility of motoneurons of different sizes. *J. Neurophysiol.*, **28**: 599–620.
Horridge, G. A. (1964). The giant mitochondria of ctenophore comb plates. *Q. J. micr. Sci.*, **105**: 301–310.
Illing, G. (1905). Verschiedene histologische Untersuchungen über die Leber der Haussäugetiere. *Anat. Anz.*, **26**: 177–193.
Kaufman, L. (1930). Innere und äussere Wachstumfaktoren. Untersuchungen an Hühnern und Tauben. *Arch. EntwMech. Org.*, **122**: 395–431.
Krishnamoorthi, B. & Krishnaswamy, S. (1967). Some considerations of the osmotic and ionic regulation in Polychaetes. *Pol. Arch. Hydrobiol.*, **14**: 7–20.
Kruszyński, J. & Boothroyd, B. (1964). Ultrastructural aspects of the nucleo-cytoplasmic relationship of invaginated nuclei in the epidermal cells of the axolotl. *Acta anat.*, **56**: 79–92.
Levi, G. (1925). Wachstum und Körpergrösse. Die strukturelle Grundlage der Körpergrösse bei vollausgebildeten und im Wachstum begriffenen Tieren. *Erg. Anat. EntwGesch.*, **26**: 86–342.
Marrable, A. W. (1965). Cell numbers during cleavage of the zebra fish egg. *J. Embryol. exp. Morph.*, **14**: 15–24.
Messier, B. & Leblond, C. P. (1960). Cell proliferation and migration as revealed by radioautography after injection of thymidine-H^3 into male rats and mice. *Am. J. Anat.*, **106**: 247–295.
Morgulis, S. (1911). Studies of inanition in its bearing upon the problem of growth. 1. *Arch. EntwMech. Org.*, **32**: 169–268.
Nakajima, Y., Pappas, G. D. & Bennett, M. V. L. (1965). The fine structure of the supramedullary neurons of the puffer with special reference to endocellular and pericellular capillaries. *Am. J. Anat.*, **116**: 471–492.
Plenk, H. (1911). Über Änderungen der Zellgrösse im Zusammenhang mit dem Körperwachstum der Tiere. *Arb. zool. Inst. Wien*, **19**: 247–288.
Rensch, B. (1954). *Neuere Probleme der Abstammungslehre. Die transspezifische Evolution.* Stuttgart: Enke Verl.
Shoshenko, K. A. (1966). K voprosu o metodike opredelenia stepeni vaskularizatsii skeletnuikh muishts (On method of determination of vascularization degree of skeletal muscles) *Arch.Anat.Hist.Embr.Leningr.*, **51**: 102–107. In Russian.
Smith, H. M. (1925). Cell size and metabolic activity in Amphibia. *Biol. Bull. Woods Hole*, **48**: 347–378.
Smith, H. W. (1953). *From fish to philosopher.* Boston: Little, Brown.
Szarski, H. (1964). The structure of respiratory organs in relation to body size in Amphibia. *Evolution*, **18**: 118–126.
Szarski, H. & Cybulska, R. (1967). Liver cell size in *Protopterus dolloi* Blngr. (Dipnoi). *Bull. Acad. polon. Sci. Cl. II*, (B) **15**: 217–220.

Szarski, H. & Czopek, G. (1966). Erythrocyte diameter in some amphibians and reptiles. *Bull. Acad. polon. Sci. Cl. II*, (B) **14**: 433–437.
Szarski, H. & Czopek, J. (1965). Liver cell size in some species of Amphibia. *Zool. Polon.*, **15**: 51–64.
Teissier, G. (1939). Biométrie de la cellule. *Tabulae Biol.*, **19**: 1–64.
Uzell, T. M. (1964). Relations of the diploid and triploid species of the *Ambystoma jeffersonianum* complex (Amphibia, Caudata). *Copeia*, **1964**: 257–300.
Vernberg, F. J. (1955a). Hematological studies on salamanders in relation to their ecology. *Herpetologica*, **11**: 129–133.
Vernberg, F. J. (1955b). Correlation of physiological behavior indexes of activity in the study of *Plethodon cinereus* (Green) and *Plethodon glutinosus* (Green). *Am. Midl. Nat.*, **54**: 382–393.
Wallace, H. (1963). A case of somatic polyploidy. *Nature, Lond.*, **199**: 1115–1116.
Zhorno, L. J. & Ovchinnikova, L. P. (1964). A cytophotometric study of DNA in the cells of stratified squamous epithelium. *Folia Hist. Cytoch.*, **2**: 173–180.
Zubina, E. V. (1963). Tsitofotometricheskoe opredelenie soderzhania DNK v iadrakh gigantskikh kletok trofoblasta (Cytophotometrical determination of the DNK content in the nuclei of giant trophoblastic cells). *Dokl. Acad. Sci. USSR*, **153**. 1428–1431. In Russian.

Evolutionary significance of the "prolactin" cells in teleostomean fishes

By Ragnar Olsson

Department of Zoology, University of Stockholm, Stockholm, Sweden

The rostral *pars distalis* is generally a well-defined portion of the anterior pituitary gland of different fishes. It is characterized by its rostral position in the adeno component as well as by the cell types which constitute its parenchyma. Typical cells of this region are a strongly erythrosinophile cell type which, in some specimens, on functional grounds has been compared to the mammalian prolactin cells, a second acidophile type which is probably the ACTH-producer and, sometimes, a basophile type which is considered to be responsible for TSH-synthesis.

There is, however, a considerable variation in morphological pattern of the rostral portions of different fish groups. One type of rostral pars distalis will here be referred to as the *compact type*, because the tissue is made up of closely packed cells which give no room for cavities. This is the most common type among fishes and it is also the one of the tetrapod rostral pars distalis.

In the *follicle type* some of the rostral cells are arranged as epithelia bordering cavities. These cavities may vary in size and number from a large number of tiny structures to one single, giant follicle which occupies most of the rostral region. The follicle type has been described from various fishes, most of them belonging to such groups which the taxonomists consider as "primitive".

A third type of rostral pars distalis organization may be called the (*hypophysial*) *duct type*, because the rostral cavity communicates with the oral cavity through a duct which ontogenetically is the persistent Rathke's pouch. The classic example of this type is the brachiopterygid *Polypterus* but a similar condition has also been described from some larval and immature teleosts.

Material and methods

This investigation is based on the following material:
Elopiformes: *Elops machnata*; 2 immature animals, 60 mm. Two brains from adults.
    Albula vulpes; 3 brains from immature animals, 31–35 cm.
Clupeiformes: *Clupea harengus*; 2 larvae, 33 mm.
    Pellonula vorax; 2 larvae of each 16, 32 and 40 mm. 3 adults.

Osteoglossiformes: *Heterotis niloticus*; 1 adult brain.
  Notopterus afer; 3 adult brains.
Mormyriformes: *Gnathonemus gambiensis*; 1 adult brain.
Salmoniformes: *Salmo salar*; Larvae 9.5, 12, 18, 24 and 27 mm.
Gonorhynchiformes: *Chanos chanos*; 4 larvae, 13 mm, 2 immature animals 23–31 cm.
  Kneria sp.; 1 adult head.
  Grasseichthys gabonensis; 1 adult head.
  Phractolaemus ansorgei; 1 adult head.

Furthermore several series of *Polypterus senegalus* at different stages.

The taxonomic system adopted in this paper is that proposed by Greenwood, Rosen, Weitzmann & Myers (1966).

The heads or brains were cut as paraffin sections, usually in sagittal series. The following staining techniques were used: Chrome alum hematoxylin, chrome alum gallocyanine (Bock, 1966), paraldehyde fuchsin / fast green-orange II, Herlant's tetrachrome stain (Herlant, 1960), periodic acid-Schiff / orange II / alcian blue, and azocarmine / orange II - aniline blue (Heidenhain).

Observations

Polypterus. The polypterids, *Polypterus* and *Calamoichthys*, have pituitaries of the duct type at all stages (references in Gérard & Cordier, 1936; Wingstrand, 1966; Kerr, 1967; Lagios, 1968).

The cavity in the rostral gland forms a few tubular diverticula which open into a common cavity which, in its turn, is ventrally extended as the buccohypophysial duct. The cells which line the cavity are tall and have elongated nuclei and a weakly basophile cytoplasm. The apical cell region contains some material which reacts strongly with periodic acid-Schiff, alcian blue, chrome hematoxylin and paraldehyde fuchsin. It is no doubt that this material is released into the cavity. The material in the cavity and the duct may occur in large amounts. It consists of a finely granulated coagulation with embedded nuclei and other structures which may be cell debris.

The cells of the buccohypophysial duct are considerably lower than the tall cavity cells, but they display the same pronounced staining reactions in the apical region. In the vicinity of the oral epithelium these cells are ciliated.

Elops. The pituitary gland of *Elops machnata* is clearly divided in three portions; the neurointermediate lobe, the proximal pars distalis and the rostral pars distalis (Fig. 1). The rostral part contains a large, funnel-shaped cavity which ventrally forms a duct. The available material of large animals was unfortunately only the brains with surrounding tissues and did not include the oral roof.

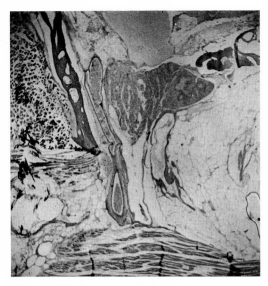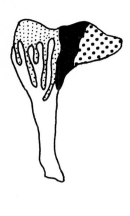

Fig. 1. *Elops machnata*, adult. Sagittal section through the pituitary gland; periodic acid—Schiff—alcian blue; 5 μm; × 38.
Coarsely dotted: neurointermediate lobe; black: proximal pars distalis; finely dotted: rostral pars distalis; white: rostral cavity.

The duct could, however, be followed widely open as far as to the parasphenoid bone in one large animal.

The cells of the rostral pars distalis are arranged as epithelial folds in the rostral cavity. The folds are composed of two cell layers separated by a delicate membrane of loose connective tissue. This tissue is richly vascularized and it is also innervated by nerves which do not carry secretory material with the preoptic staining properties.

The majority of the fold cells are tall and stain strongly acidophile. This staining reaction seems to be caused by coarse granules which are so accumulated in the basal cell pole that this region appears homogeneously acidophile. The nucleus is round and it is situated almost centrally in the cell. The apical cell pole (towards the cavity) stains reddish blue with Herlant's tetrachrome, but it may also contain coarse granules which are strongly erythrosinophile.

Another cell type in the folds is considerably smaller and occurs scattered or in small groups. These cells are always found close to the central nutritive lamina of the folds and they do not seem to reach the lumen border by means of processes. Their cytoplasm gives a strong reaction with alcian blue, paraldehyde fuchsin and periodic acid-Schiff.

At the 60 mm stage the rostral pituitary is of the duct type (Olsson, 1958). The fold formation is not yet as pronounced as in the larger animals but the cytological appearance is much the same in the two. Several cells can be seen

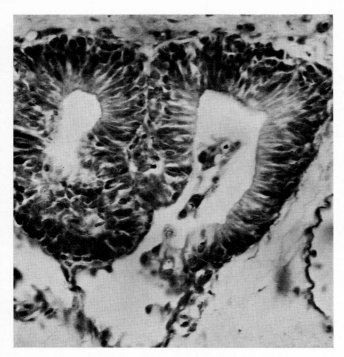

Fig. 2. *Elops machnata*, 60 mm. Transverse section through the rostral cavity. Note the cells in the lumen, some of which contain vacuoles with stainable material. Chrome hematoxyline—floxine; 5 μm, ×800.

in the cavity as well as in the duct in these young animals. These cells are apparently not disintegrating and several of them contain a large vacuole which encloses a chrome hematoxyphile substance (Fig. 2). The cells are released from the cavity epithelium and the same kind of activity seems to occur also in the duct.

Albula. The *Albula* pituitary gland is dominated by the large neurointermediate lobe (Fig. 3). The proximal pars distalis is small and gives the impression of being pushed dorsalwards by the neurointermediate lobe and the rostral cavity. The rostral pars distalis consists of dorsally situated cells and below them a very large cavity which is ventrally continued as a duct. The available material did not allow a complete investigation of this duct but as far as the sections reach, the canal is quite open without signs of obliterations.

The rostral cells do not form two-layered curtains as in *Elops* but rather a system of tubes which communicates with the main cavity. The wall cells of the tubes are very like those in the folds of *Elops*. In this case, however, they often show the same acidophile reaction in the apical cell pole as in the basal one. Also the basal, alcian blue-positive cells, are strikingly like those of *Elops*.

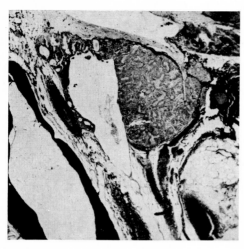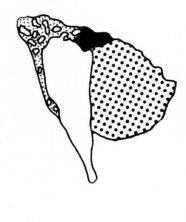

Fig. 3. *Albula vulpes*, immature. Sagittal section; Herlant's tetrachrome; 5 μm; × 38. Symbols as in Fig. 1.

In *Albula* they are found scattered in the tube peripheries. Their nuclei are slightly ovoid and the cytoplasm is sometimes extended in the form of a narrow process.

Some material is often seen in the lumina of the tubes. This material consists of a finely granulated blue substance containing some nucleus-sized bodies which stain blueish with a red core with Herlant's tetrachrome technique. The same kind of material is also found abundantly in the rostral cavity and in the duct and it is very likely that the tube material is emptied into the cavity.

Several nerve fibres in the hypophysial stalk penetrate the rostral pars distalis where they can be seen mingling with the tubes. These fibres do not contain chrome hematoxyphile secretion.

Chanos. The pituitary gland of *Chanos chanos* was studied at different stages by Tampi (1951; 1953), who described a duct in larvae and a closed cavity at older stages.

It is possible to delimit the hypophysial regions already in the 13 mm larvae. The neurointermediate lobe is a small evagination of nervous tissue which is covered by a thin medullary layer of juvenile pituitary cells. In the proximal pars distalis the rather large cells tend to form discrete strands. The rostral pars distalis is a cell tube which is caudally closed but rostrally opens into an extremely thin-walled cavity. This cavity communicates with the oral cavity through the rostro-ventrally directed hypophysial duct.

Most of the cells have the undifferentiated "chromophobe" appearance at this early stage. Very striking exceptions are, however, some cells in the rostral cell tube which display an intense and homogeneous erythrosinophilia. These

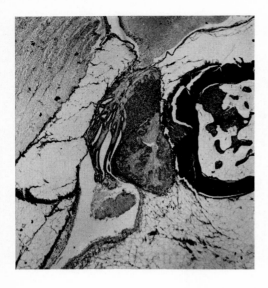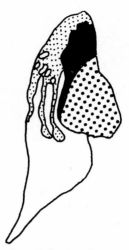

Fig. 4. *Chanos chanos*, immature. Sagittal section; Herlant's tetrachrome; 5 μm; ×38. Symbols as in Fig. 1.

precocious cells appear to release stainable material into the cavity, but very little material can be seen in the cavity as well as in the duct. A weak periodic acid-Schiff reaction could be seen at the lumen border, partly in the form of tiny droplets.

The glands of the investigated immature animals are very like those of the large *Elops* specimens (Fig. 4). The rostral part is made up of a large cavity and epithelial folds. The fold formation is obscured in the lateral parts of the gland where several tubes appear, a condition which recalls the tubular *Albula* gland.

Also in this case the folds and tubes are composed of tall acidophils together with scattered basophils (Fig. 5). Most of the acidophils are very elongated, as the equivalent cells in *Albula* and *Elops*, but in some thin folds they may be almost isoprismatic. They contain a large and almost round nucleus which is situated in the centre of the cell. The cytoplasm contains coarse erythrosinophile granules which are usually most crowded at the basal, nutritive, pole.

The cytoplasm of the basophils is finely granulated or even homogeneous. It is stained blue with Herlant's technique and alcian blue, black with gallocyanine and violet with paraldehyde fuchsin. These cells are scattered in the central portion of the folds where they may also appear in small clusters. Some of them carry short processes which may reach for a short distance between the acidophils, but it has never been seen that such a process reaches the lumen border.

Many peculiar, erythrosinophile flakes can be seen in or on the apical tips of

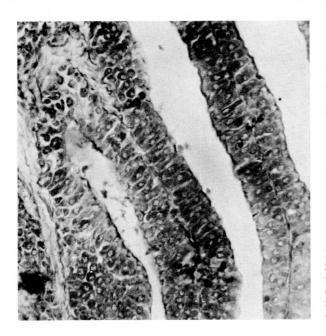

Fig. 5. *Chanos chanos*, detail from Fig. 4, showing the tall, strongly erythrosinophile cells which form the folds of the rostral pars distalis; × 800.

the erythrosinophil cells. It is likely that this is some stage of a material which is released into the cavity. Such flakes are also seen in the cavity and together with them a finely granulated material which has basophile properties.

The dorsal region of the rostral pars distalis has a brighter appearance than the folds because the affinity to the acid dyes is much less pronounced here. The cells of this region are comparatively small with small nuclei and their cytoplasm is best characterized as "chromophobe".

Both the neurointermediate lobe and the rostral pars distalis are richly innervated. The former receives fibres with an abundance of gallocyaninophile material while the numerous bundles which penetrate the fold tissue show no secretory material with the techniques used here.

Clupea and **Pellonula.** The pituitary gland of *Clupea* was carefully investigated at different stages by Buchmann (1940), who described a duct in larval fish and a follicle-condition after metamorphosis. His results could, in all fundamental parts, be confirmed by my material and it will be added here only that the 3 mm *Clupea* larvae are remarkably like the 13 mm *Chanos* larvae in pituitary morphology.

The rostral distal part of *Pellonula* contains one large cavity which gives off a few short, tubular branches. (Fig. 6). The cavity epithelium is as usual formed by tall, acidophile cells which have a special affinity to erythrosine but also stain with orange. Some of these cells contain a few basophile inclusions at the basal

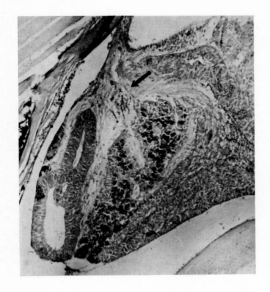

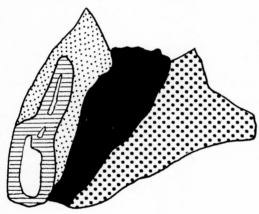

Fig. 6. *Pellonula vorax*, adult. Sagittal section. Several nerve fibres penetrate the pars distalis and some of them carry paraldehyde fuchsinophile material (arrow). Periodic acid—Schiff; 5 μm; × 500. Symbols as in Fig. 1.

cell pole. Several small rounded basophils can be seen also in this specimen in, or close to, the follicle epithelium.

Several nerve fibres penetrate different parts of the *Pellonula* hypophysis. One of the preparations indicates that some fibres which carry paraldehyde fuchsinophile material pass close to the follicle epithelium and may have a functional relationship to the follicle cells.

It was not possible to find an open duct in any of the investigated *Pellonula* larvae. The cavity is, however, in the young animals ventrally extended and forms a narrow canal. At some distance below the cavity this canal is transformed into a strand of loose connective tissue. None of the investigated specimens contained much material in the cavity, but most of them contained a sparse finely granulated and weakly basophile material.

Salmo. Several authors have investigated the pituitary gland of different salmonids and described follicles in the rostral pars distalis. A few observations only will be added here which are relevant to the problem of follicle structure and formation.

Kawamoto (1967) has described that the rostral cavity of the clupeid *Harengula zunasi* is ventrally open, although it is not continued in the form of a duct. A similar condition could be observed in my immature *Pellonula* specimens, but the cavity is here not open; it is ventrally covered by an extremely thin, squamous epithelium. This region seems to be the remnant of the obliterated duct. Also in *Salmo* some of the superficial follicles are covered by a region of thin and apparently non-secretory epithelium which faces the surrounding tissue (Fig. 7). These conditions have been mentioned in some detail here because the morphologic requisites seem to exist at this stage for a hormone diffusion from the cavity to the surrounding tissue (and its vessels). If such a process could be proved to take place, it could be considered as an intermediate stage between secretory release through a duct and release from a well-drained, compact endocrine organ.

It has not been possible to establish an open communication between follicles and the buccal cavity at any of the *Salmo* embryo stages. The anlage of the adeno component is solid and the first traces of follicles appear *de novo* at the 9.5 mm stage, where they have tiny, if any, cavities. At the 12 mm stage the cells are actively secreting material into the follicle cavities.

Gonorhynchiform fishes. Greenwood, Rosen, Weitzman & Myers (1966) have, for several reasons, placed *Chanos* as a member of the group Gonorhynchiformes in their tentative system of teleost fishes. Because of its peculiar pituitary morphology, which closely reminds that of the elopid fishes, it has already been treated here.

The available *Phractolaemus*-specimen had not been preserved for a microscopical investigation and did not allow a cytological study. It is however clear that this fish has neither follicles, cavities nor duct in the rostral pars distalis. The same is the case with *Kneria* and *Grasseichthys*, although the former has a few rosette arrangements of the elongated rostralis cells.

Osteoglossomorph fishes. The osteoglossomorph fishes have retained several features which are considered to be primitive. For this reason the pituitary morphology was investigated in three members of this group, viz. *Heterotis*, *Notopterus* and *Gnathonemus*. The glands of these fishes were, however, all of the compact type.

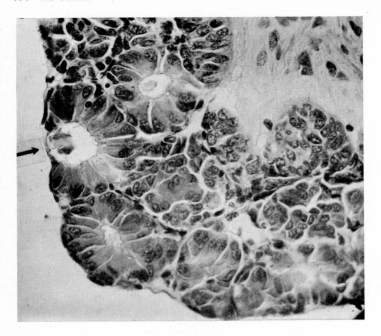

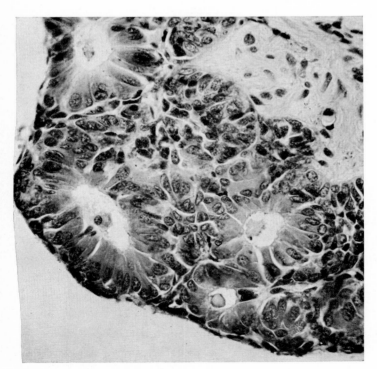

Fig. 7. *Salmo salar*, 75 mm. Two sagittal sections through the rostral pars distalis. Above: one follicle is partly covered by a very thin epithelium (arrow); Below: one follicle with a free cell in the lumen. Chrome hematoxyline; 4 μm; ×800.

Discussion

Morphological variations in the rostral pituitary. The duct type of rostral pars distalis (Fig. 8 A) contains a large cavity which is in communication with the oral cavity through a duct. This is the condition which is found at all stages of the brachiopterygids *Calamoichthys* and *Polypterus*. A pituitary duct was also described in large but immature *Elops machnata* (*E. saurus*) (Olsson, 1958) and in premetamorphic clupeids (Buchmann, 1940) and *Chanos chanos* (Tampi, 1951; 1953).

A gland of the "follicle type" may contain one single, large cavity (*Pellonula*), a few medium-sized cavities (*Salmo*) or several minute ones (*Acipenser*). We have a fairly good idea about the distribution of glands of this type in the fish system as a result of several reports:

Ganoids: *Amia calva, Lepisosteus osseus, Acipenser fulvescens* (Kerr, 1949).

Teleosts: *Clupea harengus* (Buchmann, 1940). *Engraulis telera, Gadusia chapra, Hilsa ilisha* (Misra & Sathyanesan, 1959). *Harengula zunasi* (Kawamoto, 1967).

Anguilla anguilla (Hagen, 1936; Olivereau, 1963; 1965; Knowles & Vollrath, 1966; Vollrath, 1967). *A. japonica* (Honma, 1966; Kawamoto 1967). *A. rostrata* (Scruggs, 1939). *Conger conger* (Knowles & Vollrath, 1966). *Oxyconger leptognathus, Astroconger myriaster, Muraenesox cinereus* (Kawamoto, 1967).

Salmo salar (Woodman, 1939). *S. trutta* (Scruggs, 1939, Kerr, 1940). *Oncorhynchus tschawytscha* (Scruggs, 1939). *O. rhodurus* (Kawamoto, 1967). *Salvelinus fontinalis* (Scruggs, 1939).

Chanos chanos, from the 53 mm stage (Tampi, 1951; 1953).

Fugu rubripes, F. poecilonotus (Kawamoto, 1967). *Limanda herzensteini* (Honma, 1960; not confirmed by Kawamoto, 1967).

The rostral cavities. Wingstrand (1966) has recently summarized the available reports on different kinds of cavities in the pituitary glands of various fishes. Significant differences exist, however, between the epithelium-lined cavities of the rostral pars distalis and a hypophysial cavity such as arises in for example dipnoans (Kerr, 1940). The former always form follicles which are consistently lined by tall cells with a characteristically acidophile cytoplasm. They are also always restricted to the frontal region of pars distalis.

It is likely that a genetic relationship exists between these different cell-lined cavities in the rostral pituitary, in spite of the great variations in appearance. The duct-condition is retained during the whole life-cycle of the recent brachiopterygids (Fig. 8 A). In clupeids and *Chanos* a transformation takes place during the ontogenesis from the juvenile duct-condition (Fig. 8 A) to an adult follicle stage (Fig. 8 B). Salmonids, anguillids and some other fishes have follicular

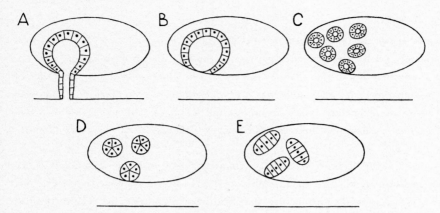

Fig. 8. Diagrammatic representation of the arrangement of the erythrosinophile epithelia cells in the rostral pars distalis of various fishes. *A*, "duct type" in communication with the oral cavity; *B*, *C*, "follicle types"; *D*, *E*, "compact types", in *D*, forming rosettes without cavities, in *E*, with only traces of epithelia-like arrangement.

pituitaries at all stages (Fig. 8 C). It is possible that we have indications even of a transformation from the follicle stage to the compact one (Fig. 8 D). Scruggs (1939) thus observed that the rostral acidophils in the cyprinid *Notemigonus* tend to be arranged in small circles, although without cavities. Kerr (1940) found tiny cavities in the rostral pituitary of a few weeks old *Perca*, but in this case the regular orientation of cells around them was lacking. The compact type, finally, occurs in the majority of fishes, but also in this case the tall acidophile cells are characteristically arranged in an epithelium-like manner. (Fig. 8 E).

The different cavities in the rostral gland which have been considered here are either the remainder of the original invaginated hypophysial (Rathke's) pouch (*e.g. Polypterus*) or schizocels which appear *de novo* in a solid anlage (*e.g. Salmo*). These different ways of formation do not speak against a phyletic kinship of the cavities. The ependyma-lined brain cavities are formed in one of these two ways and nevertheless it is no doubt that the brain ventricles are homologous structures in all craniates.

Phyletic considerations. In his illustration showing the evolution of different chordate endocrines Hoar (1965: 194) derived the duct pituitary from a simple compact gland like that of lampreys. It is, however, more likely that the hollow condition is primitive (Wingstrand, 1966); perhaps the series in Fig. 8 illustrates these evolutionary changes. This view is supported not only by ontogenetic evidence but also by the distribution of the different types among recent fishes. Fig. 9 is based on the provisional system of Greenwood, Rosen, Weitzman & Myers (1966) and illustrates in which groups pituitaries with rostral cavities may be found.

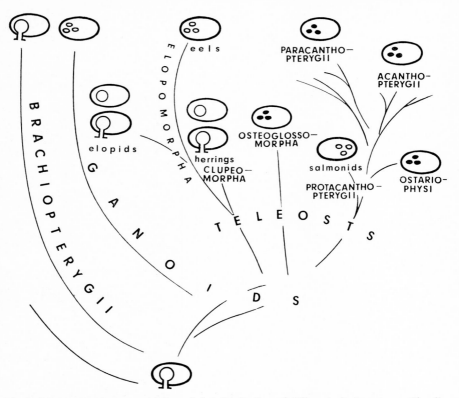

Fig. 9. Diagrammatic representation of the distribution of different pituitary types. The diagram shows in which groups a certain type has been demonstrated but does not indicate that all members of each group apply to the type which is illustrated by the symbol. *Chanos* and *Fugu* have not been considered in the diagram. If *Chanos* is not considered as an elopid, the follicle type is not yet established for this group.

Glands with a duct are found in the brachiopterygids (*Calamoichthys* and *Polypterus*) and in at least young stages of some of the most primitive teleosts, viz. elopids and clupeids. The recent ganoids *Amia*, *Lepisosteus* and *Acipenser* have reached the follicle stage and so have the anguillids on the elopomorph and salmonids on the acanthopterygid stem of teleost evolution. Compact glands are found in the more advanced protacanthopterygid, paracanthopterygid and acanthopterygid fishes.

Chanos has not been considered in the diagram. It was placed by Greenwood, Rosen, Weitzman & Myers as a protacanthopterygid in the order Gonorhynchiformes. As regards pituitary morphology it is, however, not like the gonorhynchiforms *Kneria*, *Grasseichthys* and *Phractolaemus* which all have compact glands. Instead, the rostral pars distalis of *Chanos* is strikingly like that of *Elops* and *Albula*.

Kawamoto (1967) has described "follikelartige Rosettenanordnung der Drüsenzellen" in two of the four investigated *Fugu* species (Tetraodontidae).

It is not clear from her description if these structures can be compared to the follicles of other teleosts. If this is the case, the occurrence of follicles in some tetraodontids may be an argument in favour of those taxonomists who consider these fishes as primitive. It may also be an example of how the follicle stage, for some reason, is retained in an advanced group.

The osteoglossomorph fishes are generally considered to be primitive and it may surprise that only compact glands have been found here. But it must be remembered that even though these fishes have retained several osteologic and other features of certain primitiveness, they have also acquired many specializations like electric organs, advanced brain development etc.

The fact that the recent cyclostomes possess no apparent cavities is not very surprising. If they have the cells which in fishes line the cavities, they may, of course, for functional reasons, have passed to the compact stage, although their pituitaries in most other respects have retained very generalized features. It must also be noted that follicles have been described in the pro- and mesoadenohypophysis of lampreys (Rühle & Sterba, 1966), but too little is known about these structures to allow any comparisons with the fish follicles.

Functional considerations. It is known from several investigations that the erythrosinophile cells in the rostral pars distalis of "compact" teleost pituitaries are the equivalents of the mammalian prolactin cells (Olivereau, 1963; cf. Bern, 1967 for the extensive literature on fish "prolactin", omitted below). These cells produce a prolactin-like substance (paralactin; Ball, 1965a) which has been attributed a variety of effects in fishes (Bern, 1967). One important function of this hormone seems to be to limit sodium loss and thus to prevent a dangerous fall of plasma osmolality if the osmolality of the surrounding water decreases (Ball & Olivereau, 1964; Olivereau & Ball, 1964; Ball, 1965b; Ball & Ensor, 1965; Pickford, Robertson & Sawyer, 1965). According to these investigations the hormone of the erythrosinophile cells is essential for fresh water survival of some euryhaline fishes (*Fundulus, Poecilia*).

It seems to be these "paralactin" cells which form the follicles in eels and salmonids (Olivereau, 1963) and in this case release their secretory substance into cavities. Experimental studies seem to be lacking so far in salmonids, but it is known that the pituitary gland of the Atlantic eel is not essential for freshwater survival (Fontaine, Callamand & Olivereau, 1949). Nevertheless, a removal of the paralactin cells causes electrolyte loss also in the eel (Maetz, Mayer & Chartier-Baraduc, 1967), although this effect is much less drastic than in the investigated poecilids. Morphologic studies show that the erythrosinophile cells in the follicles respond to a change in osmolality of the surrounding medium (Olivereau, 1967a; 1967b) even if they do not show the hyperactivity in demineralized water (Olivereau & Chartier-Baraduc, 1966) which is typical for the poecilids. We may thus conclude that also the eel produces a hormone which, to a certain extent, prevents sodium loss and that this hormone is produced by the erythrosinophile follicle cells (paralactin cells).

The corresponding cells of the clupeid and elopid follicles and cavities are structurally very like those of salmonids and eels. It is probable that they release substances which are chemically related to the eel paralactin and it is even possible that they have similar properties. Experimental research is here highly desirable for it might demonstrate if the corresponding principle has osmoregulatory properties even at the duct stage and if this principle is released as an exocrine secretion through the buccohypophysial duct. It may be of some interest to note that most of the fishes with a hypophysial duct are pronouncedly euryhaline. Important exceptions to this rule are the polypterids which are purely freshwater animals. The rostral cavity cells of these fishes are basophile and they produce a substance which is probably of mucoid nature (Kerr, 1967). It is, however, reasonable to believe that these cells and their product(s) either reflect a very primitive evolutionary condition, or have evolved in another direction than the corresponding structures of the other recent actinopterygids.

The basophile cells which are consistently seen in the vicinity of the cavity epithelium in primitive fishes are probably thyrotrops. If this is the case, these cells are not exocrine at the duct stage. Only later, when the rostral pars distalis has reached the follicle stage, can the thyrotrops be considered as normal members of the epithelium (cf. Hagen, 1936; Olivereau, 1965; Knowles & Vollrath, 1966).

Final remarks. Several authors (Nicoll & Bern, 1964; 1965, Nicoll, Bern & Brown, 1966; Meites & Nicoll, 1966; Bern, 1967) have pointed out the usefulness of the "prolactins" during the functional evolution of the craniates. Their ability to direct different epithelial structures has been used in various ways and a great number of these structures have been reported as "prolactin" (paralactin) target organs (see Bern, 1967). Besides the incorporation of new targets the evolution has also, at a certain degree, affected the principle itself. The "basic prolactin" of fishes (Nicoll & Bern, 1964) is not identical to that of mammals but nevertheless a part of the molecule is still common to the two (Ball, 1965b; Emmart, Pickford & Wilhelmi, 1966).

It is reasonable to believe that this "skin function hormone" was originally a mucous substance which was produced by secretory cells in the oral epithelium. The function of this substance could have been mechanical and chemical protection of the gastrointestinal mucosa cells. Much like the subpharyngial thyroxin-producers these cells have been accumulated and invaginated as a gland, the secretory duct of which was the hypophysial pouch. The next step in this evolution was the transformation to a follicular endocrine gland, a stage which the thyroid never passed.

If this conjectured mucoid substance had properties which included an osmotic protection of epithelial cells it is probable that such a mechanism worked autonomously in the early Devonian actinopterygids as well as in the recent polypterids which have, as far as we know, never abandoned fresh water. Those early fishes, however, which entered waters with an increased electrolyte content faced quite new problems. The protein portion of the original muco- or glycoprotein, which had the "fresh water properties", was retained and refined, but its release had to be controlled in a hypertonic medium. The activity of the secretory cells was thus subordinated to the central nervous system which had to act as an inhibitor. Another evolutionary step was the integration of the gland as a member (together with the adrenocortical cells, the corpuscles of Stannius and probably also urophysial and pseudobranch cells) of the complex mechanism which regulates water and electrolyte composition of the fish body.

This study was supported by grant No. 2124–19/23 from the Swedish Natural Science Research Council (Statens naturvetenskapliga forskningsråd), Stockholm. Material was placed at my disposal by Dr. R. Nishioka, Berkeley, Calif. (*Albula, Chanos*); Dr. P. H. Greenwood of the British Museum (Natural History), London (*Kneria, Grasseichthys, Phractolaemus*): and Dr. O. Nybelin, Gothenburg (*Clupea*). Most of the African fishes studied were obtained by the Swedish Gambia Expedition of 1950.

References

Ball, J. N. (1965a). A regenerated pituitary remnant in a hypophysectomized killifish (*Fundulus heteroclitus*): Further evidence for the cellular source of the teleostean prolactin-like hormone. *Gen. comp. Endocrinol.*, **5**: 181–185.

Ball, J. N. (1965b). Partial hypophysectomy in the teleost *Poecilia*: Separate identities of teleostean growth hormone and teleostean prolactin-like hormone. *Gen. comp. Endocrinol.*, **5**: 654–661.

Ball, J. N. & Ensor, D. M. (1965). Effects of prolactin on plasma sodium in the teleost *Poecilia latipinna*. *J. Endocrinol.*, **32**: 269–270.

Ball, J. N. & Olivereau, M. (1964). Rôle de la prolactine dans la survie en eau douce de *Poecilia latipinna* hypophysectomisé et arguments en faveur de sa synthèse par les cellules érythrosinophiles de l'hypophyse des téléostéens. *C. R. Acad. Sci. Paris*, **259**: 1443–1446.

Bern, H. A. (1967). Hormones and endocrine glands of fishes. *Science*, **158**: 455–462.

Bock, F. (1928). Die Hypophyse des Stichlings (*Gasterosteus aculeatus*) unter besonderer Berücksichtigung der jahreszyklischen Veränderungen. *Z. wiss. Zool.*, **131**: 645–710.

Bock, R. (1966). Über die Darstellbarkeit neurosekretorischer Substanz mit Chromalaun-Gallocyanin im supraoptico-hypophysären System beim Hund. *Histochemie*, **6**: 362–369.

Buchmann, H. (1940). Hypophyse und Thyroidea im Individualzyklus des Herings. *Zool. Jb. (Anat.)*, **66**: 191–262.

Emmart, E. W., Pickford, G. E. & Wilhelmi, A. E. (1966). Localization of prolactin within the pituitary of a cyprinodont fish, *Fundulus heteroclitus*, by a specific fluorescent antiovine prolactin globulin. *Gen. comp. Endocrinol.*, **7**: 571–583.

Fontaine, H., Callamand, O. & Olivereau, M. (1949). Hypophyse et euryhalinité chez l'anguille. *C. R. Acad. Sci. Paris*, **228**: 513–514.

Gérard, P. & Cordier, R. (1936). Sur la persistance d'une connexion bucco-hypophysaire chez les crossoptérygiens adultes. *Annls Soc. R. zool. Belg.*, **67**: 87–90.

Grant, W. C. & Pickford, G. E. (1959). The presence of the red eft water-drive factor prolactin in the pituitaries of fishes. *Biol. Bull. Woods Hole*, **116**: 429–435.

Greenwood, P. H., Rosen, D. E., Weitzman, S. H. & Myers, H. S. (1966). Phyletic studies of teleostean fishes, with a provisional classification of living forms. *Bull. Am. Mus. nat. Hist.*, **131**: 341–455.

Hagen, F. v. (1936). Die wichtigsten Endokrinen des Flussaals. 3. Hypophyse. *Zool. Jb. (Anat.)*, **61**: 467–538.

Herlant, M. (1960). Etude critique de deux techniques nouvelles destinées a mettre en évidence les différentes catégories cellulaires présentes dans la glande pituitaire. *Bull. micr. appl.*, **10**: 37–44.

Hoar, W. S. (1965). The endocrine system as a chemical link between the organism and its environment. *Trans. R. Soc. Canada*, (4) **3**: 175–200.

Hoar, W. S. (1966). Hormonal activities in the pars distalis in cyclostomes, fish and Amphibia. In *The Pituitary Gland*, ed. Harris, G. W. & Donovan, B. T., vol. **1**: 242–294. London: Butterworths.

Honma, Y. (1960).—Cited from Kawamoto, 1967.

Honma, Y. (1966). Notes on the catadromous eels obtained from off the coast of Niigata, the Sea of Japan, with reference to the histology of some of the organs. *La Mer*, **4**: 241–260.

Kawamoto, M. (1967). Zur Morphologie der Hypophysis Cerebri von Teleostiern. *Arch. Hist. Japon.*, **28**: 123–150.

Kerr, T. (1940). Histogenesis of some teleost pituitaries. *Proc. R. Soc. Edinb.*, **60**: 224–240.

Kerr, T. (1949). The pituitaries of *Amia*, *Lepidosteus* and *Acipenser*. *Proc. R. Soc. Lond.*, **118**: 973–983.

Kerr, T. (1967). Histology of the pituitary of *Polypterus*. *Gen. comp. Endocrinol.*, **9**: 464.

Knowles, F. & Vollrath, L. (1966). Neurosecretory innervation of the pituitary of the eels *Anguilla* and *Conger*. 2. The structure and innervation of the pars distalis at different stages of the life-cycle. *Phil. Trans. R. Soc.* (B) **250**: 329–342.

Lagios, M. D. (1968). Tetrapod characteristics of pituitary gland of *Calamoichthys calabaricus*. *Nature. Lond.*, **217**: 473.

Maetz, J., Mayer, N. & Chartier-Baraduc, M. M. (1967). La balance minérale du sodium chez *Anguilla anguilla* en eau de mer, en eau douce et au cours de transfert d'un milieu à l'autre: Effets de l'hypophysectomie et de la prolactine. *Gen. comp. Endocrinol.*, **8**: 177–188.

Meites, J. & Nicoll, C. S. (1966). Adenohypophysis: Prolactin. *Ann. Rev. Physiol.*, **28**: 57–88.

Misra, A. B. & Sathyanesan, A. G. (1959). On the persistence of the oro-hypophysial duct in some clupeoid fishes. *Proc. int. Congr. Zool. 15th Session, London 1958*: 999–1000.

Nicoll, C. S. & Bern, H. A. (1964). "Prolactin" and the pituitary glands of fishes. *Gen. comp. Endocrinol.*, **4**: 457–471.

Nicoll, C. S. & Bern, H. A. (1965). Pigeon crop-stimulating activity (prolactin) in the adenohypophysis of lungfish and tetrapods. *Endocrinology*, **76**: 156–160.

Nicoll, C. S., Bern, H. A.,& Brown, D. (1960). Occurrence of mammotropic activity (prolactin) in the vertebrate adenohypophysis. *J. Endocrinol.*, **34**: 343–354.

Olivereau, M. (1963). Cytophysiologie du lobe distal de l'adenohypophyse des agnathes et des poissons, à l'exclusion de celle concernant la function gonadotrope In

Cytologie de l'adénohypophyse, ed. Benoit, J. & Da Lage, C.,: 316–329. Paris: Cent. natn. Rech. Scient.

Olivereau, M. (1965). Action de la métopirone chez l'anguille normale et hypophysectomisée, en particulier sur le système hypophyso-corticosurrénalien. *Gen. comp. Endocrinol.*, **5**: 109–128.

Olivereau, M. (1967a). Reactions observées chez l'anguille maintenue dans un milieu privé d'électrolytes, en particulier au niveau du système hypothalamo-hypophysaire. *Z. Zellf.*, **80**: 264–285.

Olivereau, M. (1967b). Observations sur l'hypophyse de l'anguille femelle, en particulier lors de la maturation sexuelle. *Z. Zellf.*, **80**: 286–306.

Olivereau, M. & Ball, J. N. (1964). Contribution à l'histophysiologie de l'hypophyse des téléostéens, en particulier de celle de *Poecilia* species. *Gen. comp. Endocrinol.*, **4**: 523–532.

Olivereau, M. & Chartier-Baraduc, M. (1966). Action de la prolactine chez l'anguille intacte et hypophysectomisée. 2. Effects sur les éloctrolytes plasmatiques (sodium, potassium et calcium). *Gen. comp. Endocrinol.*, **7**: 27–36.

Olsson, R. (1958). A bucco-hypophysial canal in *Elops saurus*. *Nature, Lond.*, **182**: 1745.

Pickford, G. E., Robertson, E. E. & Sawyer, W. H. (1965). Hypophysectomy, replacement theraphy, and the tolerance of the euryhaline killifish, *Fundulus heteroclitus*, to hypotonic media. *Gen. comp. Endocrinol.*, **5**: 160–180.

Rühle, H.-J. & Sterba, G. (1966). Zur Histologie der Hypophyse des Flussneunauges (*Lampetra fluviatilis*). *Z. Zellf.*, **70**: 136–168.

Scruggs, W. M. (1939). The epithelial components of the teleost pituitary gland as identified by a standardized method of selective staining. *J. Morph.*, **65**: 187–213.

Tampi, P. R. S. (1951). Pituitary of *Chanos chanos* Forskål. *Nature, Lond.*, **167**: 686–687.

Tampi, P. R. S. (1953). On the structure of the pituitary and thyroid of *Chanos chanos* (Forskål). *Proc. natn. Inst. Sci. India*, **19**: 247–256.

Vollrath, L. (1966). The ultrastructure of the eel pituitary and the elver stage with special reference to its neurosecretory innervation. *Z. Zellf.*, **73**: 107–131.

Vollrath, L. (1967). Elektronenmikroskopische Untersuchung über die neurosekretorische Innervation der Adenohypophyse von Jungaalen. *Anat. Anz. (Ergänzh.)*, **120**: 205–211.

Wingstrand, K. G. (1966). Comparative anatomy and evolution of the hypophysis. In *The Pituitary Gland*, ed. Harris, G. W. & Donovan, B. T., vol. **1**: 58–126. London: Butterworths.

Woodman, A. (1939). The pituitary gland of the Atlantic salmon. *J. Morph*, **65**: 411–435.

Application of phylogenetic principles in systematics and evolutionary theory

By Lars Brundin

Section of Entomology, Swedish Museum of Natural History, Stockholm, Sweden

Today, 100 years after Darwin, phylogenetic evolution is accepted as a fact. The species of the recent biota are conceived as the result of innumerable speciation processes upwards through time. We realize that the species of the past have either been eliminated by extinction or ceased to exist by splitting and that the recent species are the expressions of continuous but incessantly forking gene flows in time, starting at the latest with the ancestral species of each phylum. We are agreed that the variability of the gene pool of each species is fettered by a rigid system of genetic homeostasis and that the evolutionary potential of the gene pool is able to break through mainly in connection with spatial isolation of a peripheral population, meaning escape from the homeostasis of the mother population and the acquirement of new characters and reproductive isolation. In other words, different properties evolve at different rates, and every speciation process is a confirmation of this.

The principles of phylogenetics and phylogenetic systematics advanced by Hennig (1950; 1966) are logical consequences of the meaning of the speciation process as outlined above. The intimate connection between those principles and the results of population genetics is valid not only for the definition of phylogenetic relationship and monophyly and the all-important conclusion that every recent species and every recent group must have a sister group (species) of the same absolute age in the recent biota, but also for the principles of phylogenetic arguing, using concepts as the deviation rule, synapomorphy, and geographical replacement.

However, surveying the modern literature dealing with the evolutionary process we are struck by the fact that the discussions of supraspecific groups and transpecific evolution are strongly influenced by typological thinking. Largely neglecting the nature of the speciation process and its phylogenetic meaning the authors have not perceived the conclusiveness of that process for a realistic interpretation of transpecific evolution, for proper phylogenetic arguing, and for our ability to reconstruct the history of life in time and space, i.e. biogeography. Instead the evolutionsts have largely concentrated on the role and nature of natural selection, evolutionary trends, adaptations, and grade developments. Lacking proper knowledge of sister-group relationships these discussions are often contradictory and unrealistic.

Only a limited number of students have as yet realized the far-reaching implications of phylogenetic systematics, as set out by Hennig. The few critics appearing till now, for example, Mayr (1965) and Cain (1967), have all missed an essential point, namely that phylogenetics is the search for the sister group.

There is, however, increasing interest in the problems of phylogenesis and phylogenetic systematics, especially among younger zoologists. But the situation is still bewildering. It may suffice to refer to recent papers representing different gradations of "numerical" or "phenetic" taxonomy, a mathematical approach doomed a priori because of its unsound basic principles.

Since the theory of evolution may lay claim to be considered the greatest unifying theory in biology, the controversial situation within vital sections of that theory stands out as a strange phenomenon in biological sciences of today. What is the background to this development and which are more precisely the weak points of the evolutionary theory?

Phylogenetics and classification

Few evolutionists seem to draw the necessary conclusions from the fact that the plant and animal system still is largely typological and thus unable to function as a general reference system. Most of the comparatively primitive groups are paraphyletic and based on superficial over-all similarity, i.e. on the common possession of primitive characters which are always worthless as indicators of phylogenetic relationship. When discussing anagenesis (development of transformation series, i.e. "progressive evolution") and phenomena connected with anagenesis the evolutionists operate with "groups", "lines", or "lineages". These aggregates are freely borrowed from the existing system, and it seems to be taken for granted, or at least hoped for, that they do represent strictly monophyletic groups. This method is convenient but not very critical, since the system in reality is highly provisional, a mixture of monophyletic and non-monophyletic groups.

The provisional state of the present "natural" system is partly a consequence of the immense variedness of the biota and the difficulties to obtain sufficient knowledge of existing species and their characteristics. An important factor is also the strongly varying accessibility of different groups for proper phylogenetic analysis. To this we have to add the fact that the taxonomists still are largely occupied with purely descriptive work of the classical type. Such pioneer work has to be done and will remain actual for a long time. Regrettable is, however, the circumstance that even monographs and discussions of relationships by acknowledged specialists still generally are expressions of typological thinking,

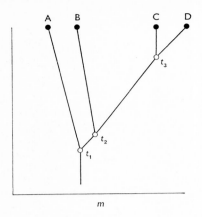

Fig. 1. Schedule explaining the definition of degree of phylogenetic relationship; the dimension *m* symbolizes the morphological differences; *t* time. (From Hennig, 1957, redrawn.)

meaning that the hierarchial grouping of the species treated is founded on degree of morphological similarity. The shortcomings of such a mode of procedure are obvious if we bear in mind that morphological similarity relationships have a reticular structure. Even in systematics the application of phylogenetic principles is still in its infancy.

Phylogenetic principles. In the phylogenetic system the degree of phylogenetic relationship is the decisive point. The structure of that system and the mutual relationships between the groups can be adequately expressed in a schedule of the type shown in Fig. 1, and only in such a schedule. Consequently the *degree of phylogenetic relationship* has to be defined as follows:

A species *x* is more nearly related to an arbitrary species *y* than to an arbitrary species *z* if, and only if, it has at least one ancestral species in common with species *y* which is not at the same time the ancestor of species *z*.

Decisive for the establishment of degree of phylogenetic relationship is thus the time for the separation from a common ancestral species. An example is given in Fig. 1. If we go back in time, investigating which of the species A, B, C, and D first meet each other, we find that D first comes across species C (point of time: t_3). These two species are *sister species* and have thus to be coordinated in one group. Further back in time (t_2) the species B will meet C/D. Consequently B and C+D are *sister groups*, meaning that they have to be coordinated in a group of higher rank. Finally (at t_1) species A joins B/C/D, thus forming the sister group of B+(C+D). In spite of the circumstance that the development of species B is much closer in time to the development of species A than to that of the species C and D, the species B is more nearly related to C and D than to A. For B has an ancestral species in common with the species C and D (t_2) which is not at the same time the ancestor of A.

The definition of a *monophyletic group* is a direct consequence of the defini-

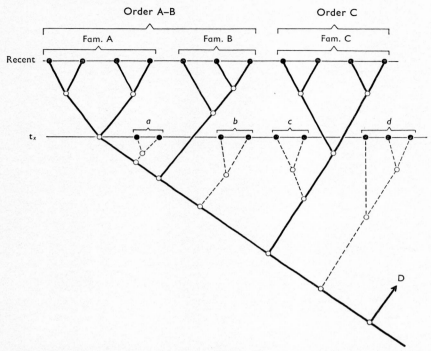

Fig. 2. Diagram demonstrating that proper knowledge of the phylogenetic relationships of the recent groups is a prerequisite for the interpretation of those of the fossil groups. In the case illustrated the contemporary fossil groups $a/b/c$ could be properly placed, which means important information as to the age of the related recent groups. The indicated position of the fossil group d presupposes that the (recent) sister group (D) of the recent supraorder $A + B + C$ is known. (From Brundin, 1966.)

tion of phylogenetic relationship: Monophyletic is every group of the system fulfilling the demand that any species belonging to it is more closely related to any other species likewise belonging to the group than to any species that does not belong to it. It is also important to note that a monophyletic group must comprise all species which are derivable from a common ancestral species.

A logical consequence of the foregoing is the circumstance that every group (down to a single species) of the phylogenetic system has and must have a sister group (which likewise can be a single species). This is valid for groups of all categorical ranks. Speciation, i.e. the splitting of an ancestral species into two daughter species, and the development of sister groups are identical processes.

We must of course calculate with the possibility that the sister group of a recent group is extinct. That is certainly often the case. In Fig. 2 the fossil group a is the sister group of the recent group A. In spite of that it holds good that group A has a sister group among the recent animals, namely group B. This implies only a sister-group relationship on a higher level.

The speciation process presupposes progression in space and development

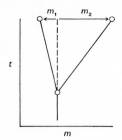

Fig. 3. Diagram explaining the rule of deviation. The dimension m symbolizes the morphological differences; t time.

of spatial isolates. Since phylogenetic evolution forms a sequence of processes performed on species, prevailingly in combination with speciation, there is an intimate connection between phylogenetic relationship and geographic distribution. In other words, sister species and sister groups do often display allopatry, i.e. *spatial replacement* (vicariism). The study of this important phenomenon is still poorly developed, but it is evident that the widespread oversight of the common existence of vicarious sister groups of higher rank is a consequence of misinterpretations of the phylogenetic relationships.

For a more detailed account of the phylogenetic principles, see Hennig, 1966.

Phylogenetic reasoning. It is tempting to suppose that closeness of phylogenetic relationship is proportionate to the number and conspicuousness of similarities. A close investigation demonstrates, however, the insufficiency of that hypothesis. The hypothesis is not quite false, as shown by the phylogenetic validity of many pre-Darwinian groups, but it leads often to false conclusions or confronts us with puzzles which are insoluble because of their very nature. There are as a matter of fact many groups comprising species whose relationships as to morphological similarity simply cannot be expressed by a hierarchial arrangement. It is thus clear that "similarities" or "characters in common" do mean something but that such concepts are in some way too diffuse to function as reliable criteria of phylogenetic relationship.

Phylogenetic reasoning starts from the phenomenon that has been named the "*rule of deviation*" by Hennig. It states that by the splitting of a species into daughter species (i.e. the development of sister groups) one of the species normally remains on the whole unchanged, while the other species clearly deviates from the ancestor (Fig. 3). This is valid for single characters and the general design alike. The former, primitive condition corresponds to Hennig's term *plesiomorph*, the latter, derivative condition to his term *apomorph*.

It is apparent that the rule of deviation and concepts like plesiomorphy and apomorphy, indeed all the concepts and principles of phylogenetic systematics, are and have to be based on the speciation process, its premises and phylogenetic meaning. Applying the principles of population genetics, the apomorphic daughter species is thus identical with the spatially isolated peripheral population which, thanks to successful escape from the rigid system of genetic homeostasis of the mother population, has been able to benefit by the development of new adjustments and to acquire reproductive isolation.

For the discussion and interpretation of phylogenesis it is of fundamental importance to note that the speciation process must be looked upon as a splitting of an ancestral species into daughter species, and not as a branching off of daughter species from a persisting ancestral species. Accordingly, when the isolated daughter population acquires reproductive isolation and the status of a separate species, then the mother species ceases to exist. We have to hold on to this interpretation even if the ancestral species, as may often be the case, should remain practically unchanged. That is, after all, a logical consequence of the definition of degree of phylogenetic relationship, and that definition is self-evident.

Because of the time perspective involved, a direct proof of the action of the deviation rule cannot be offered, but its validity follows from the remarkable fact that the cross-section of the phylogenetic tree, as displayed by the recent organisms, is on the whole in agreement with the construction of the tree.

Trying to find the place in the system which has to be assigned to a group, we are always faced with a problem similar to the rule of three. Our comparisons are of course meaningful only if they are based on homologous characters. But it is decisive that we know the quality of those characters, then only agreements in apomorphic characters, *synapomorphies*, speak in favour of a closer relationship between two groups. It is obviously sound to conclude that apomorphic agreements between two groups have been inherited from the common ancestral species of just those two groups. The use of plesiomorphic characters, on the other hand, would lead to creation of non-monophyletic aggregates called *paraphyletic* groups.

Phylogenetic analysis has to start with members of the recent biota, on the basis of a broad knowledge of comparative morphology, ecology, and distribution. We will further do wise in starting with a group that for obvious reasons stands out as apomorphic. Asking which group is more closely related to our initial group, A, than to any other group because of the common possession of at least one unique specialization (synapomorphy), we are able to establish that two groups, A and B, seemingly stand out as sister groups. Our conclusion is confirmed if there is more than one synapomorphy available and if the two groups display geographical replacement. If this procedure is continued, we will arrive step by step at a hierarchial system of sister groups comprising all existing species of the major group that we did intend to analyze. If we have worked properly, without compromise, we will now dispose not only of a phylogenetic system and a fair knowledge of the main trends in the biogeography of

Application of phylogenetic principles 479

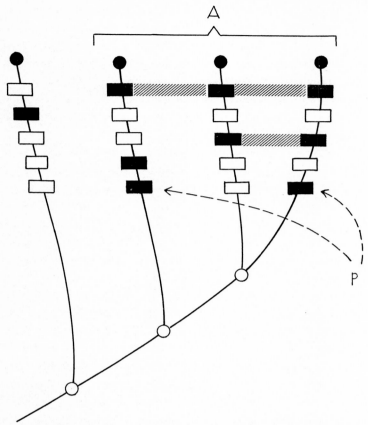

Fig. 4. Parallelism (*p*) in a group (A) comprising 3 species. Part of a schedule of arguing for the establishment of sister species (groups) in accordance with Hennig. Black rectangles symbolize the apomorphic grade of a character, white rectangles the plesiomorphic grade.

the group, but also of the chief outlines of a phylogenetic diagram where all pertinent fossil groups can be properly placed, provided that they are well enough preserved to show those apomorphic characters whose decisive importance has been demonstrated by the foregoing analysis of the recent groups (Fig. 2). This is a consequence of the fact that every fossil group must have a sister group in the recent biota. Hence it is not possible to discuss the relationships of fossil groups in a realistic way without knowledge of the sister-group system formed by the recent groups. Overlooking these causal connections the paleontologists have wasted much time in futil search for the ancestor of different groups.

However, constructing our phylogenetic system step by step, we will observe that several apomorphic characters cannot be used as direct proofs of sister-group relationship. This applies to such specializations that do not occur in all members of at least one of two sister groups (Fig. 4). This false synapomorphy is *parallelism*. But if the parallelism is unique, i.e. occurring only within one pair of sister groups, it is a confirmation of the soundness of the foregoing analysis. Parallelisms are also of importance when it comes to decide about the relative plesiomorphy (primitiveness) of two sister groups. If a unique apomorphic character is present in all species of one group, but only in one or a few comparatively apomorphic species of the other group, this is an indication that the latter group is the more conservative (plesiomorphic) one. (Several examples in Brundin, 1966.)

Some biologists seem to be dubious concerning dichotomic branching as a major feature in phylogenesis. They have to consider, however, that such an omnipresent phenomenon as parallelism (cf. below) does presuppose dichotomy (cf. Fig. 4), while multiple splitting because of its very nature excludes occurrence of parallelism. Then assumption of multiple splitting, i.e. true contemporary multiple speciation from a single ancestral species, must presuppose that no species of a group, apart from the group characters, has any apomorphic character in common with any other species of the group.

It may be added that if phylogenetic systematics starts out from a dichotomic differentiation of the phylogenetic tree, this is primarily no more than a methodological principle (cf. Hennig, 1966).

The nature of a general reference system. It is an inevitable consequence of the meaning of the speciation process that phylogenesis, anagenesis, and stasigenesis (preservation of primitiveness) are closely connected evolutionary phenomena and, further, that species and strictly monophyletic groups are the real units of evolution. Nevertheless even modern evolutionists take a sceptical or purely negative attitude towards the demand for a consequent phylogenetic system. The contradictoriness of that attitude should be obvious.

It is significant in this connection that the most ardent opponents of a phylogenetic system seem never to have made any serious personal effort to construct the sister-group system of a major, world-wide group on the basis of a phylogenetic step-by-step analysis. If they had ever tried, they would have learned that phylogenetic analysis is not subjective and speculative, as often pretended, but based on objective arguing in accordance with strict principles. They would then also appreciate that the arguing is based on all that is known of structural and ecological anagenesis, and of geographical distribution. They would learn, finally, that much in anagenetic evolution that stands out as problematic at the start will be clarified at the end, as a consequence of the phylogenetic analysis, because the whole forms an orderly pattern in space and time, as testified by mutual clarification and mutual confirmation. This latter point is essential. I refer here to my paper (1966) on chironomid midges, dealing with the connection between phylogenetic relationship, anagenetic evolution, transantarctic dispersal, and amphitropical distribution in groups from the Jurassic to the present.

The demand for a consequent phylogenetic system "neglects", according to Huxley (1958), "the difficulties raised by the facts of parallel and convergent evolution". This is a deep-going misunderstanding of the situation since parallelism, instead of being a hindrance, stands out as an important source of help in the phylogenetic analysis (see above), while convergence hardly will cause any difficulties in properly studied groups. To the typologist, however, the occurrence of parallelism is an ever present source of trouble.

Many biologists are adherents of a mixed classification. They feel that a purely phylogenetic system obscures too much of the anagenetic aspect, which seems to stand out as the essential thing for the general student of evolution. Mackerras (1967), for example, "can see no logical reason for insisting on consistency (or any other abstract principles) if it obscures an aspect of nature that we are endeavouring to understand". He finds support for his point in the following three sets of circumstances:

(a) "The most significant is that, the further one has to go back in time to find the origin of a major group of animals, the clearer do grades become and the more obscure becomes the phylogeny. The converse is equally true. In the relatively younger, lower ranking taxa, phylogeny can often be deduced with considerable confidence, whereas anagenesis can be seen but dimly."

However, the circumstance that our possibilities to compare fossil and recent groups decrease the further we are going back in time, as a consequence of increasing incompleteness of the fossil specimens and the fossil record, so that many fossil groups have to be treated dimly as grades with unknown phylogenetic background, is indeed no relevant reason for a mixed classification of the recent groups. Further, the very task stressed by Mackerras, namely the finding of the "origin" of a major group, does first and foremost represent a typological problem. It is overlooked that phylogenetics is the search for the sister group and that a broad knowledge of anagenesis is a prerequisite for phylogenetic reasoning. Fossil grade groups may be very helpful, it is true, because they demonstrate direction of change, but knowledge of the ancestral species is not essential for the establishment of phylogenetic relationships of groups of any size and age. If we had a proper knowledge of the sister-group system represented by the recent Placentalia and their plesiomorphic sister group, the Marsupialia, it would be no problem to determine the position in the phylogenetic diagram of every extinct placental group, provided of course that the fossil specimens are so well preserved that we are able to study those apomorphic characters whose decisive importance has been demonstrated by the foregoing analysis of the recent groups (cf. Fig. 2).

The details of phylogenetic evolution will always be more obscure the more we go back in time, as stated by Mackerras. He overlooks, however, that the essential point is our ability to establish the degree of phylogenetic relationship. That ability is not necessarily delimited by the age of the groups treated. Moreover, we should never forget that the recent animal kingdom does on the whole represent a recapitulation of the performed phylogenesis, and that *every* recent sister-group pair, from lowest to highest rank, is a more or less complex manifestation of relative stasigenesis (plesiomorphy) and relative anagenesis (apomorphy). This is the reason why the construction of a consequent phylogenetic hierarchy must be a point of aim for a deeper understanding of the meaning and history of the numerous organisational patterns that we have before us. The statement of Mackerras that anagenesis can be seen but dimly in the younger, low-ranked groups is not correct.

(b) "The second consideration is that this situation has, in fact, been incorporated, consciously or not, in widely accepted classifications, which at least some evolutionary taxonomists seem to find more illuminating and intellectually satisfying than a rigidly phylogenetic classification. Conservatism is no more a criterion of scientific validity than is convenience, but the opinions of serious students of evolution should at least be examined" (Mackerras, 1967).

The weakness of this point may be demonstrated by a hypothetical example (Fig. 5). If there is a group, A, in the system whose elements are kept together by a grade of a special trend or trend-complex, we have to keep the group as it is, if there is no reason to suspect that the group is anything else than monophyletic. However, let us assume that group A is represented by subgroups in Laurasia, Patagonia, and Australia. Let us assume, further, that it is shown later on by phylogenetic analysis that the subgroups of Patagonia and Australia together form the apomorphic sister group (C) of a plesiomorphic group B in New Zealand, whose members lack the actual trend characters, and that the subgroup of Laurasia (D) is the apomorphic sister group of a plesiomorphic group (E) in South Africa that also lacks the trend of group A. Starting from scratch and applying the rule of deviation we have been able to show that the group A is a heterogenous assemblage, consisting of subgroups the characters of which are the results of trend developments which have had a different origin and history, and have taken place in different continents and under different conditions of life. Thanks to the phylogenetic step-by-step analysis and the elimination of the grade group A we have got a fair insight into the history in space and time not only of all subgroups of the extensive group B/C+D/E, but also of the integrated phylogenetic and anagenetic processes. Moreover, if we have worked properly, all new knowledge of the transformation in time of ecological, ethological, and physiological characters of our group B/C+D/E, added by the research of the future, will fit into the picture and deepen our understanding of the pathways of transpecific evolution. Would it then be "more

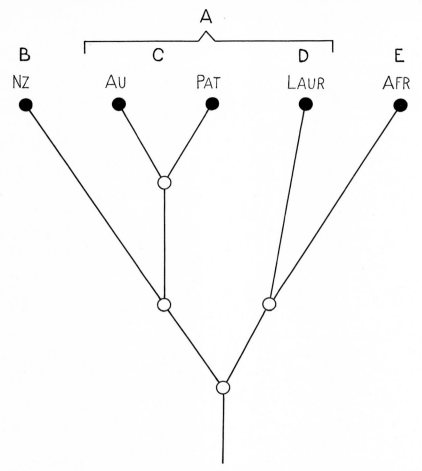

Fig. 5. Phylogenetic relationships of the elements belonging to a grade group (A) based on false synapomorphy (parallelism).

illuminating and intellectually satisfying" to keep the grade group A in a phylogenetic system?

According to Mackerras, grade groups are "often, perhaps usually, polyphyletic in origin". There is, however, little doubt that the majority of the non-monophyletic grade groups of the present animal system are paraphyletic.

In his recent paper on some principles of phylogenetic systematics Tuomikoski (1967) stresses that paraphyletic groups are sharply separated from typical polyphyletic groups and that they are more acceptable than the latter because they are often "almost" monophyletic. He is ready to give up the demand for strict monophyly and to accept paraphyletic taxa in contemporary classification "in special cases because of their greater information content and hence also better applicability to different branches of biology beyond the purely

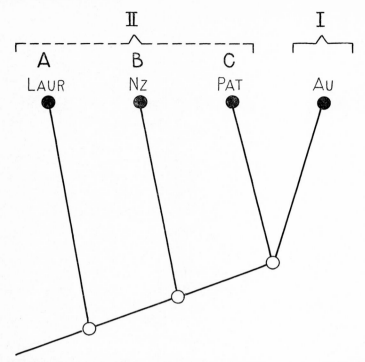

Fig. 6. Phylogenetic relationships of the elements belonging to a (paraphyletic) group (II) based on common possession of plesiomorphic characters (symplesiomorphy).

historical ones". It is important to note that Tuomikoski presumes that the "phylogeny" of accepted paraphyletic groups is "reasonably well known".

Let us assume the presence in the system of two groups, here named I and II, which are considered "closely related" (Fig. 6). Group I occurs in Australia; it is distinguished from group II by some unique specializations. Group II, on the other hand, which has an amphitropical distribution in Laurasia, New Zealand, and Patagonia, is treated as a unit simply because its members lack the specialized characters of group I. Founded on the possession of certain plesiomorphic characters, group II stands out as a probably paraphyletic and comparatively primitive grade group, while group I, though also being a grade group, is strictly monophyletic because it is founded on synapomorphy.

A phylogenetic step-by-step analysis will reveal that certain members of the amphitropical group II as to propinquity of descent are more closely related to members of group I than to the other species of group II. More precisely, the Patagonian group C is the plesiomorphic sister group of group I, which means that it has an ancestral species in common with group I that is not the ancestor of A or B. Group B of New Zealand is in turn the plesiomorphic sister group of the monophyletic group C + I, and group A of Laurasia is the plesiomorphic

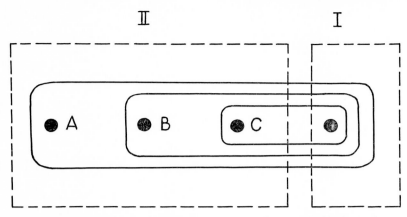

Fig. 7. Other representation of the relations between phylogenetic kinship and similarity relationship shown in Fig. 6.

sister group of the aggregate B + C + I. Working in accordance with the model given by the speciation process we have thus been able to "open" a major group to further analyses and different kinds of syntheses. That is a great step forwards.

Hence it is clear that acceptance of the view maintained by Tuomikoski (and many others), and the consequential approval of groups I and II as coordinate groups, means the keeping of a fundamentally false hierarchy. The resulting typological classification cuts straight across the phylogenetic connections (Fig. 7). Based on a subjectively chosen fraction of the involved mosaic process of evolution such a typological classification will burst from the expanding pressure of increasing knowledge of the anagenetic pattern. Moreover, every qualified study of different kinds of parallelism within the discussed complex does presuppose the use of a known sister-group constellation that invalidates the recommended typological classification.

But according to Tuomikoski it is appropriate, in spite of the above considerations, to accept paraphyletic groups because of their pretended greater information value and applicability to biological branches "beyond the purely historical ones". What is the meaning of this? We know for certain that every character, every species, every group, every ecosystem, and every habitat and distribution pattern have a history. Is it then possible for a scientific biological mind working comparatively above the registrating and descriptive alpha level ever to disregard the historical aspect, or, if we still do not have sufficient knowledge to be able to apply the time perspective, not to work consciously towards a corresponding extension of our insight?

(c) Mackerras' third consideration deals with the purpose of classification and is summarized in the statement that a mixed classification serves all needs, "because the general student of evolution, who cannot neglect grades, is usually concerned with higher taxa, zoogeographers and others who depend on phylogenetic relationships are predominantly concerned with levels at which those relationships are most clearly expressed, and the nature of the system does not matter for information storage so long as it is mechanically efficient". The reasons why this statement stands out as objectionable are largely obvious from what has been pointed out above. But in addition, it may be emphasized that the apparent difficulty of getting at the phylogenetic relationships of high-ranked groups is most probably simply a consequence of the fact that we lack a proper step-by-step analysis of the low-ranked groups making up the higher groups. Without that basic work we lack the necessary requisites for a realistic interpretation of the grand patterns of phylogenesis and geographical distribution as well as anagenesis.

We cannot reconstruct the whole story, but it is certainly possible to deepen the insight in an essential way if we join in consequent application of strict principles and in open-minded cooperation free of traditional thinking.

Concluding remarks. After the breakthrough of Darwinism and the beginning of the development of a proper synthetic theory of evolution the elucidation of the phylogenetic connections of the species and the construction of a consequent phylogenetic system stand out as fundamental and unrejectable tasks. This conclusion is a direct consequence of the meaning and mechanism of the evolutionary process. The real units of evolution are the species and the strictly monophyletic, supraspecific groups. Let us then not delay the development and deepening of the evolutionary studies unnecessarily, by arguments directed against the need for a consequent phylogenetic system.

We ought to consider that all differences between the species go back to and are contained in the phylogenetic connections, since the latter are proceeding in time. Objective consideration and integration of all existing differences is possible only in connection with the phylogenetic system. Moreover, the phylogenetic connections of the species and the supraspecific groups are exactly measurable in principle, since they go on only in one dimension, the time dimension. As a general biological reference system the phylogenetic system is superior to all other systems; indeed, it is the only system which can fulfil that purpose, without contradictions and without compromises.

Different typological systems might be fully justified, in addition to a phylogenetic system. But they are bound to be more or less one-sided and to serve special purposes. Subsidiary classifications illustrating different aspects of grade development, as proposed by Huxley (1958), will certainly contribute to the understanding of anagenetic evolution. The limitations of the gradal aspect are, however, all too often forgotten. The "systématique idéale" of Gisin (1967) is one of the most recent examples of this.

It seems appropriate to end this section with the following statement of Simpson (1953: 269): "Of course any group of objects can be arranged in graded series according to some criterion. Unless the criterion is correctly and solely phylogenetic, the result has no relevance whatever as to how the characteristics of the objects arose."

Phylogenesis and the course of evolution

Modern evolutionists agree that the evolutionary process has been more or less directional and that evolutionary change has been governed by natural selection. Mutation is considered chiefly random, merely providing the raw material for evolution by natural selection.

However, natural selection is conceived not only as directional, but as something far more deep-going: "... because it is directive, it has a share in evolutionary creation. Neither mutation nor selection alone is creative of anything important in evolution; but the two in conjunction are creative" (Huxley, 1942: 28). As we shall see below, this is a comparatively moderate view of the role of selection. On p. 466 of the cited work Huxley states that all that natural selection can ensure is survival, but also that selection is able to "produce" adaptation which is a different thing. On p. 474 he writes that progressive adaptations involving a number of separate steps cannot have arisen without the operation of some agency "which can gradually accumulate and combine a number of contributory changes: and natural selection is the only such agency that we know".

Huxley's conception of natural selection, i.e. the censorship of the environmental factors, as a creative force (though always in conjunction with mutation and recombination) appears objectionable. It is about the same as saying that the art critique evaluating the production of an artist is creative. The exhibition can be met by disapproval or more or less laudatory approval. In the latter case the artist is encouraged to go on, and further encouragement will probably lead to successive progress. If the artist happens to be many-sided, encouragement of one of his different "styles" will probably lead to a corresponding specialization. The critique and the buying public are directional and may also increase the creative rate, but they are indeed not creative. Creative is only the artist.

Creative in evolution are solely the hereditary mechanisms, via mutation and recombination. But they are also directive, as a consequence of canalized and species-specific evolutionary potentials. Every new character is an experiment meeting approval or disapproval by the environment. The censorship performed by the latter is an omnipresent directive force; but it is not creative.

The above thesis will be further discussed below. The important point to be stressed here is the tendency of Huxley (and others) to exaggerate the influence of natural selection relative to that of the genetic systems. Asking why, we will have no difficulty in finding at least one reason. Like many other evolutionists Huxley is manifestly unwilling to admit that orthogenesis in any form has played a prominent role in determining the course of evolution. Arguing against the theory of orthogenesis in its strict sense (see below), Huxley continually involves himself in inconsistencies demonstrating the presence of a prominent directional force of genetic nature collaterally with natural selection.

It is obvious that old hypotheses of creationism, divine or vitalistic guidance, and the extremer forms of orthogenesis do not conform with present scientific ideas. Huxley is certainly right when stating (l.c.: 465) that strict orthogenesis, "as implying an inevitable grinding out of results predetermined by some internal germinal clockwork", removes evolution out of the field of analysable phenomena. But it is quite another thing to say that evolution is predetermined to proceed within certain definite limits, which is equal to the postulate that mutation and variation are to a marked degree non-random. Huxley (l.c.: 509) asserts, however, that there is no evidence whatever of such a pattern as far as long-continued adaptive trends are concerned. But in the much-discussed case of the repeated over-coiling and extinction of subgroups of the fossil lamellibranch *Gryphaea* group he is forced "provisionally" to face an explanation in terms of (predetermined) orthogenesis. If this apparently non-adaptive tendency to over-coiling is due to orthogenesis, we have, according to Huxley, "no inkling of any mechanism by which it may be brought about ... Indeed its existence runs counter to fundamental selectionist principles."

However, would it not be reasonable to suspect that the *Gryphaea* case brings up some fallacy in the accepted selectionist principles? Indeed, the main pattern of speciation and phylogenetic evolution has not changed in time, and there is hardly any reason to think that the *Gryphaea* case represents something unusual. It seems obvious that *Gryphaea* exemplifies the ever present canalization of the evolutionary potential of species and groups. To the *Gryphaea* repertoir belonged the starting of a trend that turned into a deleterious grade. The trend was probably non-adaptive and during a long time indifferent to natural selection. Genetic change means experiment and differential success. The genetic system of *Gryphaea* failed, as so many other genetic systems earlier and later. *Gryphaea* is of special interest only because the reason for extinction stands out so clearly to the human mind.

We are then facing the problem raised by the occurrrence of evolutionary trends of undoubted adaptive nature. Stressing the apparent orthogenesis behind adaptive trends, Huxley (l.c.: 497) writes: "The only feature inviting orthogenetic explanation is the directive character of evolutionary trends, their apparent persistence towards a predetermined goal. But on reflexion this too is seen to be not only explicable but expected on a selectionist viewpoint." Adaptive radiation is according to Huxley "essentially a product of selection, not the outcome of any intrinsic tendency" (l.c.: 496).

This seems to be a depreciation of the influence of the genetic systems on the course of evolution. New trends are by necessity direct products of the genetic systems, experiments in new directions meeting the censorship of selection. Which is to be considered most important, the ability of the genetic systems to start new trends and to hang on in the started directions by means of successive change, or the ability of selection, to which mutations are random, to increase evolutionary rates and to favour, with high effectiveness, the adaptive proposals of the genetic systems? To me it seems meaningless to try to make a decision.

Huxley (l.c.: 510) admits that what he calls "subsidiary orthogenesis" is common enough; and he includes under that heading phenomena "which in the first place are of an orthogenetic nature in that they limit the freedom of variation and therefore of evolutionary change, and in the second place are subsidiary in that they merely provide limits within which natural selection still plays the main guiding and shaping role". Pointing out, further, that corresponding mutational effects are likely to turn up independently in various related species of a group (l.c.: 511, 514), Huxley is indeed on the point of admitting that mutations are to a marked degree non-random and that every genetic system is characterized by special "intrinsic tendencies".

One of the things which impressed me most during my phylogenetic studies of the world fauna of some groups of the dipteran order (Brundin, 1966) was the common and regular occurrence of parallelisms within the different sister-group systems. Basing my opinion on this experience I think it is realistic to say that among higher animals parallelisms have to be suspected theoretically to occur in all sister-group pairs comprising 3 or more species. The phenomenon called parallelism belongs to the major features of evolution, and it seems to be a proper conclusion that viable mutations normally are non-random in relation to the species. Many lethal mutations, on the other hand, stand out as genetic accidents, i.e. as by-products at random of the dynamic processes within a local population.

Turning from Huxley to G. G. Simpson (1953) we meet the same tendency to exaggerate the influence of natural selection relative to that of the genetic systems. Significant is Simpson's definition of selection as "anything tending to produce systematic, heritable change in populations between one generation and the next" (l.c.: 138). Hence selection is conceived as a creative factor. Simpson asserts further (l.c.: 267) that selection is "the only nonrandom or antichance evolutionary factor that is objectively demonstrated to exist". Before that he has, however, written in the same paper:

"What is inherited, mostly but not exclusively through the genotype, is known to be an organized physico-chemical system with a definite growth tendency or pattern" (l.c.: 60).

"Given a certain hereditary type of developmental pattern, the changes that can occur in it and their effects upon the structures developed are strictly limited, and alternative changes are not introduced in exactly equal numbers; but in almost every case the change can be and is in at least two, frequently more, different directions" (l.c.: 136).

The contradictoriness of Simpson's arguing is stressed further by the following statement (l.c.: 272):

"Another aspect of the same sort of limitation is seen in the fact that a population always has a certain range of realized or expressed phenotypic and genotypic variation, another much larger range of existing potential variation stored in the genetic pool of variability, and a third prospective range involved in mutations not yet fixed in the pool. Selection can act only on expressed variation."

It is thus admitted more or less indirectly by Simpson that the genetic systems are in the possession of strictly canalized and species-specific evolutionary potentials and variation patterns and hence represent a creative and directive non-random factor. Why then try to deny or go around the presence and significance of a phenomenon that is an obvious consequence of observed facts and proper inference and, moreover, opens the door to a balanced and more realistic interpretation of the evolutionary process? The reason seems to be an exaggerated fear of getting something metaphysical involved in the explanation, an attitude which in turn goes back to the crusade against the theory of orthogenesis. That is, however, to fight a ghost of the past since there are hardly any modern biologists believing that trends are able to continue indefinitely and undeviatingly in one direction independently of environmental factors. Denying that anything in the genetic processes and their outcome is oriented or predetermined we are throwing the egg away with the shell.

Mayr (1963) has recently given a comprehensive survey of vital parts of the theory of evolution. His arguments concerning the relative influence of genetic variation and natural selection on the course of evolution conform closely with those set out by Simpson, and the role of selection as helpful experimenter and purposeful creator is stressed quite as strongly. A close study shows, however, that Mayr cannot escape involving himself in contradictions supporting another interpretation. Dealing with general aspects of the evolutionary theory Mayr writes (l.c.: 8):

"The basic framework of the theory is that evolution is a two-stage phenomenon: the production of variation and the sorting of the variants by natural selection."

If Mayr's discussions and conclusions had been in general agreement with this basic thesis, there would not be any need for further discussion. It would only be desirable to remark that the concept "production of variation" would be far more stringent and adequate by addition of the adjective canalized. But the situation calls for further comments.

According to Mayr (l.c.: 7–8) mutations do not guide evolution. This is followed by the statement that "the kinds of mutations and recombinations that can occur is severely restricted". The oversight of the fact that the former statement is contradictory to the latter, and that restriction or canalization of evolutionary potential and change must be an orienting factor, leads automatically to overestimation of the influence of chance and selection.

Mayr stresses (l.c.: 214) that evolutionary change in populations is a two-factor process and writes:

"One stage consists in the generation of genetic variation. It is on this level that chance reigns supreme. The second stage is concerned in the choosing of the genotypes that will produce the next generation. On this level natural selection reigns supreme and chance plays a far less important (although not altogether negligible) role."

These conclusions are drawn by Mayr in connection with discussion of selection and chance. He illustrates the importance of chance by referring (l.c.: 203) to the relation between the number of gametes produced during the life time of a human couple (many billions in the male, many hundreds in the female) and the number of their children (at best only about a score). "It is largely a matter of chance which among the countless gametes will form the few successful zygotes". The relevancy of this example with respect to the course of evolution is, however, very limited since chance is treated in relation to the individuals and not to the species.

"Perhaps the greatest single source of randomness and indeterminacy in evolution" is, according to Mayr (l.c.: 212–213), "the selective equality of genotypes". Then it happens fairly often that different genotypes produce phenotypes that react in an identical manner to a given selection pressure. "It is this indeterminacy of the selective aspects of genotypic recombination that introduces the greatest element of chance in evolution."

However, the randomness of the discussed situation is strongly reduced if we avoid estimating chance solely in relation to selection and look upon the evolutionary process as a developmental experiment performed by a hierarchy of organisms, where the units consist of species and supraspecific groups with strictly canalized evolutionary potentials. Within each of these units the evolutionary change is predetermined to keep to a definite repertoir. The multitude of organisational patterns represented by the contemporary biota is a consequence of the particular constitution of the genetic systems and thus essentially non-random. Then selection is neutral to differences as such. The occurrence of particular genetic change is largely non-random with respect to the organisms as members of definite genetic flows in time, but random with respect to mechanical natural selection. The fate of genetic change and its phenotypic expression, on the other hand, is largely non-random with respect to selection but random in relation to the organisms. Hence randomness is a highly relative concept.

Broadly speaking, the course of evolution is determined by the interaction of largely non-random directional variation and largely non-random directional selection.

Selection is not a creative force working according to a certain plan for the "purpose" of achieving superior gene combinations, as set out by Mayr and Simpson. Stressing that selection is creative, Mayr (l.c.: 20) asks :"Is not a sculptor creative, even though he discards chips of marble?" Certainly, but the question is irrelevant in this connection since it does not exemplify the function of selection. Having to work as a "sorter" (cf. Mayr, l.c.: 8) selection deals with series of completed "sculptures" displaying variedness and different degrees of fitness. Irrelevant for the same reason is the illustration of selection as a creator given by Simpson (1947) and cited as argument by Mayr (l.c.: 202). Alluded to is an example where Simpson compares natural selection to a man trying to draw the letters c, a, and t from a pool of all letters of the alphabet "in order to achieve a purposeful combination" of the three letters into the word "cat".

It is also hardly correct to say that selection is "strongly opportunistic" and that it "induces evolutionary change" (Mayr, l.c.: 203, 613). The constructive and opportunistic experimenter is life itself, via the evolutionary dynamics of the genetic systems. Evolutionary change is induced by the latter via mutation and recombination and is exposed to selection in its phenotypic consequences, as differential fitness. How can selection induce change when it constitutes a "sieve" (Mayr, l.c.: 203) and "supervises" (id., l.c.: 202) the bringing together of genes by the genetic systems?

Our discussion of evolutionary change leads back to the problem concerning generation of genetic variation and Mayr's view (cited above) that "chance reigns supreme" on that level. Indeed, it seems hard to find a more convincing disproof of this view than the following statements set out by Mayr himself:

(1) "A well-knit system of canalization tends to narrow down evolutionary potential quite severely. It accounts for parallel evolution..." (l.c.: 281).

(2) "There are usually numerous pathways open to achieve a certain biological end and it will depend on the particular genetic constitution of the incipient species which particular pathway is chosen" (l.c.: 592).

(3) Only part of the inherent differences between groups "can be explained by the differences in selection pressures to which the organisms are exposed;

the remainder are due to the developmental and evolutionary limitation set by the organisms' genotype and its epigenetic system" (l.c.: 608).

(4) "Every group of animals is 'predisposed' to vary in certain of its structures, and to be amazingly stable in others... The characters that vary in *Drosophila* are totally different from those that vary, let us say, in grasshoppers" (l.c.: 608).

(5) Mayr (l.c.: 614) speaks also of the evolutionary potential of a "well-buffered epigenetic system" and of the "well-buffered system of developmental canalization".

All this is a confirmation of the views maintained in the present paper.

It seems important, finally, to cite a further statement of Mayr (l.c.: 11) that is in conflict with a balanced theory of evolution: "The origin of new species, signifying the origin of essentially irreversible discontinuities with entirely new potentialities, is the most important single event in evolution." This is to overlook what we have learned about phylogenetic evolution. Speciation is the most important opportunity for breakthrough of further and often entirely *new aspects* of the inherent evolutionary potential of a monophyletic group. The evolutionary potential is continuous and hierarchial and has never been greater than in the first species occurring on the globe, and never less than in the species of the present.

Concluding remarks. My views on the evolutionary process differ in some respects from those maintained by leading evolutionists. What seems worthwhile to stress in this connection is the circumstance that my conclusions stand out as a logical consequence of personal experience gained through extensive study of the connections between structural and ecological character patterns, geographical distribution patterns, and phylogenetic relationships in some old groups with world-wide distribution. The experience that the whole forms an orderly pattern in time and space, conforming with facts and theories from different fields, was very impressive. It was clear, among other things, that our possibilities to interpret the causal connections involved in the history of life are far better than generally assumed, and, further, that the involvement of chance has been less than often imagined. However, a consequent application of phylogenetic principles stands out as the fundamental prerequisite for a proper treatment of the evolutionary phenomena. Trying to build up a synthetic theory of evolution we cannot pass over such a state of matters; and the general oversight of this is obviously the basic reason for some deficiencies of the synthetic theory of today. These can be precised as follows:

(1) Overestimation of the influence of chance in the generation of genetic variation and the consequent underestimation of the directional influence of genetic variation on the course of evolution.

(2) The disinclination to conceive mutation and recombination as canalized and species-specific and variation as directional is closely connected with the reluctance to appreciate the nature and dynamics of the hereditary processes as the sole creative force in the evolutionary process.

(3) The current conception of selection not only as directional sorter but also as purposeful creator.

(4) Failure to realize that strictly monophyletic, supraspecific groups by necessity have individuality and reality and that the categories of typological systems are "timeless abstractions" (Woodger, cf. Hennig, 1966). The oversight of this state of matters has probably meant a greater handicap than perceived at the first moment.

(5) Too uniform application of population thinking.

The organisms are in the possession of directional intrinsic tendencies, the evolutionary potentials, which are governed and canalized by particular genetic constitution and definite physico-chemical laws of growth. In every species and supraspecific group the mutation-based variability, as expressed in the phenotype, is, therefore, predisposed to keep to a definite repertoir as to range and direction. The dynamic genetic systems are thus not only creative but also markedly directive. Further vital evolutionary qualities of the life processes are not only the ability of the individual organisms to propagate, the very existence of cohesive groups called species, and the ability of the species to split by speciation, but also the inherent endeavour of the species to extend their range in space. The general rule that stepwise progression in space is a prerequisite for speciation belongs to the main features of the evolutionary process.

The results of the evolutionary process are not predetermined and not predictable. Every new character is an experiment meeting the mechanical censorship of natural selection. Some of the new non-deleterious characters, especially preadaptations, may be neutral to selection, while others are adaptive and open to selection pressure. The latter characters will be favoured in proportion to degree of adaptiveness. However, since selection, in spite of all its importance, is solely a sorting and directional force setting up demands for over-all fitness, the broader meaning of selection is differential success in reproduction *and speciation*.

The cleavage process, i.e. the splitting of an ancestral species into daughter species, is the clue to interpretation of transpecific evolution. The process means not only acquirement of new characters and reproductive isolation by a spatially isolated peripheral population, but also continued existence of a conservative population preserving the characters of the ancestral species more or less unchanged. This dual meaning of speciation elucidates certain aspects of the evolutionary process. The preservation of relative primitiveness (stasigenesis) in one of the two sister species means reinsurance of persisting possibilities for breakthrough of the evolutionary potential of the ancestral species, if the experiment symbolized by the specializing apomorphic sister species proves unsuccessful. However, during extreme conditions adaptiveness is often the only way out and conservatism a danger. But long-continued specialization, however adaptive, is in itself a threat and will often mean only temporary success. Relative primitiveness is often extremely successful, and in the recent biota there are indeed many primitive groups (generally of microscopic size) displaying variedness and, above all, enormously high abundance. This circumstance is obviously of high general importance for the maintenance of basic food-chains and the rates of organic decomposition.

In sister-group pairs of every rank there is hence a dual trend towards conservatism and specialization. That means double reinsurance of survival. This state of matters underlines still more the need for a consequent application of phylogenetic principles in evolutionary biology.

References

Brundin, L. (1966). Transantarctic relationships and their significance, as evidenced by chironomid midges, with a monograph of the subfamilies Podonominae and Aphroteniinae and the austral Heptagyiae. *K. svenska VetenskAkad Handl.*, (4) **11**: 1–472.
Cain, A. J. (1967). One phylogenetic system. Review of Hennig (1966): Phylogenetic systematics. *Nature Lond.*, **216**: 412–413.
Gisin, H. (1967). La systématique idéale. *Z. zool. Syst. Evolutionsf.*, **5**: 111–128.
Hennig, W. (1950). *Grundzüge einer Theorie der phylogenetischen Systematik*. Berlin: Deutscher Zentralverl.
Hennig, W. (1957). Systematik und Phylogenese. *Ber. Hundertjahrf. dt. ent. Ges. Berl.*, **1956**: 50–71.
Hennig, W. (1966). *Phylogenetic systematics*. Urbana: University of Illinois Press.
Huxley, J. (1942). *Evolution, the modern synthesis*. London: Allen & Unwin.
Huxley, J. (1958). Evolutionary processes and taxonomy with special reference to grades. *Uppsala Univ. Årsskr.*, **1958**: 21–39.
Mackerras, I. M. (1967). Grades in the evolution and classification of insects. *J. Aust. ent. Soc.*, **6**: 3–11.
Mayr, E. (1963). *Animal species and evolution*. Cambridge: Harvard University Press.
Mayr, E. (1965). Numerical phenetics and taxonomic theory. *Syst. Zool.*, **14**: 73–97.
Simpson, G. G. (1947). The problem of plan and purpose in nature. *Sci. Monthly*, **64**: 481–495.
Simpson, G. G. (1953). *The major features of evolution*. New York: Columbia University Press.
Tuomikoski, R. (1967). Notes on some principles of phylogenetic systematics. *Annls ent. fenn.*, **33**: 137–147.

Aspects of vertebrate phylogeny

By Erik Jarvik
Section of Palaeozoology, Swedish Museum of Natural History, Stockholm, Sweden

General considerations

When they first appeared in the fossil record, in the Ordovician or more commonly in the Upper Silurian and the Devonian, the Craniata or as they are usually called the Vertebrata had already passed through the most important phases in their history. The early forms were much advanced and divided into several distinct groups (Jarvik, 1960; 1964). It is not easy to say exactly how many principal groups were represented in the early Palaeozoic but they were probably more numerous than indicated in Figure 3; also the number of subgroups is uncertain and difficult to establish. These difficulties are due not only to the incompleteness of the fossil record but also to the fact that only a small fraction of the fossils buried in the rocks have been discovered and studied by palaeontologists. We have also to consider the fact that fossil forms which externally differ but little from each other sometimes are internally very different and only distantly related.

Because of similarities in the shape of the cephalic shield several species of cephalaspids were earlier referred to the genus *Cephalaspis*. However, studies of the internal structure of this shield revealed (Stensiö, 1958; 1964) that these species belong to two different subgroups (ranked by him as orders) of the Osteostraci. Investigations of pteraspids gave similar results and anatomical studies of arthrodires (Stensiö, 1959; 1963a; 1968; and others) have demonstrated many unexpected differences in the structure of the endoskeletal shoulder girdle and the neural endocranium. Early lungfishes are externally somewhat suggestive of contemporaneous osteolepiforms and porolepiforms but differ fundamentally from all of these in internal structure (Jarvik, in this Volume). Similarly, osteolepiforms and porolepiforms agree fairly well in the pattern of dermal bones but in other anatomical regards are specialized in different ways.

These examples demonstrate the importance of a thorough knowledge of internal structure and anatomical conditions for classification and discussions of relationship and phylogeny; conclusions in these regards based chiefly on external features tend to be unreliable, and because many fossil forms are represented only by more or less incomplete exoskeletal remains, the grouping of early vertebrates and fossil vertebrates in general often is a difficult procedure.

It is fortunate, however, that both the exo- and the endoskeleton in early vertebrates, in particular in Devonian forms, often is well developed and strongly ossified. Consequently, it has been possible to make out the internal anatomy, in many instances with a great wealth of detail, in early representatives of several of the principal groups of vertebrates and as a rule early forms are better known than forms from later geological periods. Characteristically all these well-known groups represented in the early Palaeozoic are distinct and easy to distinguish from each other, each being characterized by a great number of specific structural features.

Of great importance for studies of evolution is the fact that these groups, except for the effects of the retrogressive development of the skeleton which may influence the external appearance (see Jarvik, 1960: 57–61, 92; 1964) and other general trends (reduction of basipterygoid process, loss of pineal foramen, etc.), as a rule have changed very little during their known geological history. In a previous paper (1964) many examples were given of the high degree of specialization in early vertebrates and the amazing conservatism displayed in phylogeny (see also Jarvik, 1967a).

In this connection a new example of this remarkable inertia may be given. In order to demonstrate the close agreement between extant dipnoans and their Devonian forerunners in the structure of the lower jaw and the snout I have figured the anterior part of the heads of *Dipterus* and *Neoceratodus* in ventral aspect. However, according to these figures (Jarvik, 1964, fig. 14; 1967a; fig. 1) there is a minor difference between the Devonian and the extant form that may be easily overlooked: the former has a small paired opening (*ro.o*) on the lower side of the upper lip, not shown in the latter. It may be of interest to mention that a paired opening, situated exactly as that in *Dipterus* and other Devonian dipnoans (Ørvig, 1961; White, 1965) has recently been discovered in a young specimen of *Neoceratodus*. The specimen has not been dissected but the opening (*ro.o*, Fig. 1) seems to be the most anterior pore of the supraorbital sensory canal.

The facts that very early in phylogeny, before their first appearance in the fossil record, the various groups of vertebrates acquired their characteristic, anatomical features and that they then have changed very little during vast periods of time lead to the following considerations:

(1) Of particular interest are those groups (dipnoans, coelacanthiforms, etc.) that have survived to the present time. By direct comparison of early Palaeozoic and recent representatives it is possible to gain important information about the effects of the evolutionary processes having operated within a group during a long period of time, in favourable cases amounting to 400 million of years or more. The comparisons are facilitated by the fact that early vertebrates, as mentioned above, often are well ossified, displaying a great number of structural details.

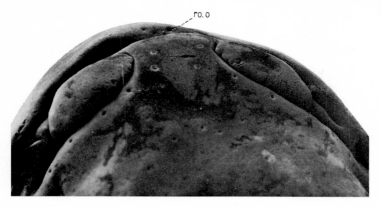

Fig. 1. *Neoceratodus forsteri*. Photograph of the anterior part of the head of young specimen (ca. 30 cm) in ventral view to demonstrate the presence of a paried opening (*ro.o*) on the ventral side of the upper lip. This small opening has persisted, obviously without change, since the Devonian.

(2) Proceeding backwards in time to the Devonian or Silurian the main evolutionary lines do not approach each other to the extent that might be expected from some current views on vertebrate evolution, but run more or less in parallel (Fig. 3). A Devonian coelacanthiform is in all important respects a typical coelacanthiform and in anatomical regards it differs from contemporaneous dipnoans in about the same way as *Latimeria* differs from *Neoceratodus*.

(3) Because in the Devonian coelacanthiforms were typical coelacanthiforms, dipnoans typical dipnoans, and sharks just sharks (Jarvik, 1964: 39, fig. 11) it seems possible that a corresponding conservatism is true also for other groups, and that for instance true holocephalians, also, existed during this period. From this theoretical point of view, at least, it is therefore unlikely that holocephalians have evolved from bradyodonts, which is the view held by Patterson (1965; and in this Volume), or from ptyctodontids, as suggested by Ørvig (1962). Thus if related to ptyctodontids they more likely share with them a common ancestor, as proposed alternatively by Ørvig.

(4) If we want to refer a new fossil form to a certain distinct group we have to show that, as far as is known, it has the specific structural features characteristic of that group. Within each group we may of course allow for some variation but we cannot always elect for membership forms that share some of its significant characters but differ widely in others. For establishment of membership in a group all known facts generally must fit and conform to the basic plan (and to the definition, if such is given) of the group.

If we want to prove relationship between two distinct fossil groups, we have to disregard all structural features that characterize each of the two groups and distinguish them from each other. We have also to disregard characters that have arisen independently because of similar habits, environments, etc. or because of the retrogressive development of the skeleton or of other general trends. Nor can we use with any confidence characters that are suspected to be primitive or known to be shared by other groups. In this case we have to rely mainly on significant common specializations inherited from a common ancestor. To find such characters, not present in any other groups, as may be gathered from my article on dipnoans (in this Volume) is not easy. Or, to take another example illustrating these difficulties, we can argue that porolepiforms and osteolepiforms are related because in contrast to other fishes they have a choana. However, the fenestra endochoanalis is different in the two groups (Jarvik, 1942) and because the origin of the choana is still obscure (see Panchen, 1967) it may very well have arisen independently.

These principles for establishment of (a) membership in a group and (b) affinities between distinct groups, are the same as used for extant vertebrates and may appear self-evident. However, the history of Palaeontology provides many examples, some even of recent date, of disregard or confusion of these simple rules. A strict application of these principles is important when we deal with forms that differ widely in external appearance, for instance because of great changes in environment (e.g. from water to land or vice versa) or as a consequence of the retrogressive development of the skeleton.

(5) The transition from fish to tetrapod (see Jarvik, 1960: 49–56; 1964: 85–86), like the change from life on land to life in the sea (whales, mosasaurs, etc.) certainly brought about considerable modifications. However, just as whales in their anatomy are typical mammals, the anatomical characters of tetrapods have no doubt been inherited from their piscine ancestors. This must mean that the possibilities for phyletic modifications were conditioned by the structure of the ancestral fish and most likely the prerequisites for life on land were in all essentials laid down already at the fish-stage. It is therefore to be expected that tetrapods in anatomical regards conform closely to their piscine ancestors, which must mean that they belong to the same principal group and are to be classified together with them. It is also to be expected that the modifications that have occurred in phylogeny due for instance to changes in environmental conditions or to general evolutionary trends are reflected in their ontogeny.

Consequently if we want to characterize the piscine ancestors of an extant group of tetrapods, or establish relationship between other recent groups (e.g. the petromyzontids) and their presumed ancestors, we have to use criteria of the same kind as those used when we argue for membership in a group. We have to demonstrate, as far as that is possible by fossil material, that the postulated ancestral group, and only that group, shows just the specific features that characterize the recent group and distinguish it from other groups. Moreover, when there are differences, we have to show that the fossil group has the prerequisites for the formation of the structures found in its presumed recent descendants, and, finally that the ontogenetic development of the descendants must be in accord with the postulated evolution.

Only when these rigorous conditions are fulfilled, only when all available data, morphological as well as embryological, fit into the picture are we justified to claim relationship; and having established such a relationship the recent group should be classified together with its ancestors.

(6) Finally, it is to be emphasized that not only the many intricate structures common to all vertebrates, but also the numerous structural features that characterize each of the early groups must have arisen before they appeared in fossils in the Devonian or earlier (see Figs. 3, 4).

In view of this fact and the remarkable conservatism which appears to be a general phenomenon in vertebrate evolution (as regards plants, see Jarvik, 1962) the earliest known vertebrates contribute less than might be expected toward elucidating relationship between the various groups of vertebrates. If we want to make out these relationships, which is a difficult procedure (see above), and unravel the course of events that in the remote past (cf. Fig. 4) led to the origin of the diversified and highly advanced fish-fauna of the early Palaeozoic, we still have to rely mainly on anatomical and embryological investigations of recent vertebrates. However, such studies carried out for more than a century have not yet brought the solution of our problems. We still cannot say anything with certainty about the systematic position and origin of the Dipnoi (Jarvik, in this Volume) and the origin and affinities also of many other main groups remain obscure. However, in spite of these difficulties the studies of early vertebrates and comparisons with recent forms have shed new light on many problems concerning the anatomy and evolution of vertebrates and as is now clear (Jarvik, 1960; 1964; 1965a) some current conceptions of vertebrate phylogeny in fundamental regards lack foundation and are in need of revision. However, before proceeding to a discussion of certain problems concerning the phylogeny of vertebrates it may be justified to give an example of the confusion that an adherance to the generally accepted conceptions may cause.

Students in many fields of research in Biology and Medicine often need information about vertebrate phylogeny. The easiest way to get such information is to consult modern textbooks of Comparative Anatomy or Vertebrate Palaeontology, and such sources are frequently utilized. This is, for instance, the case in several articles in the newly issued volume "On the evolution of the forebrain" (ed. R. Hassler & H. Stephan, Stuttgart, 1966). Schober, who in that volume treats the subject "Vergleichende Betrachtungen am Telencephalon niederer Wirbeltiere" starts with the presentation of a family tree from a well known textbook. This family tree tells (Schober, 1966: 20) that petromyzontids "zu den niedersten recenten Vertebraten gehören" and that evolution has proceeded from primitive cyclostomes, via placoderms, Osteichthyes, Choanichthyes to Amphibia. However, after this presentation Schober turns to available facts concerning important forebrain structures and on the basis partly of his own researches (1964) on *Petromyzon* he arrives at the rather surprising conclusion (1966: 29) that the forebrain in this form "nicht den ursprünglichen Bauplan verkörpert" but that the less specialized forebrain occurs in recent Amphibia, which is just the opposite of what might be expected from the family tree. Kuhlenbeck, Malewitz & Beasley (1966) arrive at a similar conclusion concerning the amphibian forebrain stating that it "provides a most suitable frame of reference for an understanding of both comparative anatomy and presumptive phylogenetic evolution of the vertebrate forebrain". Obviously there is a considerable discrepancy between the anatomical facts and the conceptions embodied in the accepted family tree. As we now shall see the results gained by specialists on brain anatomy seem to tally better with the views on vertebrate phylogeny developed in this and previous papers (Jarvik, 1960; 1964).

The basic branching of the Vertebrata

The Vertebrata form a natural, monophyletic unit, distinguished by a great number of common characters. Their nearest relatives probably are the Acrania and if so obviously they share with them a common ancestor (A, Fig. 4). However, very early in phylogeny the Vertebrata divided into two major stems, the Gnathostomata and the Cyclostomata (the term "Agnatha" is misleading; see below and Jarvik, 1964: 15), differing fundamentally in the development of the naso-hypophysial complex, the position and structure of the gills, the relations between the visceral and axial parts of the skull, the jaw apparatus, the labyrinth, the otoliths, etc. (see Jarvik, 1964: 16–26, figs. 2–6; 1965a; see also Jollie, in this Volume). Considering these great differences and the fact that several of the important specializations characteristic of the cyclostomes were demonstrably developed in their earliest known, Ordovician and Silurian representatives it is clear that the branching of the Vertebrata must have taken place in the Cambrian or more likely in the Precambrian. Because gnathostomes and cyclostomes no doubt evolved independently from a common ancestor (B, Fig. 4) it follows that these two main groups must be of the same age and one group cannot have arisen from the other. In other words the Gnathostomata and the Cyclostomata are sister groups in the sense of Hennig (1966; and earlier papers; see also Brundin, 1966; and in this Volume).

Given this conclusion it is obviously unrealistic to regard recent cyclostomes as primitive and to start from them when discussing phylogenetic problems. Recent vertebrates are end-products of two main evolutionary lines which have been separate for more than 500 million years (see below), and if we want to connect gnathostomes and cyclostomes from a phylogenetic point of view we have to go via their common ancestor (B, Fig. 4).

According to the theories of Hennig one of the two sister groups generally retains more plesiomorph (primitive) characters and can accordinlgy be said to be more primitive than the other (see Brundin, 1966: 26). When was asked which of the two main sister groups in Vertebrata is the more primitive the answer hitherto has been that it is the Cyclostomata. However, most of the arguments for that view elaborated by the anatomists of the 19th century (cartilaginous endoskeleton, absence of exoskeleton, no paired fins) have long been known to be invalid and when cyclostomes still are regarded as primitive it appears to be mainly for the following two reasons.

The first of these reasons provides an example of the power of language upon thought. As is well known the jaws in gnathostomes are modified gill arches and it has never been doubted that jawed vertebrates have arisen from forms in which these modifications had not taken place and which accordingly had no jaws. Misled by the word "jawless" many students have taken it for granted that the jawless ancestors of gnathostomes must be cyclostomes simply because cyclostomes are generally regarded as "jawless" ("Agnatha") and therefore thought to be more primitive than gnathostomes. However, as is well known (see Jarvik, 1964: 21–27, figs. 4–6) cyclostomes very early in phylogeny acquired inwardly directed gills, which prevented the formation of jaws of the gnathostome type, and consequently had to develop a biting mechanism in a different fashion (see also Stensiö, in this Volume).

The second reason is of a geological nature. The first vertebrates to appear in the geological sequence (the Ordovician astraspids) and the dominant ones in the Silurian are cyclostomes (Figs. 3, 4) and because of this fact many palaeontologists, even today, seem to believe that cyclostomes must be more primitive than gnathostomes and ancestral to them. However, if gnathostomes and cyclostomes are sister groups they therefore are of the same age. If so, the occurrence of unquestionable cyclostomes in the Ordovician can only mean that also gnathostomes must have been in existence at the same time, although they yet have been not found as fossils.

These two reasons for the view that cyclostomes are primitive are clearly untenable and expressions such as "primitive jawless cyclostomes" often found in the literature are misleading. The cyclostomes are not "jawless"; they have a biting jaw apparatus ("rasping tongue" in recent cyclostomes) which however is different in origin, anatomy and function from that in gnathostomes, and the statement that cyclostomes are primitive is unproved. If we want to make out which of the two main sister groups in the Vertebrata is more primitive, we have to rely on data provided by comparative anatomy and embryology.

It is not always easy to say which characters in vertebrates are primitive (plesiomorph) and which are advanced (apomorph). In order to find out the basic design of the vertebrates we first have to make out the ground plan for each sister group. Considering again the forebrain we must by detailed comparisons of both larval and adult stages make out which structural features the two groups of recent cyclostomes (petromyzontids and myxinoids) have in common and from that evidence we have to try to reconstruct the basic pattern of the primitive cyclostome forebrain. Then the same intricate procedure must be performed with regard to the various types of forebrain in gnathostomes, whereupon the two basic patterns must be matched. Not until that has been done will it be possible to form an opinion of the structure of the forebrain in the common ancestor (B, Fig. 4) of all vertebrates. The information thus gained will then probably enable us to decide with more confidence which structural features are primitive (plesiomorph) and which are advanced (apomorph) in the various types of forebrain in extant gnathostomes and cyclostomes.

Detailed comparative analyses of the kind now suggested must of course be made also for other parts of the brain, other organs and organ systems. Moreover due attention must be paid to conditions in the sister group of the Vertebrata, which presumably is the Acrania (Fig. 4), before it is possible to reconstruct the ancestral vertebrate and create a reliable basis for discussions of relationships and phylogeny. This certainly is a most time-consuming task which requires cooperation of specialists in many fields of research, but as far as I can see it is the only way out of our present dilemma.

However, even now it is often possible to distinguish safely between primitive and advanced characters. Returning to cyclostomes it is easy to see that they in several regards are highly specialized and convincing arguments for the traditional view that they are more primitive than gnathostomes are difficult to find. It seems likely that the presence of premandibular gills in the Osteostraci (Stensiö, in this Volume), the "permanent separation of the ventral from the dorsal roots of spinal nerves in the Petromyzontia", the preservation of the original metameric order of the tubules of the kidneys in myxinoids (Goodrich, 1909: 38, 45) and possibly the development of the posterior oblique eye-muscle in petromyzontids (Bjerring, in this Volume) are primitive. But whether that applies also to the presence of inwardly directed gills (see Jarvik, 1964: 21–24, figs. 4–6), and to the one or two semicircular canals in cyclostomes I would not venture to decide. On the other hand, cyclostomes are no doubt much more advanced than gnathostomes in the strong modifications of the naso-hypophysial complex (see Stensiö, in this Volume) and very likely also in the absence of the intracranial juncture apparatus (see below) and in the incorporation of the dorsal parts of the visceral arches into the neural endocranium (Jarvik, 1954: 89–95; 1964: 25–26). Moreover the jaw apparatus, although of different types in the two groups of recent cyclostomes, is much more complex than the different, but on the whole fairly simple jaw apparatus of gnathostomes.

From this evidence it is certainly not easy to say if gnathostomes are more primitive than cyclostomes or vice versa and it is hardly possible to come to a decision about these intricate matters until conditions in the Acrania have been analysed and are better understood. However, one thing seems to be sure: it must be incorrect to start from cyclostomes in the way that has hitherto been done when discussing the phylogeny of the Vertebrata and we have to make a serious attempt to get out of the habitual thinking that still stamps these discussions. The gnathostomes are, at any rate in certain important regards, more primitive than the cyclostomes and until the contrary has been proved it is justified to regard them as the plesiomorph sister group of the latter.

The phylogeny of the Gnathostomata

The question which is the most primitive gnathostome group is not easy to answer either. The old opinion that this position is held by selachians is still deeply rooted and unconciously influences our way of thinking. However, if we ask why the sharks are primitive, no satisfactory answer can be given; nor is it easy to decide which other group is more primitive. That it cannot be any of the better known placoderm groups seems obvious (see Stensiö, 1968) and it is at present difficult to find convincing arguments in favour of either acanthodians, actinopterygians or dipnoans. The only gnathostomes worth serious consideration in this respect are in fact the fishes hitherto classified as the Crossopterygii (see Jarvik, in this Volume) which at any rate in certain important regards seem to be remarkably primitive.

As has long been suspected (Jarvik, 1960, and in this Volume; Bjerring, 1967, and in this Volume) "crossopterygians" with the presence of the intracranial juncture apparatus appear to have preserved more primitive conditions than any other vertebrates, cyclostomes included. Primitive are probably the following features:

(1) The persistence of the intraarcual articulation of the second or mandibular metamere.

(2) The existence of separate arcual elements (supra- and infrachordal zygal plates, Bjerring) in the cranial division of the vertebral column.

(3) The presence of a paired basicranial muscle spanning the intracranial juncture apparatus (according to Bjerring probably derived from the second somite).

(4) The presence of a wide and unconstricted notochord within the neural endocranium.

Osteolepiformes and Eutetrapoda. Among "crossopterygians" we may first turn to the Osteolepiformes which in addition display the following probably primitive features:

(5) The fact that the basicranial muscle was short (in the main confined to its metamere; Bjerring, 1967).

(6) The presence of the fissura occipitalis lateralis and the fissura oticalis ventralis (Jarvik, 1954: 7, figs. 1, 30 E) which probably are vestiges of intra-arcual articulations of posterior cranial metameres.

(7) The regular serial arrangement of the dorsal elements of the visceral arches, including those of the three prootic arches that have become incorporated into the neural endocranium (Jarvik, 1954; 1960).

(8) The corresponding regular arrangement of the dental plates associated with these elements, and the presence of separate infrapharyngohyal dental plates (paraotic dental plates, etc., Jarvik, 1954; cf. Nybelin, in this Volume).

(9) The persistence of dental plates on the outside of the hyomandibula and ceratohyal (Jarvik, 1954; 1963: 17–20).

(10) The presence of a very wide spiracular gill-slit (Jarvik, 1954).

(11) The persistence of vestiges of the prespiracular gill-slit (Jarvik, 1954).

(12) The preservation of a wide communication between the spiracular canal (housing the spiracular neuromast sensory organ) and the overlying sensory canal (Jarvik, 1954: 14).

All these facts taken together tend to show that osteolepiforms are primitive but if they are the most primitive vertebrates or even the most primitive gnathostomes is of course impossible to say. But because on present knowledge it is difficult to find better arguments for primitiveness in any other group it is justified to start with them.

Like other groups of vertebrates osteolepiforms have become specialized and of particular interest are those advanced (apomorph) characters that indicate relationship with tetrapods. Among them may be mentioned the fact that the intracranial joint—as a first step towards the tetrapod condition—has become practically immovable (see Jarvik, 1937: 112–117; 1942: 457; 1944a: 32–36; Bjerring, 1967). This presumably allowed the basicranial muscle, which occupied the large gap in the ventral side of the braincase dorsal to the paraotic dental plates, to become ligamentous to an increasing extent, serving mainly to keep the two moieties of the braincase together. At the transition from fish to tetrapod this muscle, which according to Bjerring is probably homologous to the polar cartilage, was gradually transformed into skeletal tissue and it is easy to understand how the big gap in the ventral side of the braincase was thus successively obliterated.

The structure of osteolepiforms has been analysed by the writer in four different parts: the snout (Jarvik, 1937; 1942; 1964; 1966), the intermandibular division (Jarvik, 1963), the posterior part of the head (Jarvik, 1954; 1967a) and the paired fins (Jarvik, 1964; 1965a; 1965b; 1966). These analyses have unequivocally demonstrated that most specific anatomical features characteristic of osteolepiforms occur also in the Anura and in important regards in the Amniota, but not in the Porolepiformes or other fishes and not in the Urodela either. Moreover it has been shown (Säve-Söderbergh, 1932; 1935; 1936; Jarvik, 1942; 1952; 1967b) that ichthyostegalians (ichthyostegids, acanthostegids), which are the oldest known tetrapods, and other temnospondyl stegocephalians also agree well with osteolepiforms and the Anura (Säve-Söderbergh, 1936; Shishkin, 1968). The anthracosaurs, too, in important regards are osteolepiform-like (Jarvik, 1967b).

This evidence can only mean that all these groups of tetrapods, the Eutetrapoda (Säve-Söderbergh, 1934), are closely related to osteolepiforms and descendants of them or at any rate of osteolepiform-like ancestors. However, when dealing with the phylogeny of the Eutetrapoda we have to consider the fact that they include two main stems, the Batrachomorpha and the Reptilomorpha (Säve-Söderbergh, 1934; 1935; Jarvik, 1967b; Shishkin, 1968).

(a) *Batrachomorpha*. The Batrachomorpha comprise two main groups, the Temnospondyli and the Anura.

The Temnospondyli are in the late Upper Devonian represented by two distinct types, ichthyostegids and acanthostegids (Jarvik, 1952). These two types of early tetrapods are in certain regards much specialized and none of them can have given rise to the later appearing temnospondyls or other tetrapods. Moreover, they differ widely from each other in the structure of the posterior part of the skull and obviously they have become modified in different ways. These facts show that the differentiation of temnospondyls was well on the way in late Devonian times and moreover they indicate that ichthyostegids and acanthostegids branched off from ancestral osteolepiforms at an early date and possibly independently of each other (see Fig. 3). Another remarkable fact is that several of the latest (Triassic) temnospondyls in important regards (see Jarvik, 1942: 621–640, fig. 87; 1955, fig. 8; 1964: 66, fig. 22) agree more closely with well known osteolepiforms like *Eusthenopteron* than with the Devonian and other Palaeozoic forms and very likely they have evolved independently from some *Eusthenopteron*-like forebears.

The Anura cannot be derived from any temnospondyls (Jarvik, 1967b) and obviously they form another separate line of evolution. A very close agreement has been found to exist between this recent group and their Devonian forerunners. The similarities even in minor structural details between the common frog (*Rana*) and *Eusthenopteron* for instance in the structure of the snout and the posterior part of the head are really remarkable.

As recently demonstrated (Jarvik, 1967b) there is also a closer agreement than

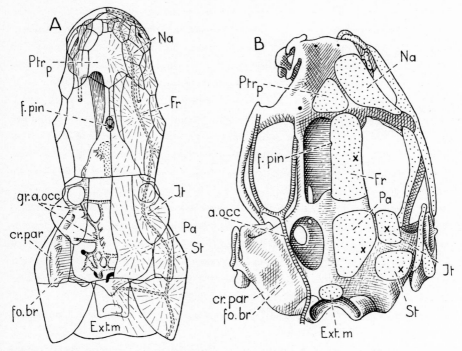

Fig. 2. Representations to demonstrate the similarities in the number, position and relations to the neural endocranium of the dermal bones of the skull roof between *A*, Devonian osteolepiforms (*Eusthenopteron foordi*), and *B*, recent anurans (composite). Modified from Jarvik, 1967 *b*.

Ext.m, median extrascapular; *Fr*, frontal; *It*, intertemporal; *Na*, nasal; *Pa*, parietal; Ptr_p, posterior median postrostral; *St*, supratemporal; *a.occ*, occipital artery; *cr.par*, crista parotica; *fo.br*, fossa bridgei; *f.pin*, pineal foramen; *gr.a.occ*, groove for occipital artery.

was previously known between anurans and osteolepiforms in the number and disposition of dermal bones of the cranial roof. At that time, however, I overlooked papers by Camp (1917) and Griffiths (1954) in which an additional bone is recorded in anurans. This bone, by Griffiths called prefrontal, arises according to him early in ontogeny "over the antero-medial edge of the auditory capsule". Because it arises at a place corresponding to the centre of radiation of the intertemporal (*It*) in osteolepiforms (Fig. 2 A) it probably is homologous to that bone. This additional bone is included in Figure 2 B, which shows also a large posterior median postrostral (Ptr_p) in a specimen of *Pelobates fuscus*.

Because of the many profound similarities between osteolepiforms and anurans and the occurrence of exactly the same specializations in both groups, because several structures found in anurans (eg. the tongue) may be easily derived from the conditions in osteolepiforms and because, finally, many remarkable changes in the ontogeny of anurans (see Jarvik, 1942: 495–554) can be understood only on the presumption that anurans are descendants of osteolepiforms, the requirements for membership in the same group are fulfilled (see above). Accordingly the Anura are to be classified together with the Osteolepiformes (see below).

The fact that even seemingly unimportant anatomical characters of the osteolepiforms persist in anurans implies that anurans in their phylogeny have been just as conservative as coelacanthiforms, dipnoans, petromyzontids, etc. This conservatism seems to be true also of the forebrain which in osteolepiforms appears to have been of the anuran type (Jarvik, 1942; Stensiö, 1963 b). In view of these facts and because osteolepiforms seem to be a primitive group of vertebrates, more primitive perhaps than cyclostomes, it is not surprising that the anuran forebrain is primitive. The results obtained by Schober (1966) and other specialists on forebrain anatomy (see above) are thus in accord with the view tentatively advanced in this paper that the anurans are in certain respects conservative descendants of one of the most primitive groups of vertebrates.

Because of the close affinities between osteolepiforms and anurans and because Palaeozoic intermediate forms are unknown, it is also obvious that discussions of the phylogeny of the Anura on a palaeontological basis must rest on direct comparisons between recent forms and their osteolepiform ancestors. Comparisons with the heterogenous and incompletely known group of late Palaeozoic tetrapods known as the Lepospondyli are of little use. These forms which according to a growing consensus of opinion (Brough & Brough, 1967) probably are reptilomorphs rather than "amphibians" do not, as far as known, show any of the specializations common to osteolepiforms and anurans and the reconstructions of the skull roof in the "ancestral lissamphibian" recently made by some students on the basis of the conditions in lepospondyls are apparently rather far from the truth (cf. Fig. 2, and Jarvik, 1967 b).

It may be true that several subgroups of the Temnospondyli (ichthyostegids. acanthostegids, Triassic stereospondyls, etc.) have evolved separately from different osteolepiforms and that the Anura represent another independent line. However, because of the close agreement between the various batrachomorphs and the well known osteolepiforms like *Eusthenopteron* it is evident that they have arisen from osteolepiforms similar to, or at any rate not deviating too much from, that form and that they have to be classified together with their piscine ancestors. As a revision of the classification of the Vertebrata lies beyond the scope of this paper we may use the term Batrachomorpha provisionally in a widened sense including also their osteolepiform ancestors (Fig. 3).

(*b*) *Reptilomorpha*. Similarly we may use the term Reptilomorpha in a widened sense including the groups hitherto classified as Reptilomorpha (anthracosaurs, reptiles, birds, mammals) and their osteolepiform ancestors (Fig. 3). However, this grouping involves certain difficulties, due mainly to our still incomplete knowledge of the variations within the Osteolepiformes.

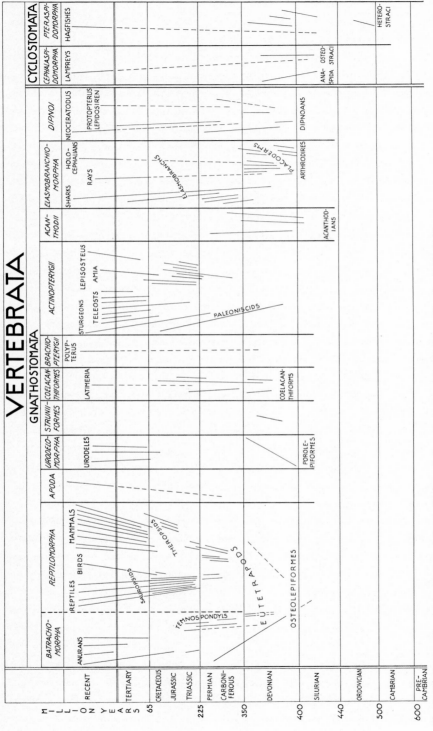

Fig. 3. Diagrammatic representation of the evolution of the vertebrates. Modified from Jarvik, 1960; 1964.

That the Reptilomorpha are closely related to the Osteolepiformes is in the first hand borne out by the following facts: the striking similarities between osteolepiforms and anthracosaurs in the patterns of the dermal bones of the skull roof and the cheek (Jarvik, 1967b); similarities in the basic structure of the palate and the lower jaw; the ontogenetic development and structure of the tongue (Jarvik, 1963) and of the paired limbs (Holmgren, 1933; Jarvik, 1964; 1965a) in amniotes; the presence of a septomaxillary developed very much as its equivalent (the lateral rostral) in the osteolepiforms (Jarvik, 1942; 625–626, fig. 86); similarities in the development of the brain (Stensiö, 1963b; Jarvik, 1965a, fig. 6).

However, in certain regards reptilomorphs differ from *Eusthenopteron* and other well known osteolepiforms and judging from the conditions in the anthracosaurs (embolomeres, seymouriamorphs), they early in phylogeny had undergone certain modifications, which have not yet been traceable back to the osteolepiform stage. The presence of embolomerous vertebrae in early reptilomorphs may possibly be of little importance in view of the variations in the vertebral column in osteolepiforms (see Westoll, 1943; Schaeffer, 1967). However, in the structure of the palate, notably the fenestra exochoanalis, the vomer and the parasphenoid, they have departed from the conditions so far known in osteolepiforms, and moreover the main part of the brain has become displaced backwards into the posterior part of the braincase which has been considerably modified (Jarvik, 1967b). Because of these modifications the brain stem—in sharp contrast to conditions in osteolepiforms and batrachomorphs—has become flexed and the pineal foramen has moved backwards to a point between the parietals. Moreover the processus ascendens palatoquadrati (epipterygoid, columella cranii) and the basipterygoid process have assumed a more posterior position in relation to the otic capsules and the parietals (see Schauinsland, 1903; de Beer, 1937) and neither these processes nor the opening for the n.opticus can be used as landmarks in homologizations of dermal bones.

None of these characters are foreshadowed in the well known osteolepiforms and the Reptilomorpha cannot be directly derived from them. Under these circumstances and because they no doubt are closely allied to osteolepiforms it may be reasonable to assume that reptilomorphs are descendants of some still unknown osteolepiform-like ancestors (cf. Kuhn-Schnyder, 1967: 335). If we want to classify them with their piscine ancestors we therefore have to postulate a separate subgroup among the Osteolepiformes, as tentatively indicated by a broken line in Figure 3.

Apoda. As is well known and sustained by the quotations from Foxon and others given below, apodans differ considerably from both urodeles and anurans and because they in certain regards are said to agree more closely with reptiles than with "amphibians" (see e.g. de Beer, 1937: 461) they appear to hold an isolated position among recent tetrapods. Earlier I had been unable to find any characters linking them either to the Porolepiformes-Urodela or to the Osteolepiformes-Eutetrapoda and I could accordingly say nothing about their affinities. However, as far as may be judged at present they most likely have evolved independently from some rhipidistid- or osteolepiform-like ancestors, and presumably they are more closely related to the Eutetrapoda than to the Urodela as tentatively indicated in Figure 3.

Porolepiformes and Urodela. In addition to the intracranial juncture apparatus (Bjerring, 1967) porolepiforms have several other, probably primitive characters in common with osteolepiforms. But the groove for the prespiracular gill-slit is more distinct than in osteolepiforms (Jarvik, 1954) and porolepiforms are the only fishes in which the presence of a large skin-flap of the mandibular gill-cover has been recorded (Jarvik, 1963: 25, figs. 10, 11). On the other hand teeth are lacking on the outside of the ceratohyal which was probably covered by muscles (Jarvik, 1963) and in this and certain other respects porolepiforms appear to be more advanced than osteolepiforms.

The porolepiforms and osteolepiforms have some advanced characters in common indicating relationship (for the choana, see above) and are generally grouped together as the Rhipidistia. However, in most regards they differ considerably and because the characteristics of porolepiforms were well established in the Lower Devonian (*Porolepis*), they must have separated from each other much earlier.

Distinct and as far as known constant differences between porolepiforms and osteolepiforms (Woodward, 1891; Jarvik, 1942; 1944b, 1948; 1950; 1954; 1955; 1959; 1960; 1962; 1963; 1964; 1965a; 1966; 1967b; and unpublished; Ørvig, 1957; Stensiö, 1963b; Jessen 1966b; 1967; Bjerring, 1967) are found in the development of the paired fins, the structure and evolution of the scales (Ørvig), the pattern of dermal bones of the head (these differences are partly due to different fusions, Jarvik, 1963; 1967b, fig. 9), the structure of the shoulder girdle and the lower jaw, the mode of connection between the fronto-ethmoidal and parietal shields and the development of the supracerebral division of the intracranial juncture apparatus as a whole (Bjerring), etc. However, the most important and profound differences refer to internal structures. The porolepiforms differ widely from osteolepiforms (and other fishes) in the two complex main divisions of the head (the snout, Jarvik, 1942; 1964; 1966; the intermandibular division, Jarvik, 1963) that have been analysed in detail in both groups, and there are also great differences in the structure of the orbitotemporal region and the mode of articulation of the palatoquadrate (Stensiö, 1963b; Jarvik, 1966). Moreover, there are distinct differences in the otico-occipital, the palatoquadrate, the hyomandibula and other parts that are well shown in the material at my disposal but not yet described.

It has also been established that porolepiforms in all those parts studied on a comparative anatomical basis show a most striking agreement even in minor and seemingly insignificant structural details with one group of recent vertebrates, the Urodela, and only with that group. Even the Lower Devonian porolepids show exactly the same partly peculiar specializations in the structure of the snout that are characteristic of urodeles and distinguish them from osteolepiforms and anurans. It has also been demonstrated that the tongue in Urodela on the one hand and extant Eutetrapoda (anurans, reptiles, birds, mammals) on the other arises in a different way and that the prerequisites for the formation of a tongue of the urodele type are present in porolepiforms, whereas that organ in the Eutetrapoda may be easily derived from the conditions in osteolepiforms. Moreover it has been demonstrated that urodeles are different in those other parts (not yet described in porolepiforms) where important similarities between osteolepiforms and eutetrapods have been established (paired limbs, Jarvik, 1965a; 1966; posterior part of head, Jarvik, 1967b). Furthermore it has been shown that the ontogenetic development of urodeles in important regards implies a recapitulation of phylogeny only on the presumption that they are descendants of porolepiforms (Jarvik, 1942: 392–424; 1963; 1964; 1966).

As the conditions for the establishment of a near relationship and membership in the same group (see above) also in this case are fulfilled the conclusion must be that the Urodela are closely related to and descendants of the Porolepiformes, and consequently they are to be classified together with them. For this systematic unit we may use, provisionally, the term Urodelomorpha (Fig. 3).

Remarks. The idea first set forth in 1942 that urodeles and anurans have evolved from two different groups of rhipidistid fishes has been criticized by students who still think that the "Amphibia" and the Tetrapoda as a whole are descendants of some hypothetical presumably Devonian proto-tetrapod and accordingly are monophyletic in origin (see comments by Thomson, and discussion by Szarski, in this Volume). This criticism has been refuted in previous papers (Jarvik, 1964; 1965a; 1966) and in this connection only the following brief remarks may be given.

(1) The attempts to find intermediate forms between osteolepiforms and porolepiforms have failed (see Jarvik, 1966). All rhipidistids known by safe and well documented descriptions are in all regards either typical osteolepiforms or typical porolepiforms.

(2) The attempts to find common characters of recent "Amphibia" ("Lissamphibia") which could serve as proof of a monophyletic origin have also been unsuccessful. The arguments that have been used are unconvincing and most

of them have been shown to be either incorrect or insignificant (Jarvik, 1965a; see also Lehman, in this Volume).

As has long been known and as may be gathered also from the following statements by modern anatomists the three groups of extant "amphibians", the Urodela, the Anura, and the Apoda, differ widely in numerous respects.

As regards the heart Foxon (1955: 217) states: "When considering Recent Amphibia it should be noted that the Urodela and the Anura differ from each other in many respects" (see also Turner, 1967). Concerning apodans he says (Foxon, 1964: 200) that "the departure from both anuran and urodelan conditions is wide".

Kuhlenbeck, Malewitz & Beasley (1966: 18) upon close comparative studies of the apodan forebrain conclude: "Again the three extant recent orders of amphibia, namely urodeles, anurans and gymnophiones, although easily comparable, show three conspicuously different transformations of the topologically invariant set of brain configuration" (see also Rudebeck, 1945).

Finally Clairambault (1963: 115), in his studies of the anuran forebrain and detailed comparisons with urodeles as described by Herrick arrives at the conclusion: "It is thus difficult to see a similarity between the telencephalon of an Urodela and that of the tadpole of an Anura".

Because of the many fundamental differences between urodeles, anurans and apodans it has been generally admitted by anatomists that these three groups have been separate for a long period of time. Stimulated by these results palaeontologists have endeavoured to find ancestral forms of these three groups among the Palaeozoic tetrapods, but hitherto without any success. The main reason for this, besides the incompleteness of the fossil record, is certainly the fact that Palaeozoic "amphibians" are either too incompletely known in anatomical regards or too advanced. If we want to discuss the origin of the recent "amphibians" we therefore have to turn to the much better known ancestral fishes, and of great importance is that the differences between recent "amphibians", notably between urodeles and anurans, are distinct also in those parts that it has been possible to study on a palaeontological basis. The fact that anurans in these parts differ from urodeles in the same way as osteolepiforms differ from porolepiforms cannot be disregarded. In the phylogeny of both the Anura and the Urodela we are clearly concerned with the same remarkable conservatism in anatomical regards that distinguishes other main groups of vertebrates and appears to be a characteristic feature of vertebrate evolution during the long periods of time that are documented by fossils.

Struniiformes. The struniiforms (Jessen, 1966a; 1966b; 1967; Jarvik, in this Volume) are in many respects an aberrant group of Devonian "crossopterygian" fishes with some peculiar palaeoniscid-like features. The internal structure is unknown and it is therefore impossible to make any definite statement about their systematic position. The fact that the skull roof is divided into fronto-ethmoidal and parietal shields in much the same way as that in rhipidistids indicates that struniiforms have a primitive intracranial juncture apparatus but cannot be taken as evidence of relationship. However, because they in certain other characters resemble rhipidistids it seems likely that they are more closely related to them than to any other fishes.

Coelacanthiformes. The coelacanthiforms have hitherto been grouped with osteolepiforms, porolepiforms and struniiforms as Crossopterygii. However, as this classification rests almost entirely on characters (the intracranial juncture apparatus) that probably are primitive the term Crossopterygii has to be rejected (see Jarvik, in this Volume). If we still want to speak of crossopterygians we can only do so if we mean with this designation fishes in which the intracranial juncture apparatus persists.

When they appear as fossils in the middle of the Devonian period coelacanthiforms had already acquired the rostral organ and other characteristic features and no doubt the group was established much earlier and had evolved independently for a long period of time (Fig. 3). With regard to the intracranial juncture apparatus the earliest known coelacanthiforms were more advanced than porolepiforms and osteolepiforms (secondary elongation of basicranial muscle, forward position of supracerebral articular division; Bjerring, 1967; Jarvik, 1967b, fig. 9). In other anatomical regards they differ very much from both of these groups and from tetrapods. Because they differ widely also from brachiopterygians and apparently are more distantly related to them than there was earlier reason to assume (see Jarvik, in this Volume) we have to admit that the origin and affinities of coelacanthiforms are obscure. However, as far as may be judged at present it may, after all, be among the other "crossopterygians" and brachiopterygians that we have to look for their closest relatives. The presence of lungs may be an important common character of "crossopterygians" and brachiopterygians (see Jarvik, in this Volume) and moreover these groups share some similarities in the pattern of the dermal bones that may indicate relationship.

Brachiopterygii. The brachiopterygians, comprising only the two extant genera *Polypterus* and *Calamoichthys*, are certainly no palaeoniscid derivatives and they cannot possibly be referred to actinopterygians as is generally done (see Jarvik, in this Volume). Nor can they be regarded as primitive. It may be true that they in forebrain anatomy are less advanced than actinopterygians (Rudebeck, 1945; Nieuwenhuys, 1963; 1966) but the parasphenoid is more intricate in structure than in any other fishes (Jarvik, 1954) and in this and other regards they are obviously highly specialized. As they differ considerably from all extant and fossil fishes, brachiopterygians, too, are to be looked upon as an isolated group of fishes and as recently emphasized also by Daget, Bauchot, Bauchot & Arnoult (1964) they very likely have evolved independently since pre-Devonian times.

Actinopterygii. Recent members of this group agree in two probably important characters. They are the only fishes that have an air-bladder (see Jarvik, in this Volume) and the forebrain is specialized in much the same way (Niewenhuys, 1963, figs. B–F; 1966). These facts indicate that actinopterygians are a natural group derived from a common ancestor, but they do not necessarily imply that they all are descendants of early palaeoniscids which is the current view. In their externally discernible features, well known in particular in *Moythomasia* (Jessen, 1968), the early (Devonian) palaeoniscids differ considerably from the contemporaneous groups of fishes and cannot be very closley related to any of them. Unfortunately the endoskeletal structures are still practically unknown in early palaeoniscids and in these regards we have to rely mainly on the Triassic forms (see Jarvik, 1964: 47). A remarkable feature in them is that there are both a fissura occipitalis lateralis and a fissura oticalis ventralis developed very much as those in osteolepiforms (see above and Jarvik, 1954) and like in them probably vestiges of intraarcual articulations of posterior cranial metameres. In this character and the presence of a comparatively short parasphenoid (see Jarvik, 1954) the Triassic palaeoniscids appear to be primitive. The presence of a well developed myodome, on the other hand, implies a specialization and in this regard they are probably more advanced than recent sturgeons, in which a myodome is lacking. However, as recently shown by Jessen (1968) *Acipenser* agrees closely with *Moythomasia* in the detailed configuration of the endoskeletal shoulder girdle and the endoskeleton of the pectoral fin, and because there are also other similarities it seems likely that the sturgeons are descendants of some primitive palaeoniscids in which the myodome had not yet developed. Whether *Amia* and *Lepisosteus*, too, are palaeoniscid derivatives remains uncertain and also the origin of the Teleostei seems to be obscure.

Remarks. The groups of gnathostome fishes now considered (osteolepiforms, porolepiforms, struniiforms, coelacanthiforms, brachiopterygians, actinopterygians) have some qualities in common in which they seem to differ distinctly from the groups (acanthodians, elasmobranchiomorphs, dipnoans) to be treated below. They have an outer dental arcade, formed in the upper jaw by two bones (premaxillary, maxillary) but in the lower jaw by only one bone (dentary). Moreover there is generally an inner dental arcade formed in the upper jaw by three bones (vomer, dermopalatine, ectopterygoid; see e.g. Nybelin, in this Volume, fig. 1) and in the lower jaw likewise by three bones (coronoids). Furthermore it has been demonstrated (Jarvik, 1967a) that the lower jaw as a whole is very similar in these groups and fundamentally different from that in dipnoans and elasmobranchiomorphs. These striking similarities in jaw structure suggest that "crossopterygians", brachiopterygians and actinopterygians form a natural unit for which the term Teleostomi could be used. However, as tetrapods, which in early forms have very similar jaws, are to be classified together with their piscine ancestors (osteolepiforms, porolepiforms) this term cannot be used as a systematic unit in classification, unless also the tetrapods are included.

Acanthodii. As recently demonstrated by Miles (1965; and in this Volume) acanthodians in several regards agree more with "bony fish" than with elasmobranchiomorphs and he suggests that they may be grouped as a separate class in the superclass Teleostomi as opposed to the superclass Elasmobranchiomorpha (cf. Schaeffer, in this Volume). However, acanthodians lack the outer dental arcade and the lower jaw (see also Gross, 1967; Ørvig, 1967) is of a special build not easily comparable to that in teleostomes. Moreover acanthodians according to Gross (1967: 121) in their dentition agree with placoderms "but differ from other fishes by the ontogenetic development and replacement of their teeth and by the histology of the functional elements (teeth, cutting edges of the jaws)". Because furthermore acanthodians in gill arch structure seem to agree with elasmobranchiomorphs rather than with teleostomes (Nelson, in this Volume) the most reasonable thing to do for the moment appears to be to place the Acanthodii as a separate group between "teleostomes" and elasmobranchiomorphs.

Elasmobranchiomorpha. This major group comprises extant and fossil elasmobranchs and the Arthrodira or Placodermi which was the dominant group of fishes in the Devonian. During the last few years our knowledge of fossil elasmobranchiomorphs, notably arthrodires (see Stensiö, 1968), but also elasmobranchs (see Patterson 1965; and in this Volume) has increased very much but is still incomplete particularly with regard to internal structures. It is therefore not easy to make a satisfactory subdivision into subgroups and no generally accepted classification exists. Because the elasmobranchiomorphs include many mutually very different forms, such as sharks, holocephalians, rhenanids, antiarchs, etc. it is also very difficult to find common characters clearly distinguishing them from other gnathostome groups and it may be asked if it is justified to group them together.

However, as demonstrated by Stensiö and others (see Stensiö, 1959; 1963a; 1968) the various arthrodires in many regards agree with extant elasmobranchs and most students now agree that holocephalians are either descendants of early arthrodires (Ørvig, 1960; 1962; Stahl, 1967; Stensiö, 1968) or at any rate share with them a common origin (Patterson, 1965: 213). The rays may perhaps also be descendants of primitive, probably rhenanid-like arthrodires (Jarvik, 1964: 37–39, fig. 11). The origin of sharks, on the other hand, is still unknown, but according to Stahl (1967: 193; cf. Patterson, 1965: 206) selachians and holocephalians, although different in several important respects, have so many features in common that it is reasonable to assume that they share a common ancestor.

This evidence, although partly indirect and on the whole not very strong, suggests that the various forms included in the Elasmobranchiomorpha (Fig. 3) are in some way related and probably of a common origin. However, considering the presence of several mutually different, highly specialized subgroups in the Devonian and the great differences between, for instance, a Devonian shark like *Cladodus* and a contemporaneous antiarch like *Bothriolepis* it is clear that their presumed common ancestor must have been in existence much earlier, and not until the relations between this ancient ancestral form and the ancestors of dipnoans and acanthodians have been made out it is possible to decide if the retention of the Elasmobranchiomorpha as a systematic unit is justified.

Dipnoi. The origin and affinities of dipnoans or lungfishes are still obscure. As shown elsewhere (Jarvik, in this Volume) they differ widely from osteolepiforms and it is impossible to find any significant specializations indicating relationship either with them or with any other "teleostome" fishes; because tetrapods are descendants of porolepiforms and osteolepiforms the similarities with "amphibians", notably urodeles, almost certainly are due to parallelism. The lungs in dipnoans may have arisen independently and the current view that the absence of the outer dental arcade in dipnoans is secondary (see e.g. Bertmar, in this Volume) is unproved and has no support in embryological evidence. Because dental plates MdX and MdY in the lower jaw of Devonian dipnoans are retained in larval stages of *Neoceratodus* (Jarvik, 1967a), whereas in the ontogeny of recent dipnoans there are not the slightest traces of equivalents of the premaxillary, maxillary and dentary of the outer dental arcade or of the dermopalatine, ectopterygoid and coronoids of the inner arcade in teleostomes it is doubtful if such elements have ever been developed in dipnoans. At any event in the absence of the outer and at least the main part of the inner dental arcades dipnoans agree with elasmobranchiomorphs and because they also in the structure of the lower jaw as a whole and in many other respects, too (see Jarvik, in this Volume), agree with extant elasmobranchiomorphs, notably holocephalians, there is reason to believe that they are more closely related to elasmobranchiomorphs than to teleostomes. However, the origins of both these main groups and of acanthodians and their mutual relations are still so obscure that it is impossible to make a decision; and if we ask which of the recent vertebrate groups is the sister group of dipnoans (holocephalians? coelacanthiforms?) no answer can be given.

The phylogeny of the Cyclostomata

This subject has been discussed above and in previous papers (Jarvik, 1960; 1964; 1965a; see also Stensiö, in this Volume). The available data warrant the following conclusions:

(1) The Cyclostomata and the Gnathostomata are sister groups, derived from a common ancestor (B, Fig. 4). The traditional view that cyclostomes are more primitive than gnathostomes is unproved (see above). Which of these two groups is more primitive cannot be decided until the conditions in the Acrania (Acraniata), which is the sister group of the Craniata (Vertebrata), have been analysed. However, because gnathostomes in important characters are more primitive than even the earliest known cyclostomes they may tentatively be regarded as the plesiomorph (more primitive) sister group of the two.

(2) The cyclostomes, extant and fossil, have many characters in common (see Stensiö, in this Volume) indicating a common ancestry (C, Fig. 4), and probably this main group as a whole is a monophyletic unit.

(3) Recent cyclostomes comprise two distinct groups, petromyzontids (lampreys) and myxinoids (hagfishes). As has been well known since the days of Johannes Müller and sustained by a great number of later students (see Jarvik, 1965a) these two groups differ fundamentally from each other in the anatomy and function of their jaw apparatus (the "rasping tongue") and in numerous other regards. These differences in practically all organ systems can only mean that these two groups have been separate for vast periods of time.

(4) As shown by Stensiö (1927; 1958; 1964; and in this Volume) cephalaspids (Osteostraci) in important regards show the same specializations that are characteristic of petromyzontids and distinguish them from myxinoids. On the basis of this and other evidence Stensiö groups petromyzontids, cephalaspids and anaspids as Cephalaspidomorpha and claims (1964: 169) that these three groups have evolved independently of each other from primitive cephalaspidomorphs. Provided the new interpretation (Jarvik, 1965a) of the "electrical fields" of the cephalaspids is correct, and the five ventral and the single dorsal sac-like evaginations of the membraneous labyrinth in *Petromyzon* are vestiges of more extensive evaginations of the membraneous labyrinth in cephalaspids, contained in the five ventral and the single dorsal canals for "electric nerves", petromyzontids are probably more closely related to cephalaspids than to anaspids in which "electrical fields" are lacking (cf. also Ørvig, in this Volume, concerning the corium in cephalaspids). It seems thus likely that petromyzontids and cephalaspids (Osteostraci) have evolved from a common ancestor (G, Fig. 4) independently of anaspids.

(5) The myxinoids do not show the specializations of the naso-hypophysial complex that are characteristic of all cephalaspidomorphs and they differ also in other important respects from both cephalaspids and petromyzontids. This

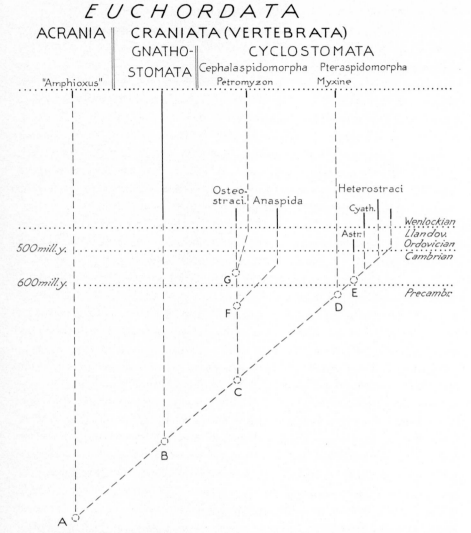

Fig. 4. Diagram illustrating certain important dates in the early evolution of the vertebrates. The phylogenetic relationships indicated are tentative.

can only mean that myxinoids are descendants of early Palaeozoic cyclostomes other than the petromyzontids and their immediate ancestors and in this sense recent cyclostomes are diphyletic in origin (Jarvik, 1965a).

(6) When discussing the phylogeny of myxinoids (for review see Heintz, 1963) we have to consider the fact that the early Palaeozoic fauna in addition to cephalaspidomorphs (cephalaspids, anaspids) includes another main group of cyclostomes, namely the Heterostraci, and the question if myxinoids are more closely related to them or to cephalaspidomorphs has been lively debated during the last few decades.

In support of the view, held by many students (see Heintz, 1963), that myxinoids are related to cephalaspidomorphs, it has been claimed that they, like petromyzontids, have a "rasping tongue" and that such a highly specialized organ can have arisen only once. Moreover it has been maintained that myxinoids like cephalaspidomorphs, but in contrast to the Heterostraci, have a single olfactory organ and a single median nostril. On this assumption myxinoids have been grouped with cephalaspidomorphs as the Monorhina, whereas the Heterostraci have been classified as the Diplorhina.

Serious objections may be raised against these arguments. As is well known myxinoids do not have a rasping tongue. They have a highly specialized, biting jaw apparatus which both in anatomy and function differs fundamentally from the jaw apparatus of petromyzontids (see Jarvik, 1965a). Nor do they have a single olfactory organ. As demonstrated by Stensiö (in this Volume), and as has long been suspected because of the presence of paired olfactory nerves, the olfactory organ in all cyclostomes is paired, as in gnathostomes.

Because neither the "rasping tongue" nor the alleged monorhine condition can be used as arguments for relationship there appears to be nothing linking myxinoids to cephalaspidomorphs. Under these circumstances and because there are several facts indicating relationship between myxinoids and heterostraceans (Stensiö, 1958; 1964; and in this Volume) the only possibility, on present evidence, is to accept Stensiö's view and classify the Myxinoidea together with the Heterostraci as the Pteraspidomorpha. However, since myxinoids as far as may be judged at present cannot be derived from any heterostraceans it is to be assumed that they share with them a common ancestor (D, Fig. 4).

(7) We know by fossils recently discovered in Arctic Canada that typical anaspids existed in the middle of the Silurian period, more precisely in late Llandoverian or early Wenlockian (Thorsteinsson, 1967). Consequently cephalaspids (Osteostraci) must have existed at the same time and because of the great differences between cephalaspids and anaspids (see Stensiö, 1964) it is to be assumed that the common ancestor of the Cephalaspidomorpha (F, Fig. 4) was considerably older.

We also know by fossils (Ørvig, 1958) that the Heterostraci were represented by several types (astraspids, eriptychiids) in Middle Ordovician and probably present already in lowermost Ordovician. Accordingly the common ancestor of the Heterostraci (E, Fig. 4) must have occurred in the Cambrian or earlier.

Because the Pteraspidomorpha demonstrably were well established in the Ordovician it is also obvious that their sister group, the Cephalaspidomorpha, must have been present at the same time; and the common ancestor of all cyclostomes (C, Fig. 4) must have preceded the common ancestors (D, F) of both the Pteraspidomorpha and the Cephalaspidomorpha. Moreover it is clear that the common ancestor (B) of the Gnathostomata and the Cyclostomata must have been in existence still earlier and that in its turn it was preceded by the common ancestor (A) of the Acrania (Acraniata) and the Craniata (Vertebrata).

Concluding remarks

With regard to the classification and phylogeny of the vertebrates there is still considerable diversity of opinion, and although most students agree that the traditional, typological (horizontal) arrangement into hierarchic classes is to be replaced by a phylogenetic (vertical) classification little in this direction so far has been done. As early as in 1927 Stensiö grouped the cyclostomes according to phylogenetic principles and in 1934 Säve-Söderbergh proposed a new classification that abolished the old classes, but as recently emphasized by Brundin (1966: 19) "very few have cared". Since 1934 our knowledge of early vertebrates has greatly improved and it has appeared more and more desirable to make a clean sweep of all the unproved statements and preconceived ideas that still form the basis of current conceptions of vertebrate phylogeny and to try to get at a grouping more consistent with available facts. The sister group concept recently introduced by Hennig (1966; and earlier papers; see also Brundin, 1966; and in this Volume) is certainly a useful tool for this purpose, because it enforces a phylogenetic way of thinking and a careful consideration of the evidence, but as far as the vertebrates are concerned it is at present not so easy to handle. We have to remember that only the youngest fragments of the evolutionary successions are available for study, and our knowledge of the internal structure of the fossil forms and the anatomy and embryology of most recent vertebrates is still too incomplete to enable us to make out the relationships between the various main groups with any degree of certainty. We can safely say that the Gnathostomata is the sister group of the Cyclostomata and that the Petromyzontida is the sister group of the Myxinoidea. But which recent group is most closely related to, and accordingly the sister group of, for instance, dipnoans, coelacanthiforms, apodans, chelonians or insectivores? We cannot at present answer these questions and the new grouping of the Vertebrata suggested in this paper (Figs. 3, 4) is therefore in many respects tentative.

As is well known (see above and Jarvik, 1960; 1964; 1967a) the main vertebrate groups, both those that have remained fishes and those that have evolved into tetrapods, very early became highly specialized and have changed very little in anatomical regards during the last 400–500 million years. This means that the main evolutionary lines run almost in parallel and if prolonged backwards in time they would hardly ever meet (cf. Fig. 3). This is clearly unreasonable and with Brough (1958: 34) we have to assume that "Evolution in the more distant past must have been a much more accelerated process than anything we know at present". In this respect phylogeny offers an interesting parallel to ontogenetic development which in the early embryonic period proceeds so rapidly that, for instance, in the chick the basic vertebrate morphology is attained at only about 50 hours after fertilization. With regard to early phyletic evolution we may conclude that gnathostomes and cyclostomes were established and well separated in the beginning of the Ordovician about 500 million years ago. From this we can further conclude that their common ancestor (B, Fig. 4) must have been living in the Cambrian or earlier, but how long before the end of that period this and other postulated ancestral forms (A, B, C, etc.) occurred is impossible to say and the direction of line A–E in Figure 4 is conjectural.

We have to admit that we do not know anything about the course of events in the remote past when the vertebrates arose and passed through the most important phases in their history. Nor can we say anything with certainty about the rate of evolution or of the environmental conditions, and we cannot explain why no vertebrate fossils have been found in Cambrian or earlier deposits. We have to admit also that our knowledge of the early vertebrates is still incomplete particularly with regard to internal structures. However, in the museums fossil material suited for anatomical studies is not yet anywhere nearly wholly utilized. Substantial sequences of Palaeozoic rocks are still incompletely searched for fossils and vast, promising areas are unexplored. It is therefore to be hoped that new discoveries and further palaeoanatomical and palaeohistological studies will shed new light on the early history of the vertebrates and contribute to the analysis of the origin and relationships of the various groups. But Vertebrate Palaeontology alone cannot solve the problems. We strongly need much more detailed information about anatomy, histology, embryology, etc. of extant vertebrates. It is only in cooperation with other branches of Biology and Natural Sciences as a whole that progress can be made.

References

Bjerring, H. C. (1967). Does a homology exist between the basicranial muscle and the polar cartilage? *Colloques int. Cent. natn. Rech. Scient.*, **163**: 223–267.
Brough, J. (1958). Time and evolution. In *Studies on Fossil Vertebrates*, ed. Westoll, T. S.,: 16–38. London: Athlone Press.
Brough, Margaret C. & Brough, J. (1967). Studies on early tetrapods. 1. The Lower Carboniferous microsaurs. 2. *Microbrachis*, the type microsaur. 3. The genus *Gephyrostegus. Phil. Trans. R. Soc.* (B) **252**: 107–165.
Brundin, L. (1966). Transantarctic relationships and their significance, as evidenced by chironomid midges, with a monograph of the subfamilies Podonominae and Aphroteniinae and the austral Heptagyiae. *K. svenska VetenskAkad. Handl.*, (4) **11**: 1–472.
Camp, C. L. (1917). An extinct toad from Rancho La Brea. *Univ. Calif. Publ. Geol.*, **10**: 287–292.

Clairambault. P. (1963). Le tèlencéphale de *Discoglossus pictus* (Oth.). Étude anatomique chez le têtard et chez l'adulte. *J. Gehirnf.*, **6**: 87–121.
Daget, J., Bauchot, M.-L. & R., & Arnoult, J. (1964). Développement du chondrocrâne et des arcs aortiques chez *Polypterus senegalus* Cuvier. *Acta zool. Stockh.*, **46**: 201–244.
de Beer, G. R. (1937). *The development of the vertebrate skull.* Oxford: Oxford University Press.
Foxon, G. E. H. (1955). Problems of the double circulation in vertebrates. *Biol. Rev.*, **30**: 196–228.
Foxon, G. E. H. (1964). Blood and respiration. In *Physiology of the Amphibia*, ed. Moore, J. A.: 151–209. New York & London: Academic Press.
Goodrich, E. S. (1909). Cyclostomes and fishes. In *A Treatise on Zoology*, ed. Lankester, E. R. Vol. 9, Vertebrata Craniata, fasc. 1. London: A. & C. Black.
Griffiths, I. (1954). On the "otic element" in Amphibia Salientia. *Proc. zool. Soc. Lond.* **124**: 35–50.
Gross, W. (1967). Über das Gebiss der Acanthodier und Placodermen. In *Fossil Vertebrates*, ed. Patterson, C. & Greenwood, P. H., *J. Linn. Soc. (Zool.)*, **47**: 121–130.
Heintz, A. (1963). Phylogenetic aspects of myxinoids. In *The Biology of Myxine*, ed. Brodal, A. & Fänge, R.,: 9–21. Oslo: Universitetsforl.
Hennig, W. (1966). *Phylogenetic systematics.* Urbana: University of Illinois Press.
Holmgren, N. (1933). On the origin of the tetrapod limb. *Acta zool. Stockh.*, **14**: 185–295.
Jarvik, E. (1937). On the species of *Eusthenopteron* found in Russia and the Baltic states. *Bull. geol. Inst. Upsala*, **27**: 63–127.
Jarvik, E. (1942). On the structure of the snout of crossopterygians and lower gnathostomes in general. *Zool. Bidr. Uppsala*, **21**: 235–675.
Jarvik, E. (1944a). On the dermal bones, sensory canals and pit-lines of the skull in *Eusthenopteron foordi* Whiteaves, with some remarks on *E. säve-söderberghi* Jarvik. *K. svenska VetenskAkad. Handl.*, (3) **21**: 1–48.
Jarvik, E. (1944b). On the exoskeletal shoulder-girdle of teleostomian fishes, with special reference to *Eusthenopteron foordi* Whiteaves. *K. svenska VetenskAkad. Handl.*, (3) **21**: 1–32.
Jarvik, E. (1948). On the morphology and taxonomy of the Middle Devonian osteolepid fishes of Scotland. *K. svenska VetenskAkad. Handl.*, (3) **25**: 1–301.
Jarvik, E. (1950). Middle Devonian vertebrates from Canning Land and Wegeners Halvø (East Greenland). 2. Crossopterygii. *Medd. Grønland*, **96**: 1–132.
Jarvik, E. (1952). On the fish-like tail in the ichthyostegid stegocephalians with descriptions of a new stegocephalian and a new crossopterygian from the Upper Devonian of East Greenland. *Medd. Grønland*, **114**: 1–90.
Jarvik, E. (1954). On the visceral skeleton in *Eusthenopteron* with a discussion of the parasphenoid and palatoquadrate in fishes. *K. svenska VetenskAkad. Handl.*, (4) **5**: 1–104.
Jarvik, E. (1955). The oldest tetrapods and their forerunners. *Sci. Monthly*, **80**: 141–154.
Jarvik, E. (1959). Dermal fin-rays and Holmgren's principle of delamination. *K. svenska VetenskAkad. Handl.*, (4) **6**: 1–51.
Jarvik, E. (1960). *Théories de l'évolution des vertébrés reconsidérées à la lumière des récentes découvertes sur les vertébrés inférieurs.* Paris: Masson.
Jarvik, E. (1962). Les porolépiformes et l'origine des urodèles. *Colloques int. Cent. natn. Rech. Scient.*, **104**: 87–101.
Jarvik, E. (1963). The composition of the intermandibular division of the head in fish and tetrapods and the diphyletic origin of the tetrapod tongue. *K. svenska VetenskAkad. Handl.*, (4) **9**: 1–74.

Jarvik, E. (1964). Specializations in early vertebrates. *Annls Soc. R. zool. Belg.*, **94**: 11–95.
Jarvik, E. (1965a). Die Raspelzunge der Cyclostomen und die pentadactyle Extremität der Tetrapoden als Beweise für monophyletische Herkunft. *Zool. Anz.*, **175**: 101–143.
Jarvik, E. (1965b). On the origin of girdles and paired fins. *Israel J. Zool.*, **14**: 141–172.
Jarvik, E. (1966). Remarks on the structure of the snout in *Megalichthys* and certain other rhipidistid crossopterygians. *Ark. Zool.*, (2) **19**: 41–98.
Jarvik, E. (1967a). On the structure of the lower jaw in dipnoans: with a description of an early Devonian dipnoan from Canada, *Melanognathus canadensis* gen. et sp. nov. In *Fossil Vertebrates*, ed. Patterson, C. & Greenwood, P. H., *J. Linn. Soc.* (*zool.*), **47**: 155–183.
Jarvik, E. (1967b). The homologies of frontal and parietal bones in fishes and tetrapods. *Colloques int. Cent. natn. Rech. Scient.*, **163**: 181–213.
Jessen, H. (1966a). Struniiformes. In *Traité de Paléontologie*, ed. Piveteau, J., **4:3**: 387–398.
Jessen, H. (1966b). Die Crossopterygier des Oberen Plattenkalkes (Devon) der Bergish-Gladbach-Paffrather Mulde (Rheinisches Schiefergebirge) unter Berücksichtigung von amerikanischem und europäischem *Onychodus*-Material. *Ark. Zool.*, (2) **18**: 305–389.
Jessen, H. (1967). The position of the Struniiformes (*Strunius* and *Onychodus*) among crossopterygians. *Colloques int. Cent. natn. Rech. Scient.*, **163**: 173–180.
Jessen, H. (1968). *Moythomasia nitida* Gross und *M.* cf. *striata* Gross, devonische Palaeonisciden aus dem oberen Plattenkalk der Bergisch-Gladbach-Paffrather Mulde (Rheinisches Schiefergebirge). *Palaeontographica*, (A) **128**: 87–114.
Kuhlenbeck, H., Malewitz, T. D. & Beasley, A. B. (1966). Further observations on the morphology of the forebrain in Gymnophiona, with reference to the topologic vertebrate forebrain pattern. In *Evolution of the Forebrain*, ed. Hassler, R. & Stephan, H.,: 9–19. Stuttgart: Thieme Verl.
Kuhn-Schnyder, E. (1967). Paläontologie als stammesgeschichtliche Urkundenforschung. In *Die Evolution der Organismen*, hrsg. Heberer, G., vol. **1**: 238–419. Stuttgart: Fischer.
Miles, R. S. (1965). Some features in the cranial morphology of Acanthodians and the relationships of the Acanthodii. *Acta zool. Stockh.*, **46**: 233–255.
Nieuwenhuys, R. (1963). The comparative anatomy of the actinopterygian forebrain. *J. Gehirnf.*, **6**: 171–192.
Nieuwenhuys, R. (1966). The interpretation of the cell masses in the teleostean forebrain. In *Evolution of the Forebrain*, ed. Hassler, R. & Stephan, H.,: 32–39. Stuttgart: Thieme Verl.
Ørvig, T. (1957). Remarks on the vertebrate fauna of the Lower Upper Devonian of Escuminac Bay, P. Q., Canada, with special reference to the porolepiform crossopterygians. *Ark. Zool.* (2) **10**: 367–426.
Ørvig, T. (1958). *Pycnaspis splendens*, new genus, new species, a new ostracoderm from the Upper Ordovician of North America. *Proc. U.S. nat. Mus.*, **108**: 1–23.
Ørvig, T. (1960). New finds of acanthodians, arthrodires, crossopterygians, ganoids and dipnoans in the Upper Middle Devonian calcareous flags (Oberer Plattenkalk) of the Bergisch Gladbach-Paffrath trough. 1. *Paläont. Z.*, **34**: 295–335.
Ørvig, T. (1961). New finds of acanthodians, arthrodires, crossopterygians, ganoids and dipnoans in the Upper Middle Devonian calcareous flags (Oberer Plattenkalk) of the Bergisch Gladbach-Paffrath trough. 2. *Paläont. Z.*, **35**: 10–27.
Ørvig, T. (1962). Y a-t-il une relation directe entre les arthrodires ptyctodontides et les holocéphales? *Colloques int. Cent. natn. Rech. Scient.*, **104**: 49–60.

Ørvig, T. (1967). Some new acanthodian material from the Lower Devonian of Europe. In *Fossil Vertebrates*, ed. Patterson, C. & Greenwood, P. H., *J. Linn. Soc. (Zool.)*, **47**: 131–153.

Panchen, A. L. (1967). The nostrils of choanate fishes and early tetrapods. *Biol. Rev.*, **42**: 374–420.

Patterson, C. (1965). The phylogeny of the chimaeroids. *Phil. Trans. R. Soc.*, (B) **249**: 101–219.

Rudebeck, B. (1945). Contributions to forebrain morphology in Dipnoi. *Acta zool. Stockh.*, **26**: 9–156.

Säve-Söderbergh, G. (1932). Preliminary note on Devonian stegocephalians from East Greenland. *Medd. Grønland*, **94**: 1–107.

Säve-Söderbergh, G. (1934). Some points of view concerning the evolution of the vertebrates and the classification of this group. *Ark. Zool.*, **26A**: 1–20.

Säve-Söderbergh, G. (1935). On the dermal bones of the head in labyrinthodont stegocephalians and primitive Reptilia with special reference to Eotriassic stegocephalians from East Greenland. *Medd. Grønland*, **98**: 1–211.

Säve-Söderbergh, G. (1936). On the morphology of Triassic stegocephalians from Spitsbergen, and the interpretation of the endocranium in the Labyrinthodontia. *K. svenska VetenskAkad. Handl.*, (3) **16**: 1–181.

Schaeffer, B. (1967). Osteichthyan vertebrae. In *Fossil Vertebrates*, ed. Patterson, C. & Greenwood, P. H., *J. Linn. Soc. (Zool.)*, **47**: 185–195.

Schauinsland, H. (1903). Beiträge zur Entwicklungsgeschichte und Anatomie der Wirbeltiere. 1–3. *Zoologica, Stuttg.*, **39**: 1–168.

Schober, W. (1964). Vergleichend-anatomische Untersuchungen am Gehirn der Larven und adulten Tiere von *Lampetra fluviatilis* (Linné, 1758) und *Lampetra planeri* (Bloch, 1784). *J. Gehirnf.*, **7**: 107–209.

Schober, W. (1966). Vergleichende Betrachtungen am Telencephalon niederer Wirbeltiere. In *Evolution of the Forebrain*, ed. Hassler, R. & Stephan, H.,: 20–31. Stuttgart: Thieme Verl.

Shishkin, M. A. (1968). On the cranial arterial system of the labyrinthodonts. *Acta zool. Stockh.*, **49**: 1–22.

Stahl, Barbara J. (1967). Morphology and relationships of the Holocephali with special reference to the venous system. *Bull. Mus. comp. Zool. Harv.*, **135**: 141–213.

Stensiö, E. A. (1927). The Downtonian and Devonian vertebrates of Spitsbergen. 1. Family Cephalaspidae. *Skr. Svalbard Nordishavet*, **12**: i–xii, 1–391.

Stensiö, E. A. (1958). Les cyclostomes fossiles ou ostracodermes. In *Traité de Zoologie*, ed. Grassé, P.-P., **13:1**: 173–425.

Stensiö, E. A. (1959). On the pectoral fin and shoulder girdle of the arthrodires. *K. svenska VetenskAkad. Handl.*, (4) **8**: 1–229.

Stensiö, E. A. (1963a). Anatomical studies on the arthrodiran head. 1. Preface, geological and geographical distribution, the organisation of the arthrodires, the anatomy of the head in the Dolichothoraci, Coccosteomorphi and Pachyosteomorphi. Taxonomic appendix. *K. svenska VetenskAkad. Handl.*, (4) **9**: 1–419.

Stensiö, E. A. (1963b). The brain and the cranial nerves in fossil, lower craniate vertebrates. *Skr. norske VidenskAkad. Oslo., Mat.-naturv. Kl.*, **1963**: 1–120.

Stensiö, E. A. (1964). Les cyclostomes fossiles ou ostracodermes. In *Traité de Paléontologie*, ed. Piveteau, J., **4:1**: 96–382.

Stensiö, E. A. (1968). Les arthrodires. In *Traité de Paléontologie*, ed. Piveteau, J., **4:2**. Paris: Masson. In press.

Thorsteinsson, R. (1967). Preliminary note on Silurian and Devonian ostracoderms from Cornwallis and Somerset Islands, Canadian Arctic Archipelago. *Colloques int. Cent. natn. Rech. Scient.*, **168**: 45–47.

Turner, S. C. (1967). A comparative account of the development of the heart of a newt and a frog. *Acta zool. Stockh.*, **48**: 43–57.

Westoll, T. S. (1943). The origin of the tetrapods. *Biol. Rev.*, **18**: 78–98.

White, E. I. (1965). The head of *Dipterus valenciennesi* Sedgwick & Murchison. *Bull. Br. Mus. nat. Hist.: Geol.*, **11**: 1–45.

Woodward, A. S. (1891). *Catalogue of the fossil fishes in the British Museum (Natural History)* Vol. **2**. London: Br. Mus. (nat. Hist.).

Index of authors, genera and species

(Italic type indicates pages where illustrations appear)

Acanthodes, 109–125, 129–140, *131*.
   A. bronni, 109, *110*, *112*, 114, *115*, *121*, 129, 131, *132*, *133*.
   A. sulcatus, 124.
Acanthorhina, 167.
Acipenser, *266*, 271, *273*, *274*, 275, 423, 465, 467, 516.
   A. fulvescens, 465.
   A. sturio, *133*.
Acris, 449.
   A. crepitans, 445, 449.
Addens, J. L., 355, 356.
Aequiarchegonaspis, 44.
   A. schmidti, *39*.
Aetobatus, 194.
Agassiz, J. L. R., 195, 199, 204, 391, 393, 399, 412.
Agassizodus, 155, 157, 160, 161, 162, 163, 164, 165.
   A. sp., 156.
   A.? sp., 399, 411.
Alaspis, 382, 383, *384*, 385, 388.
   A. macrotuberculata, *378*, 382, *383*.
Albula, 431, 458, 460, 467.
   A. vulpes, 455, *459*.
Alepocephalus, 138.
   A. macropterus, *138*.
Allis, E. P. Jr., 92, 105, 116, 125, 134, 135, 136, 137, 138, 141, 232, 242, 259, 272, 280, 415, 427, 429, 431, 434, 436, 438.
Allocryptaspis, 62.
Althaspis? spatulirostris, *43*.
Alytes obstetricans, 312.
Ambystoma (*Amblystoma*), 298.
   A. tigrinum, 317, 318, 320, *322*, 325.
Amia, 238, *266*, *273*, *274*, 275, 307, 318, 414, 421, 422, 423, 424, 425, 429, 431, 432, 434, 436, 437, 467, 516.
   A. calva, *134*, 318, 343, *facing 353*, *430*, *433*, *435*, 465.
Amiurus, 273.
Amphioxus, 447.

Amphiuma, 259.
   A. means, 312.
Andrews, S. Mahala, 10, 300.
Angaraspis urvantzevi, *58*.
Anglaspis, 84.
Anguilla anguilla, 465.
   A. japonica, 465.
   A. rostrata, 465.
Anthony, J. & Robineau, D., 226, 242; see also Millot, J. & Anthony, J.
Arambourg, C. & Bertin, L., 153, 168; see also Bertin, L.
Archegosaurus, 309.
Ariëns Kappers, C. U., Huber, G. C. & Crosby, E. C., 356.
Aspidosteus, 55.
Asteracanthus, 155, 198.
Astraspis, 381, 388.
Astroconger myriaster, 465.
Ateleaspis robusta, *33*, 386.
Atz, J. W., 280.
Australosomus, 137, 434.
   A. kochi, *133*.
Ayers, H., 91, 105.

Badenhorst, Cornelia E., 328.
Baldauf, R. J., 328.
Balfour, F. M., 92, 105.
Balinsky, B. I., 90, 105.
Ball, J. N., 468, 469, 470.
Ball, J. N. & Ensor, D. M., 468, 470.
Ball, J. N. & Olivereau, M., 468, 470; see also Olivereau, M.; Olivereau, M. & Ball, J. N.
Bancroft, J. R., 331, 340.
Bdellostoma, 18, 23, 65, 102.
   B. stouti, 24, 26.
Beerman, W., 448, 451.
Bendix-Almgreen, S. E., 153–170; 10, 171, 181, 192, 193, 197, 200, 204.
Benninghoff, A., 317, 325, 329.
Bentley, P. J., 448, 451.
Berezowski, A., 446, 451.

Berg, L. S., 153, 168.
Bern, H. A., 468, 469, 470; *see also* Nicoll, C. S. & Bern, H. A.; Nicoll, C. S., Bern, H. A. & Brown, D.
Berrill, N. J., 374, 375, 393.
Bertallanfy, L. v., 446, 451.
Bertin, L., 238, 241, 242; *see also* Arambourg, C. & Bertin, L.
Bertmar, G., 259–283; 10, 68, 99, 105, 120, 125, 137, 141, 232, 237, 239, 240, 353, 357, 518.
Birgeria, 386, 432.
Bjerring, H. C., 341–357; 10, 225, 242, 504, 505, 506, 512, 515, 523.
Bock, F., 470.
Bock, R., 456, 470.
Bombina, 331, *332*, 336, *337*, 338.
  B. bombina, 331, *333*, 335.
  B. orientalis, 318, 328, 331.
Bonde, N., 10.
Boreosomus, 307.
  B. piveteaui, *390*.
Borlorocoetes taeniatus, 312.
Bothriolepis, 518.
  B. canadensis, *378*.
Bridge, T. W., 137, 141.
Brien, P., 238, 239, 242, 281.
Brien, P. & Bouillon, J., 238, 239, 242.
Brodal, A. & Fänge, R., 65, 68.
Broman, I., 262, 263, 281.
Brough, J., 522, 523.
Brough, Margaret, C. & Brough, J., 509, 523.
Brundin, L., 473–495; 10, 502, 503, 522, 523.
Buchmann, H., 461, 465, 470.
Buistrov, A. P., 82, 84, 87, 308, 381, 385, 389, 392, 393.
Buistrov, A. P. & Efremov, J. A., 308, 314.
Burckhardt, R., 331, 340.
Bush, F. M., 446, 451.

Cain, A. J., 474, 495.
Calamoichthys, 237, 238, 456, 465, 467, 515.
Callorhynchus, 184, 202, *273*.
Calyptocephalus, 317, 328.
Camp, C. L., 508, 523.
Campbell, K. S. W., 252, 257.
Casier, E., 199, 204.
Caster, K. E. & Eaton, T. H. Jr., 373, 393; *see also* Eaton, T. H. Jr.; Eaton, T. H. Jr. & Stewart, P. L.
Caturus, 434, 436.
Cephalaspis, 272, 497.
  "*C.*" *signata*, *28*.

Ceratodus, 241.
Ceratophrys, 311, 312.
  C. dorsata, 311.
Chanos, 459, 461, 463, 467.
  C. chanos, 456, 459, *460*, *461*, 465.
Chapman, B. G., 448, 451.
Cheiracanthus, 124.
Chimaera, 232.
  C. monstrosa, *235*.
Chimaeropsis, 168.
Chinlea, 210.
Chirodipterus, 211, *214*, *215*, 218, 234, 236.
  C. wildungensis, *234*.
Chlamydoselachus, 116, 123.
Chondrenchelys, 156, 157, 158, 159, 160, 161, 162, 163, 164, 165, 166, 167, 192, 193, 199, 200, 220.
  C. problematica, 156.
Church, G., 446, 451.
Cladodus, 518.
Cladoselache, 160.
Clairambault, P., 514, 524.
Clark, W. B., 355, 357.
Climatius, 118, 139.
Clupea, 266, *273*, 461.
  C. harengus, 455, 465.
Cobitis, *273*.
Cochliodus, 194, 195, 196, 197, 199, 200, 202.
  C. contortus, 193, *194*, 195, *196*, *198*: *see also* "*Streblodus oblongus*"; "*Tomodus convexus*".
  C. latus, 197.
  C. vanhornii, 195.
Coelacanthus, 210.
Coelolepis luhai (syn. *Phlebolepis elegans*), 81, 82.
Cole, F. J., 68, 92, 105.
Commoner, B., 448, 451.
Conchopoma, 255.
Conger conger, 465.
Copodus, 197, 199.
  C. spatulatus, *198*.
Corsin, J.; *see* Devillers, C. & Corsin, J.
Corsy, F., 130, 141.
Cryptobranchus, 339.
Ctenaspis, 62.
Ctenoptychius, 156, 159, 160, 161, 162, 164, 165.
  C. apicalis, 156.
Curtis, A. S. G., 421, 427.
Cyclopterus, 266.
Cyprinus, *273*.
  C. carpio, 420.

Czopek, G., 450, 451; *see also* Szarski, H. & Czopek, G.
Daget, J., 130, 141.
Daget, J., Bauchot, M.-L. & R., & Arnoult, J., 238, 242, 516, 524.
Damas, H., 22, 27, 68, 92, 98, 102, 105, 353, 357.
Daniel, J. F., 116, 117, 118, 123, 125.
Davis, J. W., 193, 195, 204.
Dawson, J. A., 65, 68.
Dean, B., 109, 112, 125, 129, 130, 131, 132, 140, 141, 153, 160, 161, 168, 171, 174, 179 199, 204.
de Beer, G. R., 98, 99, 105, 119, 123, 125, 135, 141, 204, 217, 259, 281, 511, 512, 524.
de Beer, G. R. & Moy-Thomas, J. A., 182, 204; *see also* Moy-Thomas, J. A.
Deltodus, 200.
Deltoptychius, 155, 158, 160, 161, 162, 163, 164, 165, 166, 168, 176, 177, 178, 179, 180, 181, 183, 184, 189, 191, 192, 193, 194, 195, 196, 197, 200, 201.
  D. acutus, *198*.
  D. armigerus, 156, 174, *178*, 195.
  D. moythomasi, 156, 181, 191.
Denison, R. H., 247–257; 10, 39, 68, 73, 79, 93, 94, 102, 105, 211, 221, 227, 242, 365, 366, 370, 373, 374, 375, 376, 380, 381, 382, 389, 393.
de Smet, W., 238, 242.
Desmognathus fuscus, 312.
Devillers, C., 10, 130, 139, 141, 232, 242, 307.
Devillers, C. & Corsin, J., 413–428.
Dietz, P. A., 434, 438.
Dikenaspis yukonensis, *39*, 40, 42, 44, 46.
Dineley, D. L., 10.
Diplocercides, 207, 225.
Diplovertebron, 309.
Diplurus, 210.
Dipnorhynchus, 236, 247, 252, 253, 254.
  D. lehmanni, 247, 254, 255.
  D. sussmilchi, 247, 252, 255.
Dipterus, 231, 233, 234, 236, 249, 250, 251, 252, 253, 421, 498.
  D. valenciennesi, *231*, *236*, *253*.
Discoglossus, 312.
  D. pictus, 312.
Dodson, J. W., 367, 370.
Dohrn, A., 91, 92, 105.
Dollo, L., 139, 141, 223, 228.
Doryaspis (syn. *Lyktaspis*), 55, 73.
  D. nathorsti (syn. *Lyktaspis nathorsti*), 73.

Drepanaspis gemuendenensis, *56*, *392*.
Drosophila, 493.
Duke, K. L., 445, 451.
Dunkleosteus, 145, *147*, 148, 150.
Dvinosaurus, 308.

Eaton, T. H. Jr., 118, 126; *see also* Caster, K. E. & Eaton, T. H. Jr.
Eaton, T. H. Jr. & Stewart, P. L., 326, 329.
Ectosteorhachis, *215*, 287, *288*, 289, 292.
Edds, M. V. Jr., 369, 370.
Edgeworth, F. H., 118, 122, 126, 223, 240, 434, 438.
Eglonaspis, 49, 58, 63.
  E. rostrata, *57*.
Elops, 135, 138, 429, 431, 432, 434, 436, 437, 439, 440, 441, 442, 443, 456, 458, 460, 467.
  E. hawaiiensis, *133*, *138*.
  E. lacerta, 441.
  E. machnata, 455, 456, *457*, *458*, 465.
  E. saurus, *430*, *435*, 439, *440*, *441*, *442*, 465.
Emmart, E. W., Pickford, G. E. & Wilhelmi, A. E., 469, 470; *see also* Pickford, G. E., Robertson, E. E. & Sawyer, W. H.; Grant, W. C. & Pickford, G. E.
Engraulis telera, 465.
Engström, K., 10.
Epiceratodus, 234.
Epinephelus, 137, 138.
  E. hexagonatus, *138*.
Erikodus, 155, 157, 158, 159, 160, 161, 162, 163, 164, 165, 166, 167.
  E. groenlandicus, 156.
Errolichthys, 386.
Eryops, *298*, 309.
Escuminaspis, 382, 384.
Esox, *266*, *273*, 424, 434.
"*Estheria*", 124.
Eucentrurus paradoxus, 156.
Eusthenopteron, 116, 134, 135, 137, 207, *209*, 210, *214*, 216, 225, 250, 251, 287, 289, 290, *291*, 293, *296*, 300, 301, 302, 307, 309, 434, 436, 437, 439, 440, 441, 442, 443, 507, 509, 511.
  E. foordi, *134*, 292, 293, 294, 300, 351, facing 352, *433*, *508*.
  E. saevesoederberghi, 293, *294*.

Fadenia, 155, 156, 157, 158, 159, 160, 161, 162, 163, 164, 165, 166, 167.
  F. crenulata, 156.
Fell, Honor B. & Robison, R., 417, 427.

Fleissig, J., 339, 340.
Fleurantia, 210.
Fontaine, H., Callamand, O. & Olivereau, M., 468, 471; see also Olivereau, M.
Forster–Cooper, C., 271, 281.
Fox, H., 130, 141, 221, 223, 236, 240, 242, 260, 265, 270, 272, 280, 281, 446, 451.
Foxon, G. E. H., 242, 243, 511, 514, 524.
Føyn, S. & Heintz, A., 73, 79; see also Heintz, A.; Kiaer, J. & Heintz, A.
Francis, E. T., 297, 304.
Francois, Y., 308, 314.
Friend, P. F., 79.
Friend, P. F., Heintz, N. & Moody-Stuart, M., 73, 79; see also Heintz, N.
Fugu, 467.
  F. poecilonotus, 465.
  F. rubripes, 465.
Fundulus, 468.
Fürbringer, P., 91, 106, 130, 136, 141, 241.

Gadow, H., 116, 117, 126.
Gadus, 273.
Gadusia chapra, 465.
Gallien, L. & Durocher, M., 318, 329.
Ganorhynchus, 255.
Garman, S., 130, 141.
Gasterosteus, 266.
Gaupp, E., 92, 106, 130, 135, 141.
Gauthier, G. F. & Padykula, H. A., 446, 451.
Gegenbaur, C., 89, 136, 139, 141, 223.
Gelderen, C. van, 275, 276, 281.
Gérard, P. & Cordier, R., 456, 471.
Giebel, C. G. A., 172, 204.
Gilette, R., 312, 313, 314.
Gisin, H., 486, 495.
Glimcher, M. J., 364, 370.
Glossoidaspis (syn.? *Protaspis*), 62.
Glyptolepis, 209, 225, 289, 290, 292, 302.
Gnathonemus, 463.
  G. gambiensis, 456.
Goette, A., 92, 106.
Goin, O. B., Goin, C. J. & Bachmann, K., 448, 451.
Goodey, T., 116, 126.
Goodrich, E. S., 21, 22, 27, 36, 68, 89, 92, 106, 118, 119, 126, 223, 226, 227, 228, 237, 238, 241, 243, 271, 281, 307, 314, 385, 391, 394, 429, 438, 504, 524.
Göppert, E., 68.
Górski, A., 446, 452.
Goss, R. J., 445, 452.
Goujet, D., 10.

Gould, S. J., 449, 452.
Graham-Smith, W., 374, 394.
Graham-Smith, W. & Westoll, T. S., 210, 221; see also Westoll, T. S.; Lehmann, W. & Westoll, T. S.
Grant, W. C. & Pickford, G. E., 471; see also Pickford, G. E., Robertson, E. E. & Sawyer, W. H.; Emmart, E. W., Pickford, G. E. & Wilhelmi, A. E.
Grasseichthys, 463, 467.
  G. gabonensis, 456.
Greenwood, P. H., 10.
Greenwood, P. H., Rosen, D. E., Weitzman, S. H. & Myers, H. S., 456, 463, 466, 467, 471; see also Schaeffer, B. & Rosen, D. E.
Gregory, W. K., 123, 126, 130, 141, 153, 168, 373, 394, 424.
Gregory, W. K. & Raven, H. C., 300, 304.
Greil, A., 236, 259, 281.
Griffiths, I., 508, 524.
Griphognathus, 255.
Gross, W., 10, 81, 82, 87, 145, 148, 211, 221, 227, 239, 243, 272, 281, 373, 385, 387, 389, 392, 394, 411, 412, 517, 524.
Gryphaea, 488.
Guinnebault, M., 414, 427.
Gustafson, G., 60, 68.
Gyroptychius, 249, 287.
  G. milleri, 293.
  G. sp., 292, 293.

Haas, G., 10.
Hagelin, L.-O. & Johnels, A. G., 22, 68; see also Johnels, A. G.
Hagen, F. v., 465, 469, 471.
Hamilton, W. J., Boyd, J. D. & Mossman, H. W., 239, 243.
Hammarberg, F., 137, 141.
Hanawalt, P. C. & Haynes, R. H., 448, 452.
Harder, W., 130, 141.
Harengula zunasi, 463, 465.
Harrington, R. W., 130, 137, 141.
Hassler, R. & Stephen, H., 501.
Hatschek, B., 356, 357.
Hay, Elizabeth D., 369, 370.
Heier, P., 355, 357.
Heintz, A., 145–148; 10, 29, 55, 68, 73, 81, 150, 386, 392, 394, 520, 521, 524; see also Føyn, S. & Heintz, A.; Kiaer, J. & Heintz, A.
Heintz, N., 73–80; 10, 13, 56; see also Friend, P. F., Heintz, N. & Moody-Stuart, M.
Helicoprion, 155, 171.

Helling, H., 331, 340.
Helodus, 155, 156, 157, 158, 159, 160, 161, 162, 163, 164, 165, 166, 167, 176, 182, 184, 185, 187, 192, 193, 194, 195, 196, 197, 200, 202.
H. simplex, 156, 195, *198*.
Hemicyclaspis, *384*.
H. murchisoni, *383*.
Hemmingsen, A. M., 449, 452.
Henneman, E. & Olson, C. B., 447, 452.
Henneman, E., Somjen, G. & Carpenter, D. O., 447, 452.
Hennig, W., 473, 474, 475, 477, 479, 480, 494, 495, 502, 503, 522, 524.
Hepsetus, 266, *273*, *274*, 275.
Heptranchias, 120, 123.
H. perlo, 116.
Herlant, M., 456, 471.
Herre, W., 69.
Hertwig, O., 89, 90, 106, 312, 314.
Heterodontus, 123, 232, 263.
Heterotis, 463.
H. niloticus, 456.
Hexanchus, 118.
Hibernaspis macrolepis, *57*.
Hilsa ilisha, 465.
Hime, 138.
H. japonica, *138*.
Hinsberg, V., 331, 340.
Hirella gracilis, *33*.
Hoar, W. S., 466, 471.
Hoffman, C. K., 357.
Holmesella? sp., 399, 401, *404*, *407*, 408, *409*, *410*, 411.
Holmgren, N., 22, 24, 27, 65, 69, 89, 92, 106, 114, 115, 117, 118, 119, 120, 126, 130, 134, 135, 136, 137, 139, 140, 141, 142, 153, 158, 165, 167, 168, 184, 204, 223, 243, 259, 281, 295, 296, 297, 298, 300, 304, 317, 329, 343, 357, 368, 370, 387, 394, 511, 524.
Holmgren, N. & Pehrson, T., 281; see also Pehrson, T.
Holmgren, N. & Stensiö, E. A., 14, 27, 65, 69, 91, 106, 130, 142, 227, 229, 232, 236, 271, 272, 281, 292, 304, 422, 427; see also Stensiö, E. A.
Holmqvist, O., 434, 438.
Holodipterus, 255.
Holoptychius, 437.
Holtfreter, J., 421, 427.
Homalacanthus, 124.
Homarus americanus, 363.

Homostius, 145, *146*, 147, 148, 150.
H. milleri, 146.
Honma, Y., 465, 471.
Hoppe, K.-H., 82, 87, 385, 394.
Horridge, G. A., 448, 452.
Hörstadius, S., 366, 370.
Hotton, N. 3rd., 116, 117, 123, 126.
Howes, G. B., 92, 106.
Hoyte, D. A. N., 417, 428.
Huene, F. v., 317, 329.
Hughes, G. M., 208, 221.
Hughes, G. M. & Shelton, G., 208, 221.
Hussakof, L., 233, 243.
Huxley, J., 481, 486, 487, 488, 489, 495.
Huxley, T. H., 91, 92, 106, 116, 118, 123, 126, 241.
Hyla arborea, 449.
H. crucifer, 449.
H. japonica, 318, 328.
Hylodes binotatus, 312.
Hynobius, 298, 336, *337*: see also "*Salamandrella*".
H. keyserlingi, 331, 335.
Hypsocormus, 434, 436.

Ichthyostega, *299*, *309*, 310.
Illing, G., 446, 452.
Ischnacanthus, 123, 139.
Isenberg, H., Moss, M. L. & Lavine, L., 360, 361, 370; see also Moss, M. L.; Moss M. L. & Meehan, M.; Moss, M. L. & Murchison, E.

Jaekel, O., 109, 113, 118, 126, 156, 159, 161, 168, 171, 172, 174, 178, 179, 180, 181, 182, 184, 187, 188, 195, 204.
Janassa, 156, 159, 160, 161, 162, 163, 164, 165.
J. bituminosa, 156.
Jarvik, E., 7–8, 223–245, 497–527; 10, 14, 19, 29, 53, 65, 69, 89, 90, 91, 92, 96, 97, 98, 103, 104, 106, 116, 119, 126, 130, 134, 135, 136, 137, 142, 155, 161, 165, 168, 169, 207, 208, 209, 210, 214, 216, 221, 222, 250, 251, 253, 257, 259, 260, 270, 271, 272, 277, 279, 280, 281, 282, 285, 288, 289, 290, 291, 292, 293, 294, 295, 296, 298, 299, 300, 301, 302, 303, 304, 305, 307, 309, 310, 314, 317, 318, 322, 326, 329, 368, 369, 370, 373, 375, 393, 394, 421, 424, 428, 432, 433, 434, 436, 437, 438, 439, 440, 441, 442, 443.
Jarvikia, 251.

Jefferies, R. P. S., 373, 394.
Jessen, H., 429–438; 10, 155, 169, 208, 229 237, 243, 279, 282, 512, 514, 516, 525.
Johnels, A. G., 10, 22, 27, 69, 92, 98, 99, 106, 353, 357, 385, 394; *see also* Hagelin, L.-O. & Johnels, A. G.
Johnston, J. B., 355, 357.
Jollie, M., 89–107; 10, 113, 126, 502.

Kaensche, C. C., 69.
Kaufman, L., 445, 452.
Kawamoto, M., 463, 465, 467, 471.
Kentuckia, *214*, *215*, 216, *217*.
Kerr, J. G., 223, 243.
Kerr, T., 276, 282, 312, 313, 314, 456, 465, 466, 469, 471.
Kerr, T. & van Oordt, P. G. W. J., 276, 282.
Kesteven, H. L., 223, 241, 243.
Kiaer, J., 29, 37, 69, 70, 71, 73, 78, 81, 227, 243.
Kiaer, J. & Heintz, A., 81, 82, 83, 84, 85, 86, 87; *see also* Heintz, A.; Føyn, S. & Heintz, A.
Kiaeraspis auchenaspidoides, *15*.
Kindahl, M., 223, 243.
Kisselewa, Z. N., 136, 142.
Kner, R., 124, 125.
Kneria, 463, 467.
 K. sp., 456.
Knowles, F. & Vollrath, L., 465, 469, 471; *see also* Vollrath, L.
Kölliker, A., 90, 106.
Koltzoff, N. K., 98, 106.
Kotlassia, 308.
Krefft, G., 259, 282.
Krishnamoorhi, B. & Krishnaswamy, S., 447, 452.
Kruszyński, J. & Boothroyd, B., 448, 452.
Kuhlenbeck, H., Malewitz, T. D. & Beasley, A. B., 502, 514, 525.
Kuhn-Schnyder, E., 511, 525.
Kulczycki, J., 282, 290, 305.
Kupffer, C. W. v., 22, 24, 69.

Lacerta, *296*.
Lagios, M. D., 456, 471.
Lamb, A. B., 343, 357.
Lanarkia, 81, 82, 84, 85, 86, 87.
Lankester, E. R., 73, 79.
Larsell, O., 355, 357.
Latimeria, 135, 137, *209*, 225, 226, 227, 237, 238, 389, 436, 499.
 L. chalumnae, *134*, 270, 351.
Lebedkina, Natalie S., 317–329.

Le Danois, Yseult, 422, 428.
Lehman, J.-P., 307–315; 10, 130, 134, 142, 219, 220, 222, 228, 231, 233, 243, 244, 255, 257, 282, 374, 386, 394, 434, 438, 514.
Lehmann, W. & Westoll, T. S., 254, 257; *see also* Westoll, T. S.; Graham-Smith, W. & Westoll, T. S.
Lekander, B., 414, 424, 428.
Lepidosiren, 233, 234, 259, 260, 261, *266*, 269, *279*.
 L. paradoxa, 268.
Lepisosteus, 137, *266*, *273*, 318, 467, 516.
 L. osseus, *134*, 465.
Leptosteus, 149.
Levi, G., 445, 452.
Liem, K. F., 431, 438.
Limanda herzensteini, 465.
Limulus, 373.
Lindström, M., 374, 394.
Lindström, T., 27, 69.
Lison, L., 154, 169.
Liu, Y.-H., 14, 32, 69.
Logania, 85, 86, 87.
 L. scotica, 84, 85; *see also Thelodus scoticus*.
Løvtrup, S., 10.
Lubosch, W., 22, 23, 69.
Lucilia coesar, 448.
Luha, A., 81.
Luther, A., 65, 69, 122, 126.
Lydekerrina, 309.
Lyktaspis, 55, 56, 57, 58, 59, 60, 73–79, *76*, *77*; *see also Doryaspis*.
 L. dani, 75.
 L. minor, 75.
 L. nathorsti, 56, 73, *74*, 75, *78*, 79; *see also Doryaspis nathorsti*; *Pteraspis nathorsti*; *Scaphaspis nathorsti*.

McAllister, D. E., 140, 142.
Mackerras, I. M., 481, 482, 483, 485, 495.
McMurrich, J. P., 434, 438.
Maetz, J., Mayer, N. & Chartier-Baraduc, M., 468, 471; *see also* Olivereau, M. & Chartier-Baraduc, M.
Makushok, V. M., 234, 235, 244.
Marinelli, W. & Strenger, A., 23, 27, 64, 65, 69.
Mark-Kurik, Elga, 145, 146, 147, 148, 150, 151; *see also* Obruchev, D. V. & Mark-Kurik, Elga.
Marrable, A. W., 446, 452.
Martinsson, A., 382, 394.

Mathews, M. B., 364, 370.
Matthes, E., 22, 69.
Mayr, E., 474, 490, 491, 492, 493, 495.
Medvedeva, Irene M., 331–340; 282.
Megalichthys, 227, *291*, 292.
Megalops, 431, 443.
Meites, J. & Nicoll, C. S., 469, 471; see also Nicoll, C. S. & Bern, H. A.; Nicoll, C. S., Bern, H. A. & Brown, D.
Menaspacanthus, 201.
Menaspis, 155, 158, 160, 161, 162, 163, 164, 165, 166, 168, 171, 172–193, 194, 195, 196, 197, 199, 200, 201, 202.
 M. armata, 156, 171, *173*, *175*, *177*, *180*, *182*, *183*, *186*, *187*, *189*, *190*.
Mesacanthus, 123, 124.
Messier, B. & Leblond, C. P., 446, 452.
Metopacanthus, 167, 168, 202.
 M. granulatus, *199*.
Meyer, P., 312, 315.
Miles, R. S., 109–127; 11, 129, 130, 131, 132, 135, 140, 142, 171, 204, 207, 208, 212, 222, 387, 394, 517, 525.
Millot, J. & Anthony, J., 130, 134, 137, 142, 209, 225, 226, 227, 238, 244, 355, 357, 389, 395, 436, 438; see also Anthony, J. & Robineau, D.
Mimetaspis sp., *16*, *17*.
Misra, A. B. & Sathyanesan, A. G., 465, 471.
Moona, J. C., 137, 142.
Morgulis, S., 446, 452.
Moss, M. L., 359–371; 11, 373, 375, 376, 377, 380, 395, 420, 428; see also Isenberg, H., Moss, M. L. & Lavine, L.
Moss, M. L. & Meehan, M., 362, 363, 371.
Moss, M. L. & Murchison, E., 362, 371.
Moy-Thomas, J. A., 153, 156, 157, 158, 159, 163, 169, 171, 182, 184, 185, 191, 195, 204, 386, 395, 414, 422, 428; see also de Beer, G. R. & Moy-Thomas, J. A.
Moythomasia, 436, 516.
Müller, J., 65, 69, 92, 106, 519.
Muraenesox cinereus, 465.
Mustelus, 273.
Mylopteraspis, 52.
Myriacanthus, 168, 179, 183, 184, 197, 202.
Myxine, 20, 60, 65, 92, 102.
 M. glutinosa, *64*.

Nakajima, Y., Pappas, G. D. & Bennet, M. V. L., 450, 452.
Nanpanaspis, 32.
Neal, H. V., 343, 357.

Nectaspis areolata, *34*.
Necturus, 272.
Nelsen, O. E., 341, 357.
Nelson, G. J., 129–143; 11, 120, 124, 126, 517.
Nemomyxine, 60.
Neoceratodus, 136, *209*, *217*, 218, 228, 229, 232, 233, 234, 236, 251, 260, 261, 263, 265, *266*, 269, *273*, *274*, 275, 279, 498, 499, 518.
 N. forsteri, *134*, *234*, 259, 260, *261*, *262*, *268*, *499*.
Neomyxine, 60.
Nesides, 225.
Neumayer, L., 22, 23, 26, 27, 69.
Newberry, J. S. & Worthen, A. H., 195, 197, 204; see also St. John, O. & Worthen, A. H.
Nicoll, C. S. & Bern, H. A., 469, 471; see also Meites, J. & Nicoll, C. S.; Bern, H. A.
Nicoll, C. S., Bern, H. A., & Brown, D., 469, 471.
Nielsen, E., 11, 115, 116, 117, 119, 120, 126, 130, 133, 137, 142, 154, 156, 157, 169, 199, 204, 207, 209, 213, 216, 222, 307, 315, 386, 395, 432, 434, 438.
Nieuwenhuys, R., 515, 516, 525.
Nieuwenhuys, R. & Hickey, M., 226, 244.
Noble, G. K., 311, 312, 315.
Notemigonus, 466.
Notomyxine, 60.
Notopterus, 463.
 N. afer, 456.
Notorhynchus, 120.
Nybelin, O., 439–443; 11, 431, 438, 506, 516.

Obruchev, D. V., 11, 56, 57, 58, 69, 73, 77, 79, 82, 84, 87, 148, 153, 169, 195, 204, 381, 385, 389, 395.
Obruchev, D. V. & Mark-Kurik, Elga, 392, 395.
Olbiaspis coalescens, *57*.
Olivereau, M., 465, 468, 469, 471, 472.
Olivereau, M. & Ball, J. N., 468, 472; see also Ball, J. N.
Olivereau, M. & Chartier-Baraduc, M., 468, 472; see also Maetz, J., Mayr, N. & Chartier-Baraduc, M.
Olsson, R., 455–472; 11.
Oltmanns, E., 312, 315.
Oncorhynchus rhodurus, 465.
 O. tschawytscha, 465.
Ophiacodon, *298*, *299*.

Ornithoprion, 155, 156, 157, 158, 159, 160, 161, 162, 164, 165, 192, 194, 406, 411.
O. hertwigi, 156.
Orodus, 199, 399, *400, 401, 403, 404, 405, 406*, 408, 409, 411.
Ørvig, T., 373–397; 11, 81, 82, 87, 91, 153, 154, 155, 163, 169, 171, 197, 198, 199, 204, 205, 230, 238, 240, 241, 244, 249, 257, 269, 282, 399, 401, 402, 408, 409, 411, 412, 498, 499, 512, 517, 519, 521, 525.
Osmerus, 136.
Osteoglossum, 134.
Osteolepis, 233, 249, 253, 287, *288*, 289.
Owen, R., 195, 205, 259.
Oxyconger leptognathus, 465.

Pachycormus, 434, 436.
P. macropterus, 134.
Pageau, Y., 11.
Palaeoniscus? sp., 432.
Panchen, A. L., 232, 244, 278, 308, 315, 500, 526.
Pander, C. H., 81, 87.
Parker, W. K., 91, 92, 106, 241, 328, 329.
Parrington, F. R., 11, 386, 395, 413, 415, 416, 428.
Parsons, T. S. & Williams, E. E., 306, 307, 310, 311, 315; *see also* Williams, E. E.
Paterson, N. F., 317, 329.
Patten, W., 373, 395.
Patterson, C., 171–205; 11, 153, 155, 156, 157, 158, 159, 160, 162, 163, 165, 167, 168, 169, 220, 222, 499, 517, 518, 526.
Pautard, F., 375, 396.
Pehrson, T., 11, 229, 230, 232, 236, 237, 272, 282, 318, 329, 414, 420, 422, 423, 424, 425, 428, 436, 438; *see also* Holmgren, N. & Pehrson, T.
Pellonula, 461, 462, 463, 465.
P. vorax, 455, *462*.
Pelobates fuscus, 508.
Perca, 466.
Pernkopf, E. & Lehner, J., 241, 244.
Peter, K., 22, 70.
Petromyzon, *19, 30*, 31, 36, 37, 92, 98, 102, 384, 385, 502, 519.
P. planeri, 23.
P. sp., *21*.
Phaneropleuron, 228, 250.
Pharyngolepis, 18, 37, 38.
Phlebolepis elegans, *33*, 81–87, *83*, *85*, *86*, 101.
Pholidogaster, 309.

Phractolaemus, 463, 467.
P. ansorgei, 456.
Pickford, G. E., Robertson, E. E. & Sawyer, W. H., 468, 472; *see also* Emmart, E. W., Pickford, G. E. & Wilhelmi, A. E.; Grant, W. C. & Pickford, G. E.
Pinkus, F., 288, 305.
Pipa americana, 331.
Piveteau, J., 310, 315.
Platt, Julia B., 99, 106, 343, 357.
Plenk, H., 452.
Plethodon cinereus, 446.
P. glutinosus, 446.
Pleuracanthus, 228.
Pleurodeles, 415, 416, 417, 426.
P. waltlii, 317, 318, 320, *322, 323*, 324, 325, 414, *418*.
Plourdosteus, 145, 147.
Poecilia, 468.
Poecilodus sanctiludovici, 195.
Poll, M., 238, 244.
Pollard, H. B., 223, 436, 438.
Polyacanthus, 273.
Polybranchiaspis, 14.
Polyodon, 137.
Polypterus, 223, 237, 238, 239, *266, 273*, 389, 420, 423, 424, 425, 429, 431, 432, 436, 437, 455, 456, 465, 466, 467, 515.
P. bichir, 430.
P. lapraedi, 430, 433, 435.
P. senegalus, 318, 456.
P. sp., 318.
Poraspis, *41*, 46, 54.
P. pompeckji, 51.
Porolepis, 211, 221, 225, 249, 287, *288*, 289, 290, *291*, 512.
Protacrodus vetustus, 411.
Protaspis, 62; *see also Glossoidaspis*.
Proteus, 312.
P. anguineus, 312.
Protopterus, 229, 231, 232, 234, 237, 238, 241, 260, 261, 264, *266*, 269, 276.
P. aethiopicus, 229, *231, 236*.
P. annectens, 229, *230, 236*, 260, *262, 268*.
P. dolloi, facing 232, 234, *235*.
Psammodus, 197, 199.
P. rugosus, 198.
Psammolepis, 393.
P. paradoxa, 392.
Psephaspis, 392, 393.
Psephodus, 197, 200.
P. magnus, 198.

Pteraspis, 79.
  P. nathorsti (syn. *Lyktaspis nathorsti*), 73.
Pteronisculus, 115, 116, 117, 137, 207, *209*, 216, 307.
Ptychodus, 155, 198.

Racophorus leucomystax, 312.
Radinsky, L., 153, 154, 155, 169, 197, 205.
Raja, 263.
Ramaswami, W., 328, 329.
Rana, 332, 333.
  R. catesbeiana, 446, 449.
  R. esculenta, 312, 331, *334.*
  R. pipiens, 312, *313.*
  R. ridibunda, 318, *327*, 328.
  R. temporaria, 318, 328, 331, *334.*
Ranodon, 332.
  R. sibiricus, 317, 318, *319, 320, 321*, 322, *323*, 324, *325*, 331, *334.*
Rauther, M., 238, 244.
Rawles, M. E., 367, 371.
Rayner, Dorothy H., 213, 214, 215, 216, 217, 222.
Regan, C. T., 434, 438.
Reinbach, W., 317, 328, 329.
Reis, O. M., 109, 111, 113, 114, 117, 118, 122, 127, 129, 130, 131, 132, 135, 140, 142, 171, 179, 205.
Remane, A., 259, 282, 308.
Rensch, B., 445, 446, 452.
Retzius, G., 234, 241, 244.
Rhachiosteus, 150.
Rhinodipterus, 230.
  R. ulrichi, 231.
Rhinoptera, 194.
Rhodes, F. H. T. & Wingard, P. S., 373, 396.
Rhynchodipterus, 251.
Ridewood, W. G., 439, 443.
Ritchie, A., 81–88; 11, 33, 70, 101.
Robertson, J. D., 374, 396.
Romer, A. S., 29, 70, 102, 103, 107, 203, 205, 207, 215, 216, 217, 222, 228, 244, 259, 280, 282, 292, 298, 299, 305, 309, 310, 315, 365, 366, 371, 374, 396.
Roux, G. H., 389, 396.
Rudebeck, B., 223, 241, 244, 262, 282, 514, 515, 526.
Rühle, H.-J. & Sterba, G., 468, 472.
Ruud, G., 234, 235, 244.

St. John, O. & Worthen, A. H., 195, 205; *see also* Newberry, J. S. & Worthen, A. H.

Salamandra, 299, 313.
  S. maculata, 312.
"*Salamandrella*" (syn. *Hynobius*), 296.
Salamandrina perpicillata, 312.
Salmo, 136, 261, *266, 273, 274,* 275, 414, 421, 422, 424, 425, 463, 465, 466.
  S. irideus, 423, 424.
  S. salar, 456, *464*, 465.
  S. trutta, 465.
Salvelinus fontinalis, 465.
Sandalodus, 197, 200.
Sarcoprion, 155, 156, 157, 158, 159, 160, 162, 163, 164, 165, 166, 167.
  S. edax, 156.
Sauripteris, 296, 300, *301.*
  S. taylori, 300.
Säve-Söderbergh, G., 96, 107, 211, 214, 215, 218, 222, 223, 227, 228, 229, 232, 234, 236, 244, 251, 257, 259, 282, 310, 315, 317, 329, 424, 507, 522, 526.
Scammon, R. E., 342, 357.
Scaphaspis nathorsti (syn. *Lyktaspis nathorsti*), 73, 79.
Scaphiopus couchi, 449.
Scaumenacia, 227, 230, 231, 233, 234, 236, 253.
  S. curta, 231, 236.
Schaeffer, B., 207–222; 11, 117, 118, 122, 123, 124, 127, 242, 244, 252, 260, 282, 511 517, 526.
Schaeffer, B. & Rosen, D. E., 122, 127, 213, 222; *see also* Greenwood, P. H., Rosen, D. E., Weitzman, S. H. & Myers, H. S.
Schauinsland, H., 511, 526.
Schmalhausen, I. I., 136, 143, 282, 295, 296, 298, 305, 338, 340.
Schmidt, W. J. & Keil, A., 377, 396.
Schober, W., 501, 502, 509, 526.
Schowing, J., 415, 428.
Schultze, H.-P., 11, 227, 244.
Scourfield, D. J., 374, 396.
Scruggs, W. M., 465, 466, 472.
Scyllium, 241, 263.
Sedra, S. N. & Michael. M. I., 328, 329.
Semon, R., 223, 244.
Sengel, P. & Abbot, V. K., 367, 371.
Seretaspis, 54, 62.
Sewertzoff, A. N., 22, 27, 70, 91, 92, 107, 223, 244, 282, 295, 296, 297, 305.
Seydel, O., 331, 339, 340.
Seymouria, 416.
Shipley, A. E., 70.

Shishkin, M. A., 507, 526.
Shoshenko, K. A., 450, 452.
Siberiaspis plana, 58.
Sillman, L. S., 373, 396.
Simopteraspis, 54.
  S. primaeva, *50*, *61*.
Simpson, G. G., 219, 222, 487, 489, 490, 492, 495.
Siren, 317.
  S. intermedia, 446.
Smith, B. G., 116, 123, 127.
Smith, H. M., 446, 452.
Smith, H. W., 374, 396, 448, 452.
Soederberghia, 219, 231, 251, 255.
Spencer, W. K., 373, 396.
Spiegelman, S., 420, 428.
Spjeldnæs, N., 374, 375, 376, 382, 396.
Squaloraja, 166, 167, 171, 182, 193, 201, 202.
Squalus, 274, 275.
  S. acanthias, 341, *342*, 343, *344–346*, 347, *348–350*, 353.
Squatina, 161, 164.
Stahl, Barbara S., 11, 154, 169, 518, 526.
Steiner, H., 298, 299, 305.
Stensiö, E. A., 13–71, 148–151; 11, 73, 78, 79, 80, 82, 84, 88, 91, 96, 101, 103, 104, 107, 119, 125, 127, 135, 143, 145, 147, 162, 165, 169, 170, 184, 205, 207, 208, 211, 222, 225, 226, 229, 231, 234, 237, 242, 244, 245, 270, 276, 279, 282, 285, 288, 305, 376, 381, 382, 383, 384, 385, 386, 387, 389, 396, 401, 412, 424, 497, 503, 504, 505, 509, 511, 512, 517, 519, 521, 522, 526; see also Holmgren, N. & Stensiö, E. A.
Stichaeopsis nana, 235.
Stone, L. S., 99, 107.
Strahan, R., 11, 70, 99, 100, 101, 107.
Streblodus (syn. *Cochliodus*), 195.
"*S. oblongus*" (syn. *Cochliodus contortus*), 195, *196*, *198*.
  S. colei, 195.
Stromer v., Reichenbach, E., 159, 170.
Szarski, H., 305–306, 445–451; 11, 259, 278, 282, 297, 298, 513.
Szarski, H. & Cybulska, R., 446, 452.
Szarski, H. & Czopek, G., 445, 446, 453; see also Czopek, G.
Szarski, H. & Czopek, J., 445, 446, 453.

Tampi, P. R. S., 459, 465, 472.
Tarlo, L. B. H., 73, 79, 80, 368, 371, 374, 375, 380, 381, 389, 396.

Tarlo, L. B. H. & Tarlo, Beryl J., 392, 396.
Tarpon, 443.
Tarrasius, 386.
Tatarinov, L. P., 326, 329.
Tatarko, K., 420, 428.
Teissier, G., 445, 448, 453.
Tesseraspis, 382, 388.
Thelodus, 81, 82, 87.
  T. scotius (syn. *Logania scotica*), 84.
Thomson, K. S., 285–306; 11, 116, 122, 127, 213, 217, 222, 260, 265, 283, 513.
Thorsteinsson, R., 11, 521, 526.
Thrams, O. K., 331, 340.
Thursius, 249.
"*Tomodus convexus*" (syn. *Cochliodus contortus*), 195.
Towe, K. M. & Cifelli, R., 361, 371.
Tragulus, 445.
Traquair, R. H., 252, 257, 385, 396.
Tremataspis, 272.
  T. sp., *33*.
Trematops, 299.
Tretjakoff, D. K., 91, 107, 355, 357.
Trigla, 434.
Trimerorachis, 309.
Triton, *332*, 336, *337*.
  T. cristatus karelini, 317, 318, 320, 322, 323, 324, 325, 331, 335.
  T. taeniatus, 331, *333*, 335.
Triturus, *296*, *297*.
Troitsky, W., 420, 428.
Troschel, F. H., 234, 245.
Tuomikoski, R., 483, 484, 485, 495.
Turinia, 38, 55, 60, 85, 87, 385, 386.
Turner, S. C., 514, 526.

Uranolophus, 227, 247–256.
  U. wyomingensis, 247, *248*, *249*, *250*, *251*, *252*, *253*, *254*, *255*, *256*.
Urist, M. R., 375, 396.
Uronemus, 255.
Uzell, T. M., 448, 453.

Vandebroek, G., 11.
Vernberg, F. J., 446, 453.
Vernonaspis sekwiae, *39*, 40, 42, 44, 46.
Vetter, B., 434, 438.
Vollrath, L., 465, 472; see also Knowles, F. & Vollrath, L.
Vorobyeva, Emilia, 11, 294, 317, 329.
Vorontsova, M. A., Liozner, L. D., et al., 318, 329.

Wallace, H., 448, 453.
Wängsjö, G., 382, 396.
Watabe, N. & Wilbur, K. M., 363, 371.
Watson, D. M. S., 70, 100, 107, 109, 112, 113, 114, 118, 119, 120, 123, 124, 125, 127, 129, 130, 131, 132, 140, 143, 155, 170, 210, 222, 257, 387, 397.
Watson, D. M. S. & Gill, E. L., 227, 228, 232, 255.
Weigelt, J., 156, 170, 171, 182, 192, 205.
Weiss, P., 416, 417, 428.
Westoll, T. S., 82, 87, 88, 227, 228, 229, 232, 233, 234, 236, 245, 247, 254, 271, 272, 280, 283, 374, 381, 385, 386, 391, 397, 511, 527; *see also* Graham-Smith, W. & Westoll, T. S.; Lehmann, W. & Westoll, T. S.
White, E. I., 70–71; 11, 52, 73, 79, 80, 220, 222, 228, 231, 234, 236, 237, 271, 283, 374, 397, 498, 527.
Wickbom, T., 223, 245.
Wijhe, J. W. van, 134, 136, 137, 143, 341, 357.

Williams, E. E., 307, 308, 310, 311, 315; *see also* Parsons, T. S. & Williams, E. E.
Wingstrand, K. G., 223, 245, 276, 283, 456, 465, 466, 472.
Wolff, E., 417, 428.
Woodland, W. N. F., 107.
Woodman, A., 465, 472.
Woodward, A. S., 73, 80, 153, 156, 159, 170, 171, 180, 184, 193, 195, 196, 205, 237, 512, 527.

Xenacanthus, 116.
Xenopus, 317, 328.

Zangerl, R., 399–412; 11, 156, 157, 158, 159, 160, 161, 162, 163, 170, 171, 193, 199, 205, 387.
Zangerl, R. & Richardson, E. S., 171, 205, 399, 412.
Zascinaspis sp., *42, 45*.
Zhorno, L. J. & Ovchinnikova, L. P., 448, 453.
Zubina, E. V., 448, 453.